E. JAMES POTCHEN, M.D., *Consulting Editor*

Professor of Radiology
The Johns Hopkins University School of Medicine
Baltimore, Maryland; formerly Professor of Radiology
The Edward Mallinckrodt Institute of Radiology
Washington University School of Medicine
St. Louis, Missouri

Published

GASTROINTESTINAL ANGIOGRAPHY
  Stewart R. Reuter, M.D., and Helen C. Redman, M.D.

THE RADIOLOGY OF JOINT DISEASE
  D. M. Forrester, M.D., and John W. Nesson, M.D.

RADIOLOGY OF THE PANCREAS AND DUODENUM
  S. Boyd Eaton, Jr., M.D., and Joseph T. Ferrucci, Jr., M.D.

THE HAND IN RADIOLOGIC DIAGNOSIS
  Andrew K. Poznanski, M.D.

Forthcoming Monographs

RADIOLOGIC DIAGNOSIS OF RENAL PARENCHYMAL DISEASE
  Alan J. Davidson, M.D.

THE RADIOLOGY OF VERTEBRAL TRAUMA
  John A. Gehweiler, Jr., M.D., Raymond L. Osborne, Jr., M.D., and R. Frederick Becker, Ph.D.

CHRONIC RENAL FAILURE
  Harry J. Griffiths, M.D.

FUNDAMENTALS OF ABDOMINAL AND PELVIC ULTRASONOGRAPHY
  George R. Leopold, M.D., and W. Michael Asher, M.D.

SPECIAL PROCEDURES IN CHEST RADIOLOGY
  Stuart S. Sagel, M.D.

PEDIATRIC RADIOLOGY OF THE ALIMENTARY TRACT
  Edward B. Singleton, M.D., Milton L. Wagner, M.D., and Robert V. Dutton, M.D.

ARTHROGRAPHY: PRINCIPLES AND TECHNIQUES
  Tom W. Staple, M.D.

CLINICAL PEDIATRIC AND ADOLESCENT UROGRAPHY
  Alfred L. Weber, M.D., and Richard C. Pfister, M.D.

CORONARY ARTERIOGRAPHY
  Lewis Wexler, M.D., and Ivo Obrez, M.D.

Volume 5 in the Series
SAUNDERS
MONOGRAPHS
IN CLINICAL
RADIOLOGY

RADIOLOGY OF THE ILEOCECAL AREA

ROBERT N. BERK, M.D.

Professor and Chairman
Department of Radiology
University of Texas
Southwestern Medical School
Dallas, Texas

ELLIOTT C. LASSER, M.D.

Professor and Chairman
Department of Radiology
University of California, San Diego
School of Medicine
San Diego, California

1975

W. B. SAUNDERS COMPANY • *Philadelphia* • *London* • *Toronto*

W. B. Saunders Company: West Washington Square
Philadelphia, PA 19105

12 Dyott Street
London, WC1A 1DB

833 Oxford Street
Toronto, Ontario M8Z 5T9, Canada

Radiology of the Ileocecal Area   ISBN   0-7216-1689-5

Last digit is the print number: 9   8   7   6   5   4   3   2   1

TO SONDRA AND PHYLLIS

EDITOR'S FOREWORD

"When the x-ray was discovered, a new means of investigating the alimentary tract was provided, which permitted observations to be made without interfering with the animals to any disturbing degree. This means of research was suggested to me, while still a medical student, by my teacher of physiology, Professor H. P. Bowditch in the Autumn of 1896."

—*Walter B. Cannon, 1911*

These first radiologic studies of the gastrointestinal tract clearly demonstrated the remarkable potential of contrast examinations to study the structure and function of the digestive organs in both health and disease. Cannon was the first to describe the unique physiologic activity of the ileocecal region, using bismuth subnitrate as a contrast agent for fluoroscopic studies. Diseases of this area have been recognized at least since the work of Jean Fernel when he described the clinical symptoms of appendicitis in the 16th century. The disease was not widely appreciated until Reginald Fitz coined the term "appendicitis" in 1886 and recognized the association of the clinical symptoms with the operative and postmortem findings in the ileocecal area.

It is remarkable to appreciate that within less than a century our recognition of right lower quadrant pathology has developed to the state of sophistication found in this book. The authors have compiled information on the radiology of the ileocecal region, describing the many diseases and subtle abnormalities that can be detected only by the sophisticated use of radiologic tools. This work will show those physicians who may not yet appreciate the capability of radiologic techniques to evaluate right lower quadrant abnormalities the notable impact that such studies can make on the diagnosis of diseases in this area. Many books on gastrointestinal radiology have been written, but there is none that offers greater insight into this most important and often difficult region.

Drs. Berk and Lasser are eminently well qualified as authorities in gastrointestinal radiology. After training in radiology at the Peter Bent Brigham Hospital, Dr. Berk joined Dr. Lasser at the University of Pittsburgh in 1961 where they performed basic research on contrast materials used for cholecystography. In 1968 they went to the University of California at San Diego where they continued their studies. Recently, Dr. Berk has been appointed Professor and Chairman of the Department of Radiology at the Southwestern Medical School in Dallas, Texas. Dr. Lasser served as Professor and Chairman of the Department of Radiology at the University of Pittsburgh and more recently at the University of California, San Diego. Authors of numerous publications, Drs. Berk and Lasser have made many innovative contributions to gastrointestinal radiology.

I am particularly delighted to see this manuscript published in the Saunders Monographs in Clinical Radiology series. It will provide an excellent resource for radiologists, both in training and in practice, to comprehend the potential of standard radiologic examinations in contributing to the management of common as well as rare clinical problems involving the ileocecal area.

E. JAMES POTCHEN, M.D.

PREFACE

In these days of ever increasing complexity of diagnostic modalities, in the era of ultrasound, interventional angiography and direct pancreatico-cholangiography, it is refreshing to pause for a time and concentrate again on the fundamental techniques of gastrointestinal radiology. It is rewarding to return to the plain abdominal radiograph and to barium studies of the stomach, small bowel and colon in order to re-examine their use in the diagnosis of mundane as well as unusual diseases of the abdomen. It is still the artful performance and skillful interpretation of these basic studies that provide the information that is essential for the proficient care of the majority of patients in daily practice.

The ileocecal area is an imprecise anatomical region with an artificial boundary and an arbitrary definition, yet it is convenient from clinical and radiologic standpoints to consider diseases occurring in this location together. Anatomically, the ileocecal area is remarkably uncomplicated compared to other regions such as the pancreaticoduodenal area, where multiple structures interrelate in an intricate fashion. Still, numerous and varied diseases involve the ileocecal area and provide a diversity of radiologic findings that make the ileocecal region a fascinating area for study.

The ileocecal area is one of three segments of the gastrointestinal tract where a sphincter-like mechanism separates adjacent parts of the gastrointestinal tract that are widely different in caliber. Like the esophagogastric junction and the pylorus, the ileocecal area is subject to certain diseases that specifically involve the proximal and not the distal segment, or the converse may be true. Other less common diseases cross the junction to involve both portions of the bowel without regard for the sphincter.

The goal of the Saunders series of monographs in clinical radiology is to "explore in depth some of the major problems facing the clinical radiologist." It is intended that the monographs fall between and complement review articles and the more comprehensive multidisciplinary texts. Consequently, this monograph concerning the ileocecal area is not intended to be encyclopedic or to replace the major textbooks of gastrointestinal radiology. The purpose is to review in considerable detail most of the diseases that involve the structures in the ileocecal area in an effort to increase the reader's awareness of the potentially extractable information that can be gleaned from the morphologic details detectable radiologically. It is hoped that this monograph not only will provide a reference of illustrative material for the practicing radiologist and the student of radiology but also will serve to sharpen the reader's focus and increase the effective "signal-to-noise" ratio in this important area.

ROBERT N. BERK, M.D.
ELLIOTT C. LASSER, M.D.

ACKNOWLEDGMENTS

The following is a list of friends and colleagues who have graciously permitted us to use their cases as illustrations in this monograph. Without their generous assistance, we would have been unable to demonstrate important radiologic features of many of the diseases that involve the ileocecal area.

Arnold L. Bachman
Thomas C. Beneventano
Folke J. Brahme
Klaus M. Bron
Arthur R. Clemett
Vincent T. DeVita
Wendel W. Dietz
John L. Doppman
John H. Feist
Benjamin Felson
Donald J. Fleischli
Robert S. Francis
Henry I. Goldberg
David Greenberg
Ronald D. Harris
Theodore F. Hilbish
Harold G. Jacobson
Norman Joffe
Ann M. Lewicki

George R. Leopold
James J. McCort
Joseph A. Marasco
Adrian G. Mikulicich
William L. Pogue
Thomas A. Raneur
John M. Riley
Donald J. Ritt
Clarence J. Schein
Peter M. Shimkin
Norman R. Silverman
John L. Smith
Richard C. Smith
Richard M. Taketa
Michael H. Weller
Ralph C. Wilde
Jerome F. Wiot
Bernard S. Wolf

We are indebted to Miss Marylou Ott for the fine technical quality of many of the radiographs included in this book. Her genius as a radiologic technologist and department administrator is nonpareil.

We would like to express our thanks to Miss Beverly Evans for spending many hours at the arduous task of typing the manuscript and to Mr. Jack Hanley of the W. B. Saunders Company for his valuable assistance and encouragement. Thanks also go to Mr. Raymund McMahon for his careful checking of the reference listings.

We are particularly grateful to our residents whose bright inquiries, enthusiasm and warm friendship have been a constant source of inspiration.

Finally, we would like to acknowledge the huge debt we and others owe to Richard H. Marshak, M.D. This master radiologist has aided us immeasurably through his writing, teaching and friendship over the years. His wisdom has added to the understanding of every aspect of gastrointestinal radiology, including that of the ileocecal area.

ROBERT N. BERK, M.D.
ELLIOTT C. LASSER, M.D.

CONTENTS

Chapter 1

CONGENITAL ANOMALIES OF THE ILEOCECAL AREA

Expect for Meckel's diverticula and variations in position and fixation of the cecum, there are no congenital anomalies that are peculiar only to the ileocecal area. The terminal ileum and cecum are subject to the same anomalies that occur elsewhere in the gastrointestinal tract. In this chapter the discussion of anomalies of the ileocecal area will be limited to Meckel's diverticula and its complications, enteric duplications, abnormalities of rotation and fixation of the cecum and arteriovenous malformations of the cecum and terminal ileum.

MECKEL'S DIVERTICULUM

Meckel's diverticulum is the most common anomaly of the gastrointestinal tract, occurring in from 0.3 to 3.0 per cent of autopsy series.[1,2] It is located on the antimesenteric border anywhere in the last six feet of the ileum, the most common location being one foot from the ileocecal valve. The diverticulum is usually approximately 2 cm. in length and ends blindly. The diameter of the mouth of the diverticulum often equals that of the small intestine.

Meckel's diverticulum is a remnant of the vitelline (omphalomesenteric) duct which serves as a communication between the gut and the yolk sac.[3] The duct is usually completely obliterated and detached by the fifth week of embryonic life. Closure proceeds from the umbilicus toward the intestine. When obliteration is incomplete, the enteric end of the duct persists as a diverticulum.[1] The diverticulum may remain joined to the umbilicus by a fibrous band. Rarely, the entire duct remains patent, resulting in a fistula between the ileum and the umbilicus through which fecal material may be discharged.

Histologically, the wall of the diverticulum usually remains asymptomatic throughout life. Complications occur in only 2.5 per cent of cases[4] and include bleeding, diverticulitis, perforation and peritonitis, intestinal obstruction (intussusception or volvulus), presence of a foreign body, incarceration in an inguinal hernia, umbilical discharge and tumor. Two-thirds of the complications occur in patients less than 2 years of age, and more than one-third occur in patients less than 1 year of age.[1] Bleeding is the most common complication. It is usually copious and painless and results from ulceration of ileal mucosa adjacent to heterotopic gastric mucosa. Meckel's diverticula account for one-half of all lower gastrointestinal bleeding in children.[5]

Meckel's diverticula are rarely diagnosed radiologically. Plain abdominal radiographs may show a large cyst in the abdomen, with a long air-fluid level on upright projections or a calcified enterolith in the diverticulum. Rarely, unabsorbed contrast material can be detected contained within the diverticulum in the course of performing an oral cholecystogram. When diverticulitis occurs, an abscess, intraperitoneal fluid or paralytic ileus may be present.

The diagnosis is seldom made on barium studies, because the diverticulum is difficult to distinguish from normal loops of small bowel. Careful compression spot radiographs are usually necessary to demonstrate the diverticulum unless it is large (Figs. 1–1 to 1–3). If the vitelline duct persists as a fibrous band, the diverticulum may maintain an orientation toward the umbilicus which facilitates its visualization.[1]

(Text continued on page 4)

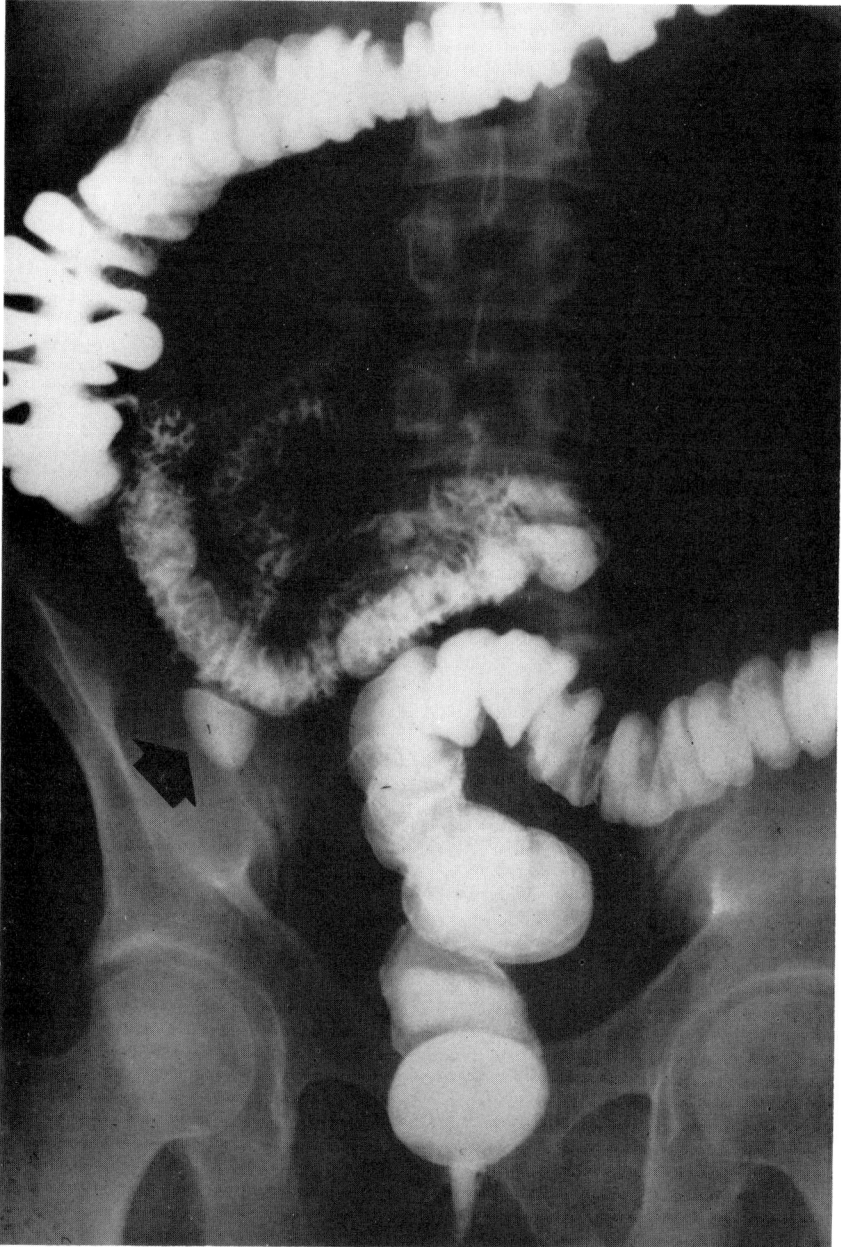

Figure 1–1. Barium enema examination showing a Meckel's diverticulum of average size (arrow).

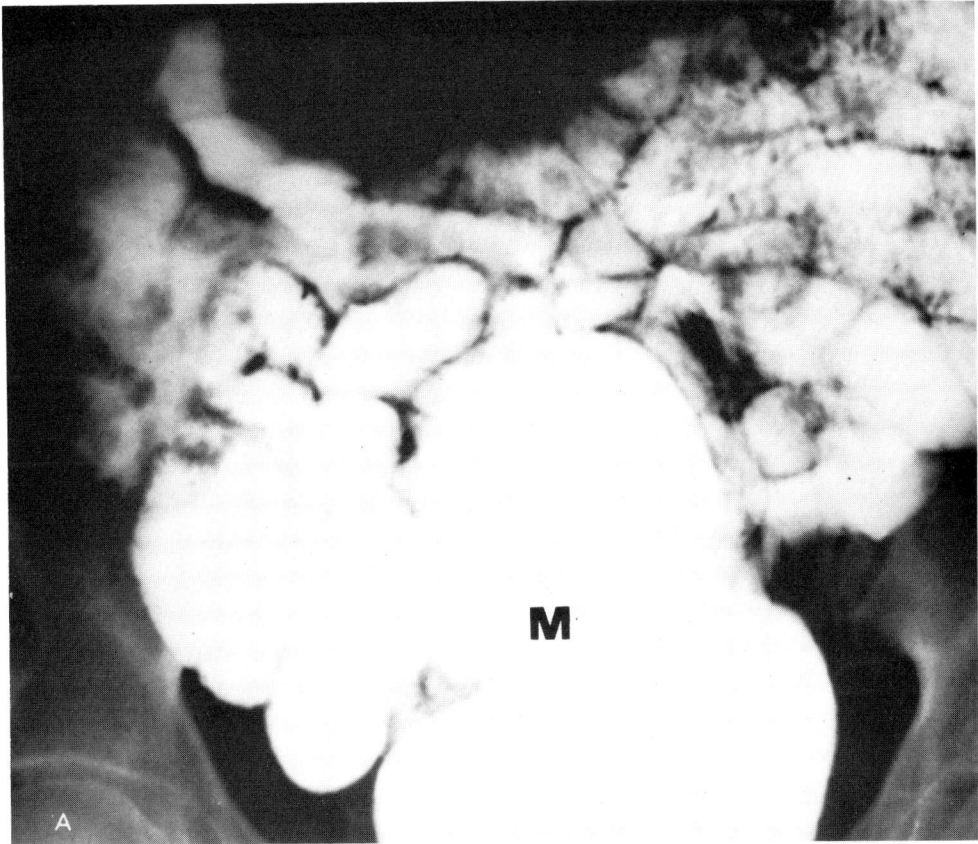

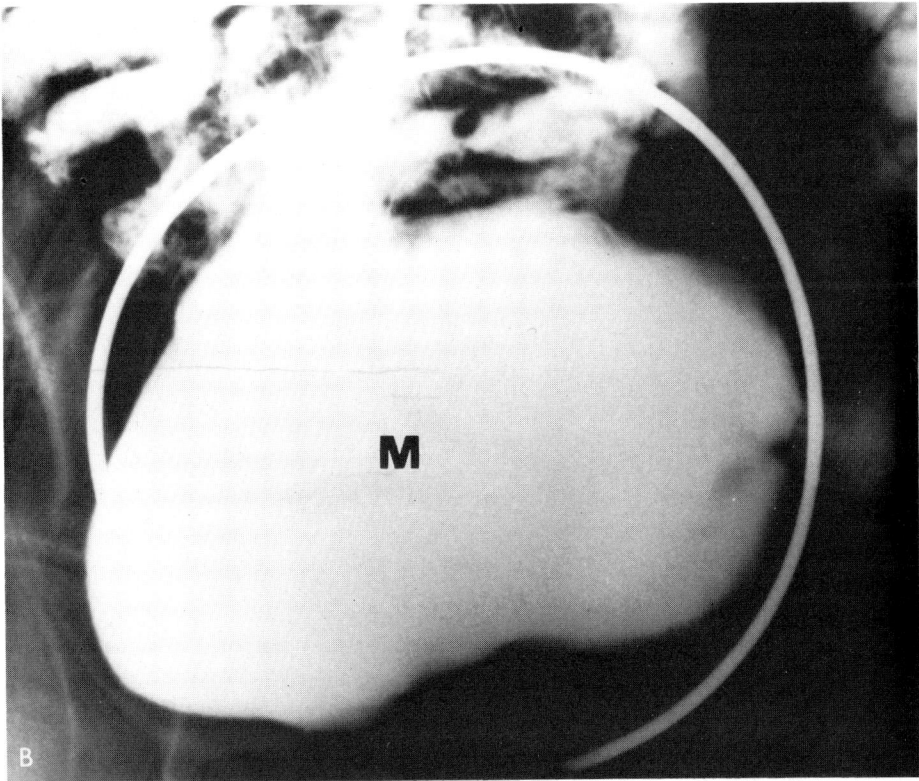

Figure 1–2. Radiograph from a small bowel examination (A) and a compression spot film (B) showing a large Meckel's diverticulum (M).

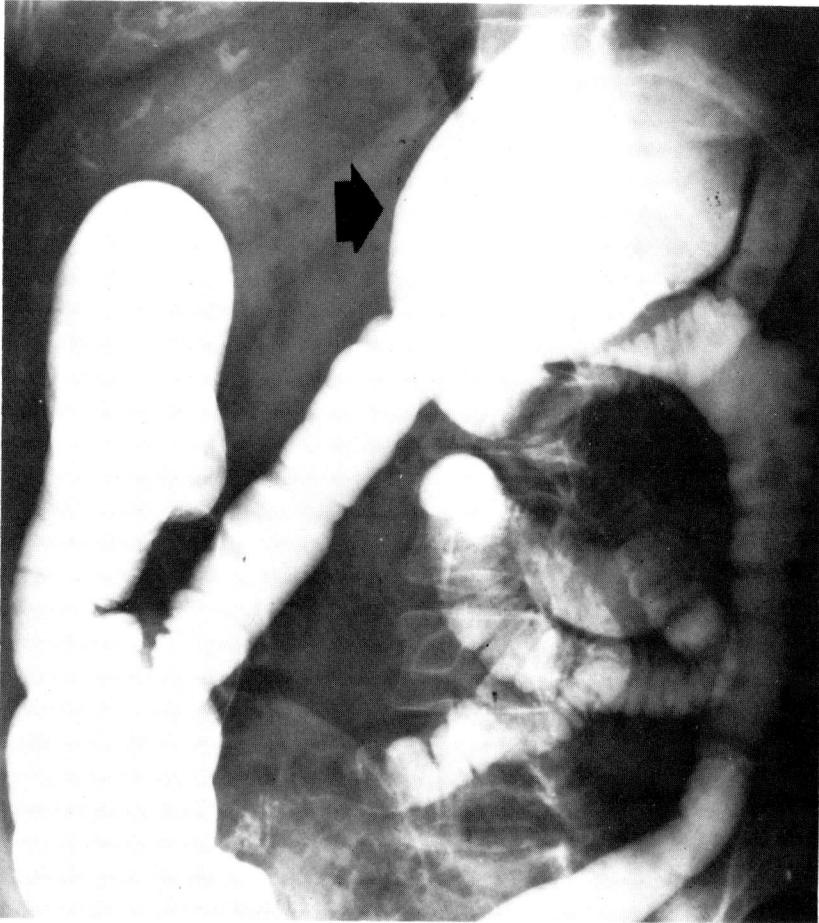

Figure 1–3. Barium enema examination showing a large Meckel's diverticulum (arrow). (From Feist, J. H., and Wilde, R. C.: Symptomatic Meckel's diverticula of unusually large size. Amer. J. Roent., 83:882, May, 1960. Courtesy of Charles C Thomas, Publisher.)

When the lumen of the duct persists throughout its length, the diagnosis can be established by injecting the umbilicus with contrast material under fluoroscopic control. If bleeding is sufficiently brisk, superior mesenteric angiography may establish the diagnosis of a bleeding Meckel's diverticulum.

Recently, isotopic scanning has been suggested as a method for establishing the presence of a Meckel's diverticulum.[6] The technique depends on the uptake of technetium^{99m} sodium pertechnetate by the heterotopic gastric mucosa in the diverticulum. The isotope is given intravenously, and the abdomen is scanned at 30 minutes and at 4 hours. Since only approximately 12 to 15 per cent of cases contain heterotopic mucosa and not all these have gastric mucosa, the scanning technique cannot be successful in more than 15 per cent of the cases.

DUPLICATION OF THE TERMINAL ILEUM

Duplications may occur anywhere along the length of the gastrointestinal tract but are most common in the small intestine. Most of those in the small bowel are in the ileum. The elongated or spherical cyst lies adjacent to and parallel with the small intestine on the mesenteric side, forming an accessory tube which usually does not communicate with the lumen of the bowel.[7]

Duplications, also termed enteric or enterogenous cysts, are lined with ileal mucosa. The anomaly is the result of abnormal recanalization of the solid core of epithelial cells which form the intestine early in fetal life.

Complications include intestinal obstruction and bleeding. Obstruction may be due to impingement of a noncommunicating duplication which is distended with secretions. The duplication may cause an intussusception or volvulus.

Barium studies of the small intestine may show an extramucosal, intramural type defect in the bowel wall.[8] Occasionally, the duplication fills with barium, in which case the diagnosis of duplication may be obvious if the barium enters a tubular structure lying parallel to the intestinal lumen. When the duplication is spherical, barium in the lumen may simulate a large ulceration or excavation similar to those that occur in leiomyosarcoma or the exoenteric form of lymphosarcoma. When a duplication of the ileum communicates with the intestinal lumen, radiologic differentiation from a Meckel's diverticulum is usually impossible (Fig. 1–4). The distinction can be made at surgery, however, because duplications involve the mesenteric side of the bowel, while a Meckel's diverticulum arises from the antimesenteric border. Rarely, a septum is identified in the lumen of the ileum or cecum, indicating a mild form of duplication (Fig. 1–5).

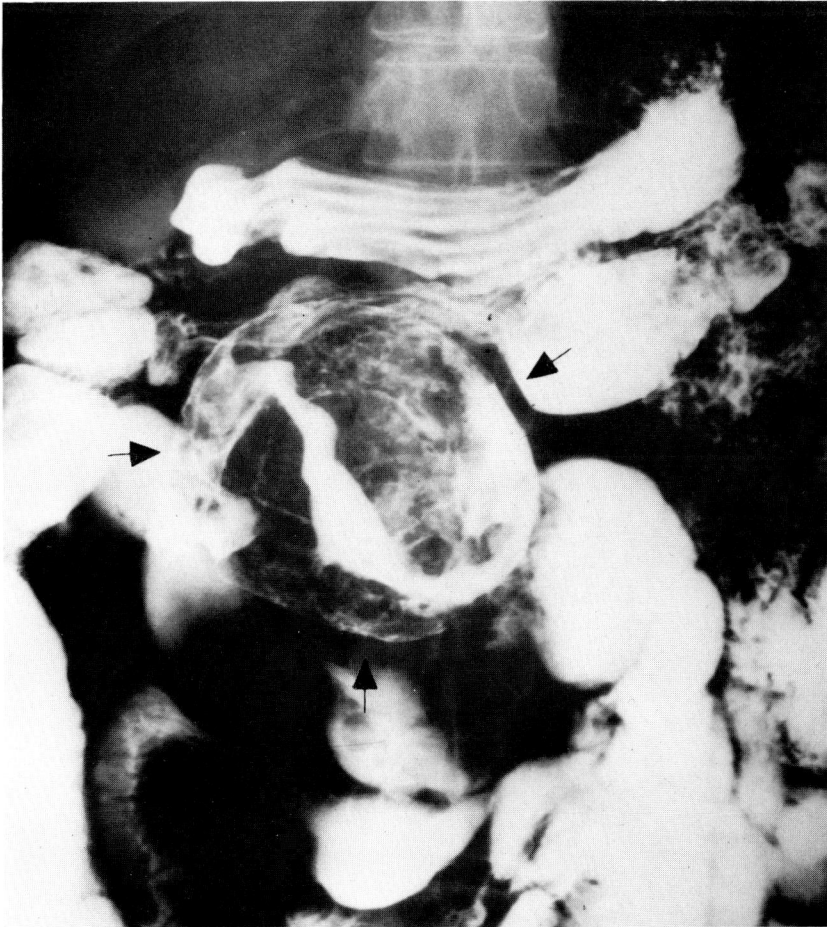

Figure 1–4. Small bowel examination showing a duplication cyst arising from the ileum and communicating with the intestinal lumen (arrows). A loop of ileum is displaced over the surface of the mass. Debris and barium are present within the cyst.

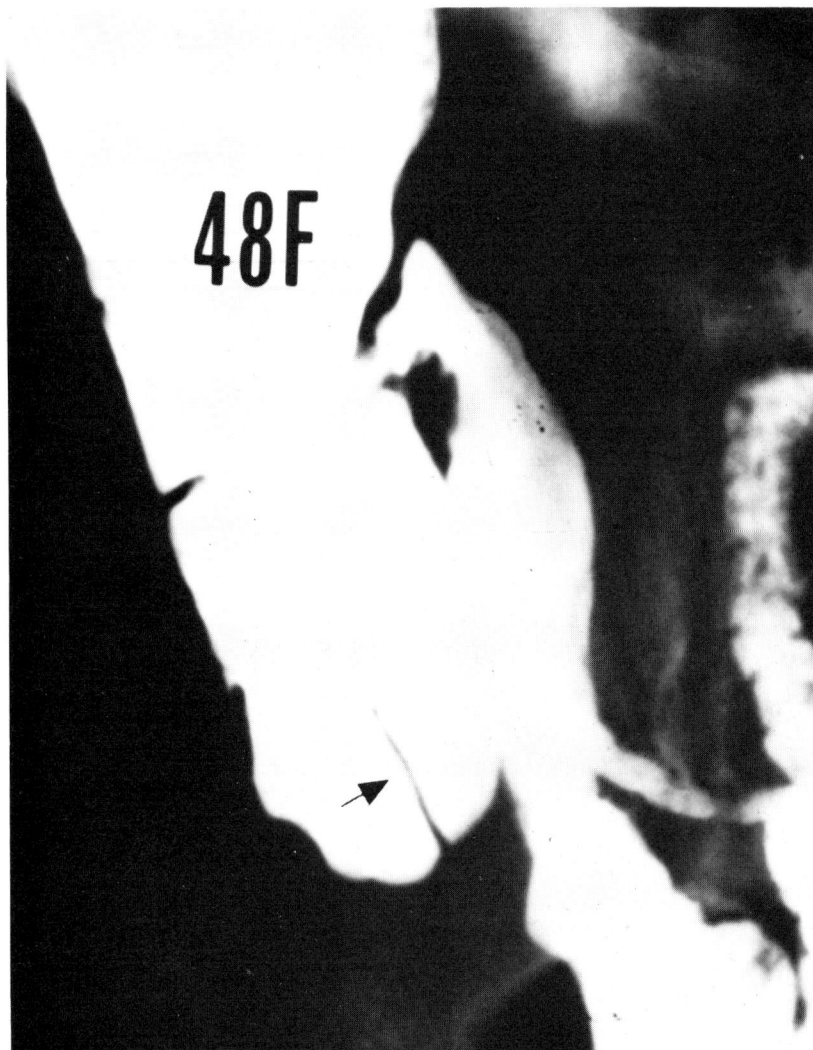

Figure 1–5. Barium enema examination showing a septum in the cecum (arrow). This is a minimal form of duplication.

ERRORS IN ROTATION AND FIXATION

The primitive digestive tube consists of the foregut and the hindgut with a wide opening of the yolk sac between, which is gradually narrowed and reduced to a small foramen leading to the vitelline duct.[8] About the sixth week a diverticulum of the gut appears just caudal to the opening of the vitelline duct and indicates the future cecum and vermiform process. The part of the loop on the distal side of the cecal diver-

ticulum increases in diameter and forms the future ascending and transverse portions of the large intestine. For a time a portion of the loop extends beyond the abdominal cavity into the umbilical cord, but by the end of the third month it is withdrawn into the cavity. The digestive tube greatly elongates and is coiled on itself, with the loops of the small intestine falling to the right while the large intestine lies on the left side. The gut then rotates upon itself in a counterclockwise direction so that the large intestine is carried over the front of the small intestine and the cecum is placed immediately below the liver. About the sixth month the cecum descends into the right iliac fossa. The mesentery of the ascending colon disappears, so that the ascending colon becomes fixed to the retroperitoneum. The cecum, which is an outpouching of the antimesenteric border of the gut, has no mesentery and is fixed posteriorly in only a small percentage of cases.

Varying degrees of failure of rotation of the cecum occur (Fig. 1–6). With partial rotation, the cecum remains in the right upper quadrant adjacent to the liver. When there is complete failure of rotation, the cecum and ascending colon remain in their primitive position in the left side of the abdomen, with the small bowel lying entirely on the right.

Failure of fixation occurs when the mesentery of the ascending colon fails to disappear. In this situation the ascending colon remains on a mesentery and is free to move within the peritoneal cavity (Figs. 1–7 to 1–10). In extreme cases, the small bowel lies lateral to the ascending colon and the cecum is in the pelvis. Cecal volvulus is likely to develop when failure of fixation occurs, because of the abnormal mobility which allows the ascending colon to twist upon itself.

(Text continued on page 12)

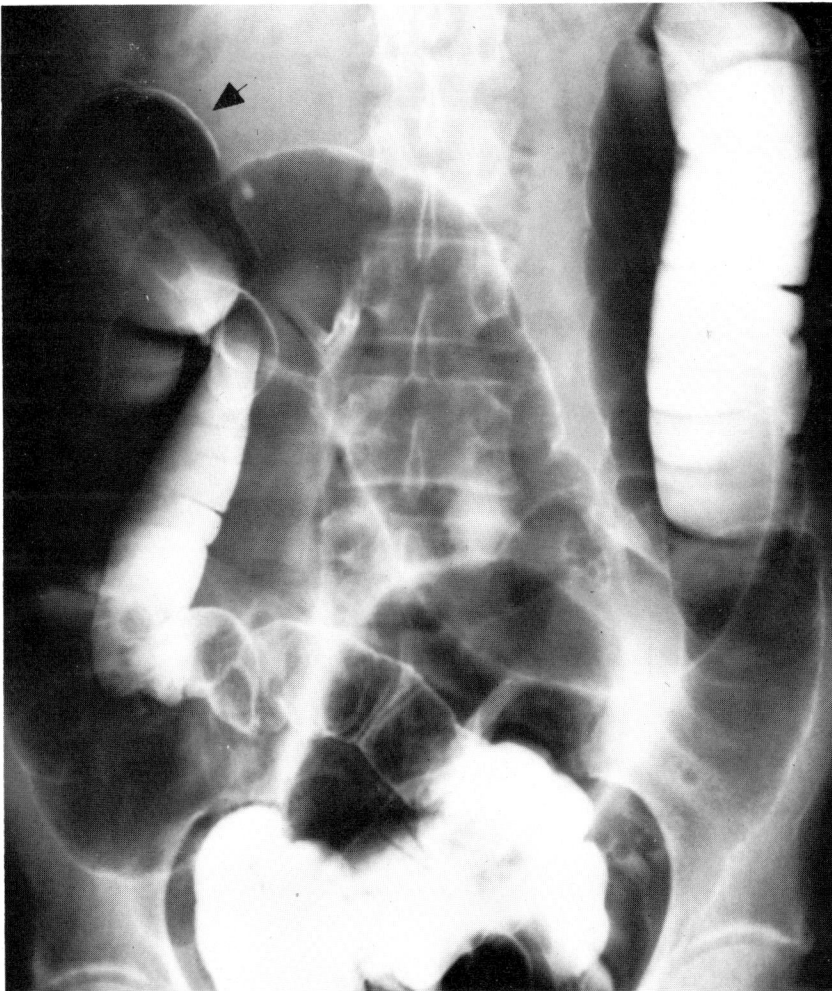

Figure 1–6. Air contrast enema examination showing malrotation of the colon. The cecum is in the right upper quadrant (arrow). The ileum is filled with barium.

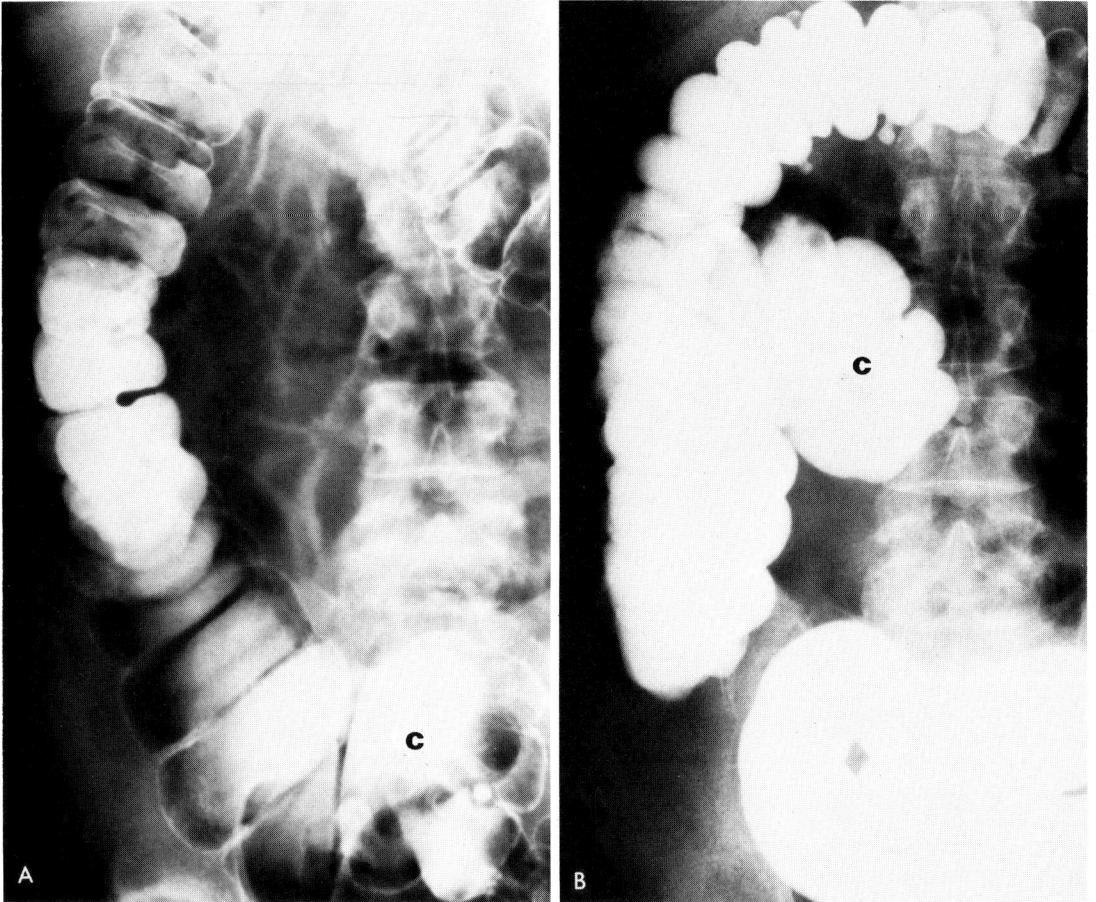

Figure 1–7. Air contrast (*A*) and barium enema examination (*B*) showing excessive mobility of the cecum (c). This represents failure of retroperitoneal fixation so the cecum remains on a long mesentery and is free to move within the peritoneal cavity.

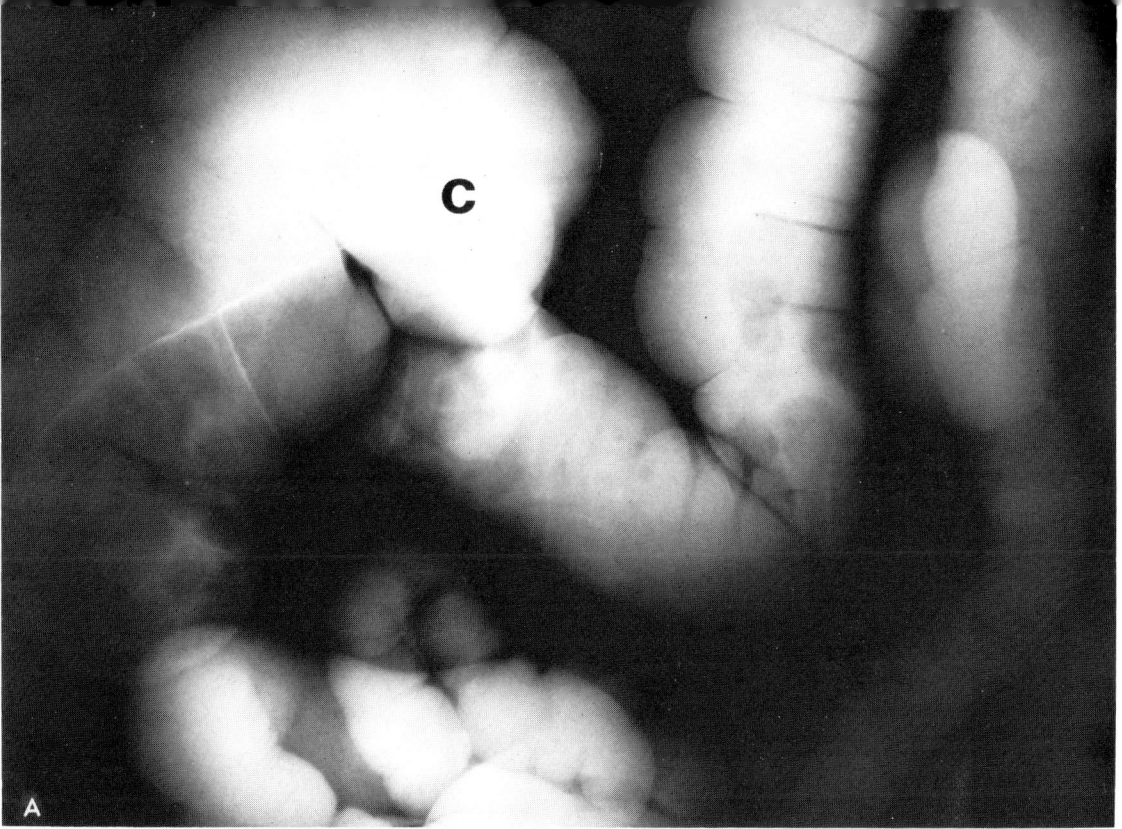

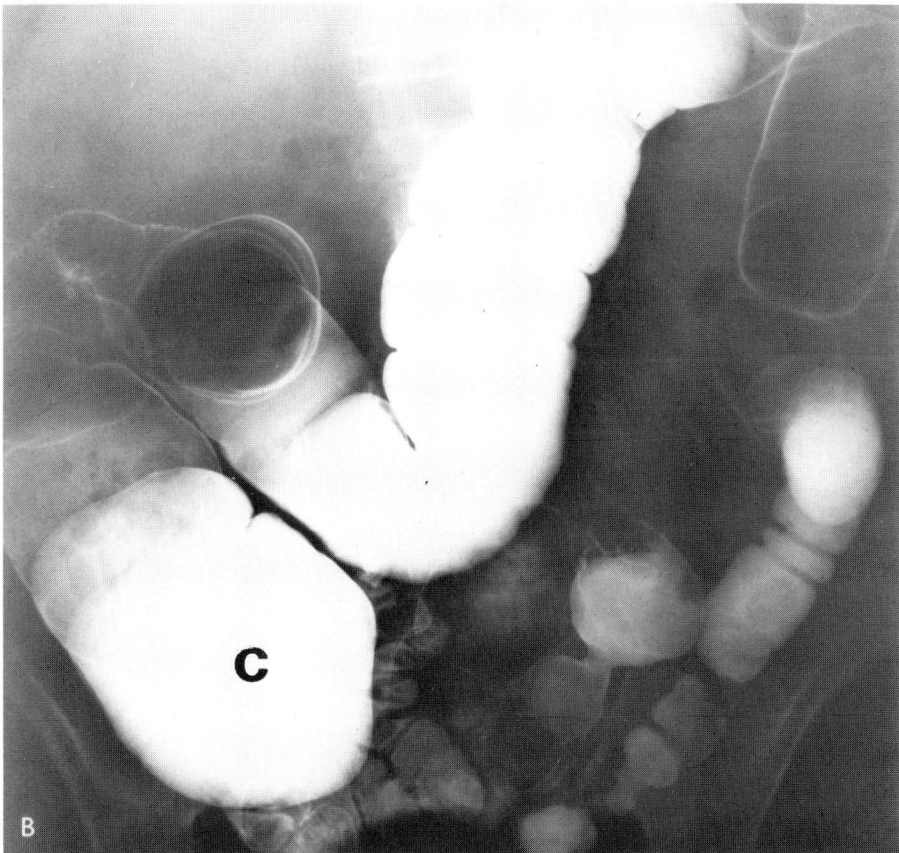

Figure 1–8. Barium enema examination showing excessive mobility of the cecum (c) which is on a long mesentery. *A*, Pre-evacuation; *B*, postevacuation.

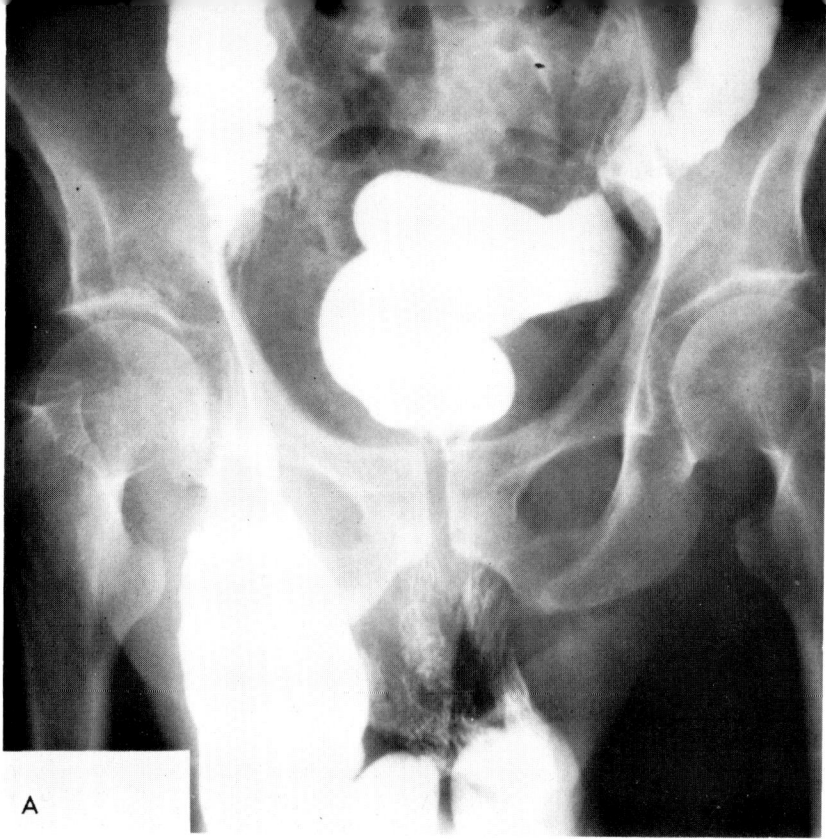

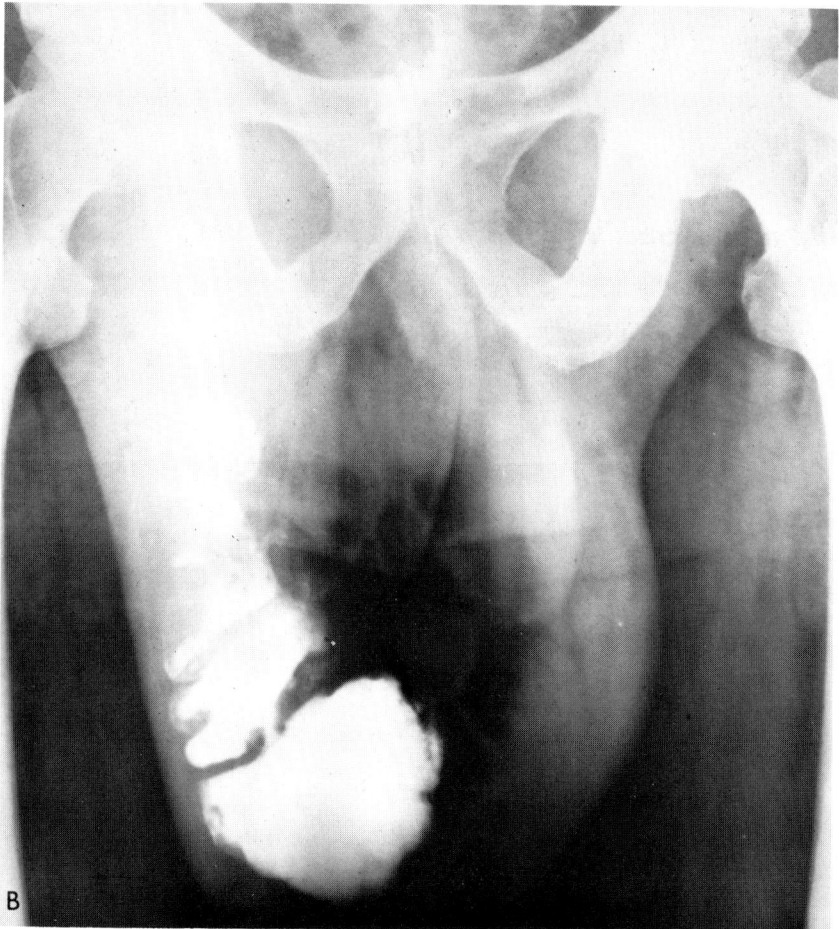

Figure 1-9. *See opposite page for legend.*

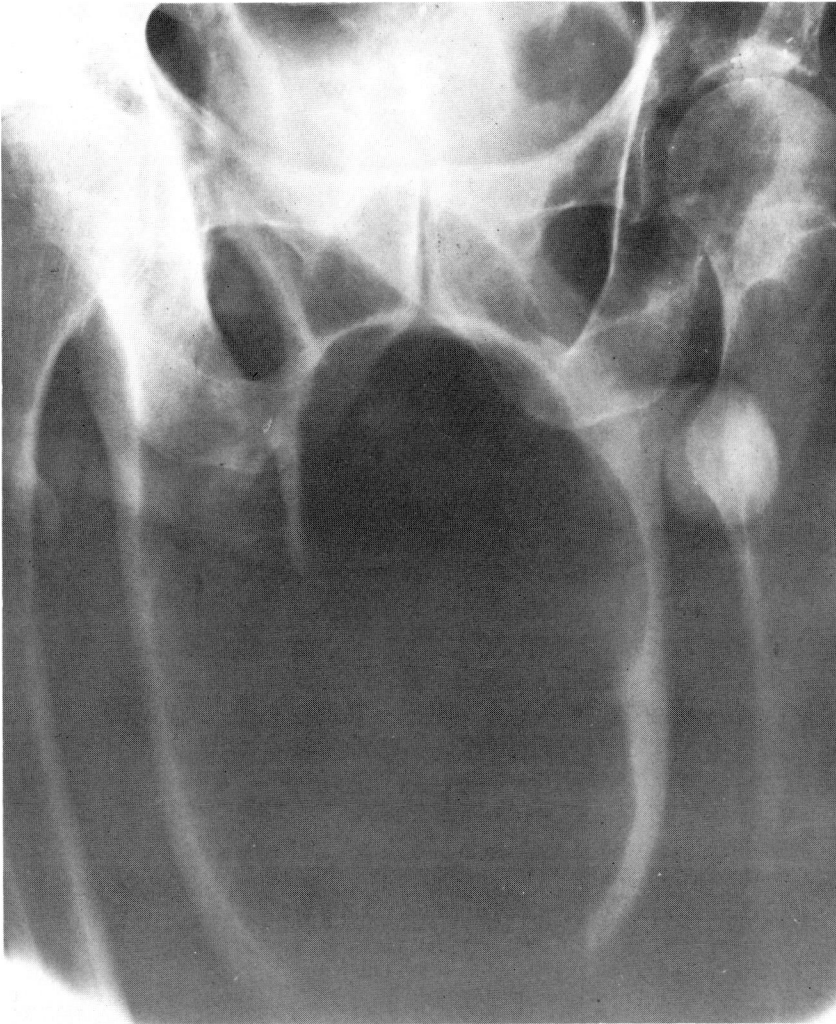

Figure 1–10. Radiograph showing marked distention of the cecum in the scrotum. The patient had a strangulated right inguinal hernia containing the cecum and terminal ileum. At operation the dilated cecum was originally dusky in color but returned to normal after relief of the strangulation and was not resected.

Figure 1–9. Barium enema examination (*A*) and postevacuation radiograph (*B*) showing the cecum, ascending colon and terminal ileum in a large right inguinal hernia.

ARTERIOVENOUS MALFORMATIONS OF THE TERMINAL ILEUM

An arteriovenous malformation, also termed an arteriovenous fistula or angiodysplasia, is any direct communication between an artery and vein. Depending on the volume of blood shunted between the artery and vein, the feeding artery dilates and becomes tortuous. The draining veins are wide and may become aneurysmal in diameter.

Arteriovenous malformations may be congenital, neoplastic, traumatic or iatrogenic.[9] The congenital variety is most common in the lung and brain. Multiple congenital malformations occur in many organs in hereditary hemorrhagic telangiectasia (Osler-Weber-Rendu disease).

In a large series studied by Gentry et al., 41 per cent of arteriovenous fistulae were in

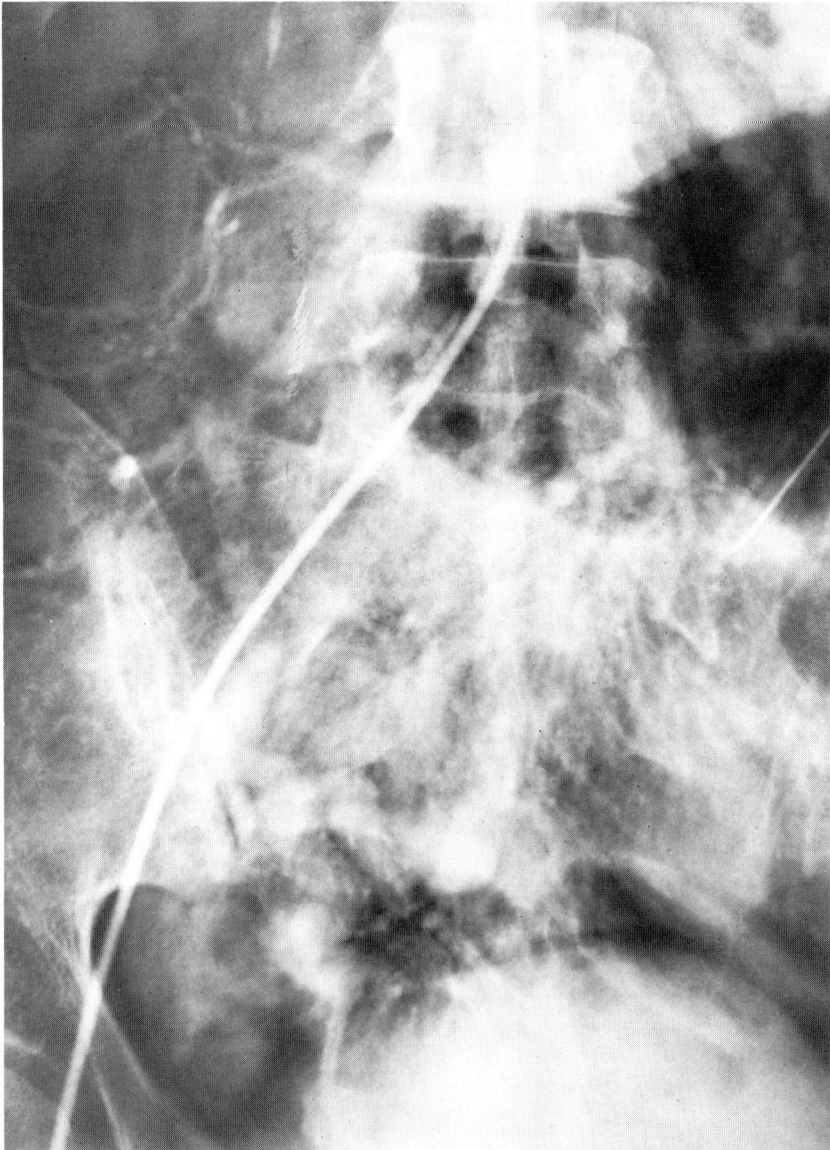

Figure 1-11. Superior mesenteric arteriogram in the venous phase showing markedly dilated ileocolic veins due to an arteriovenous malformation of the terminal ileum. Barium studies were normal.

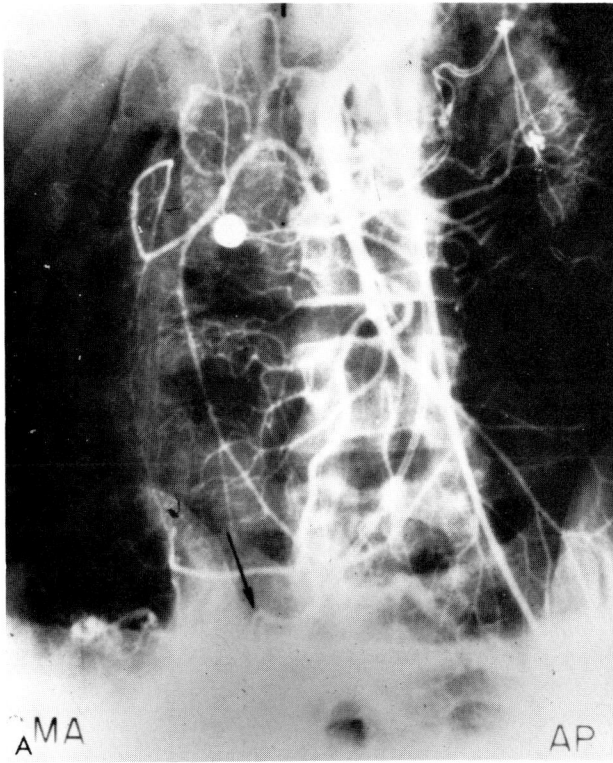

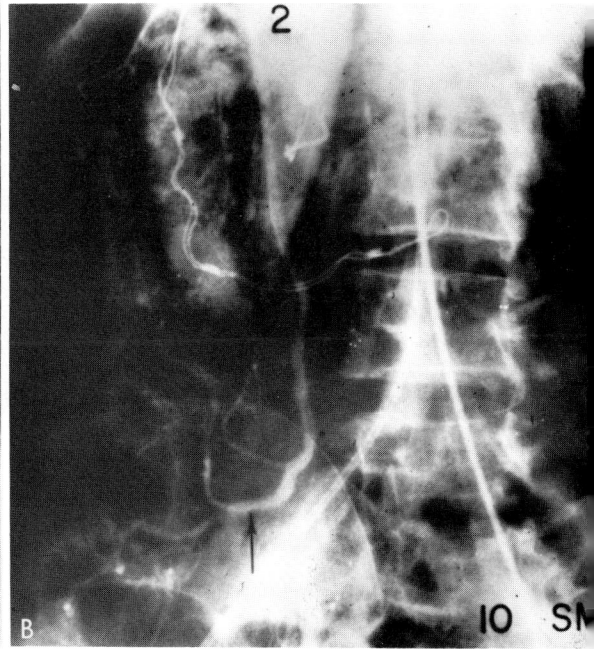

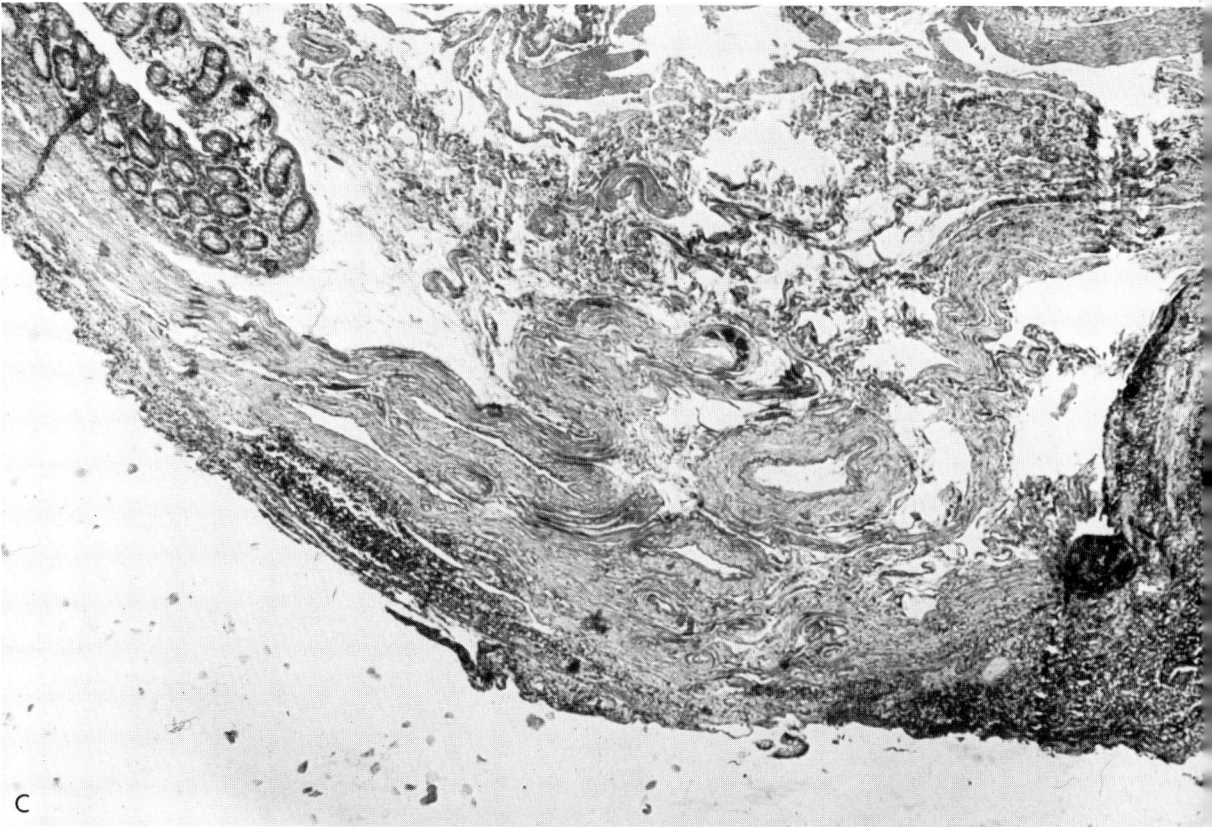

Figure 1–12. Superior mesenteric arteriogram in the arterial phase (A) and venous phase (B) showing a dilated intestinal artery and vein (arrow) due to an arteriovenous malformation in the cecum. (C) Low power histologic section showing the angiodysplasia.

the small intestine, 31 per cent were in the stomach and 26 per cent were in the colon.[10] Sixteen per cent of the lesions in the small intestine are multiple, whereas in the colon only 7 per cent are multiple.[11] Of the solitary lesions of the colon, 54 per cent are in the sigmoid colon or rectum, 27 per cent are in the cecum or ascending colon and 19 per cent are in the transverse or descending colon.[12]

Arteriovenous malformations of the small intestine are more common than was suspected before the use of mesenteric angiography. Angiography reveals the malformation either as an incidental finding or as the source of bleeding. Up to 50 per cent of patients with arteriovenous malformations of the intestine also have similar malformations in the liver.[9]

Barium studies of the intestine and in some cases laparotomy fail to reveal intestinal malformations. Mesenteric angiography demonstrates the dilated and tortuous feeding intestinal artery.[12] The associated veins fill with contrast media early and are dilated and tortuous (Figs. 1–11 and 1–12). When the patient is actively bleeding, extravasation of contrast material into the lumen may be demonstrated on the angiogram.[12,13]

REFERENCES

1. Sleisenger, M. H., and Fordtran, J. S. (Eds.): Gastrointestinal Disease. Philadelphia, W. B. Saunders Co., 1973, p. 859.
2. Berne, A. S.: Meckel's diverticulum. New Eng. J. Med., 260:690, 1959.
3. Dalinka, M. K., and Wunder, J. F.: Meckel's diverticulum and its complications with emphasis on roentgenologic demonstration. Radiology, 106:295, 1973.
4. Johns, T. N. P., Wheeler, J. R., and John, F. S.: Meckel's diverticulum and Meckel's diverticulum disease: a study of 154 cases. Ann. Surg., 150:241, 1959.
5. Rutherford, R. B., and Akers, D. R.: Meckel's diverticulum: a review of 148 pediatric patients with special reference to the pattern of bleeding and to mesodiverticular vascular bands. Surg., 59:618, 1966.
6. White, A. F., Oh, K. S., Weber, A. L., and James, A. E.: Roentgenologic manifestations of Meckel's diverticulum. Amer. J. Roent., 118:86, 1973.
7. Caffey, J.: Pediatric X-ray Diagnosis. Chicago, Year Book Medical Publishers, Inc., 1972, p. 639.
8. Marshak, R. H., and Lindner, A. E.: Radiology of the Small Intestine. Philadelphia, W. B. Saunders Co., 1970, p. 416.
9. Kittredge, R. D., Kanick, V., and Finby, N.: Arteriovenous fistulas. Amer. J. Roent., 100:431, 1967.
10. Gentry, R. W., Dockerty, M. B., and Clagett, O. T.: Vascular malformations and vascular tumors of the gastrointestinal tract. Int. Abstr. Surg., 88:281, 1949.
11. River, L., Silverstein, J., and Tope, J. W.: Benign neoplasms of the small intestine. Int. Abstr. Surg., 102:1, 1956.
12. Genant, H. K., and Ranniger, K.: Vascular dysplasias of the ascending colon. Amer. J. Roent., 115:349, 1972.
13. Whitehouse, G. H.: Solitary angiodysplastic lesions in the ileocecal region diagnosed by angiography. Gut, 14:977, 1973.

Chapter 2

THE ILEOCECAL VALVE

In the train of logic involved in the analysis of roentgen findings, successful detection of an abnormality on a radiograph often depends on the comparison of one side with the other. Thus, in the analysis of the lungs, the kidneys, the brain and the bones of the extremities, one can compare the involved organ with its opposite fellow to determine if an abnormality exists. A second technique of radiographic analysis relies on the presence of a recurring pattern that appears throughout a structure, such as the appearance of the folds of the intestinal mucosa, the arborization pattern of blood vessels and the bifurcations of the tracheobronchial tree. For those structures that are both unpaired and form no consistent continuum; pattern recognition still comes into play by recalling a mental image of the structure as it has been learned by previous experience. This is the case in the radiologic evaluation of the heart, the sella turcica, the stomach, the pyloric canal and the duodenal bulb, to mention a few examples. The ileocecal valve falls into this category. Hence, it is clear that every radiologist must establish his own repertoire of mental images of the radiologic appearance of the normal valve in order to distinguish cases in which an abnormality exists. In order to add to this experience, the first section of this chapter is devoted to the normal ileocecal valve and its variations.

The ileocecal valve is located at the junction of the ileum and colon, almost always at the level of the first complete haustral segment above the tip of the cecum. It demarcates the cecum below from the ascending colon above. The valve functions as a true sphincter in the sense of controlling the rate of ingress of ileal contents into the colon. In some individuals it undoubtedly plays a role in regulating the retrograde passage of feces and gases accumulated in the colon.

THE NORMAL VALVE: ANATOMIC APPEARANCE AND RADIOLOGIC PRESENTATION

Scrupulous attention to the details of radiographic technique is essential for the precise radiologic identification of the anatomic features of the ileocecal valve. Properly exposed radiographs made with careful graded compression under fluoroscopic control with the patient in an optimal position both prone and supine are essential. Barium distention and air contrast enema techniques are supplemental examinations, and both should be performed for maximal radiographic evaluation of the ileocecal valve. Small bowel barium studies are also valuable.

Familiarity with the wide range of normal in the size, shape and location of the valve is necessary for accurate differentiation of a normal valve from the abnormal. Otherwise, incorrect radiologic diagnoses will result in unnecessary surgery. In addition, it is important to recognize the valve as an anatomic landmark in the course of performing a barium enema examination. Unless the terminal ileum or the appendix fills with barium or unless the ileocecal valve is identified, it is impossible to be certain that the cecum has been completely visualized. Appropriate compression spot films of the cecum and knowledge that the ileocecal valve is at the level of the first complete haustral fold of the colon are useful in identifying the valve.

The ileocecal valve commonly presents as a round or oval protuberance arising from the medial and posterior wall of the colon at the junction of the cecum and ascending colon (Fig. 2–1). The size and location of the valve vary widely, so that it sometimes can be identified on the posterior or even on the lateral wall of the colon. Fleischner and

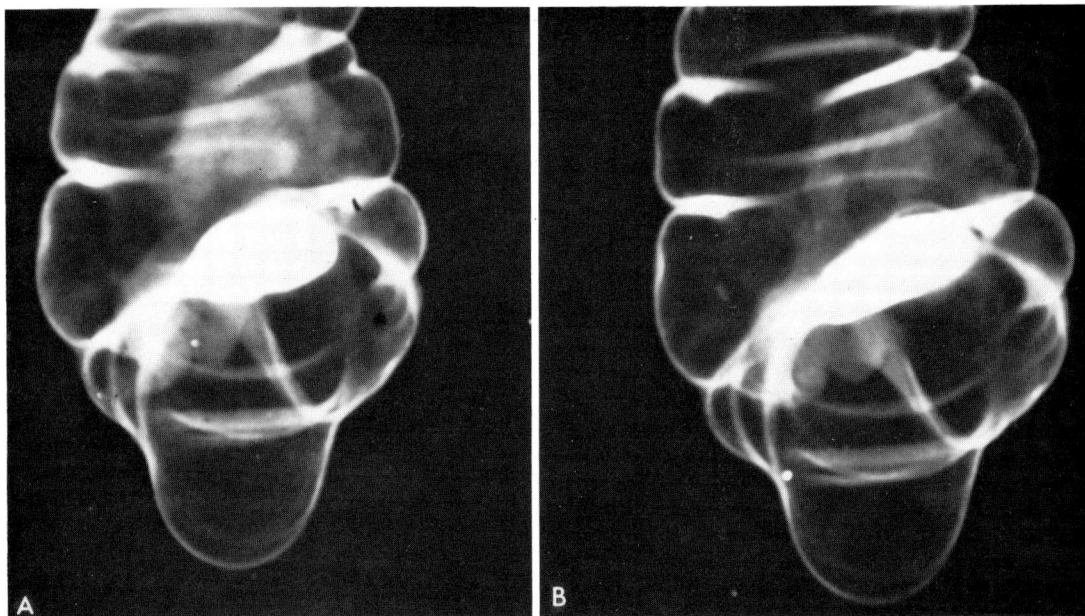

Figure 2-1. Radiograph of a specimen of a normal cecum with the mucosa and ileocecal valve coated with barium. Two degrees of distention (*A* and *B*) showing the valve en face. Note that the valve occurs at the level of the first major haustral fold distal to the tip of the cecum. The size of the valve varies with the degree of distention.

Bernstein noted posterior insertion of the valve in 7 per cent of their cases studied at autopsy.[1] Often the valve is best seen in profile by rotating the patient into the right posterior oblique projection, but because variations are common, the best position for spot-filming must be determined on an individual basis under fluoroscopic control.

While it is unusual for the valve to be larger than 3 cm. in diameter with the typical geometry of spot-filming, it is inadvisable to attach strict measurements to a maximum "normal" size. It is a common experience, for example, for a valve visualized by moderate compression in a partially distended colon to appear larger and plumper than the same valve examined in a completely distended colon without local pressure. The presence of symmetry of the valve and the recognition of a normal overlying mucosal pattern are more important determinants of normal than criteria that depend on measurements.

When the valve is viewed en face, an inconstant stellate pattern in the midportion representing the orifice of the terminal ileum can frequently be identified. The stellate appearance is a reflection of the

pliability of the mucosa overlying the valve and is in itself a good indicator that the valve is normal (Fig. 2-2). Buirge found no correlation between the size and shape of the opening and the type of valve.[2]

Fleischner and Bernstein noted fusion of the adjacent walls of the ileum and cecum from 2 to 5 cm. in length in several of their cases. This should not be regarded as an inflammatory change when it is detected radiographically.[1]

LIPOMATOUS INFILTRATION OF THE ILEOCECAL VALVE

Lipomatous infiltration of the ileocecal valve is characterized by enlargement of the valve due to submucosal infiltration with fat[3] (Fig. 2-3). Lack of a distinct capsule differentiates the fat in this condition from a true lipoma arising from the lips of the valve. A number of other terms have been used to describe the valve in lipomatous infiltration including hypertrophy of the ileocecal valve, lipohyperplasia and fatty degeneration. In most cases the ileocecal area is otherwise normal, although lipoma-

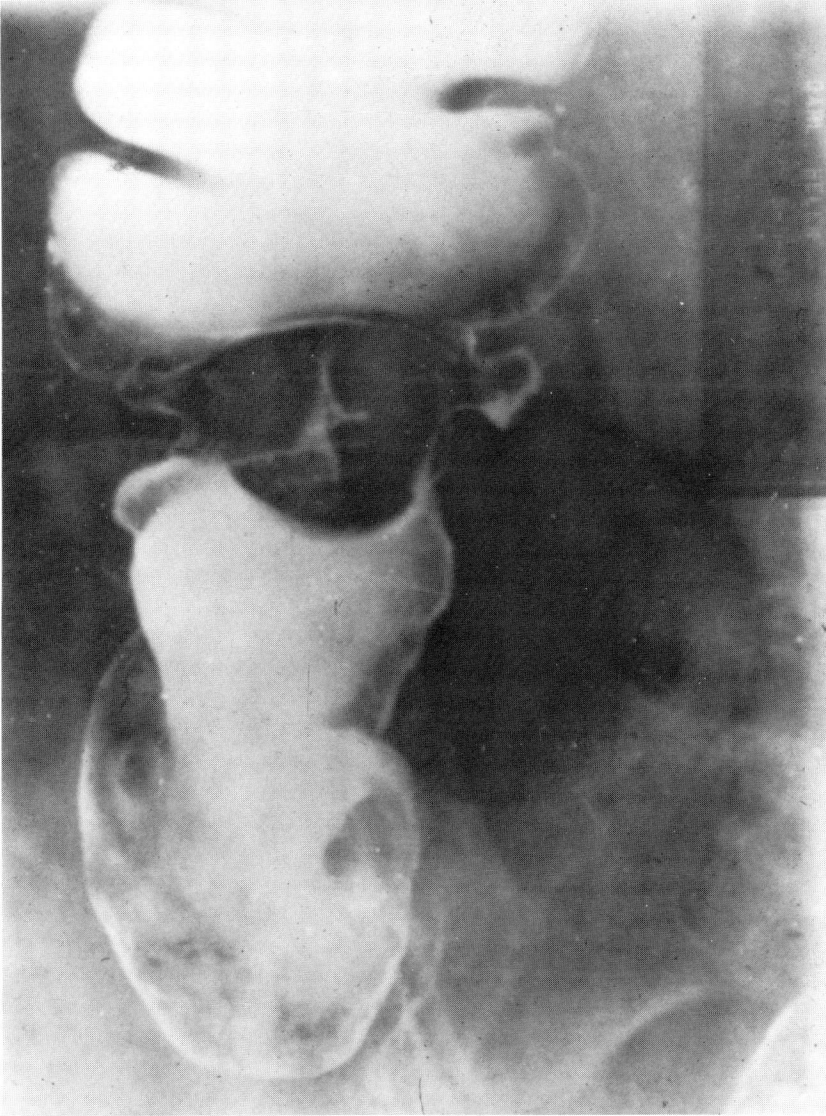

Figure 2–2. Barium enema examination showing a prominent ileocecal valve viewed en face. The stellate appearance at the center of the valve is the ileal orifice. Mucosal folds radiate from the opening, indicating that the mucosa is pliable. This suggests that the valve, while prominent, is normal.

tosis of the valve is a fairly common feature of Crohn's disease of the terminal ileum.

Miscalculations of actual valvular size occur with some frequency and depend on observer experience. As noted above, rigid measurements are not satisfactory because of variations in size due to distention of the cecum and compression and magnification. Lasser and Rigler measured normal valves and determined that the range of normal for either lip is up to 1.5 cm. in thickness.[4]

Hinkel suggested that 4 cm. is the upper limit of normal for the valve including both lips.[5]

Lasser and Rigler proposed an etiologic classification of conditions producing enlargement of the ileocecal valve.[4] This includes edema (idiopathic and post-traumatic), submucosal fat accumulation, herniation of ileal mucosa, tumors (benign and malignant) and inflammatory lesions involving the valve. The cause of edema is

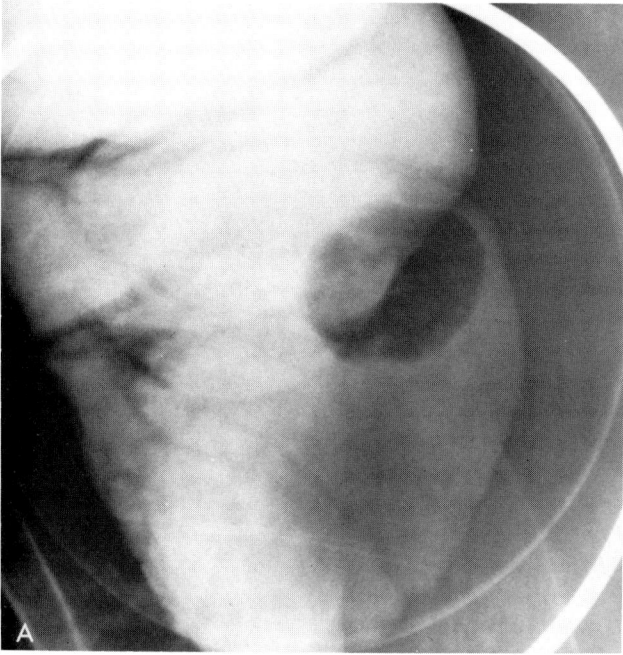

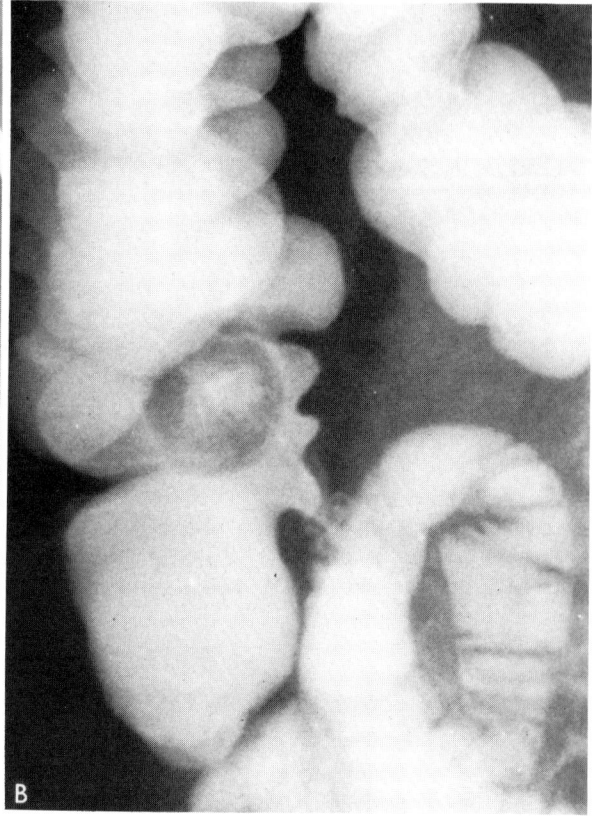

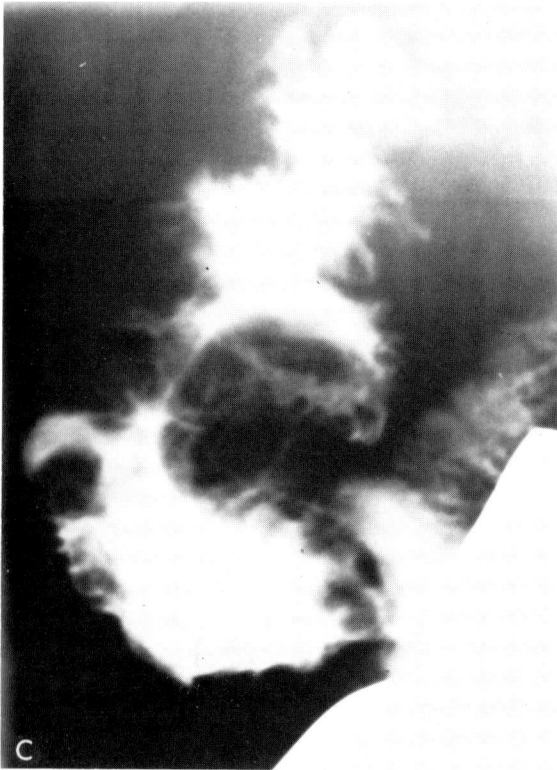

Figure 2–3. Barium enema and air contrast examinations from five different patients showing examples of lipomatous hypertrophy of the ileocecal valve (A to E). The surface of the valve is smooth, distinguishing it from tumor.

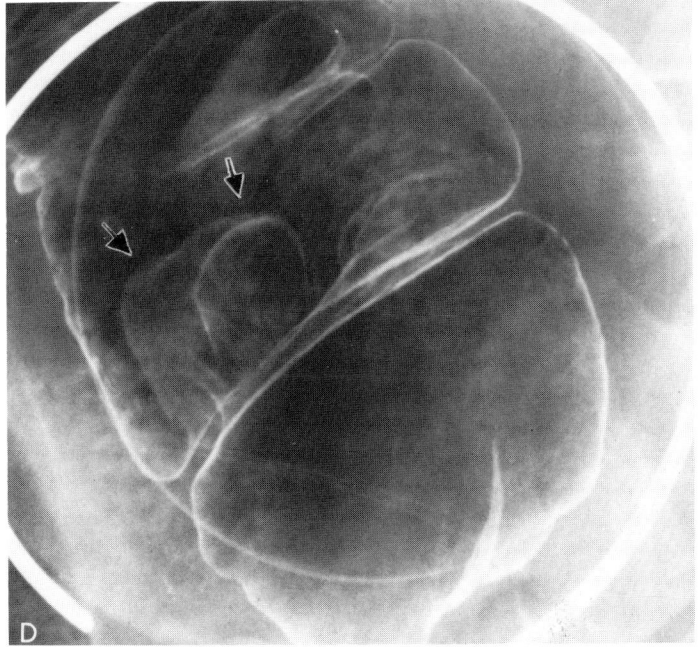

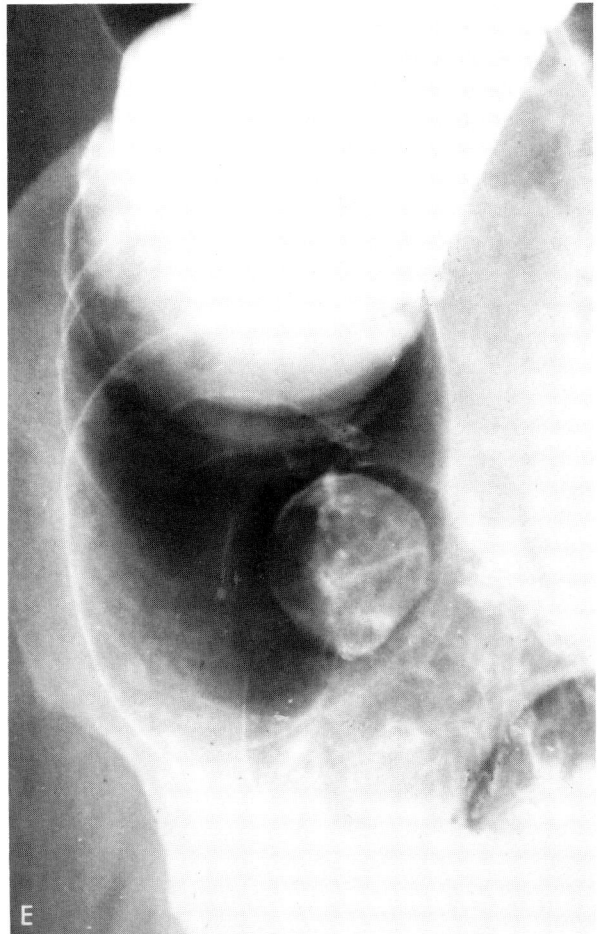

Figure 2-3. (Continued)

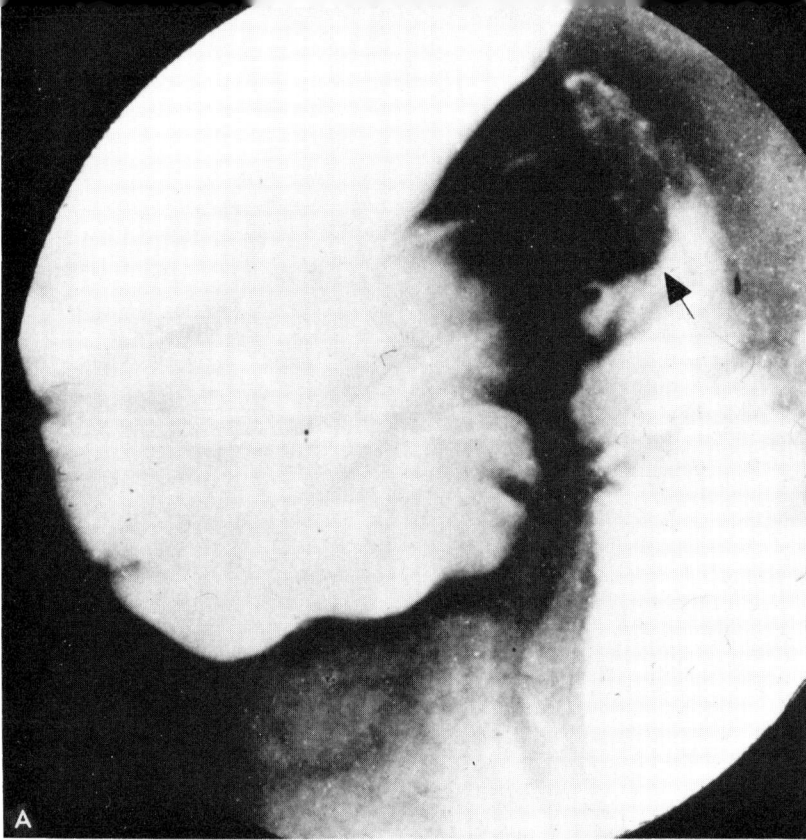

Figure 2-4. Barium enema examination (*A*) showing retrograde prolapse of the lower lip of an ileocecal valve into the ileum due to the hydrostatic pressure of the barium enema. The valve simulates a tumor or polyp in the terminal ileum (arrow). (*B*) Histologic section of the lower lip of the ileocecal valve showing submucosal edema. (From Rigler, L. G., and Lasser, E.: Prolapse of the lower lip of the ileococal valve into the terminal ileum. Amer. J. Roent. 65:878, 1951. Courtesy of Charles C Thomas, Publisher.)

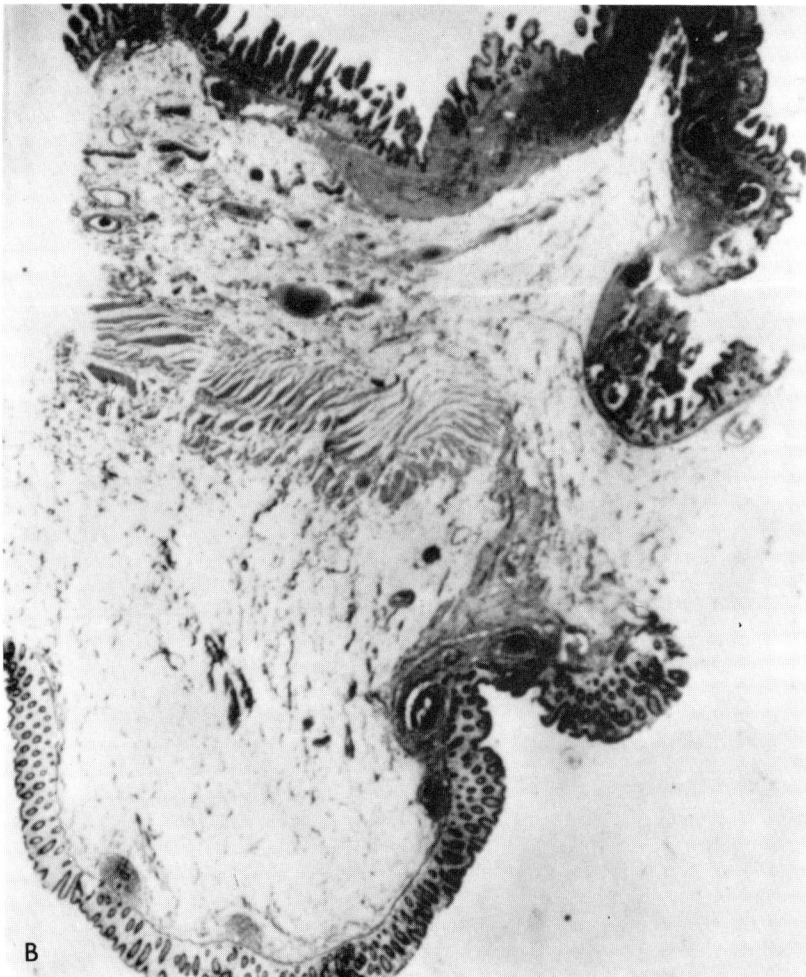

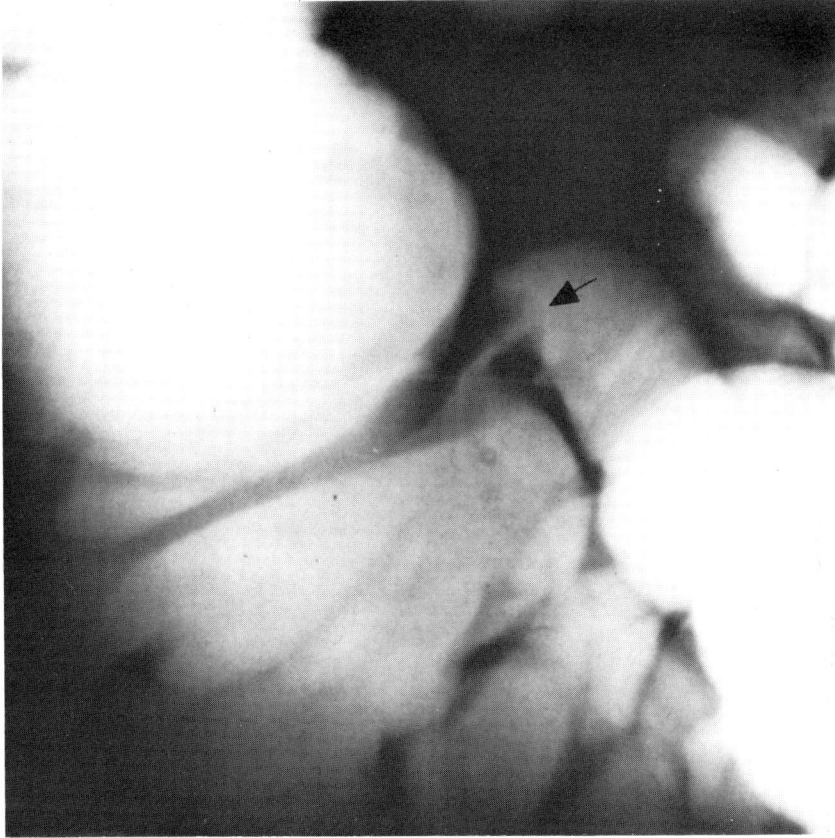

Figure 2–5.   Barium enema examination showing slight retrograde prolapse of the ileocecal valve into the terminal ileum (arrow).

obscure, although occasionally it is due to intermittent intussusception. When tumors involve the valve, a distinct mass may be identified attached to an otherwise normal valve. Idiopathic enlargement of the valve may on occasion be accompanied by retrograde prolapse of one or both lips into the ileum due to the hydrostatic pressure present in the cecum during the barium enema examination[6] (Figs. 2–4 and 2–5). In these cases, care must be taken to distinguish prolapse from a pedunculated tumor.

Lipomatosis is uncommon before the age of 40 and is usually seen between 50 and 70 years of age. The mean age in one series was 42.5 years.[7] Females predominate 2 to 1. Kelby found that 50 per cent of patients with lipomatosis have clinical manifestations[8], while 82 per cent of Lasser and Rigler's series had abdominal complaints.[4] Castleman suggests that the symptoms may be due to chronic intussusception, the enlarged valve producing intermittent coloco-

lic intussusception.[9] Lasser and Rigler propose that a cycle is established in which the enlarged valve tends to cause intussusception, and each intussusception causes further enlargement of the valve.[4] Another possibility is that the enlarged valve produces a partial obstruction at the ileocecal junction. Whether a patient's symptoms are actually related to an enlarged valve is difficult to determine. Lasser and Rigler have suggested that a syndrome may exist consisting of bloating, right-sided abdominal pain and tenderness over the ileocecal valve.[4]

When the valve is enlarged due to fatty infiltration, the radiographic appearance is that of a fairly smooth mass which is sharply demarcated from the surrounding bowel mucosa. While the surface is smooth, lobulations in the contour of the normal valve are not infrequent[10] (Figs. 2–6 to 2–8). These undulations are probably due to increased tone of the muscularis propria and

(*Text continued on page 26*)

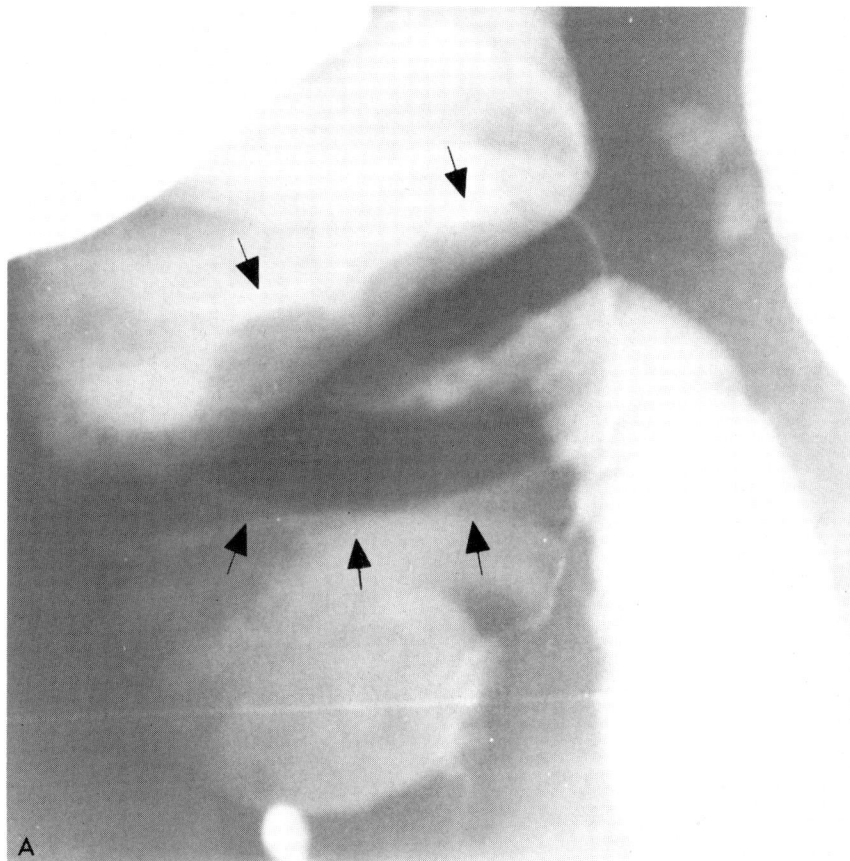

Figure 2-6. (A) Anterior-posterior and (B) left posterior oblique views of barium enema examination showing enlargement of the ileocecal valve due to fatty infiltration (arrows). Note the lobulated appearance of the valve in B due to contraction of the muscularis propria. (C) Photograph of the ileocecal valve made at surgery. The cecum has been opened, and there is diffuse enlargement of the valve (arrows). (From Berk, R. N., et al.: Lipomatosis of the ileocecal valve. Amer. J. Roent., *119*:323, 1973. Courtesy of Charles C Thomas, Publisher.)

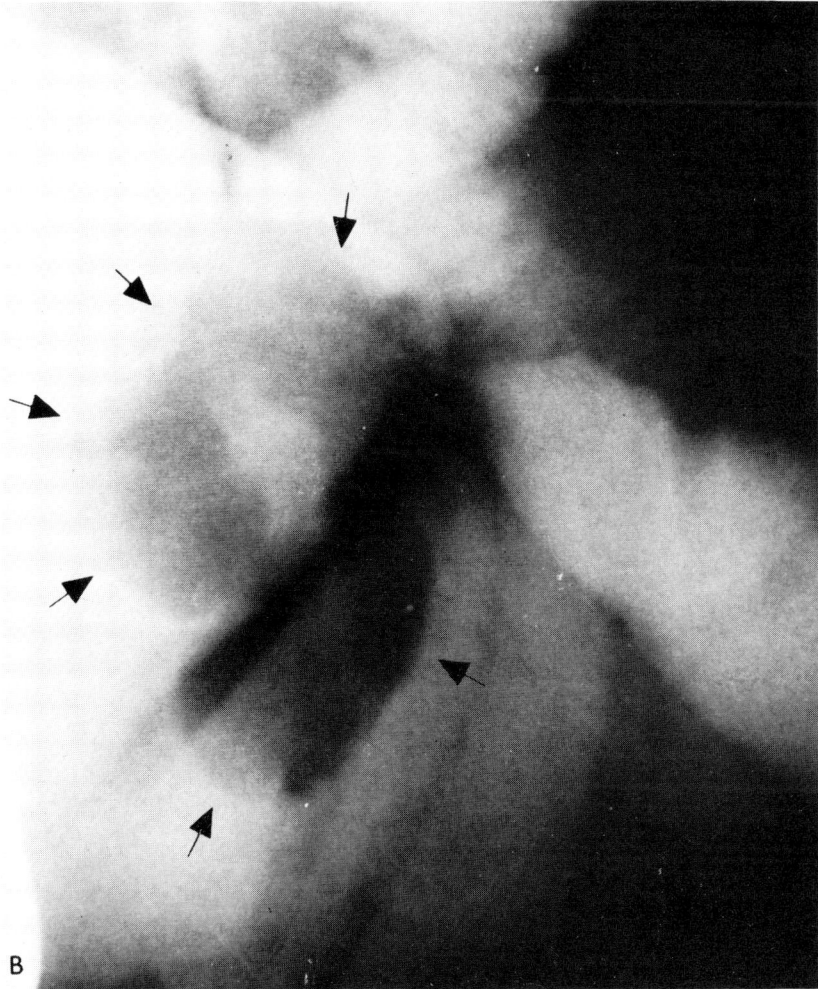

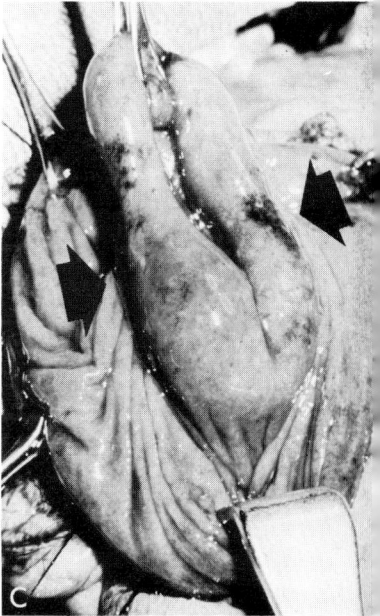

Figure 2-6. (Continued)

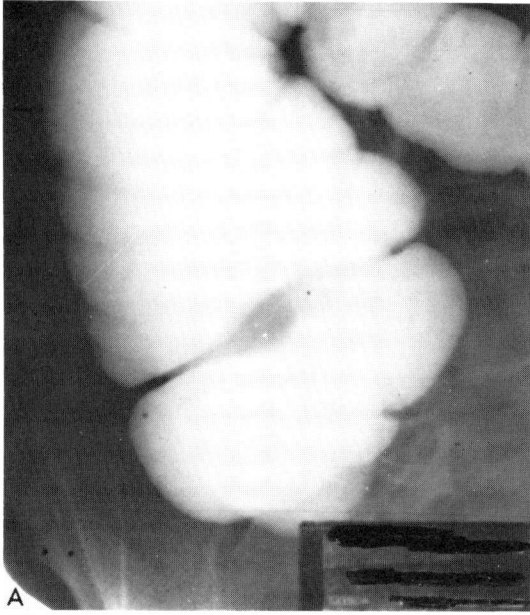

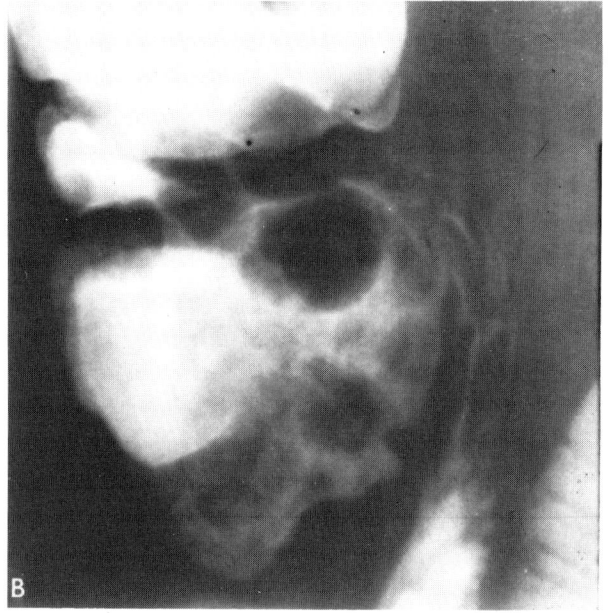

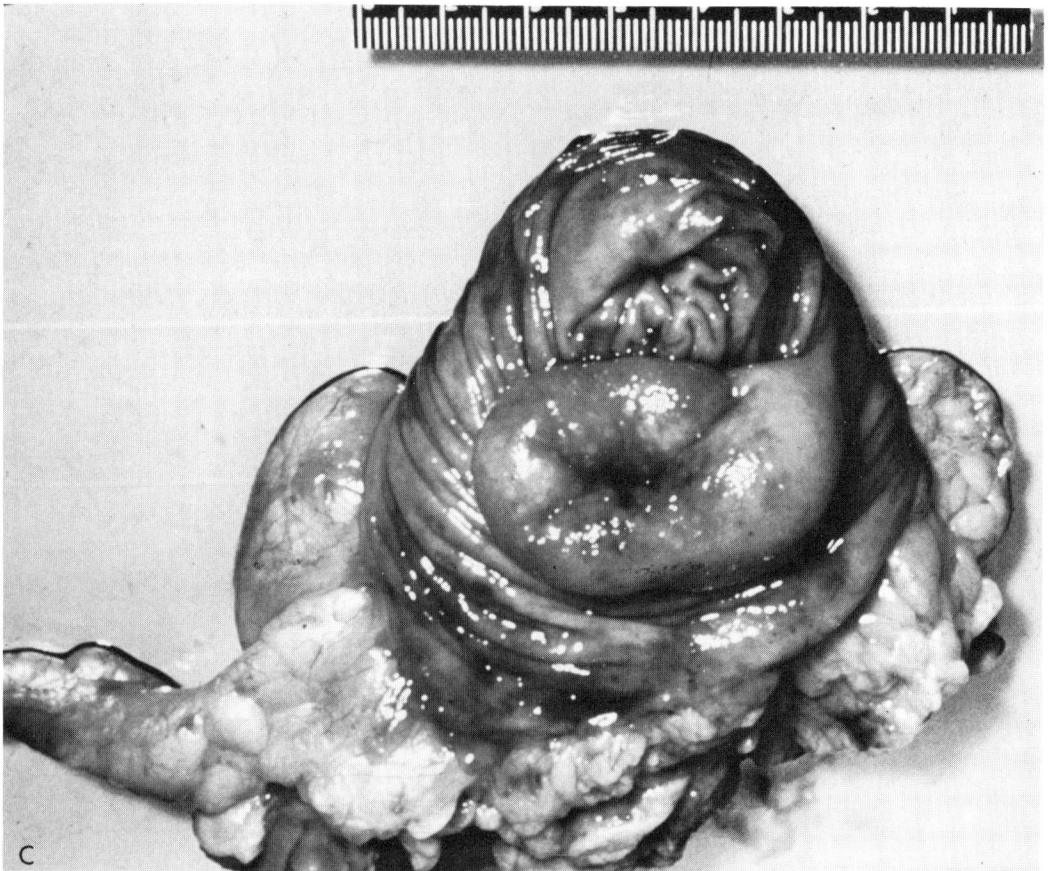

Figure 2–7. Barium enema examination (*A* and *B*) showing lobular enlargement of the ileocecal valve on the basis of lipomatous infiltration. The appearance is similar to that noted in the valve shown in Figure 2–6. (*C*) Photograph of the valve made following surgery.

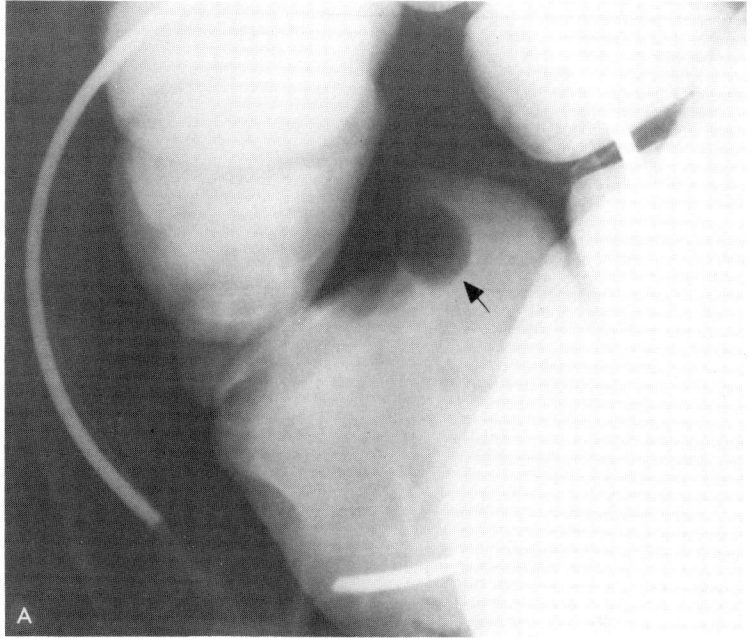

Figure 2–8. Barium enema examination (*A* and *B*) showing irregular enlargement of the ileocecal valve. The lobular character of the lower lip simulates a lipoma of the valve (arrow), but at surgery no discrete lipoma was evident.

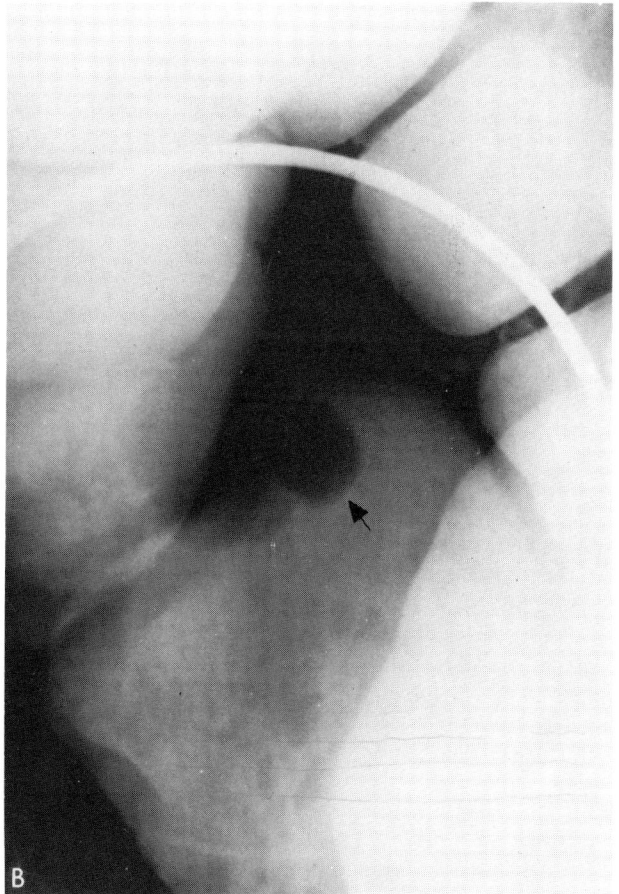

should not be mistaken for disease. Boquist et al., studying gross specimens in 10 cases of lipomatosis, noted enlargement of the valve with a polypoid appearance in several cases.[7] Histologically their cases showed diffuse fatty infiltration in the submucosa which often formed polypoid protrusions. Attention should be given to whether or not the surface of the valve is smooth or shaggy rather than to whether there is nodularity. However, when such irregularities are prominent, difficult diagnostic problems arise. In some instances an accurate diagnosis can be established only by colonoscopy or surgical exploration.

TUMORS

Benign Tumors of the Ileocecal Valve

The same variety of benign lesions that arise in the small bowel and colon occur in the ileocecal valve itself. By far the commonest benign lesion of the valve is a lipoma. Histologic differentiation of a lipoma from lipomatous infiltration depends on the presence of a true capsule in a lipoma. Radiographically, lipomas present as sharply circumscribed rounded masses arising from either lip of the valve (Figs. 2–9 to 2–14). On occasion, films of optimal quality demonstrate increased radiolucency in comparison to the adjacent water-density tissues. Neither lipomas nor lipomatous infiltration need constitute a source of clinical concern except in unique circumstances. Adenomatous polyps and villous adenomas occurring in association with lipomatous hyperplasia of the valve have been described, and excoriations or ulcerations of the mucosa overlying enlarged valves have been responsible for intestinal bleeding of considerable extent in some instances.

With the exception of the lipomas, benign lesions arising from the valve are for the most part impossible to separate one from another and are often difficult to differentiate from malignant tumors that originate on the valve. The surface of adenomatous polyps and villous adenomas is usually irregular and somewhat shaggy in comparison to lipomas and simple fatty infiltration which, although often lobulated, are invariably smooth (Figs. 2–15 and 2–16). When a discrete enlargement of one portion

of the valve is detected which cannot be clearly delineated as simple hypertrophy or a true lipoma, the nature of the mass must be verified by some method other than barium enema study to rule out the presence of a premalignant or a malignant lesion.

Carcinoid Tumors Involving the Ileocecal Valve

A carcinoid tumor localized to the valve may be indistinguishable from other tumors of the valve. Several features, however, may in some instances be helpful: carcinoids are often centered at the orifice of the valve rather than arising discretely from an upper or lower lip; they tend to elevate the mucosa rather than to infiltrate or ulcerate, and hence are smoothly demarcated; and they may on occasion produce a corrugated appearance of the adjacent cecal wall (Figs. 2–17 and 2–18). This, presumably, reflects edema and spasm of the wall and may be related to plugging of the lymphatics by carcinoid tumor cells (Fig. 2–18C). However, this appearance is not specific for carcinoid and may be found in a number of other circumstances including inflammatory lesions in the cecum (Fig. 2–19).

Adenocarcinomas and Other Malignant Tumors Arising from the Ileocecal Valve

Approximately 2 per cent of adenocarcinomas of the colon arise from the ileocecal valve. These are generally broad-based polypoid tumors with a typical irregular mucosal surface (Fig. 2–20). In some instances, however, the surface of the lesion is deceptively smooth, so that the carcinoma may be indistinguishable from benign lesions arising from the valve (Figs. 2–21 and 2–22). Contrarily, villous adenocarcinomas often have a frondlike shaggy surface that is typical of these lesions (Fig. 2–23).

Colocolic intussusception of a carcinoma arising in the ileocecal valve may occur and prevent the radiologic demonstration of the valve itself, except when the intussusception is successfully reduced by the barium enema examination (Fig. 2–24).

(Text continued on page 47)

Figure 2-9. Barium enema examination showing a lipoma of the ileocecal valve. Changeability of the shape of the lesion may permit an accurate diagnosis of lipoma.

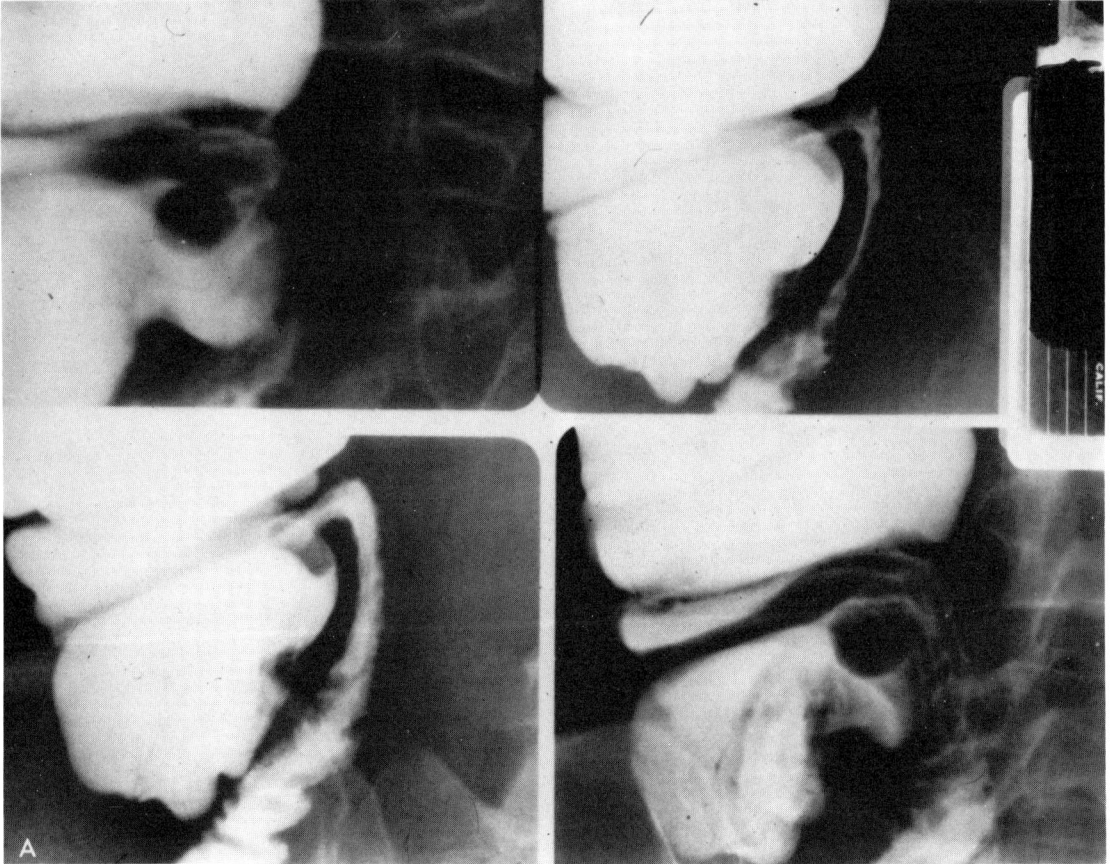

Figure 2–10. Barium enema examination (A) showing a characteristic lipoma protruding from the lower lip of an ileocecal valve.

(Legend continued on opposite page.)

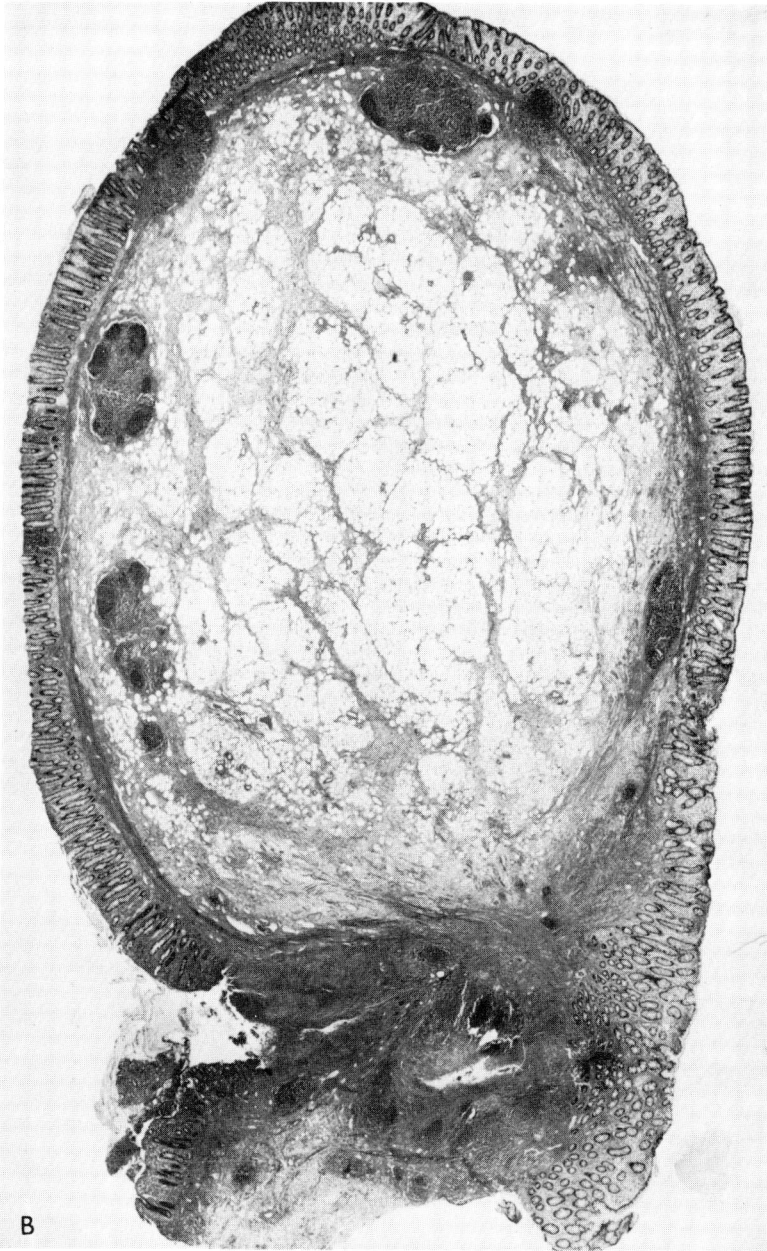

Figure 2–10. (*Continued*). (B) Histologic section of the entire lipoma showing complete encapsulation.

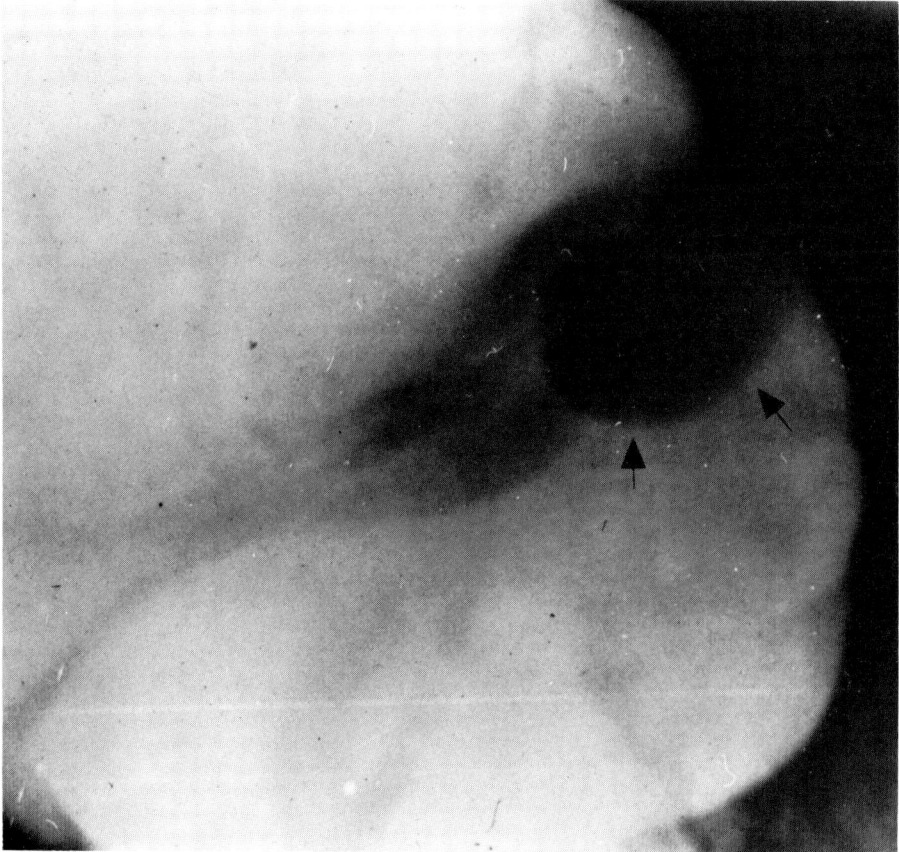

Figure 2-11. Barium enema examination showing another example of a lipoma of the lower lip of the ileocecal valve with characteristic radiologic features (arrows).

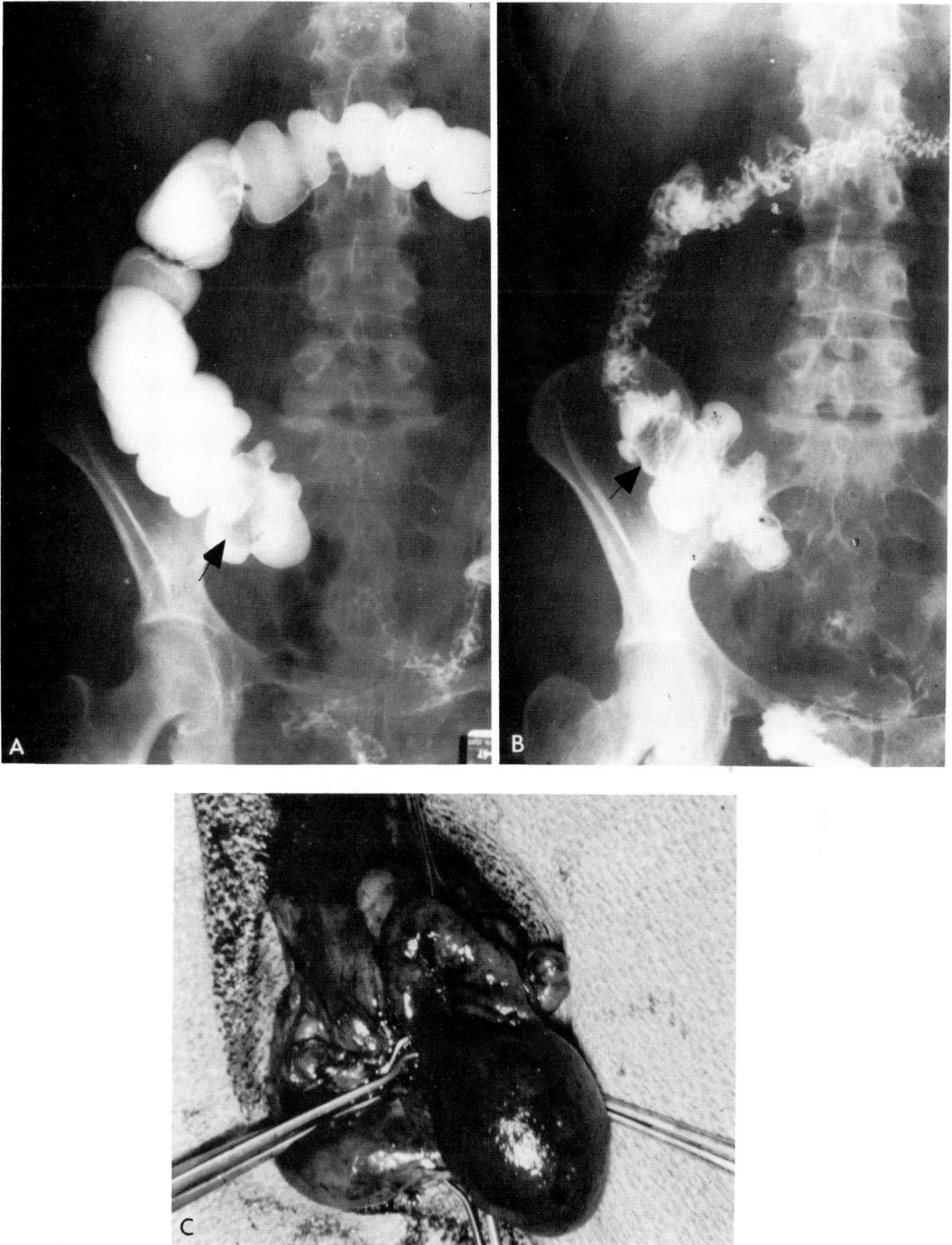

Figure 2-12. Barium enema examination (A and B) showing a lipoma of the ileocecal valve (arrows). The mass is at the level of the valve. (C) Photograph made at surgery showing a lipoma attached to the ileocecal valve. The smooth, rounded and encapsulated appearance of this benign tumor is evident.

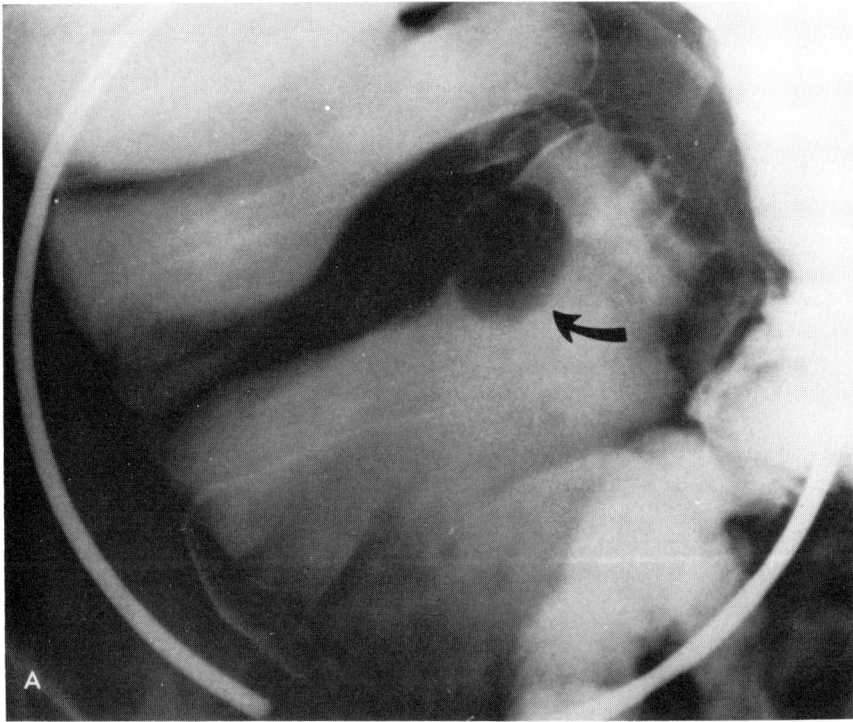

Figure 2–13. (A) Barium enema. (*Legend continued on opposite page.*)

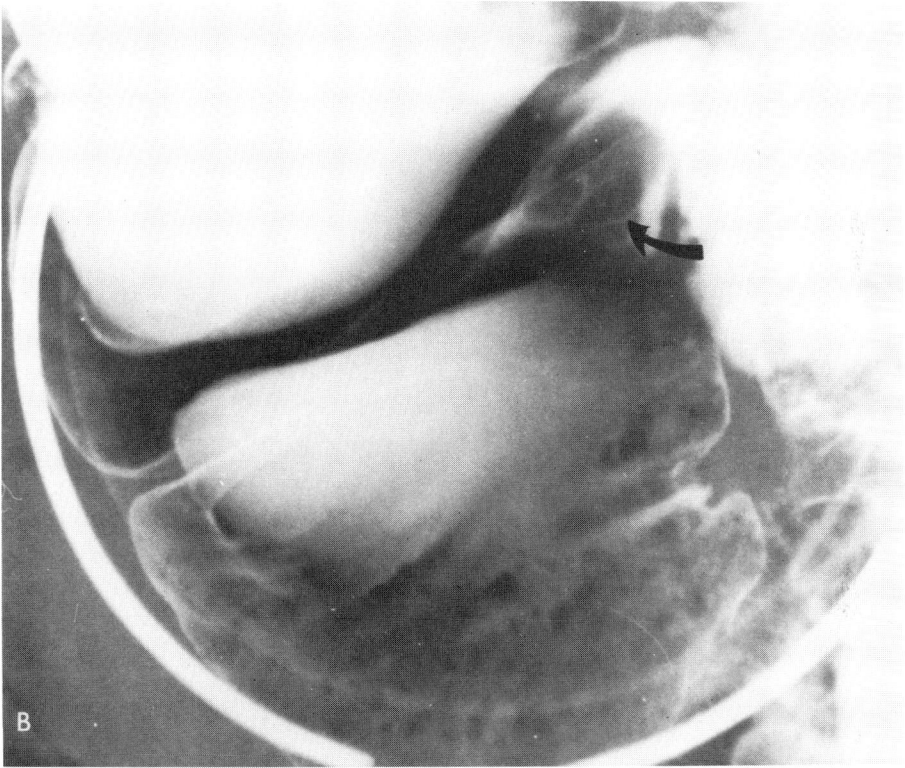

Figure 2–13. (*Continued*). (*B*) Air contrast study. Both *A* and *B* show a characteristic lipoma arising from the lower lip of the ileocecal valve (arrows). Prolapse of the lipoma into the ileum is visible in *B* (arrow).

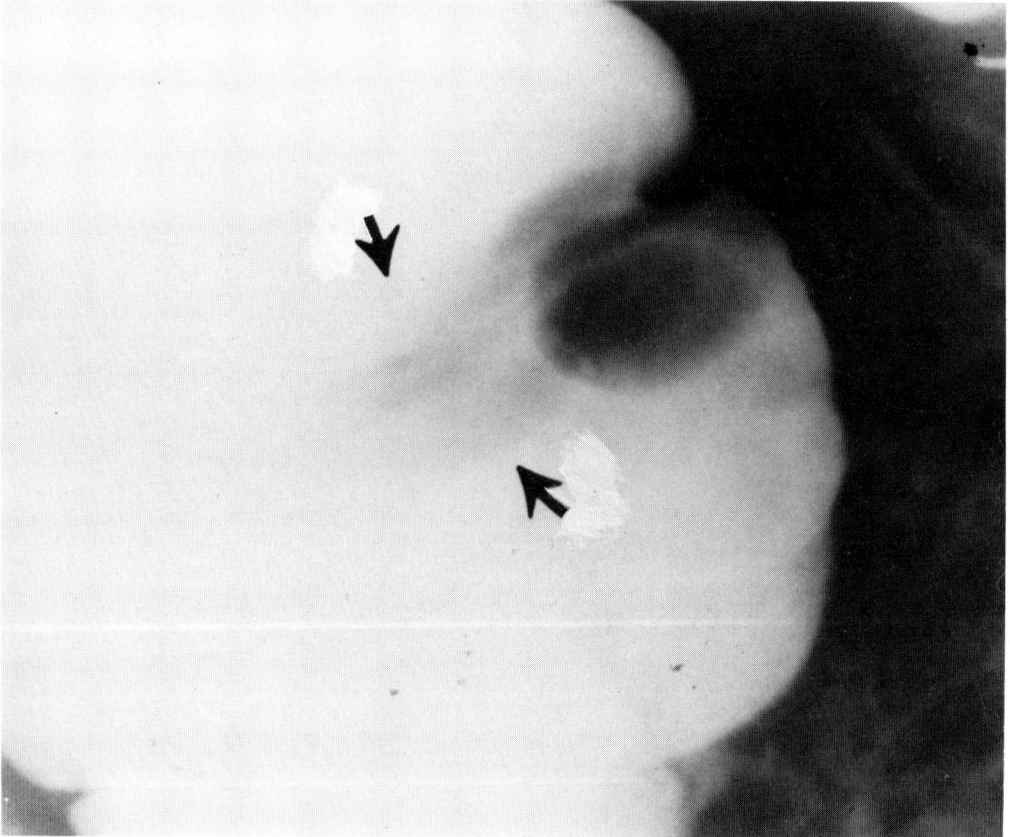

Figure 2–14. Barium enema examination showing another characteristic example of a lipoma of the lower lip of the ileocecal valve. The ileocecal valve is faintly seen (arrows).

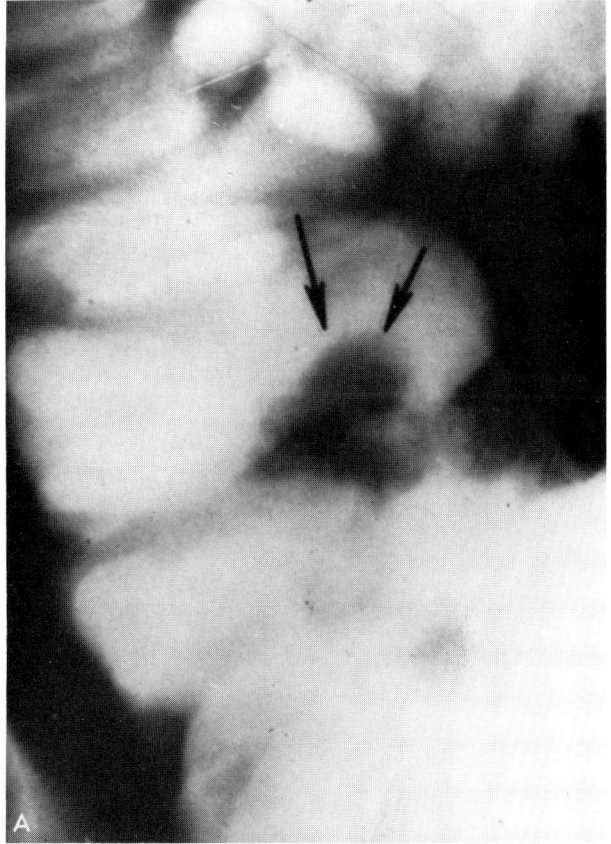

Figure 2–15. Barium enema examination (A) showing a villous adenoma arising from the upper lip of the valve (arrows). The typical frondlike surface characteristically associated with this lesion is not well seen. (B) Histologic section of the villous adenoma arising from the valve.

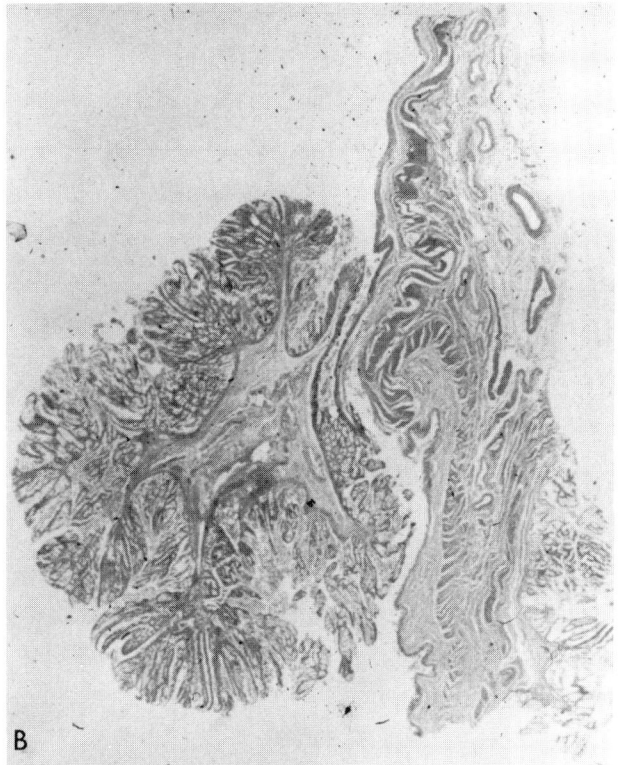

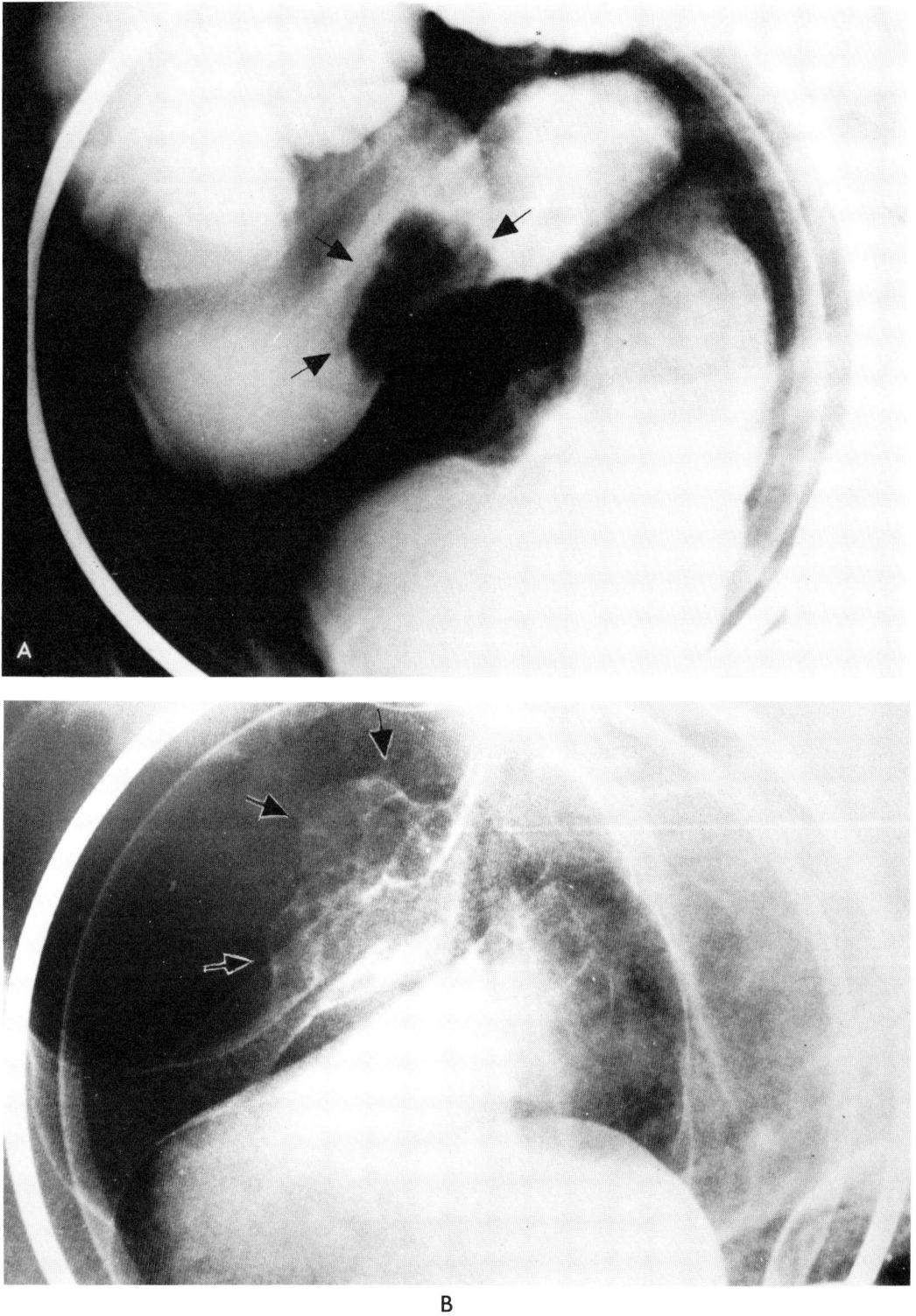

Figure 2-16. Barium enema (*A*) and air contrast examinations (*B*) showing an adenomatous polyp arising from the valve (arrows).

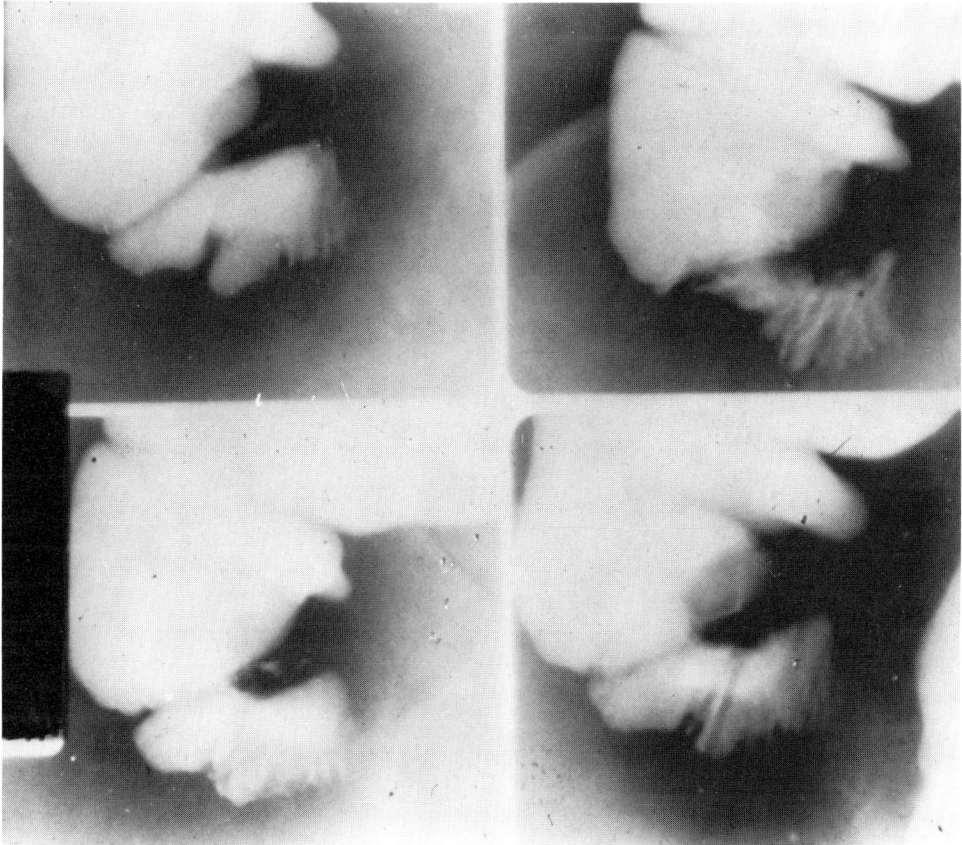

Figure 2–17. Four spot radiographs made during a barium enema examination showing a carcinoid tumor involving the ileocecal valve. The corrugated appearance on the opposite wall of the cecum is well shown.

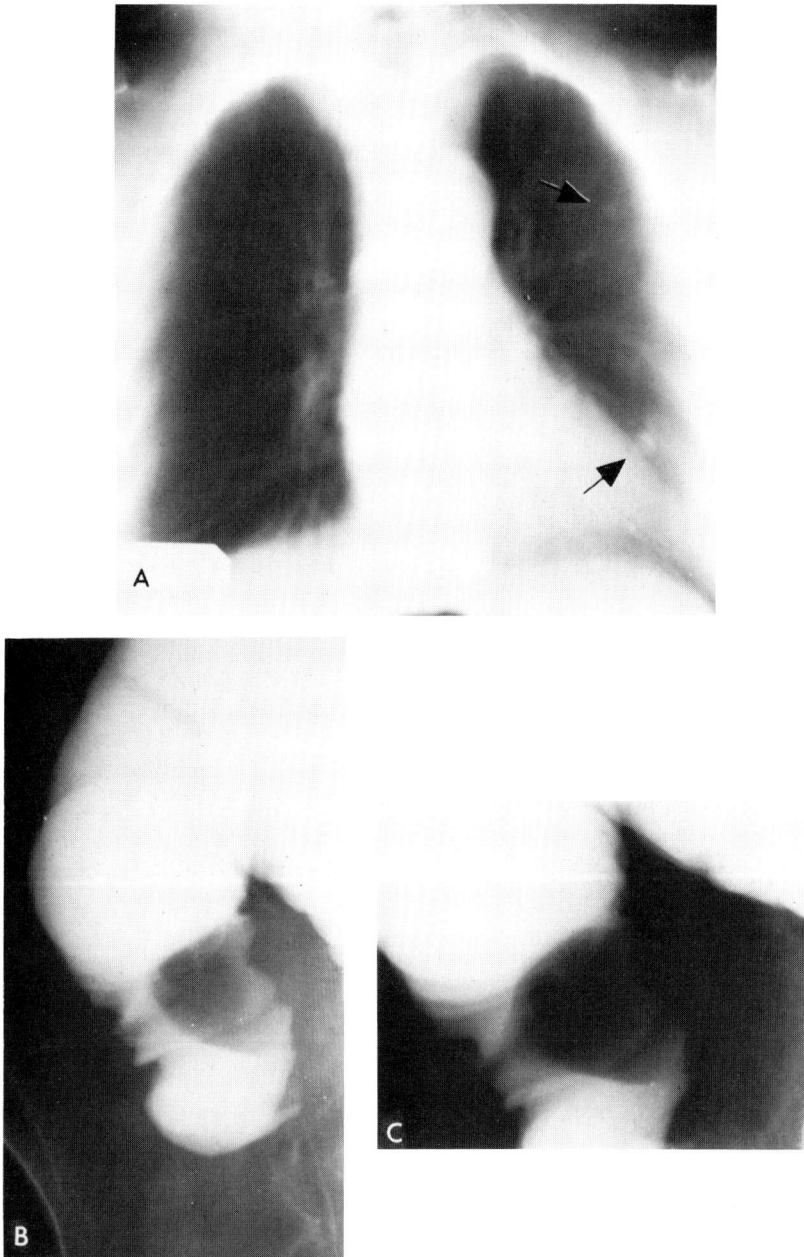

Figure 2–18. Laminogram of the chest (A) and radiographs of the ileocecal valve (B and C) in a patient with a large carcinoid tumor involving the ileocecal valve with pulmonary metastasis (arrows). The sharply demarcated edges and smooth surface of the enlarged valve are well shown. The corrugated appearance of the opposite wall of the cecum is again evident.

(Legend continued on opposite page.)

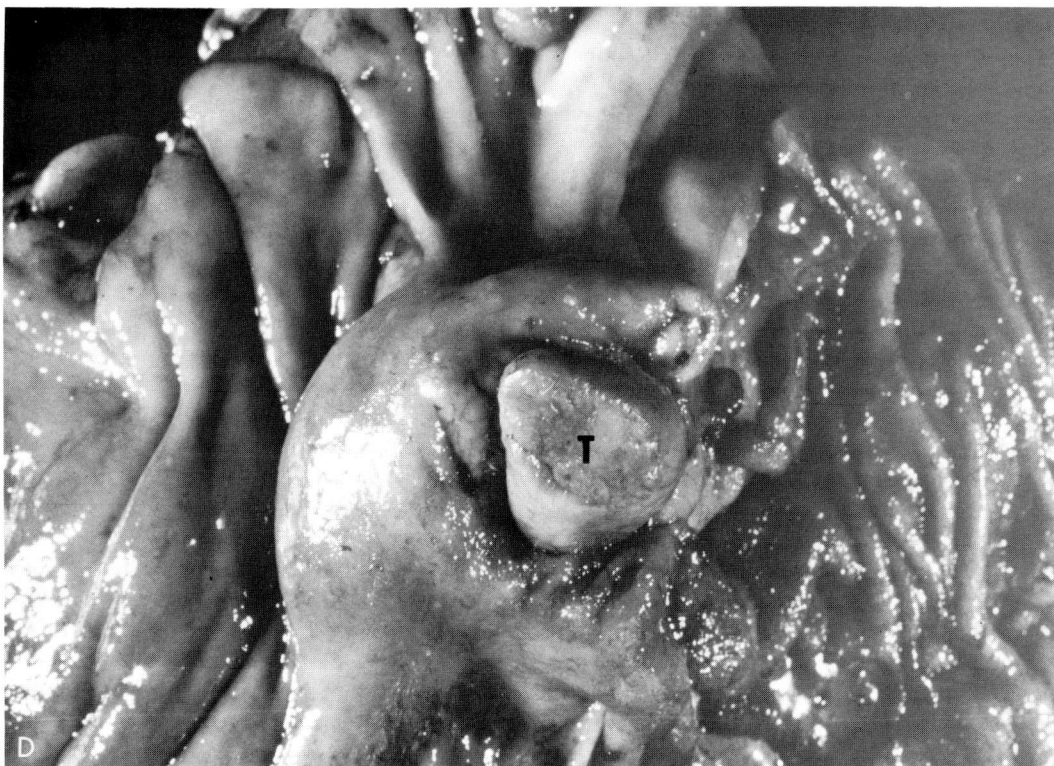

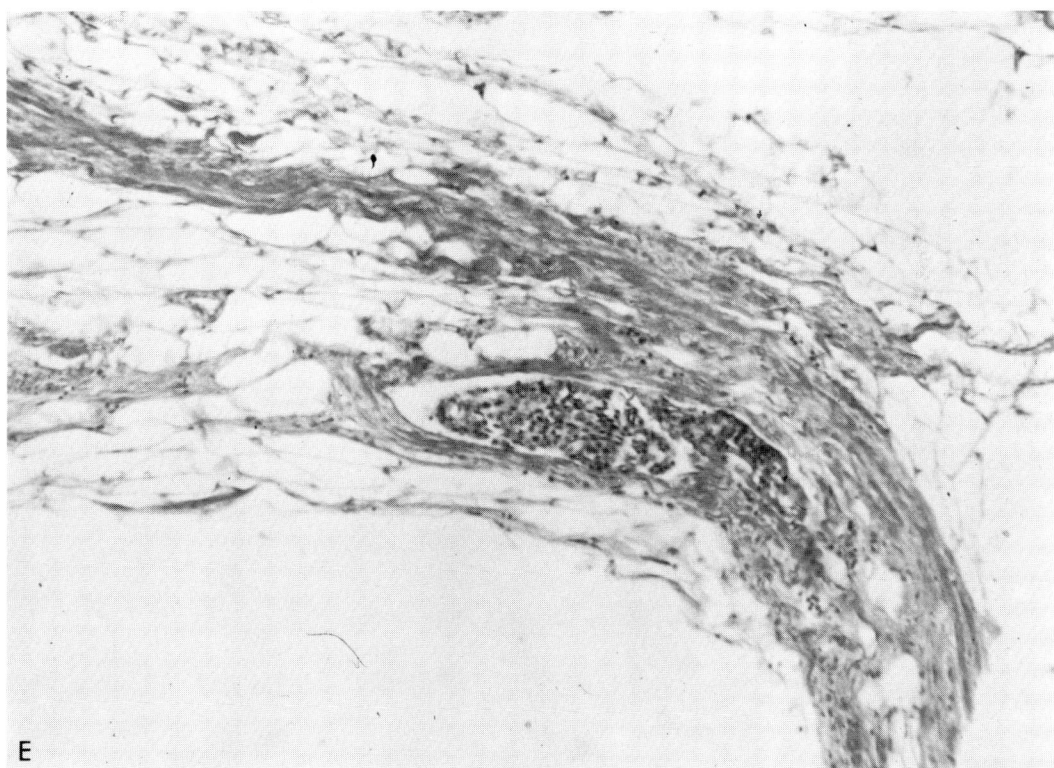

Figure 2–18. (Continued). (*D*) Gross specimen of the carcinoid. A segment of tumor protrudes through the enlarged valve (T). (*E*) Histologic section of the ileocecal valve showing nests of carcinoid within the lymphatics.

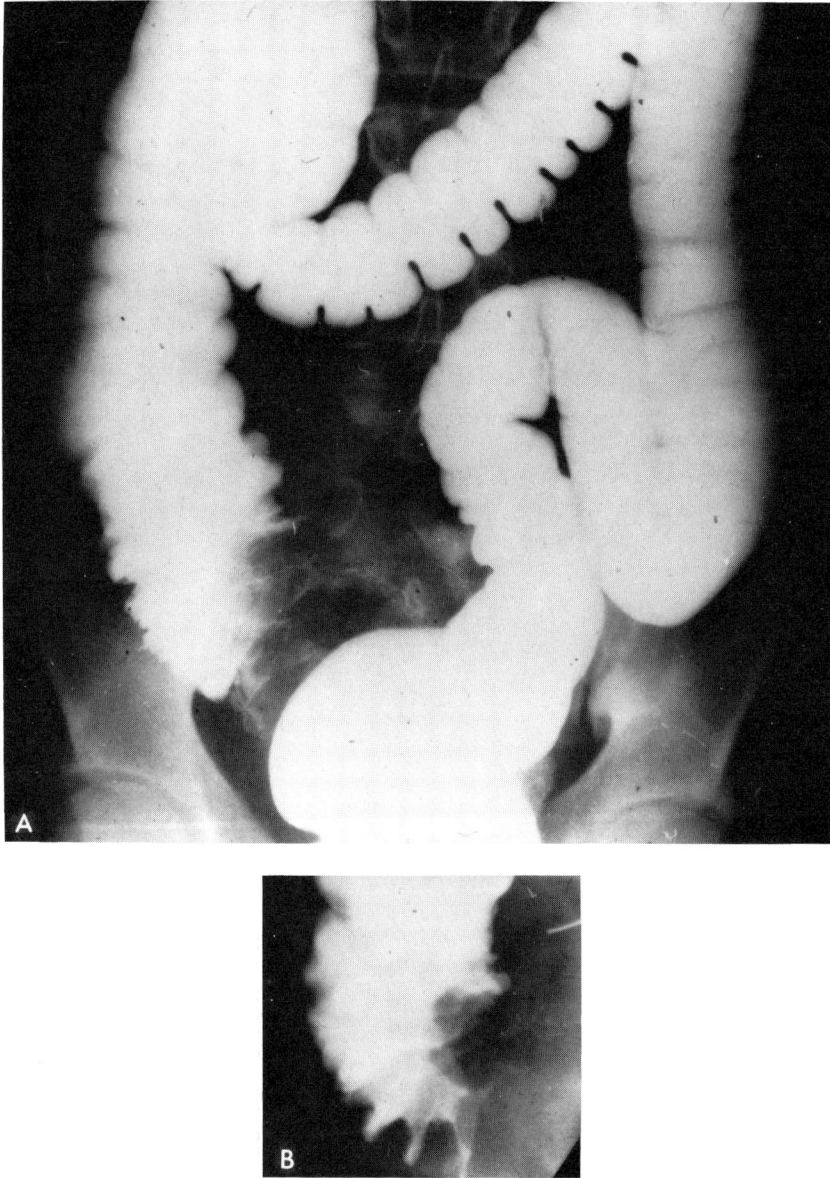

Figure 2-19. Barium enema examination (*A*) with spot radiograph (*B*) showing an appendiceal abscess with involvement of adjacent cecal wall, producing a corrugated appearance similar to that noted with a carcinoid tumor.

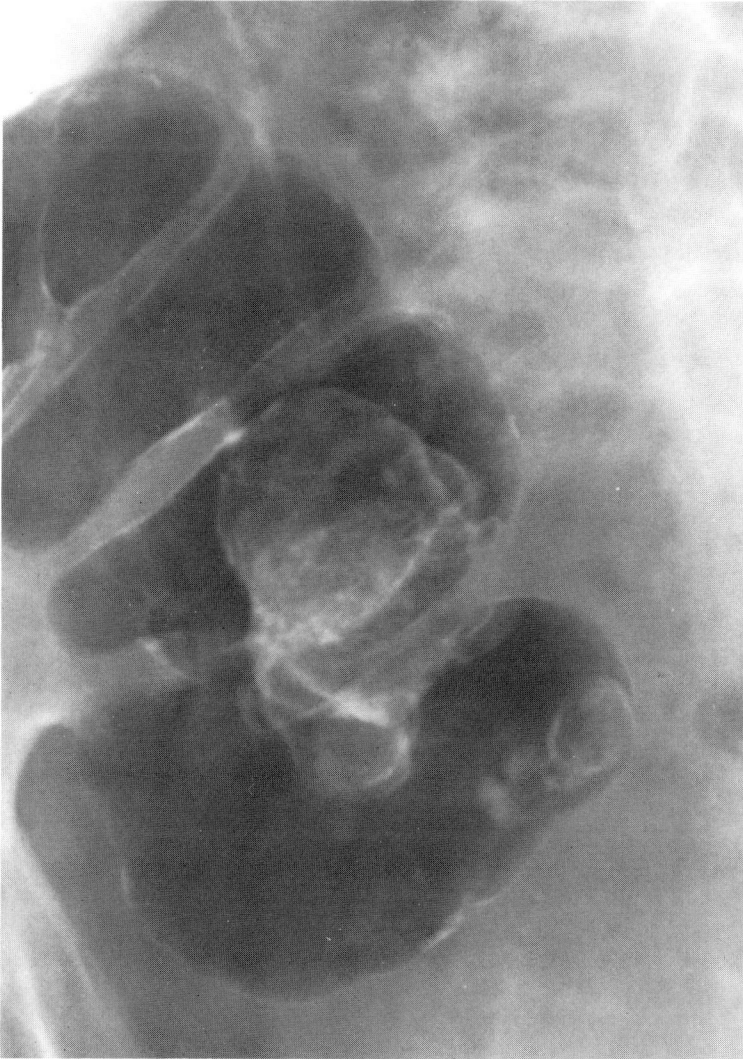

Figure 2-20. Air contrast study showing a large adenocarcinoma arising from the upper lip of the ileocecal valve. Mucosal irregularity differentiates carcinoma from a benign tumor.

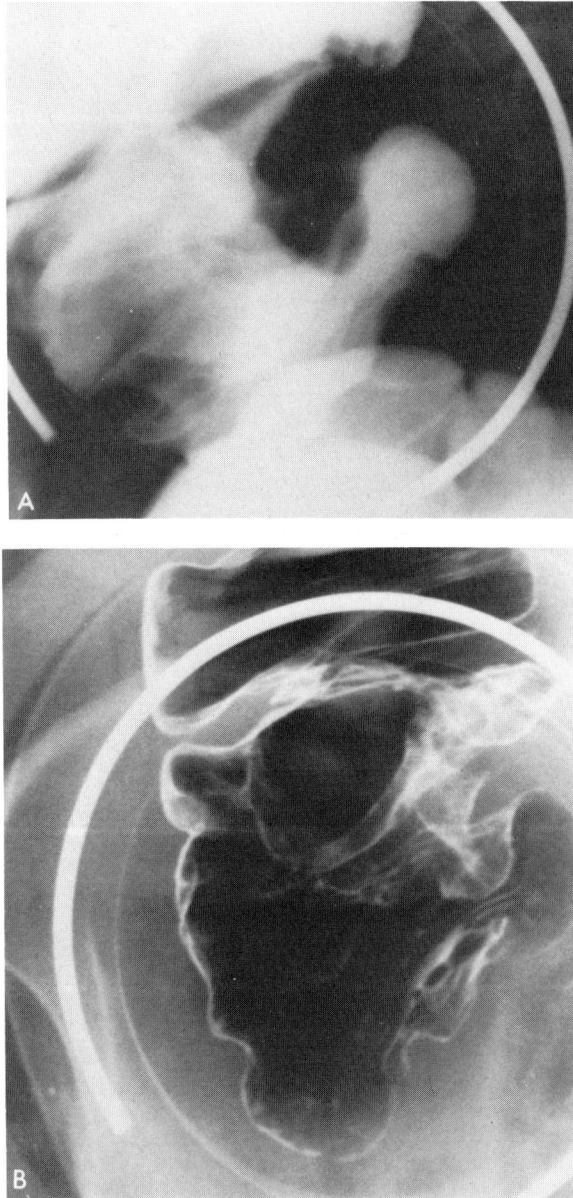

Figure 2-21. Barium enema (A) and air contrast studies (B) showing a smoothly demarcated mass arising from the lower lip of the ileocecal valve due to adenocarcinoma. The lesion is indistinguishable from a benign tumor.

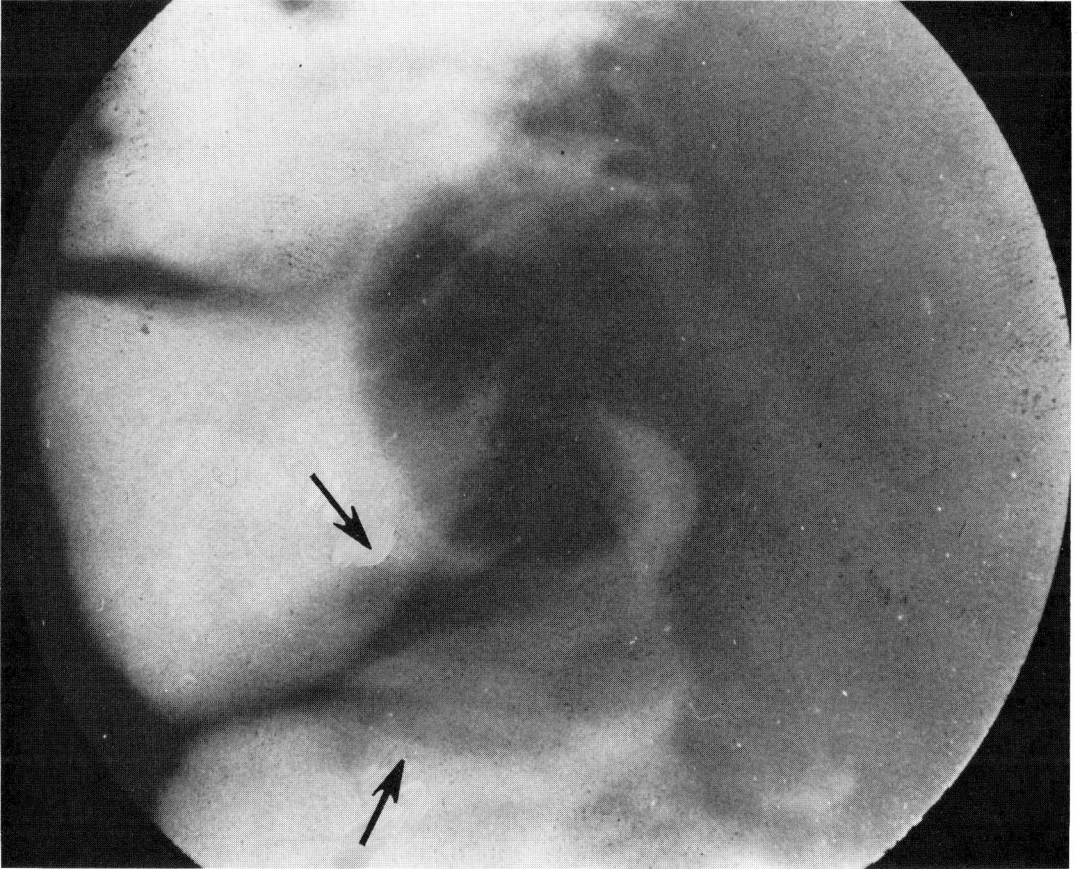

Figure 2–22. Barium enema examination showing a large, well-demarcated adenocarcinoma aris-ing from the upper lip of an ileocecal valve (arrows). Again, differentiation from a benign tumor is impossible.

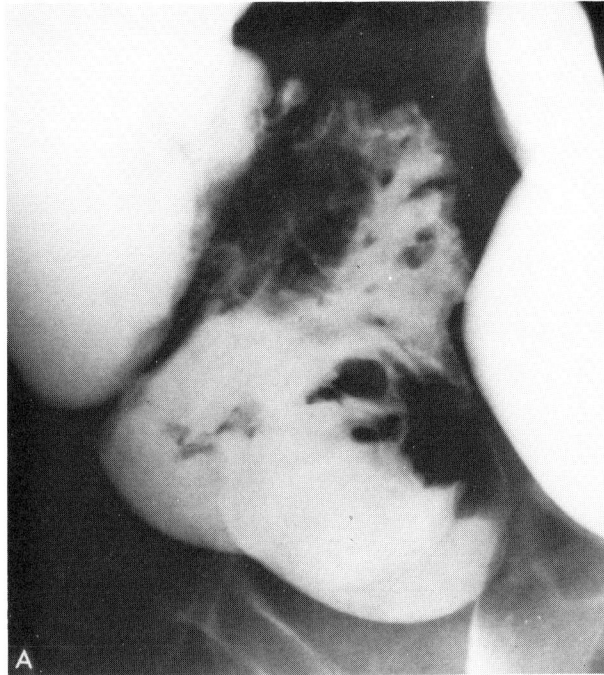

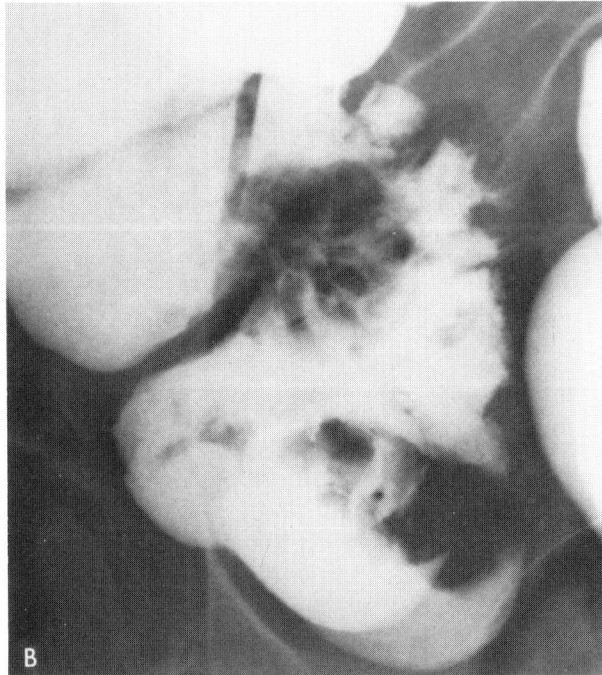

Figure 2–23. Barium enema examination showing a large irregular mass arising from the ileocecal valve due to a villous adenocarcinoma (*A* to *D*). The characteristic frondlike shaggy appearance of the surface is evident. (*E*) Another villous adenocarcinoma of the ileocecal valve showing irregularity of the surface.

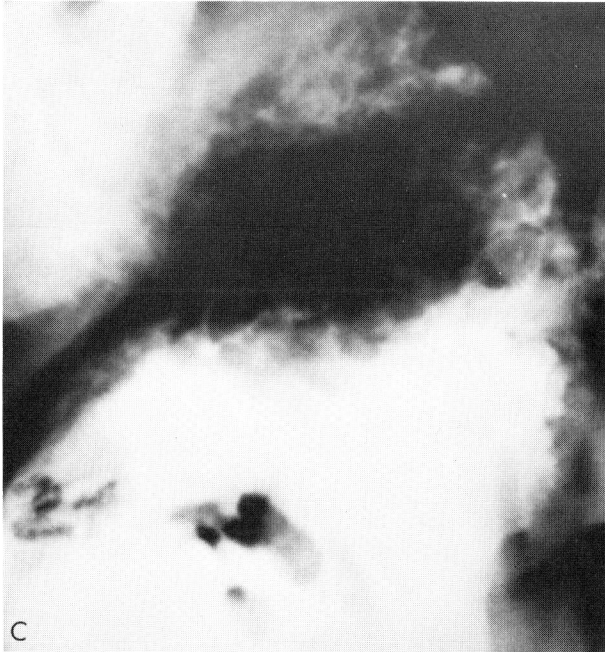

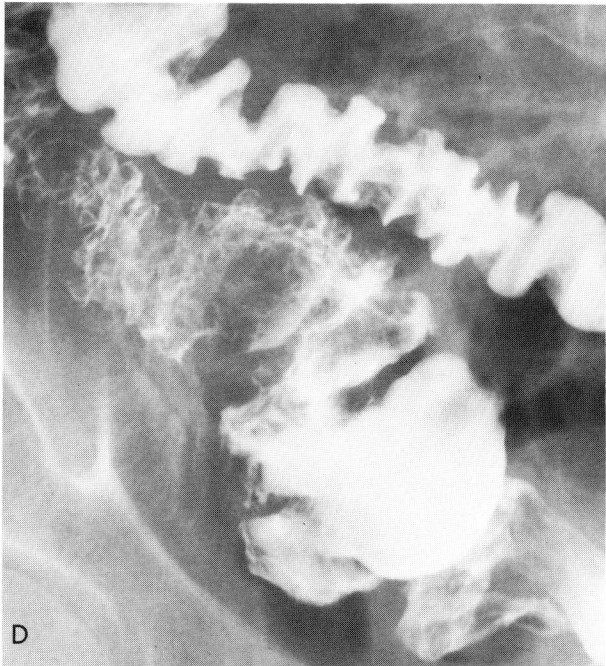

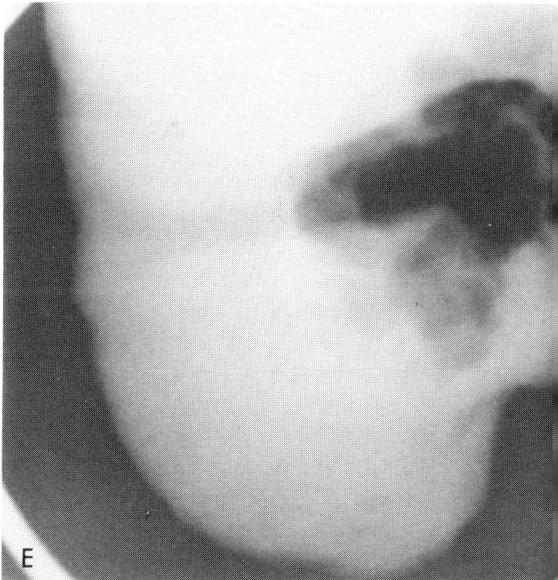

Figure 2–23. (*Continued*)

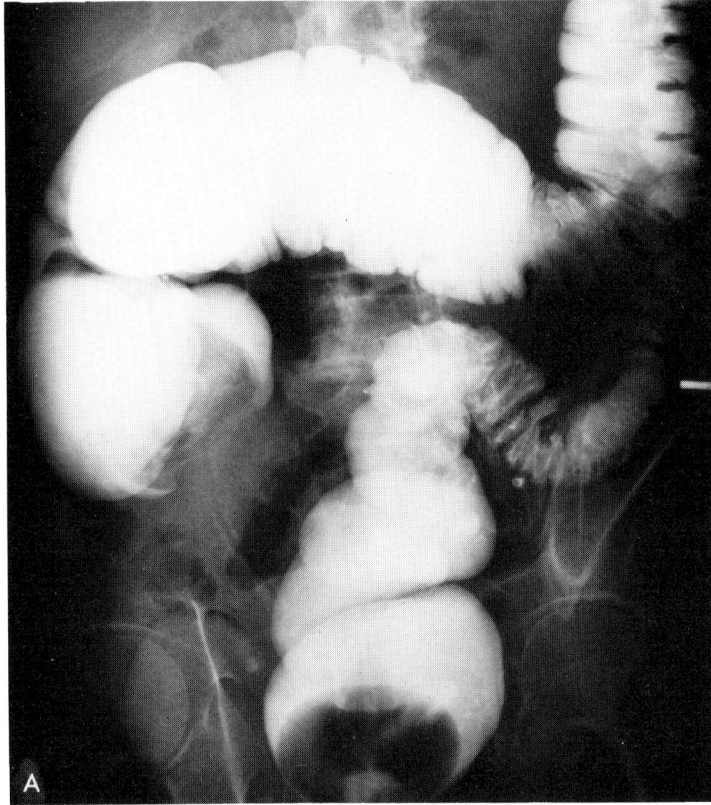

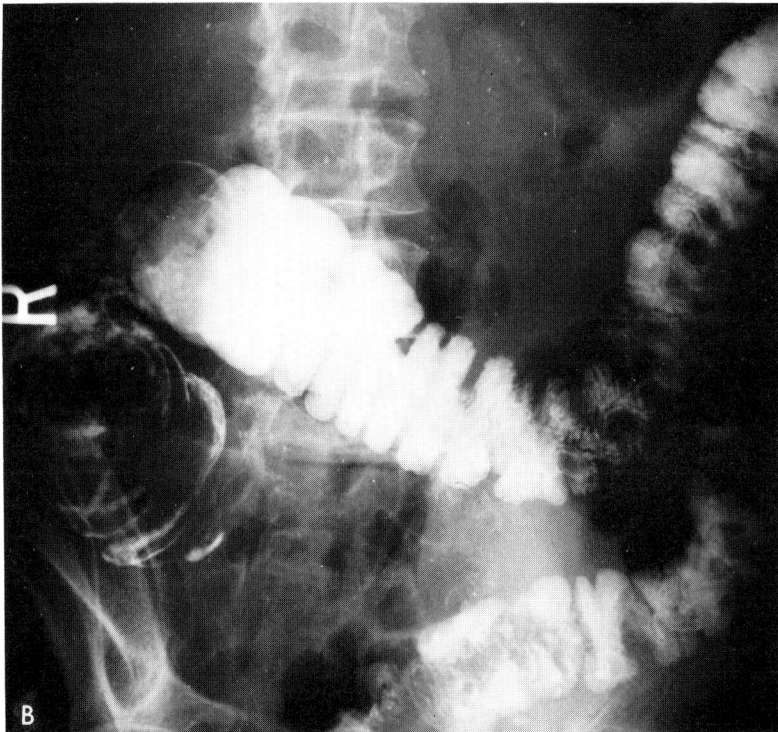

Figure 2–24. Barium enema examination (*A*) and postevacuation radiograph (*B*) showing an intussuscepting adenocarcinoma of the ileocecal valve.

Rarely, intussusception can occur owing to metastatic lesions localized to the valve (Fig. 2–25). Carcinoma of the valve may produce small bowel obstruction or appendicitis and present with the radiologic manifestations of these abnormalities.

Since the ileocecal valve contains histologic elements of both the terminal ileum and cecum, it is not surprising that malignant lesions found in the terminal ileum as well as in the cecum may occur in the valve. Thus, lymphosarcoma and other types of lymphomas may be localized to the valve. Lymphosarcoma, occurring predominantly in the ileocecal valve, often involves the terminal ileum, whereas carcinoma is less likely to do so (Fig. 2–26). Figure 2–27 is a histologic section through the upper lip of the ileocecal valve showing marked edema in a patient with lymphosarcoma who has received previous radiation therapy. Lymphomas confined to the ileocecal region produce the same radiologic manifestations as those lesions elsewhere in the intestinal tract. Multiple nodules, excavation, ulceration, infiltration and a polypoid mass are among the protean types of involvement. Because lymphosarcoma characteristically produces little desmoplastic reaction, the tumor often reaches a large size before narrowing of the intestinal lumen results in obstruction (Fig. 2–26).

ENLARGEMENT OF THE VALVE IN INTUSSUSCEPTION

The ileocecal valve may be involved in intussusception either incidentally, as in idiopathic ileocolic intussusception typical of childhood, or directly, producing a colocolic intussusception by virtue of a "lead point" produced by simple enlargement or tumors arising from the lips of the valve.

Enlargement of the valve found in association with idiopathic intussusception of infancy and childhood most likely results from edema due to mechanical trauma incurred in the course of the telescoping action of the bowel. In some instances in which the roentgen findings are otherwise equivocal, such enlargement may be an aid in recognizing a patient with a recently reduced intussusception. Idiopathic enlargement of the valve is far less common in the pediatric age group than in adults, so enlargement of the valve in a child with an appropriate history should raise the suspicion of a recent intussusception (Figs. 2–28 and 2–29).

PROLAPSE OF THE ILEAL MUCOSA

Prolapse of ileal mucosa through the orifice of the ileocecal valve is a rare occurrence. When encountered, it may simulate a nodular tumor arising from the valve and, except for direct visualization by colonoscopy, usually requires verification by surgery (Figs. 2–30 to 2–32). A changing configuration of the prolapsed mucosa may be helpful in making the proper diagnosis. This is difficult to evaluate because the appearance may vary depending on whether the valve is viewed en face or in profile, whether the cecum is fully distended and whether compression has been applied. Repeat studies with careful attention to these details may supply an answer.

INFLAMMATORY LESIONS INVOLVING THE VALVE

Inflammatory lesions due to a number of infectious agents that involve the ileocecal region may affect the ileocecal valve. Thus, tuberculosis, amebiasis, typhoid fever, anisakiasis and actinomycosis may alter the appearance of the valve. The inflammatory lesions most often involving the valve are regional enteritis and ulcerative colitis.

Since the ileocecal valve represents, in effect, a cross section of the small bowel placed in profile against the barium- or gas-filled lumen of the cecum, it is not surprising that regional enteritis frequently alters the appearance of the valve. The extensive panenteric lymphedema that occurs in Crohn's disease regularly produces enlargement of the valve. This may be in the form of direct inflammatory enlargement, lipomatous infiltration or single and multiple well-encapsulated lipomas (Figs. 2–33 to 2–36).

When ulcerative colitis involves the ileocecal valve there may be stiffening and irregularity of the mucosal contour (Fig. 2–37). Most commonly, in this disease the valve is not enlarged, and, in some cases with questionable roentgen features in the

(Text continued on page 61)

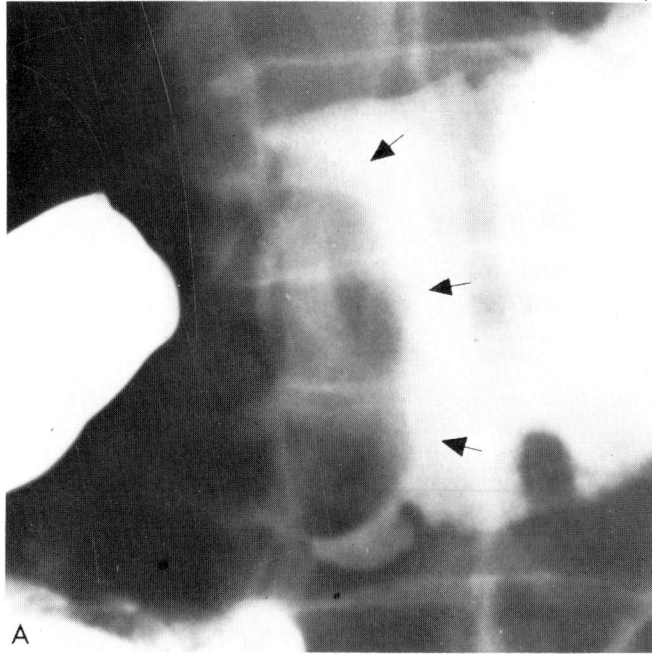

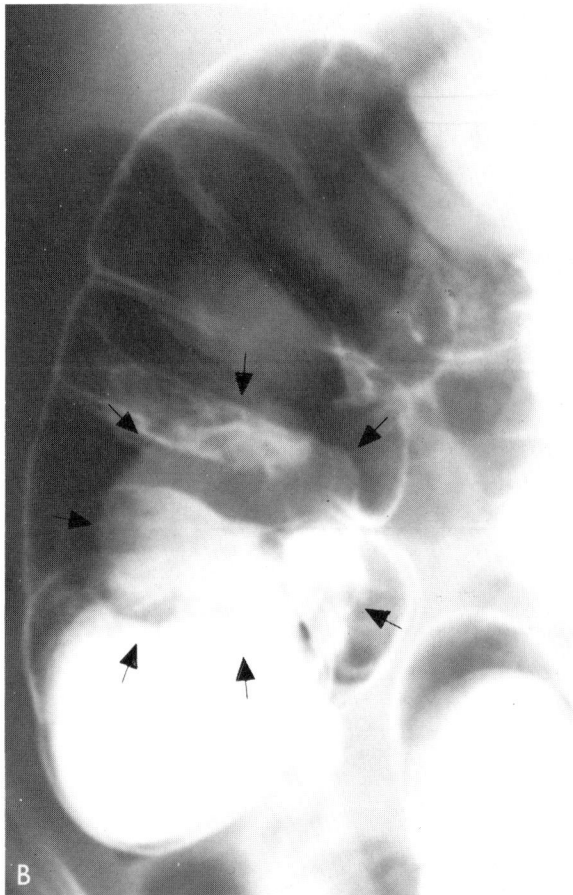

Figure 2-25. (A) Spot radiograph of an intussuscepting metastasis to the ileocecal valve made during a barium enema examination showing the intussusceptum at the hepatic flexure (arrows). (B) Air contrast examination showing the ileocecal metastasis (arrows) with the intussusception completely reduced. (*Legend continued on opposite page.*)

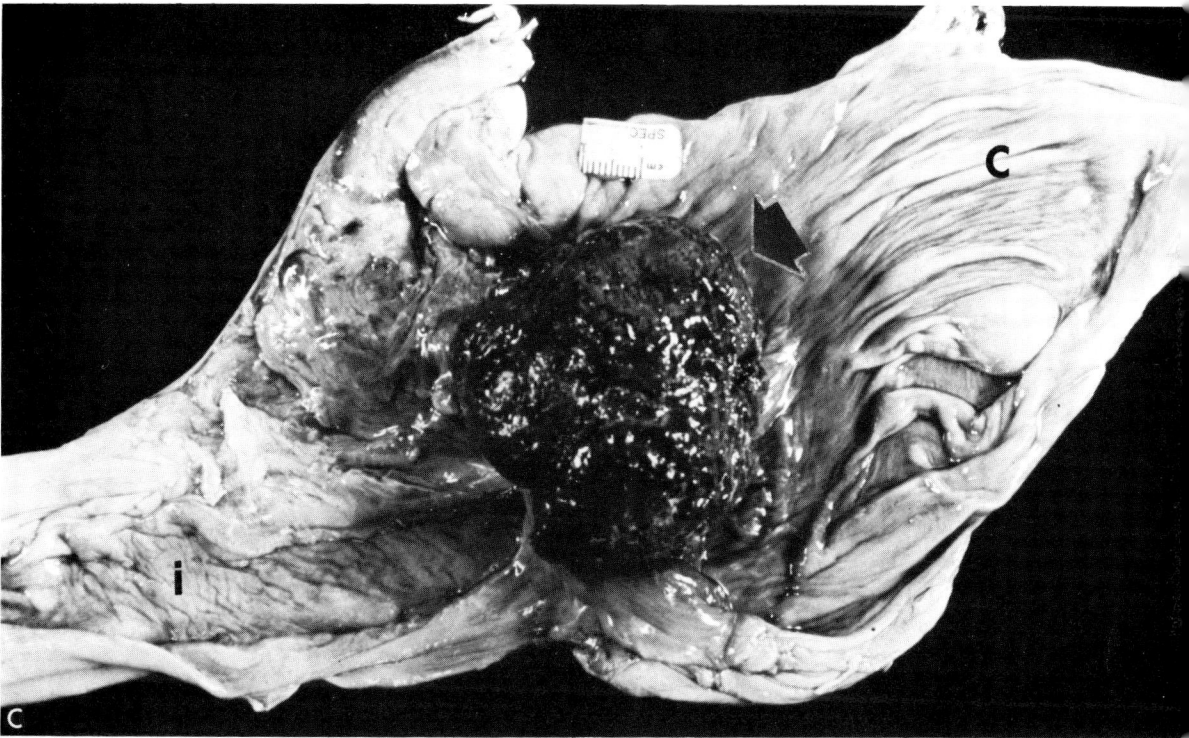

Figure 2–25. (*Continued*). (C) Gross autopsy specimen viewed from behind. The ileum (i) and the cecum (c) have been opened, showing the mass. Arrows point to the metastasis. The primary tumor was a giant cell carcinoma of the lung.

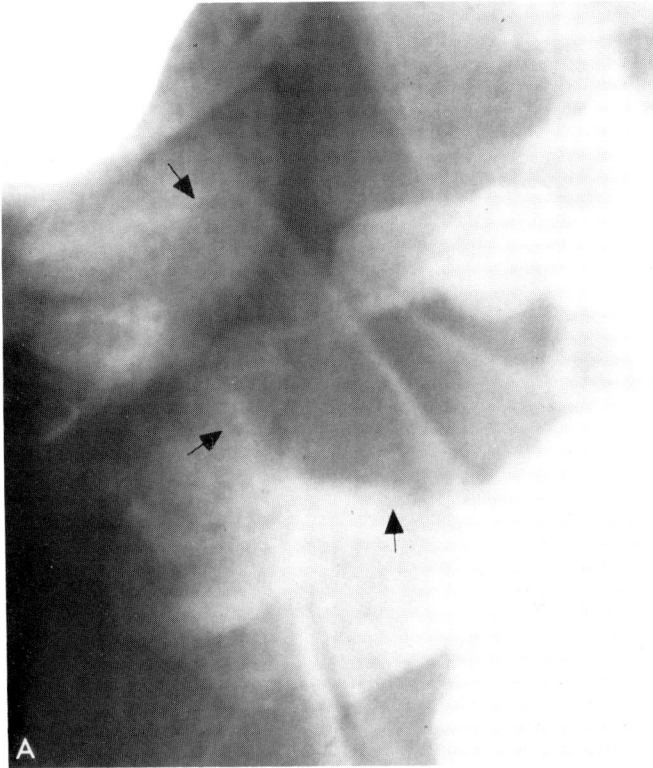

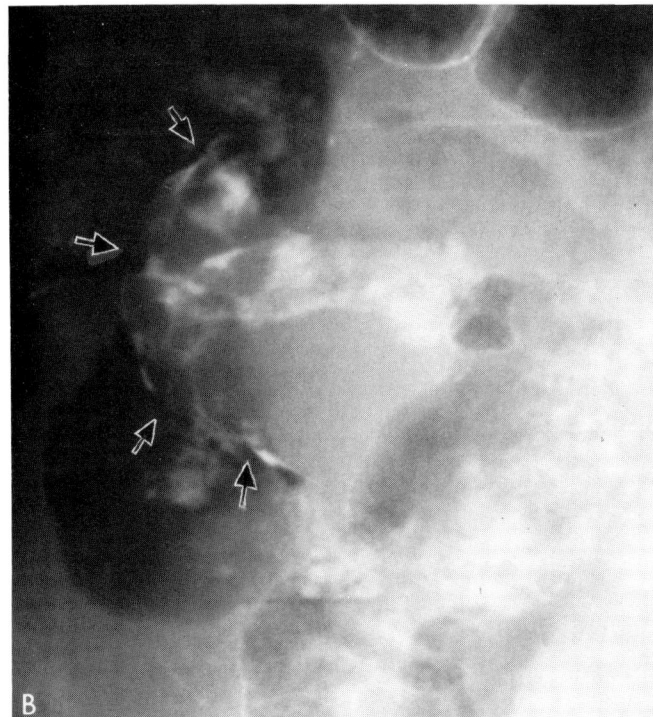

Figure 2–26. Barium enema (A) and air contrast study (B) showing lymphosarcoma involving the valve and distal segment of terminal ileum (arrows). A large polypoid tumor and extension from the ileum into the valve is characteristic of a lymphosarcoma.

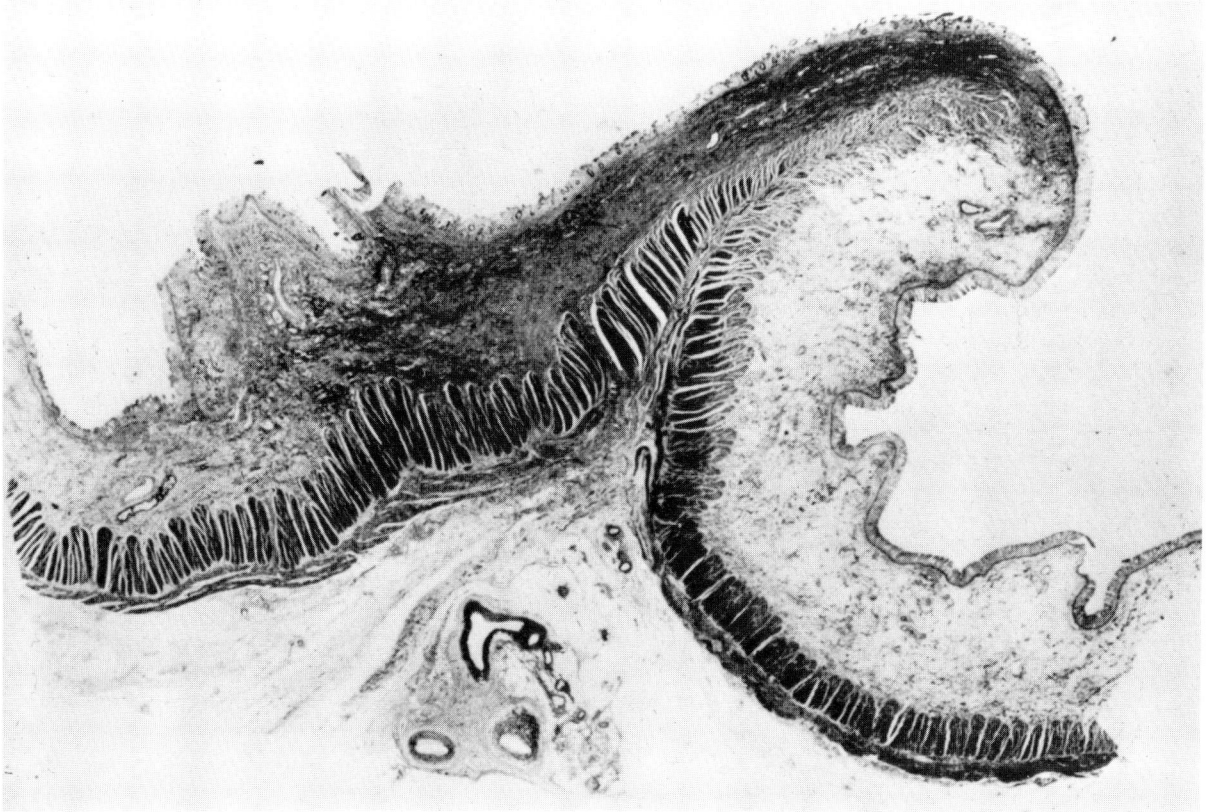

Figure 2–27. Whole-mount specimen of the upper lip of an ileocecal valve of a patient with lymphosarcoma who had received radiation therapy. Postradiation thickening and edema are evident. The specimen shows the manner in which the muscle and the wall of the ileum and the colon come together and extend for a variable distance into the valve itself. The junction between colonic mucosa and ileal mucosa occurs at the tip of the ileocecal valve.

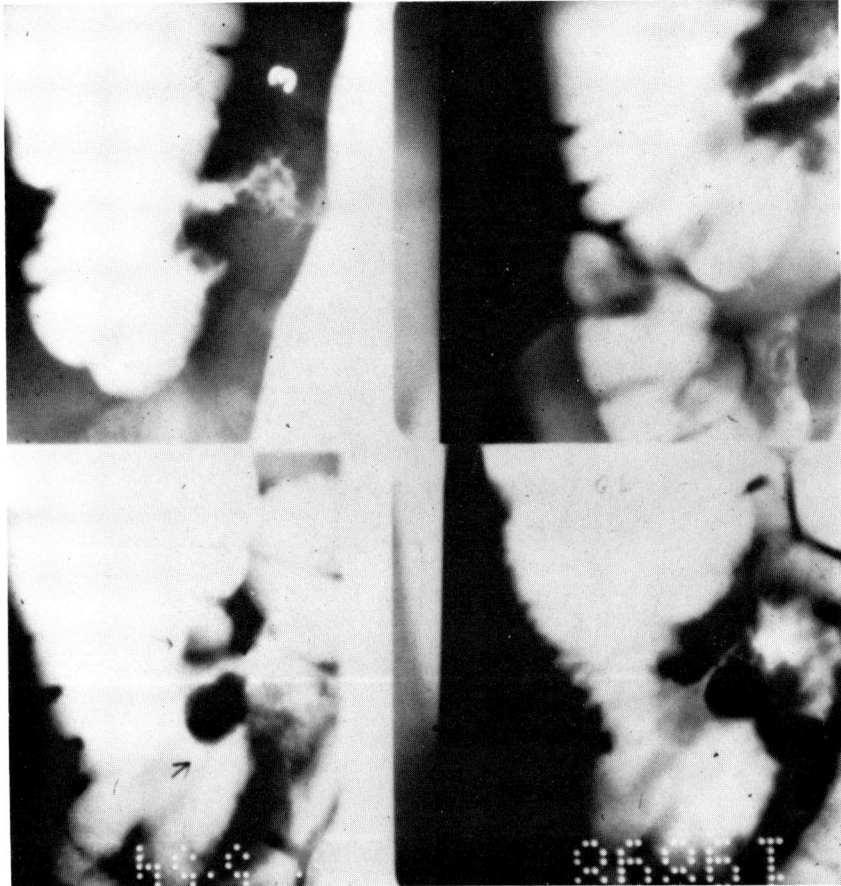

Figure 2-28. Spot radiographs from a barium enema examination showing thickened edematous lips of the ileocecal valve in a child in whom an ileocolic intussusception was reduced during the fluoroscopic examination.

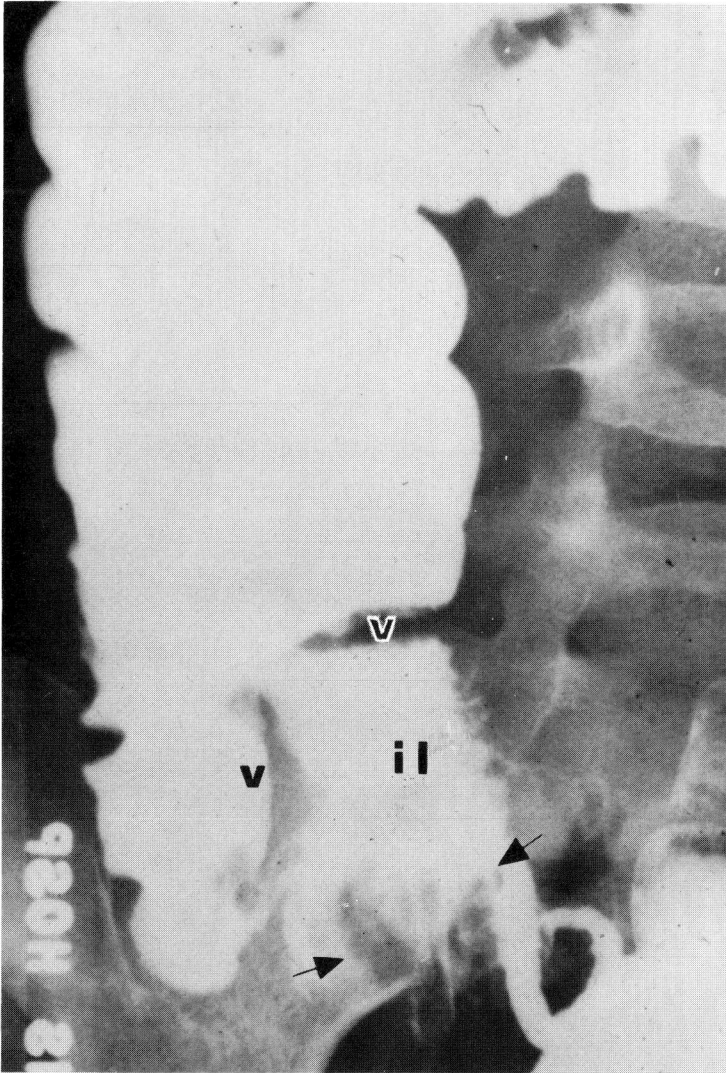

Figure 2–29. Barium enema examination showing an incompletely reduced ileocolic intussusception in a child. The remaining ileo-ileal intussusceptum (arrows) is seen outlined by barium in the
terminal ileum (il). Even moderate thickening of the lips of the ileocecal valve (v) in a child is unusual
and should suggest the possibility of a reduced intussusception or some other abnormality. (From
Lasser, E. C., and Rigler, L. G.: Ileocecal valve syndrome. Gastroenterology, 28:1, 1955. Copyright
1955, The Williams & Wilkins Company, Baltimore.)

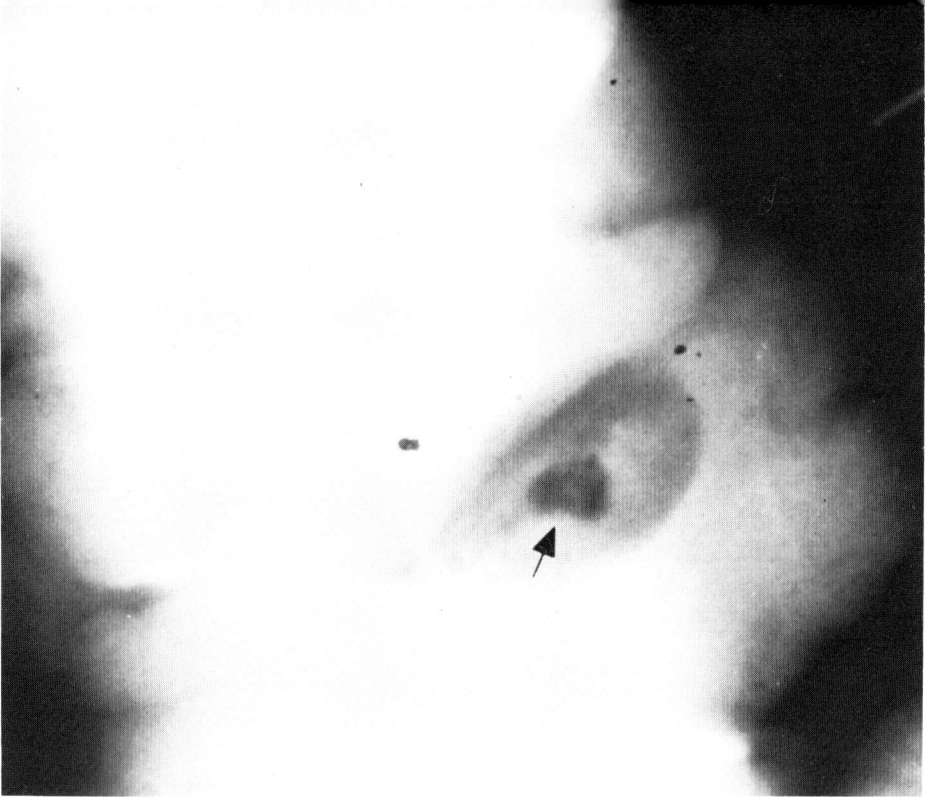

Figure 2–30. Barium enema examination showing mucosal prolapse from the terminal ileum through the orifice of the ileocecal valve viewed en face (arrow).

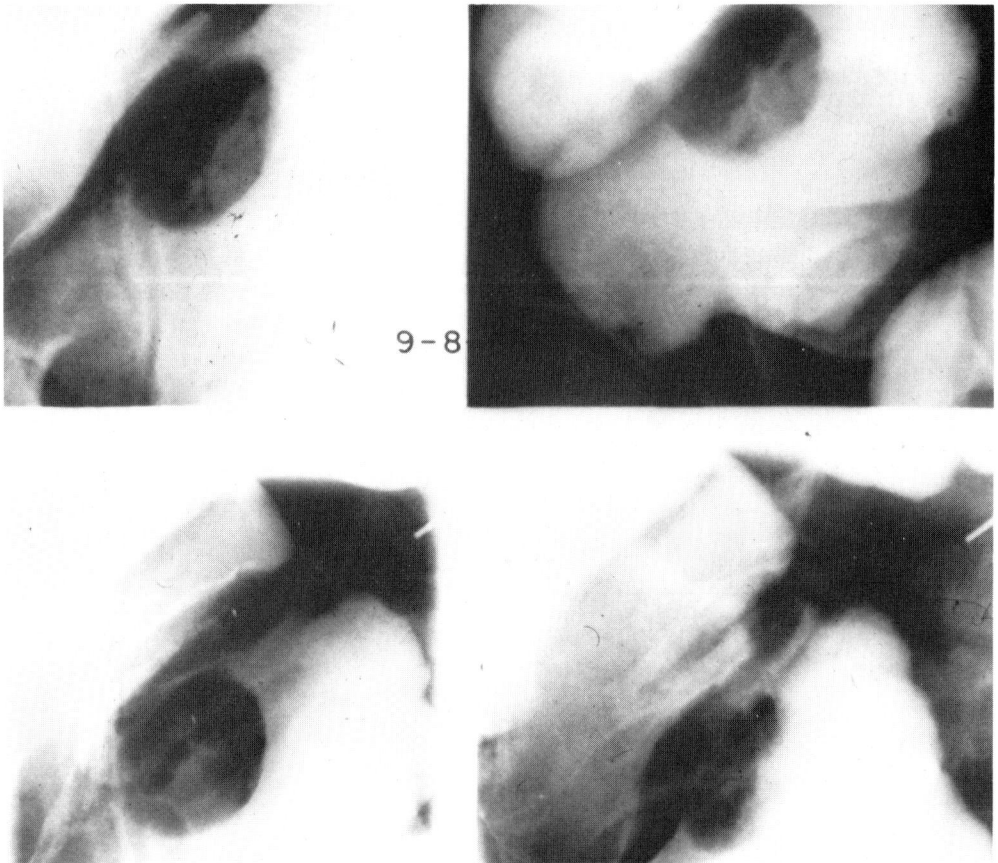

9 - 8

Figure 2–31. Barium enema examination showing the changing appearance of a large prolapse of ileal mucosa into the cecum through the ileocecal valve. The prolapsed mucosa itself simulates the valve, and it may not be possible to differentiate between the two.

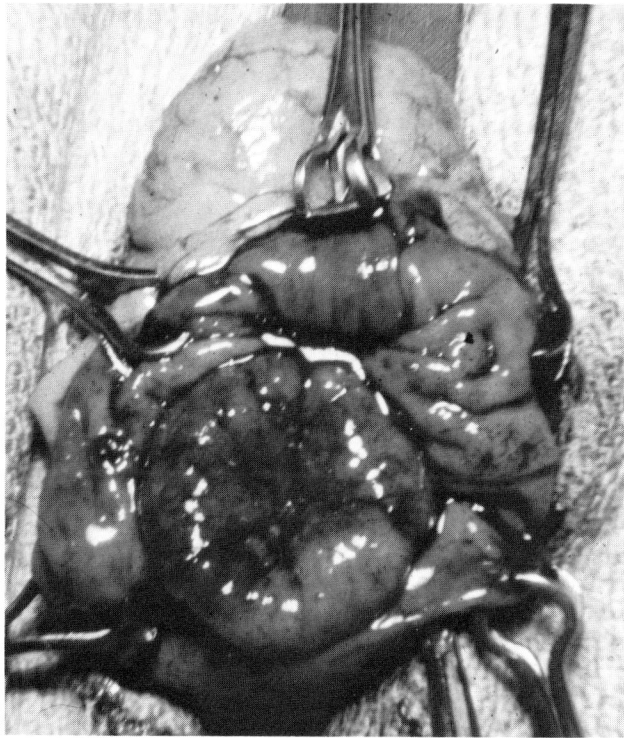

Figure 2–32. Photograph made at surgery showing the appearance of a prolapsed ileal mucosa. The forceps grasp the ileocecal valve itself, and the central rosette-like mass is caused by the prolapsed ileal mucosa.

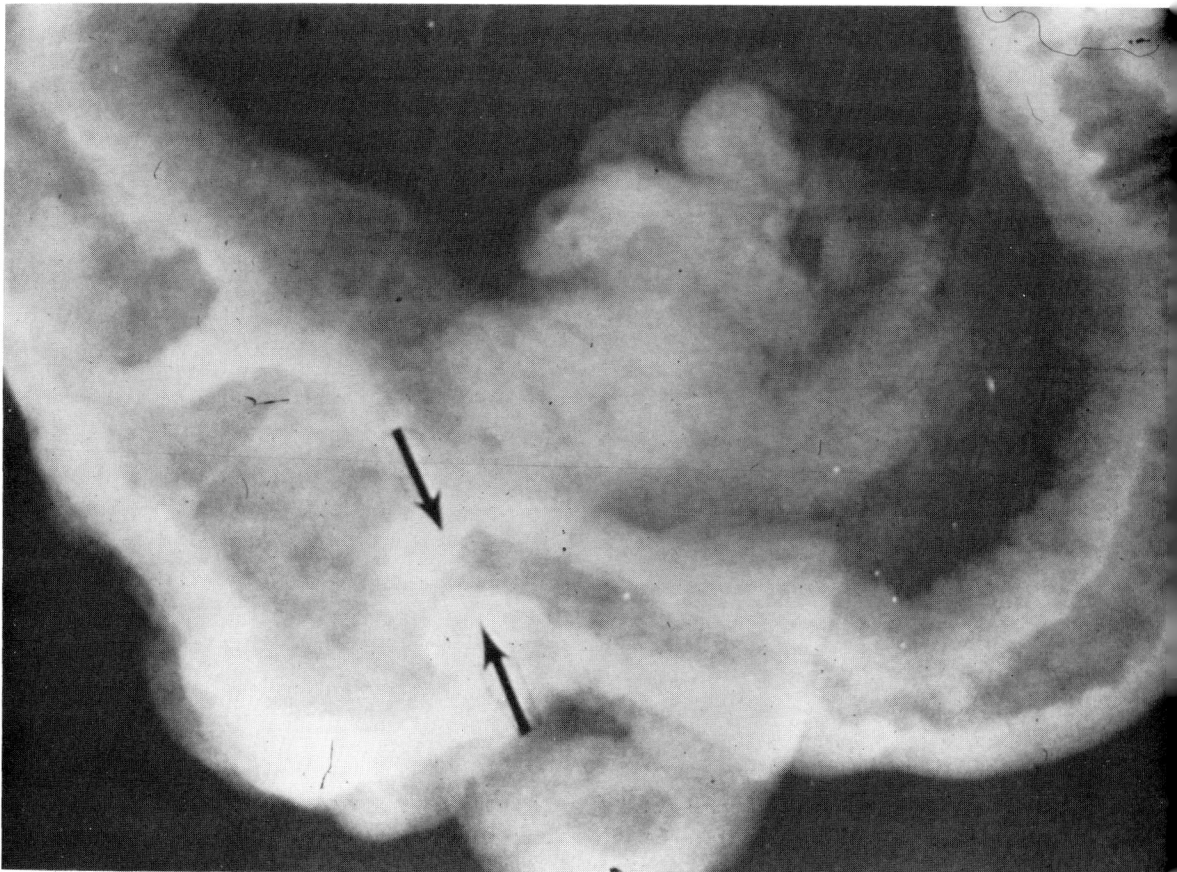

Figure 2–33. Radiograph of a gross specimen of Crohn's disease involving the terminal ileum and colon. The arrows point to the large ileocecal valve that is frequently seen in Crohn's disease. Note the marked thickening of the terminal ileum and cecum.

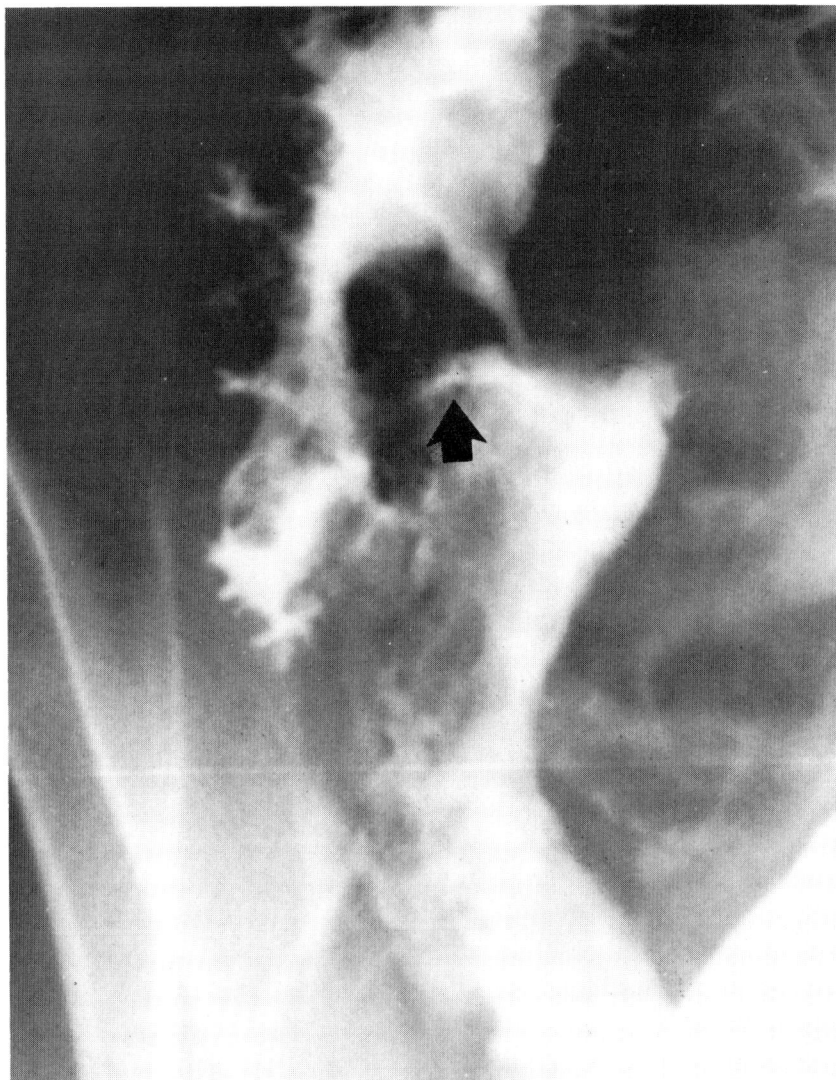

Figure 2–34. Postevacuation radiograph showing an enlarged edematous ileocecal valve in a 16-year-old girl with Crohn's disease of the terminal ileum and cecum. The arrow marks the orifice of the ileocecal valve.

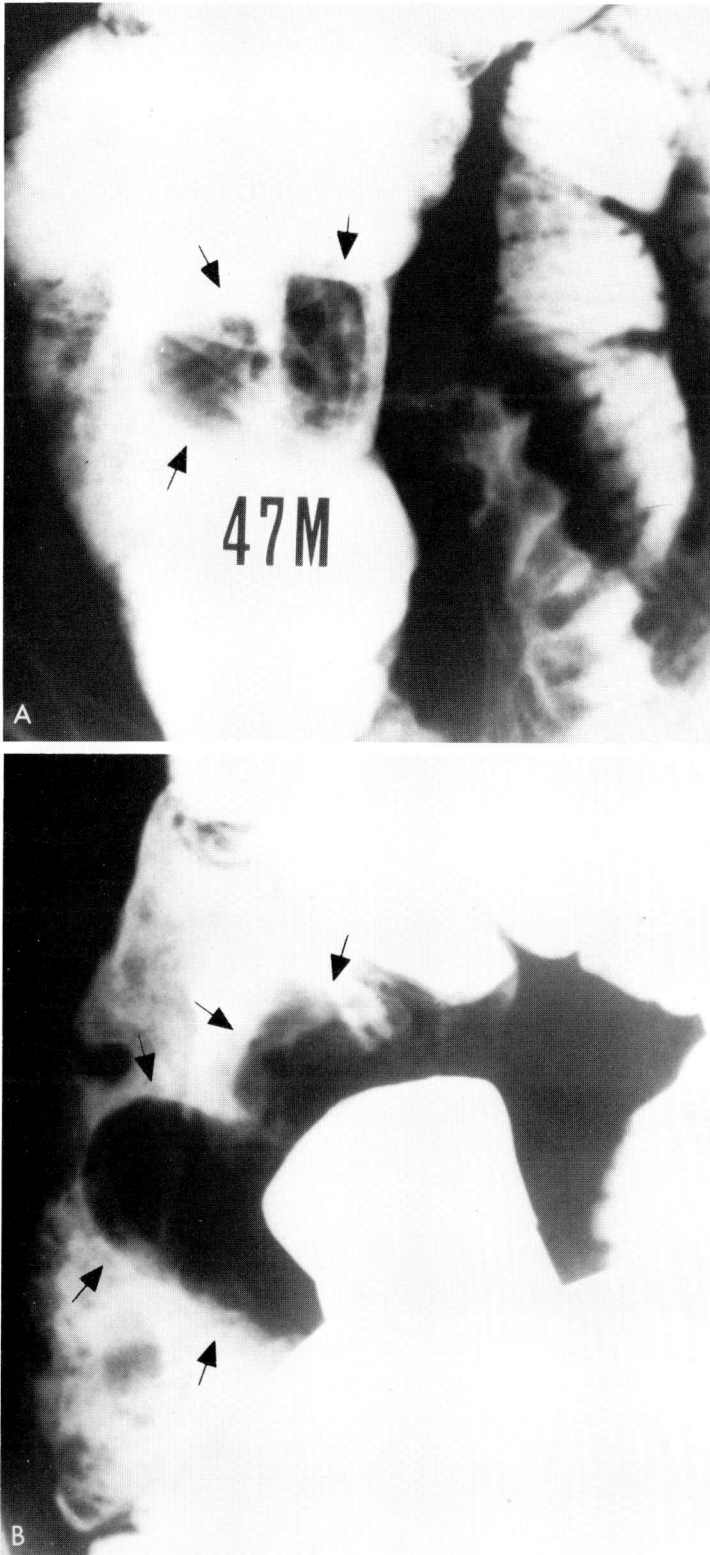

Figure 2–35. Barium enema examination (A and B) showing an enlarged ileocecal valve (arrows) and Crohn's disease of the terminal ileum. Note the edematous mucosa of the terminal ileum in A. At surgery, a lipoma of the valve was also present.

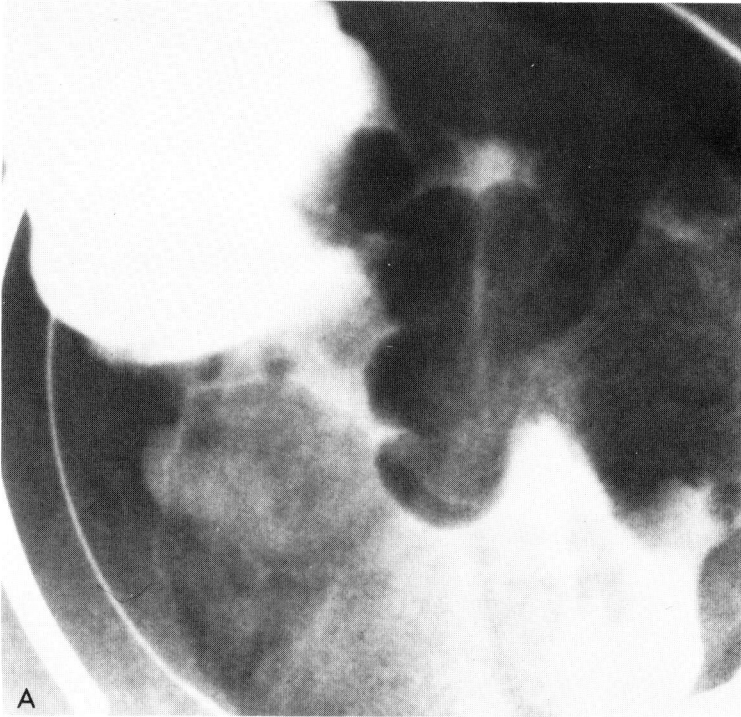

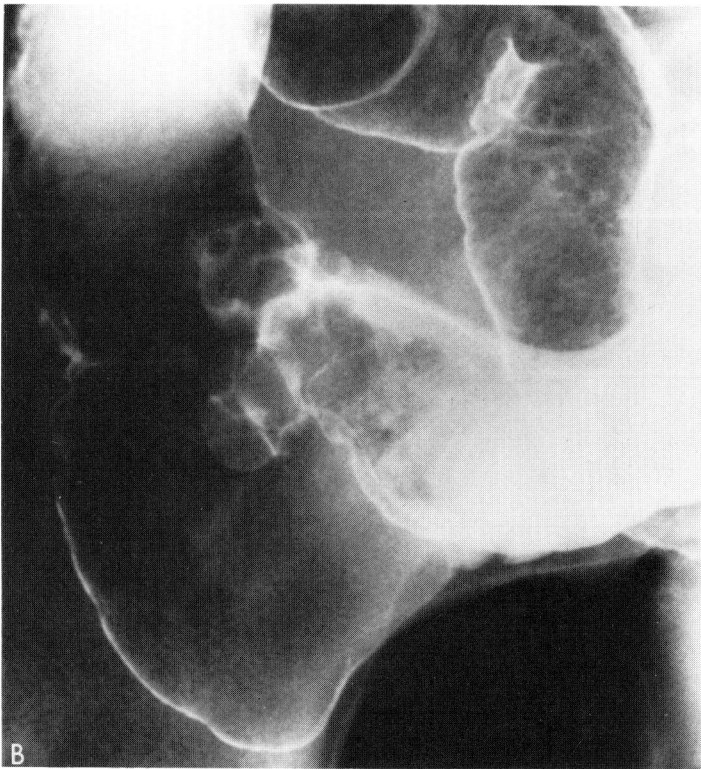

Figure 2–36. Barium enema (A) and air contrast study (B) in a patient with Crohn's ileocolitis. Multiple grapelike lipomas of the valve are present. *(Legend continued on opposite page.)*

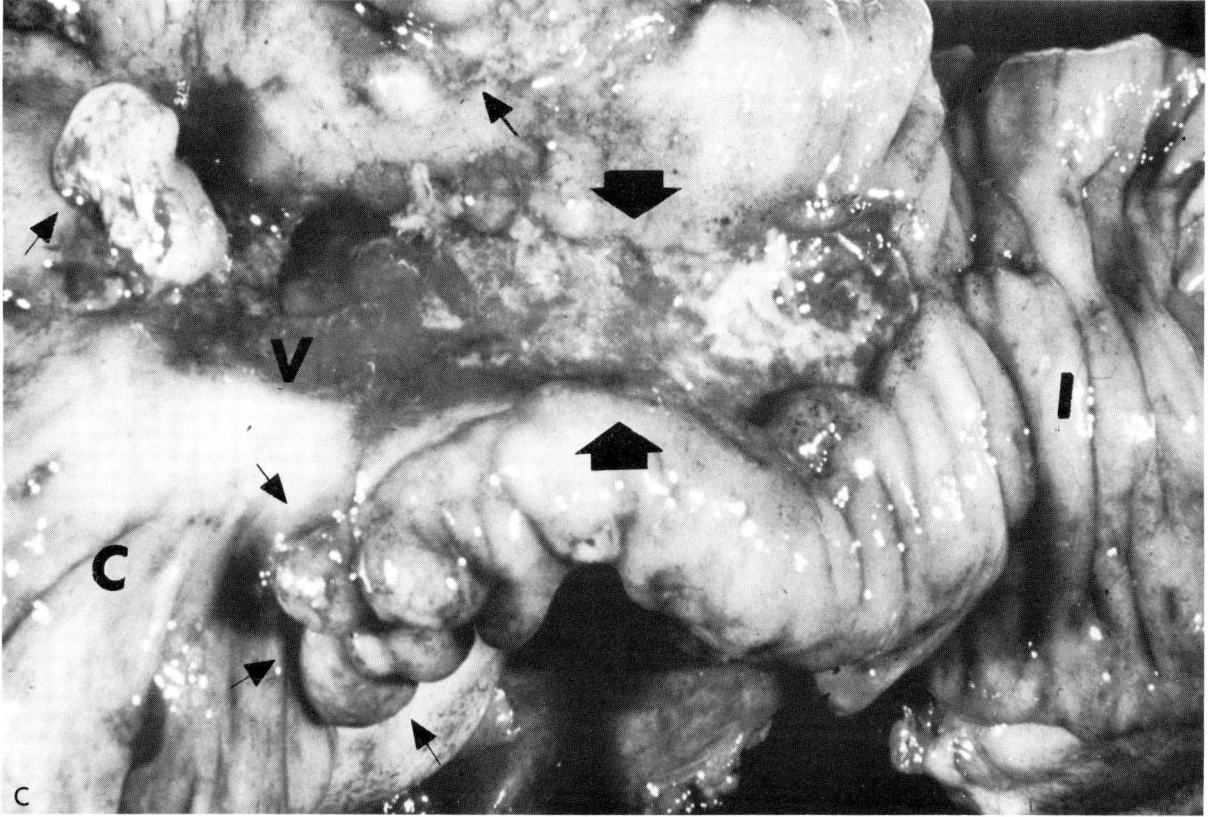

Figure 2–36. (*Continued*). (*C*) Photograph of the surgical specimen, opened to show the mucosal surface. Large arrows point to a deep ulcer of the ileum; small arrows point to a cluster of well-encapsulated lipomas (v = ileocecal valve; c = cecum; i = ileum). (From Berk, R. N., et al.: Lipomatosis of the ileocecal valve. Amer. J. Roent., *119*:323, 1973. Courtesy of Charles C Thomas, Publisher.)

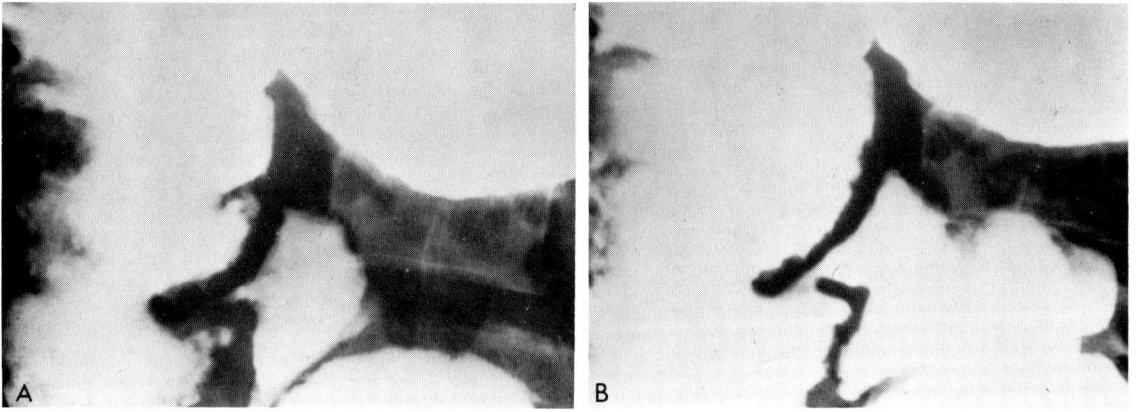

Figure 2–37. Barium enema examination (*A* and *B*) in a patient with chronic ulcerative colitis. Rigidity of the ileocecal valve is evident. (From Lasser, E. C., and Rigler, L. G.: Observations on the structure and function of the ileocecal valve. Radiology, 63:176, 1954.)

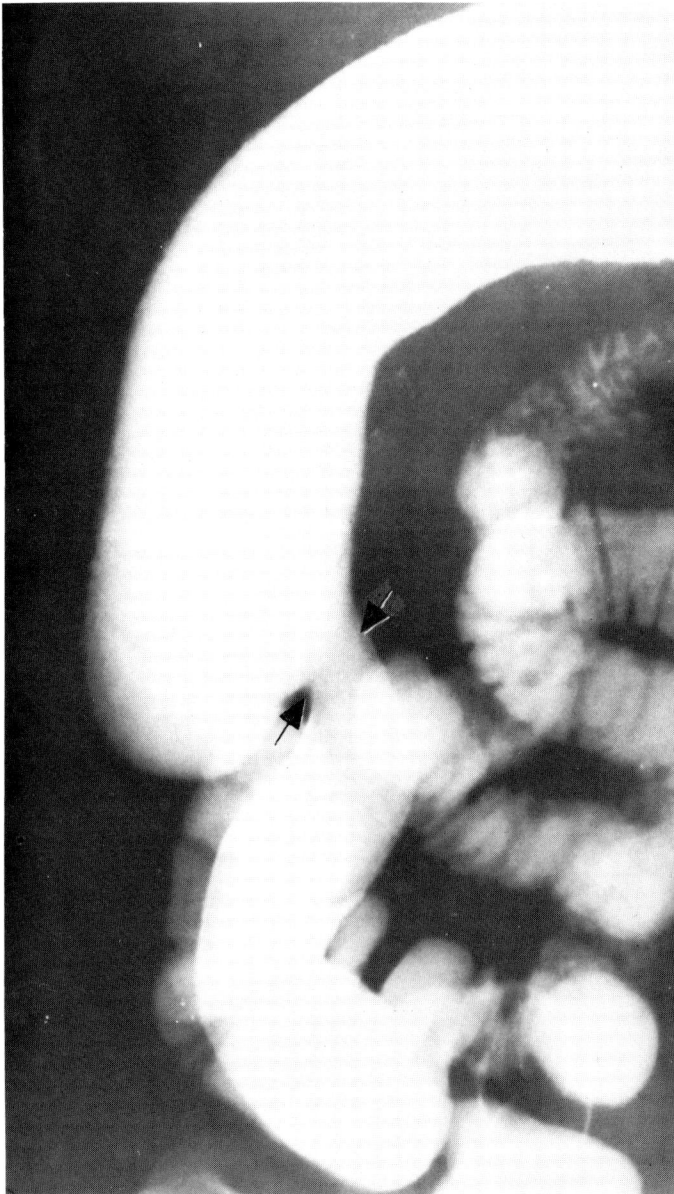

Figure 2–38. Barium enema examination showing a characteristic atrophic ileocecal valve in a patient with ulcerative colitis and "backwash ileitis" (arrows). The valve is fixed in an open position.

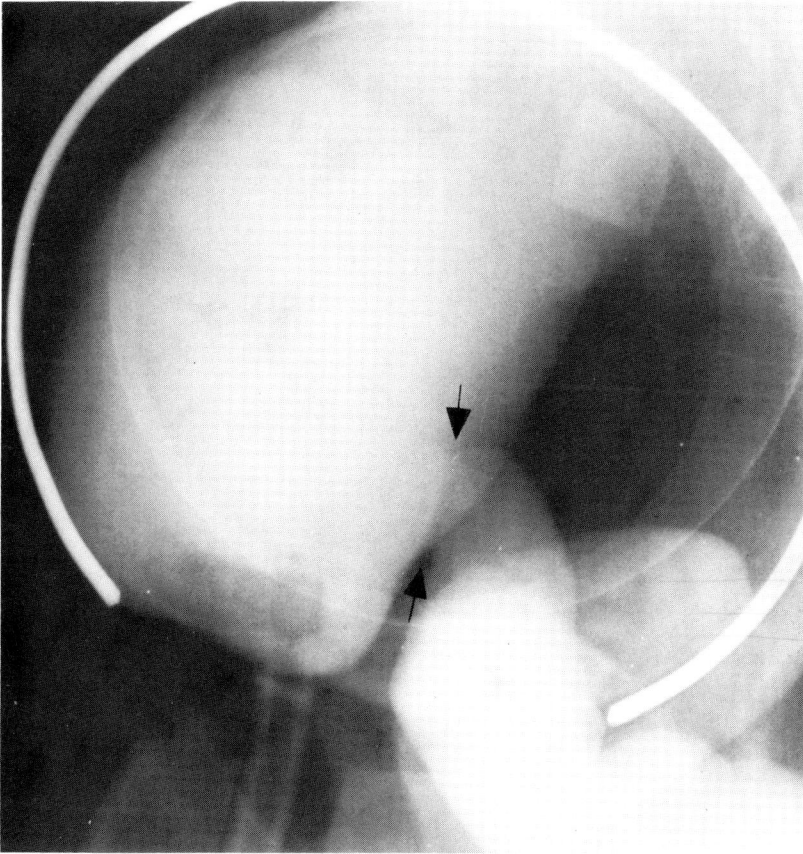

Figure 2–39. Barium enema examination showing a thin patulous valve in a patient with cathartic colon (arrows). The appearance of the ileocecal valve in cathartic colon and in ulcerative colitis is similar.

colon, this may be added evidence that one is dealing with ulcerative colitis rather than granulomatous colitis. Commonly, the valve in ulcerative colitis is patulous as distinct from that in regional enteritis, in which the valve tends to be narrowed (Fig. 2–38). The ileocecal valve is also wide open in cases of prolonged cathartic abuse (Fig. 2–39). Differentiation from chronic ulcerative colitis may be difficult on a radiologic basis alone.

THE ILEOCECAL VALVE IN INTESTINAL OBSTRUCTION

Most reports in the literature indicate that the ileocecal valve usually remains competent in cases of large bowel obstruction.[11] The reason for this is apparent from the following studies. The average pressure in the colon in 34 cases of large bowel obstruction was 14 cm. of water and the highest was 34 cm. of water.[12] Hydrostatic pressure in the colon during a barium enema examination is much higher. Reflux failed to occur in 64 per cent of patients with a pressure of 25 cm. of water or below during the barium enema examination.[13] Reflux did not occur in 16 per cent at 50 cm. or below nor in 4 per cent at 90 cm. or below. Hence, colonic obstruction produces sufficient pressure in the colon to overcome the resistance of the ileocecal valve in only approximately one-third of the cases.

In colonic obstruction there are two basic mechanisms that affect the distribution of gas in the large and small bowel.[14, 15] In the early stages of obstruction, a competent ileocecal valve permits continued entrance of small bowel contents into the large bowel and prevents retrograde reflux. As obstruction progresses and the cecum distends with gas and fluid, the tightly closed ileocecal valve causes secondary obstruction in the small bowel, with subsequent small bowel distention. When the ileocecal valve is incompetent, the gas in the colon is readily decompressed into the small bowel.

Consequently, with incompetence or relative incompetence of the ileocecal valve in large bowel obstruction, multiple distended loops of small bowel noted on plain abdominal radiographs occur owing to reflux. With a competent valve, the radiographic picture is one of a dilated colon with a large, thin-walled cecum. If the obstruction continues, small bowel distention occurs owing to a secondary obstruction of the small bowel at the ileocecal valve. The presence of air in the small bowel on abdominal radiographs, therefore, does not indicate the competence of the valve. Air may accumulate in the small intestine either because of an incompetent valve or because of secondary small bowel obstruction created by a competent ileocecal valve.

With these considerations in mind, the ileocecal valve cannot categorically be considered an "escape valve" in the face of large bowel obstruction with air in the distal small bowel. The colon in these circumstances may function as a closed-loop obstruction, and cecal perforation from distention may occur.

REFERENCES

1. Fleischner, F. G., and Bernstein, C.: Roentgen-anatomical studies of the normal ileocecal valve. Radiology, 54:43, 1950.
2. Buirge, R. E.: Gross variations in the ileocecal valve. Anat. Rec. 86:373, 1943.
3. Golden, R.: Radiological Examination of the Small Intestine. Philadelphia, J. P. Lippincott Co., 1945, pp. 210–215.
4. Lasser, E. C., and Rigler, L. G.: Ileocecal valve syndrome. Gastroenterology, 28:1, 1955.
5. Hinkel, C. L.: Roentgenological examination and evaluation of the ileocecal valve. Amer. J. Roent., 68:171, 1952.
6. Rubin, S., Dann, D. S., Ezekial, C., and Vincent, J.: Retrograde prolapse of the ileocecal valve. Amer. J. Roent., 87:706, 1962.
7. Boquist, L., Bergdahl, L., and Andersson, A.: Lipomatosis of the ileocecal valve. Cancer, 29:136, 1972.
8. Kelby, G. M.: Submucous lipomas of the ileocecal valve. Lancet, 68:301, 1948.
9. Castleman, B.: Submucous lipoma of the ileocecal valve, Cabot case 29501. New Eng. J. Med., 229:948, 1943.
10. Berk, R. N., Davis, G. B., and Cholhassey, E. B.: Lipomatosis of the ileocecal valve. Amer. J. Roent., 119:323, 1973.
11. Hodges, P. C., and Miller, R. E.: Intestinal obstruction. Amer. J. Roent., 74:1015, 1955.
12. Dennis, C.: Treatment of large bowel obstruction. Surgery, 15:713, 1944.
13. Rendleman, D. F., Anthony, J. E., Davis, C., et al.: Reflux pressure studies on the ileocecal valve of dogs and humans. Surgery, 44:640, 1958.
14. Love, L.: The role of the ileocecal valve in large bowel obstruction. Radiology, 75:391, 1960.
15. Lasser, E. C., and Rigler, L. G.: Observations on the structure and function of the ileocecal valve. Radiology, 63:176, 1954.

Chapter 3

THE APPENDIX

Of the viscera in the right lower quadrant of the abdomen, the organ most frequently involved by disease is the appendix. Acute appendicitis is most common. Abnormalities of the terminal ileum and cecum involve the appendix either because of spread in continuity or because of contiguity. The appendix is only rarely the site of a primary tumor.

The radiologic diagnosis of diseases of the appendix is almost entirely dependent on the extent to which changes are produced in adjacent structures such as the cecum and terminal ileum. Although in many cases there may be no radiologic manifestations of appendiceal disease, the abdominal plain film, barium enema examination and small intestine study frequently allow the radiologist to make an important contribution to the diagnosis.

VARIATIONS IN POSITION

The appendix becomes clearly defined at 10 to 12 weeks of uterine life when the cecum is subhepatic in location.[1] The position of the appendix varies greatly and is determined by the following factors: the degree of descent of the cecum, the degree of cecal fixation to the posterior abdominal wall, the configuration of the cecum, the length of the appendix (normally 1 to 9 inches), the degree of development of the pericecal fossae, the presence of associated adhesions (whether they arise from the appendix or associated organs), the presence and extent of submucosal lymphoid hyperplasia and the habitus of the patient.[1]

There are numerous normal variations in the position of the appendix, so that it may assume any of the positions of the radii of a circle with the center at the origin of the appendix from the cecum.

According to Treves, there are four basic anatomic variations in the origin of the appendix from the cecum which may be listed as follows:[2]

Type I. In the fetal type the appendix arises from the apex of the cecum and forms a continuation of the long axis of the colon (Fig. 3–1).

Type II. The cecum is roughly quadrilateral in shape and the appendix arises between two bulging cecal sacculi instead of at the summit of the colon, as in the fetal type (Fig. 3–2).

Type III. In this, the most common type, the portion of the cecum lateral to the origin of the appendix bulges and the base of the appendix arises from the medial wall. The apex of the cecum is positioned medially. Therefore, a false apex is formed by the highly developed portion lateral to the appendiceal origin (Fig. 3–3).

Type IV. In this form the appendiceal origin is close to the ileocecal valve without any trace of the original apex of the cecum (Fig. 3–4).

Several of the complications associated with appendiceal disease, specifically abscess formation and intestinal obstruction, are directly related to the anatomic position of the appendix.[1] In two reviews summarizing a total of 75,000 appendectomies, it is stated that the appendix is usually an intraperitoneal organ lying anterior to the cecum or in line with it.[3] There is general agreement that about 25 per cent of appendices lie in a retrocecal position (Fig. 3–5). A retrocecal appendix may be either intraperitoneal or retroperitoneal. Of the retrocecal appendices, 60 per cent are in a mobile retroperitoneal position.

Seventy-five per cent of the appendices lie anterior to the cecum and fall into three major categories: those which lie caudal to the cecum (one-third of the total), those lying medial to the cecum (one-third of the total) and those lateral to the cecum (6 per cent of the total).[1]

The appendix may be in an unusual position because of abnormalities in the position of the cecum (Figs. 3–6 and 3–7). The appendix, as an elongated variably mobile

(Text continued on page 74)

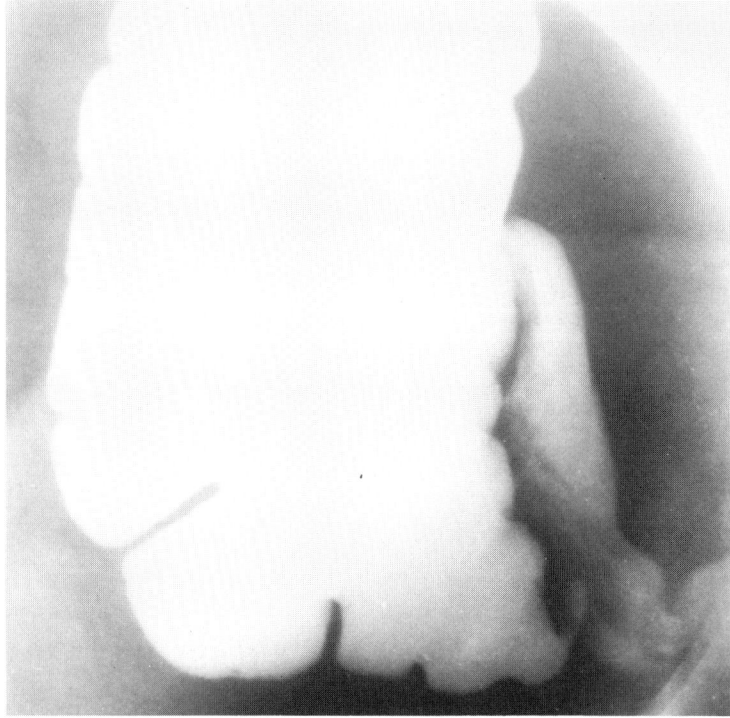

Figure 3–1. Fetal type anatomic variation in the origin of the appendix. The appendix arises from the apex of a conical cecum forming a continuum with the colon. (From Beneventano, T. C., et al.: The roentgen aspects of some appendiceal abnormalities. Amer. J. Roent., 96:344, 1966. Courtesy of Charles C Thomas, Publisher.)

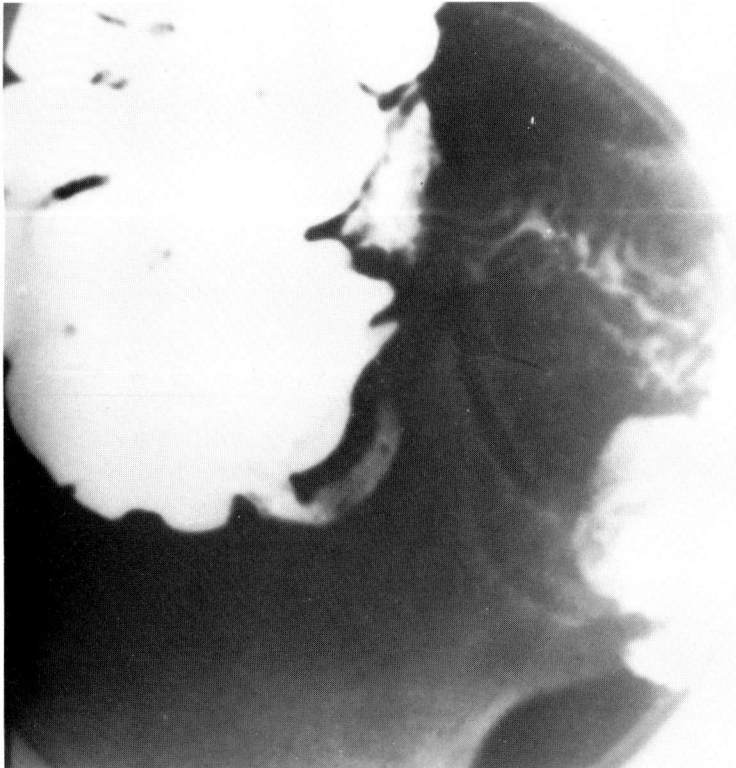

Figure 3–2. Second type of anatomic variation in the origin of the appendix. The appendix arises between two haustra of a quadrilateral cecum instead of at the summit. (From Beneventano, T. C., et al.: The roentgen aspects of some appendiceal abnormalities. Amer. J. Roent., 96:344, 1966. Courtesy of Charles C Thomas, Publisher.)

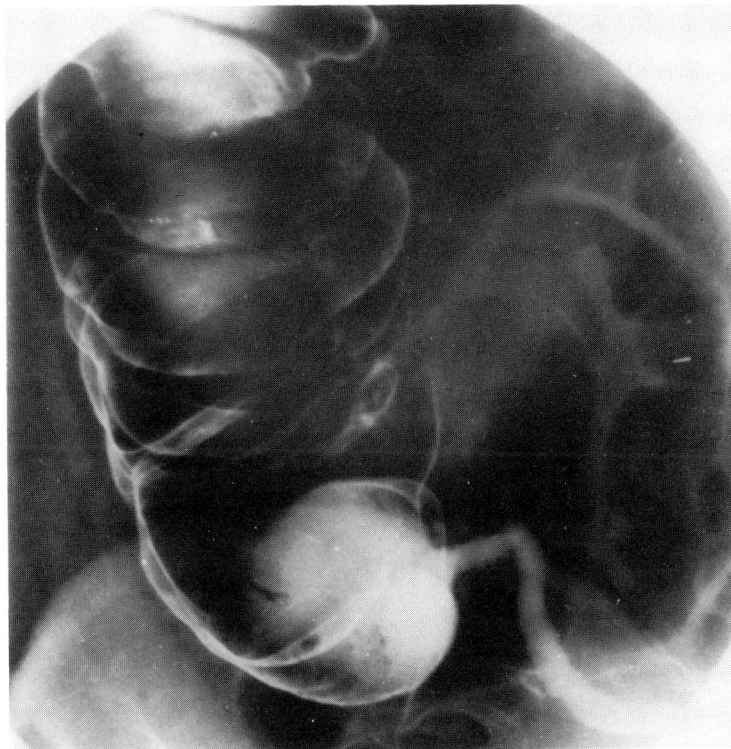

Figure 3-3. Third type of anatomic variation in the origin of the appendix. The appendix arises from the medial wall of the cecum with a prominent bulge below, which forms a false apex of the cecum. This is the most common type. (From Beneventano, T. C., et al.: The roentgen aspects of some appendiceal abnormalities. Amer. J. Roent., 96:344, 1966. Courtesy of Charles C Thomas, Publisher.)

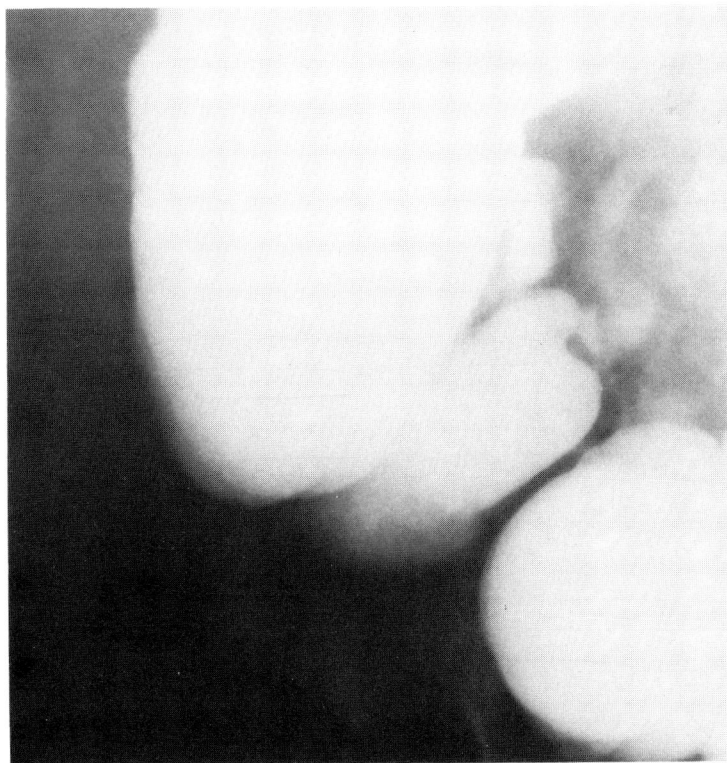

Figure 3-4. Fourth type of anatomic variation in the origin of the appendix. The appendix arises close to the ileocecal valve. (From Beneventano, T. C., et al.: The roentgen aspects of some appendiceal abnormalities. Amer. J. Roent., 96:344, 1966. Courtesy of Charles C Thomas, Publisher.)

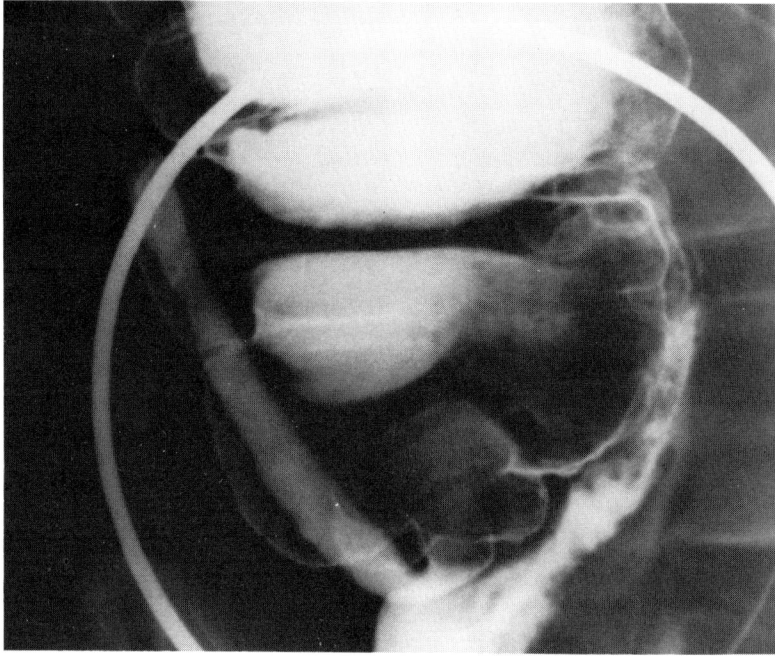

Figure 3–5. Air contrast colon examination showing filling of the appendix with barium. The appendix is retrocecal in position.

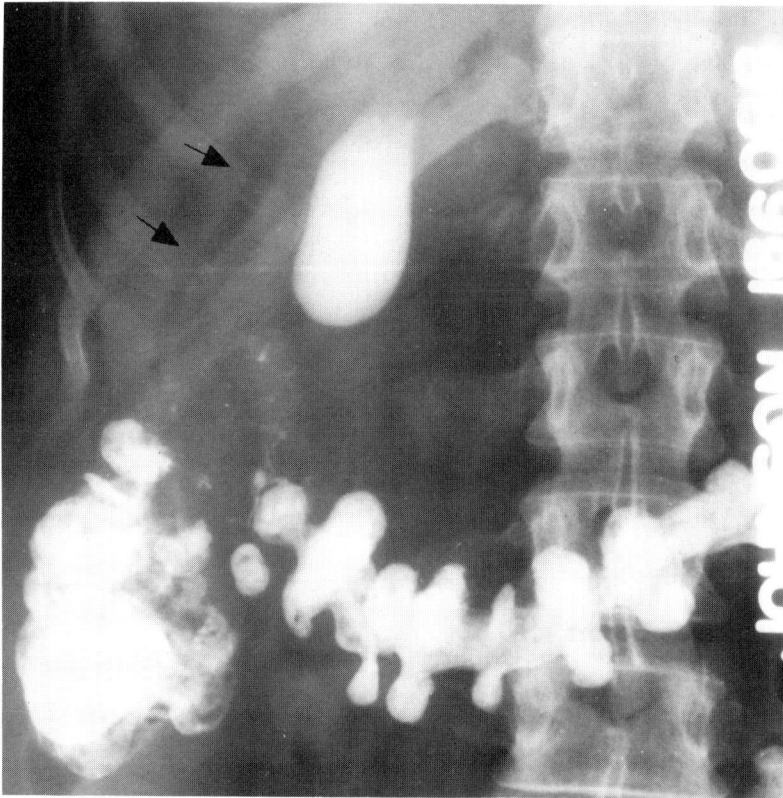

Figure 3–6. Oral cholecystogram showing a gas-filled appendix (arrows) adjacent to the liver and the opacified gallbladder. Air often collects in the appendix when it is in this location and does not indicate that the appendix is inflamed. (From Beneventano, T. C., et al.: The roentgen aspects of some appendiceal abnormalities. Amer. J. Roent., *96*:344, 1966. Courtesy of Charles C Thomas, Publisher.)

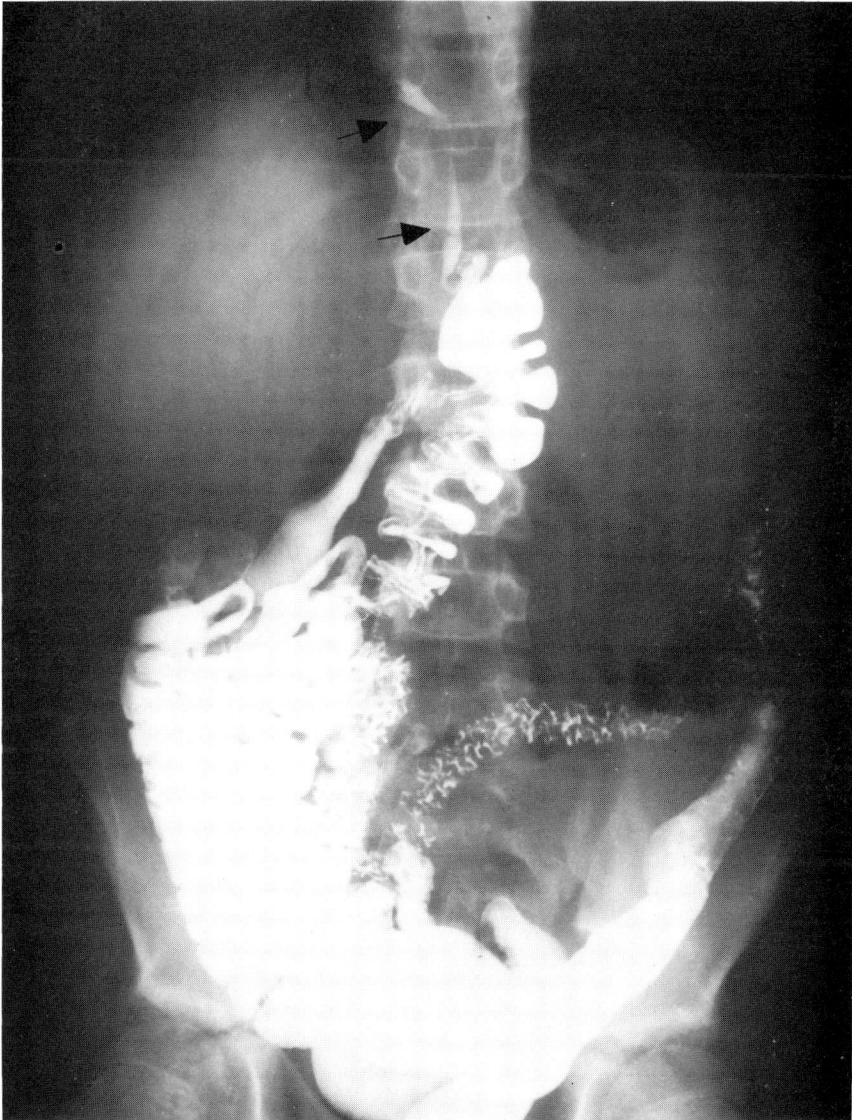

Figure 3-7. Barium-filled appendix (arrows) is in a foramen of Morgagni hernia in a patient with malrotation of the colon. (From Beneventano, T. C., et al.: The roentgen aspects of some appendiceal abnormalities. Amer. J. Roent., 96:344, 1966. Courtesy of Charles C Thomas, Publisher.)

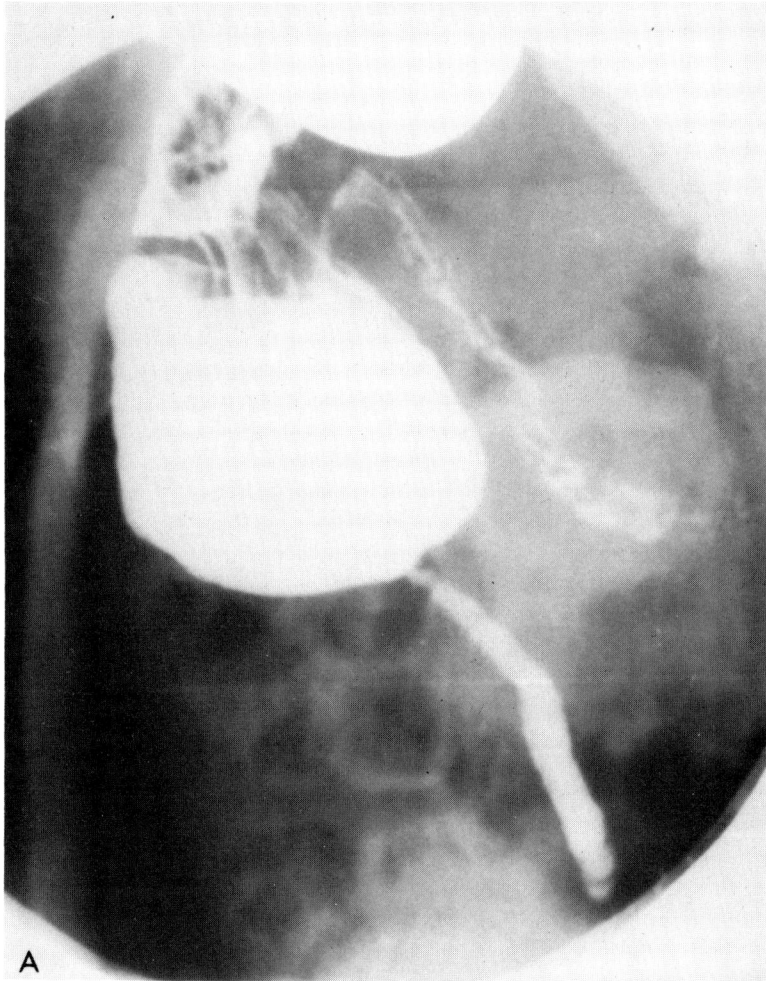

Figure 3–8. Barium enema examination (A) with air contrast study (B). (*Legend continued on opposite page.*)

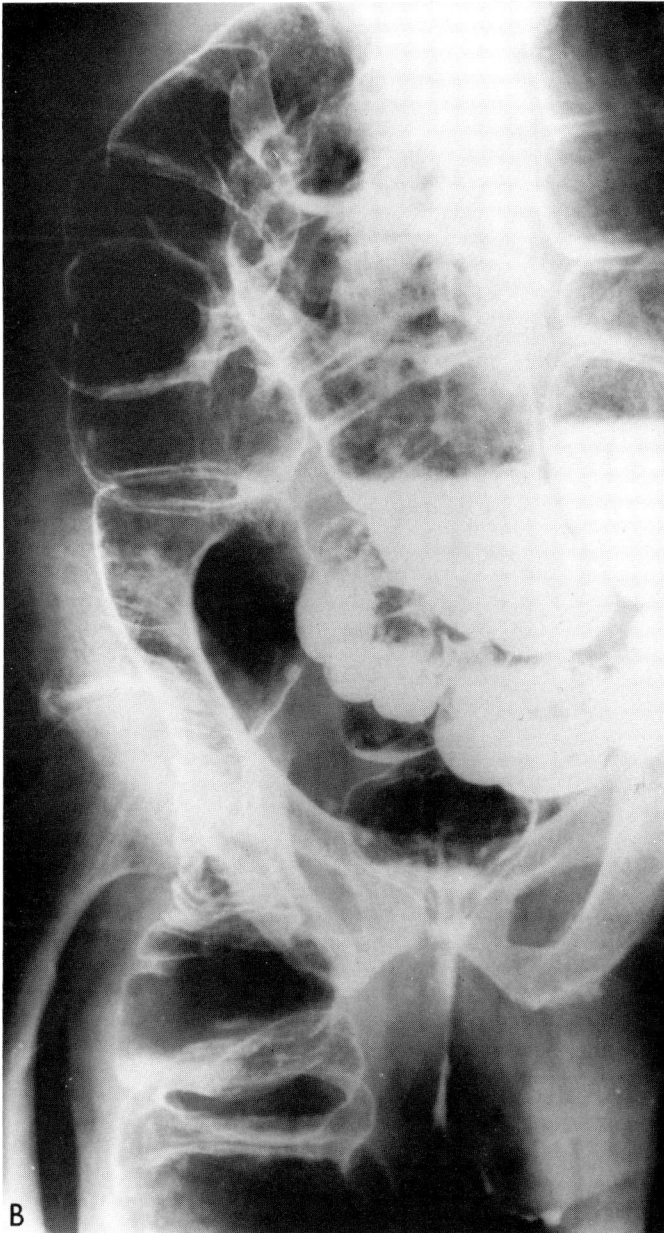

Figure 3–8. (*Continued*). Both A and B show the appendix, cecum and ileum in a right inguinal hernia. In A the right femur is visible on the observer's left. (From Beneventano, T. C., et al.: The roentgen aspects of some appendiceal abnormalities. Amer. J. Roent., 96:344, 1966. Courtesy of Charles C Thomas, Publisher.)

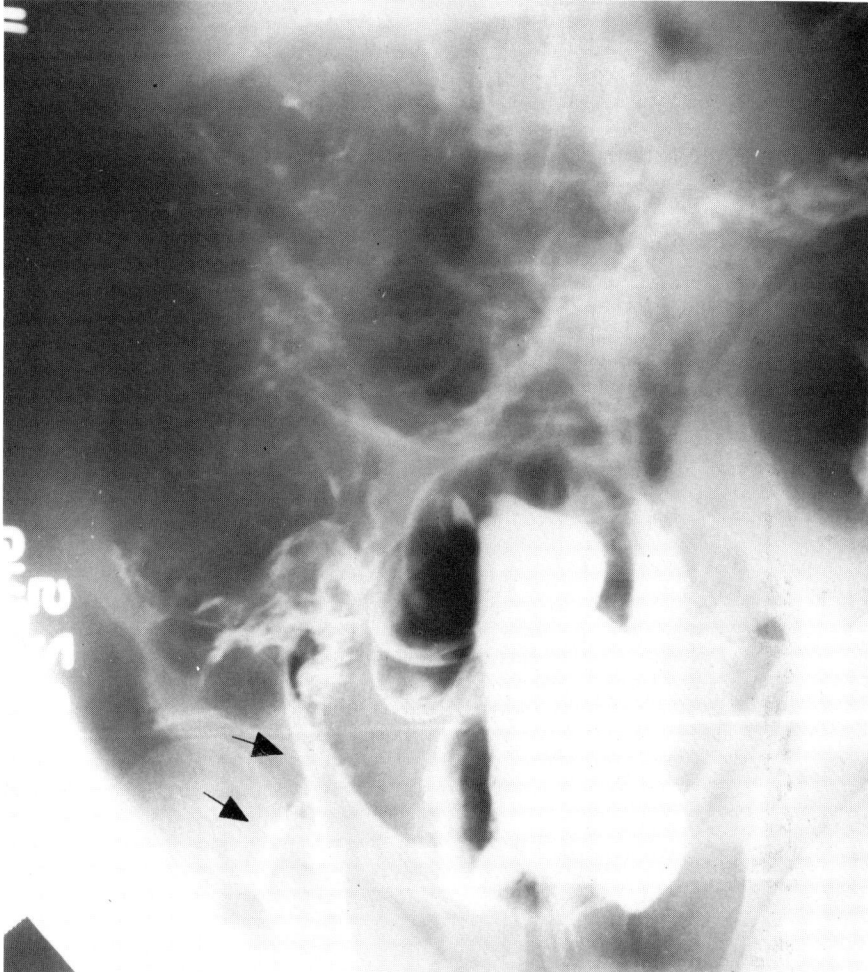

Figure 3–9. Air contrast examination showing a barium-filled appendix (arrows) in a right inguinal hernia. (From Beneventano, T. C., et al.: The roentgen aspects of some appendiceal abnormalities. Amer. J. Roent., 96:344, 1966. Courtesy of Charles C Thomas, Publisher.)

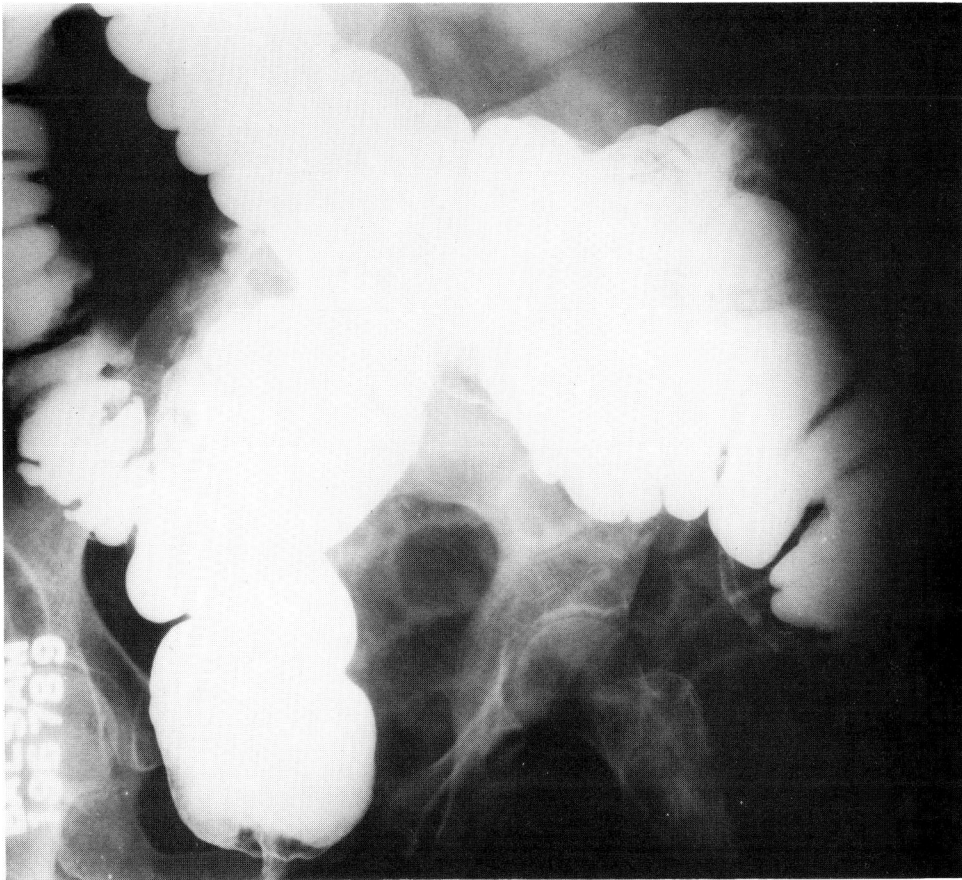

Figure 3–10.   Barium enema examination showing the appendix and cecum in a ventral hernia. (From Beneventano, T. C., et al.: The roentgen aspects of some appendiceal abnormalities. Amer. J. Roent., 96:344, 1966. Courtesy of Charles C Thomas, Publisher.)

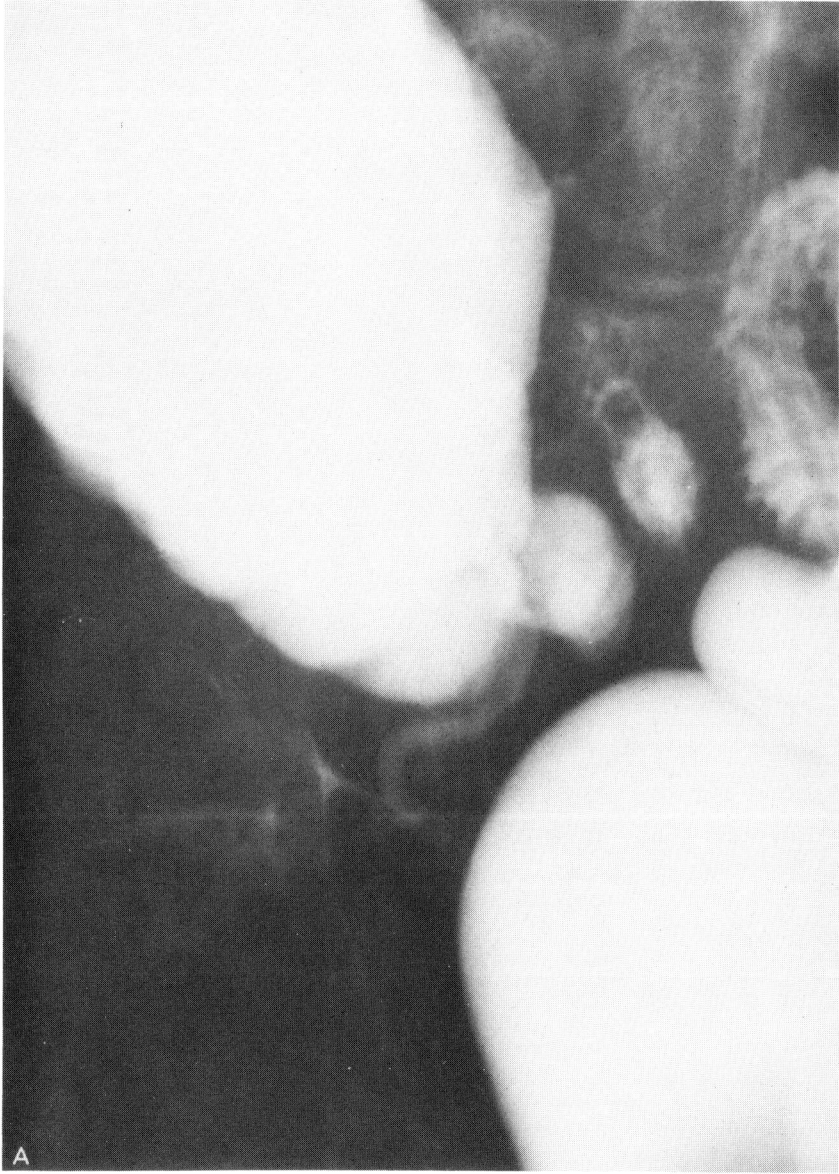

Figure 3–11. Barium enema examination showing the appendix in a right femoral hernia (A). The patient presented with an abscess in the right groin which was due to acute appendicitis. (*Legend continued on opposite page.*)

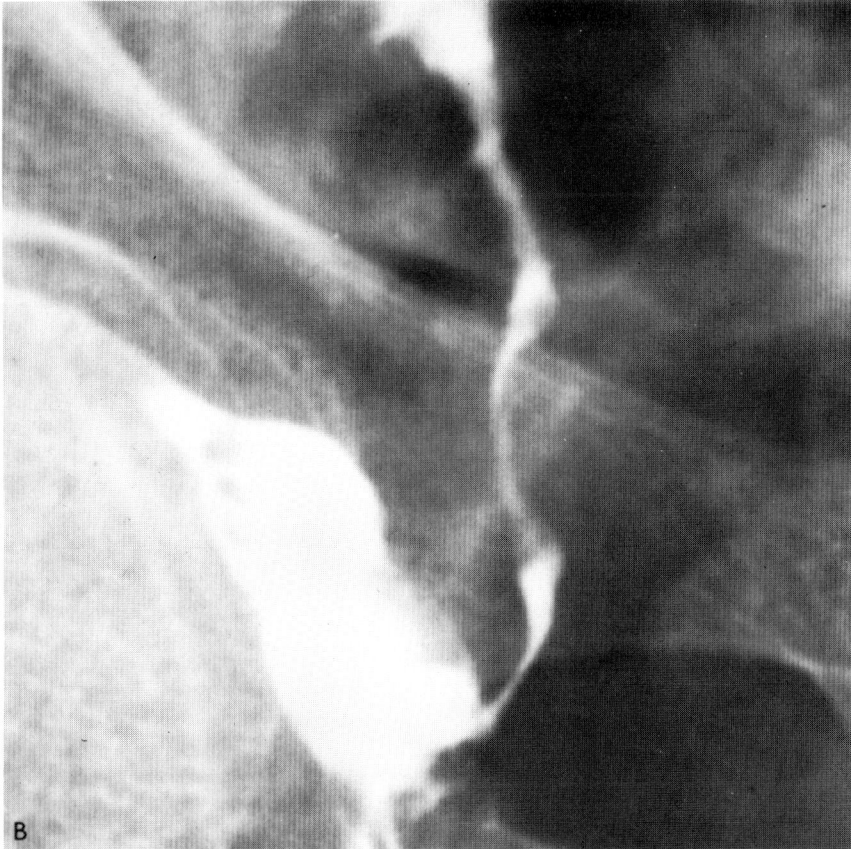

Figure 3-11. (*Continued*).   Barium entered the abscess via the appendix during the barium enema study (*B*).

structure, may present in numerous ectopic positions within the peritoneal cavity as well as in extraperitoneal spaces. Such mobility is possible in instances where the cecum and ascending colon have a well-developed mesentery or are loosely fixed to the posterior abdominal wall. Unusual locations of the appendix may cause acute appendicitis to mimic diseases of the gallbladder, the right kidney, the right lung and the sigmoid colon. Differential diagnosis in these cases may be extremely difficult.

Collins described the appendix in external hernias in 1.1 per cent of autopsy and surgical specimens[3] (Figs. 3–8 to 3–11).

DIVERTICULA

Diverticula of the appendix may be either congenital or acquired. The congenital variety are most often single and contain all layers of the appendiceal wall. Less than 30 cases have been reported.[4]

Acquired diverticula are usually multiple and consist of simple mucosal herniations through defects in the muscularis propria due either to weakness created by penetration of blood vessels or to defects produced by acute inflammatory diseases.[5] Most diverticula occur in the distal one-half of the appendix along the mesenteric border (Fig. 3–12). The outpouches are incidental findings at appendectomy and have no clinical significance. Joffe reported a walled-off perforation of the appendix presenting as a large appendiceal diverticulum[6] (Fig. 3–13).

FOREIGN BODIES

Balch and Silver reported 8 cases of foreign bodies in the appendix and reviewed 217 previously reported.[7] Various objects have been detected in the appendix on barium studies (Fig. 3–14). These include pins, seeds, bones, lead shot, glass, teeth, nails and numerous other miscellaneous objects. The first two recorded appendectomies in 1735 and 1759 were performed on patients with pins in the appendix.[7] Nearly three-quarters of the patients reported with appendiceal foreign bodies had abdominal pain. A prophylactic appendectomy is recommended when a sharp object is identified in the appendix, if it persists on subsequent examinations.

ACUTE APPENDICITIS

Acute appendicitis is a common abdominal disease which in the majority of cases presents little diagnostic difficulty. In a significant number of patients, however, the clinical findings are obscure or minimal, and prompt diagnosis is not possible. It is in these cases that the findings on plain abdominal radiograph or barium examination of the colon and small intestine may be helpful in establishing the diagnosis. The fact that radiologic examination can play a role in the diagnosis of appendicitis is not widely appreciated by practicing physicians. Prompt diagnosis is the most important factor in the management of patients with acute appendicitis. This is particularly important in older patients, in whom the interval between onset of symptoms and perforation may be short and in whom symptoms and signs are often minimal.

In a review of 4500 patients who were treated for acute appendicitis at the Massachusetts General Hospital during a 23-year period, it was shown that the preoperative diagnoses were incorrect in 18 per cent.[8] In 8.3 per cent of the cases there were no surgical findings to explain the symptoms and signs of an acute abdominal condition. The mortality ranged from 0.1 per cent when the appendix was inflamed but not gangrenous to 13.1 per cent when there was an appendiceal abscess not treated initially by appendectomy. The appendix was found to be perforated in 15 per cent of the cases.

Wangensteen showed the importance of obstruction of the appendix in the pathogenesis of appendicitis.[9] A closed-loop obstruction is formed when a fecalith occludes the lumen, resulting in an accumulation of fluid in the obstructed portion. An inflammatory process follows, often culminating in thrombosis and infarction. The wall distal to the obstruction becomes thinned by distention and its mucosa becomes ulcerated. The reaction may then extend to the serosa and the adjacent peritoneum, the omentum, the mesoappendix and the tip of the cecum. Finally, perforation may occur and lead to a local abscess or to generalized peritonitis.

(Text continued on page 79)

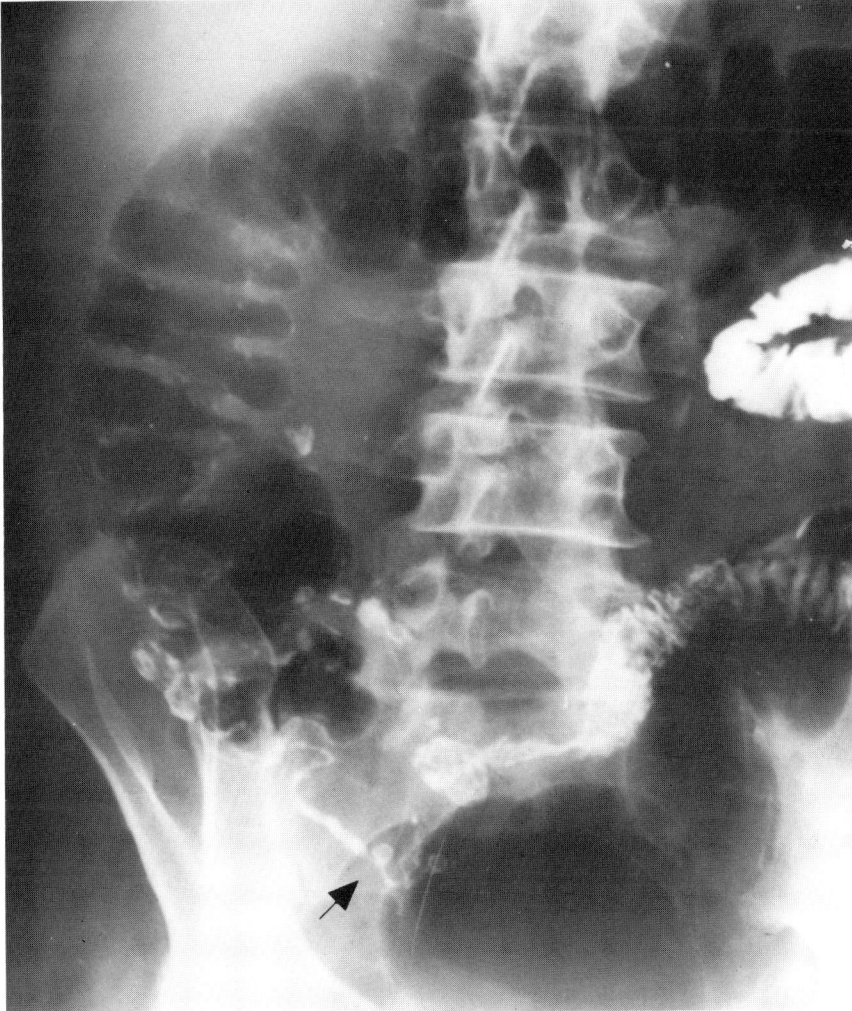

Figure 3–12. Several diverticula are present in the distal portion of the appendix (arrow). (From Beneventano, T. C., et al.: The roentgen aspects of some appendiceal abnormalities. Amer. J. Roent., 96:344, 1966. Courtesy of Charles C Thomas, Publisher.)

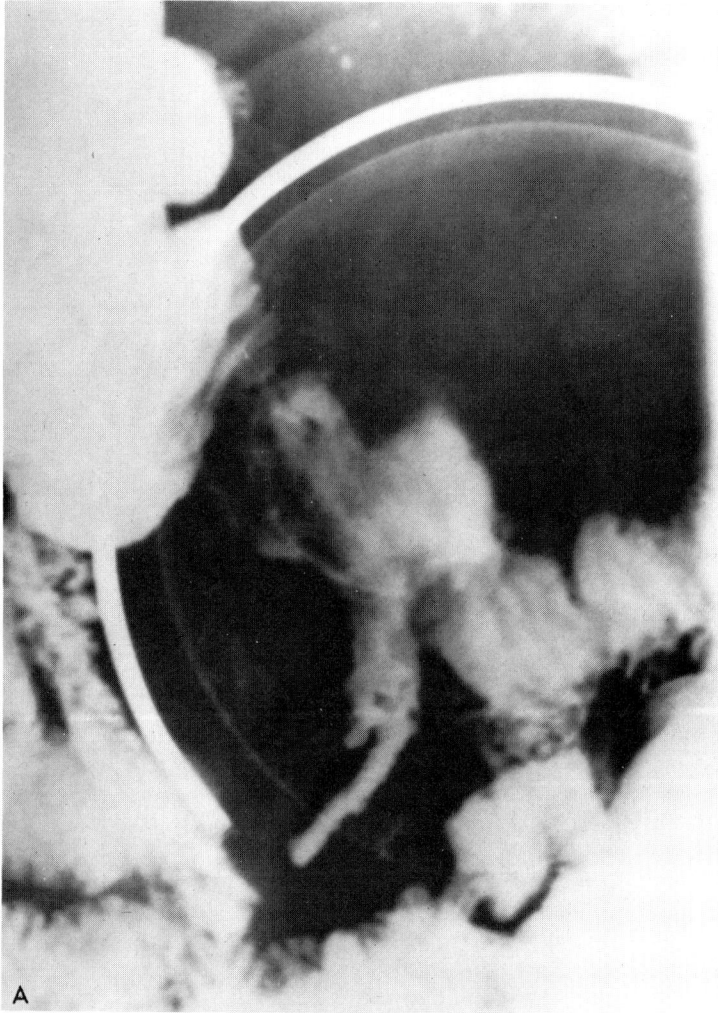

Figure 3–13. There is an irregular collection of barium communicating with the midportion of the appendix (A). (*Legend continued on opposite page.*)

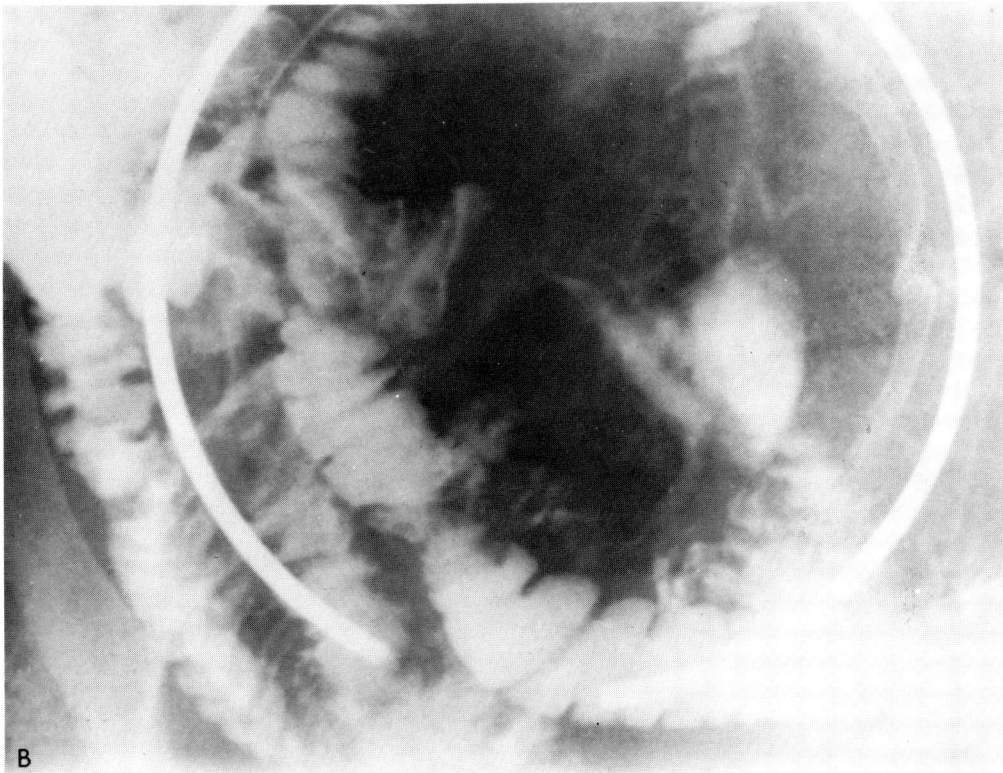

Figure 3–13. (*Continued*).   Eight weeks later the barium-filled cavity is smooth-walled and more discrete (*B*). Appendiceal diverticula may arise from walled-off localized perforation of this type. (From Joffe, N.: Some uncommon roentgenographic findings associated with acute perforative appendicitis. Radiology, *110*:301, 1973.)

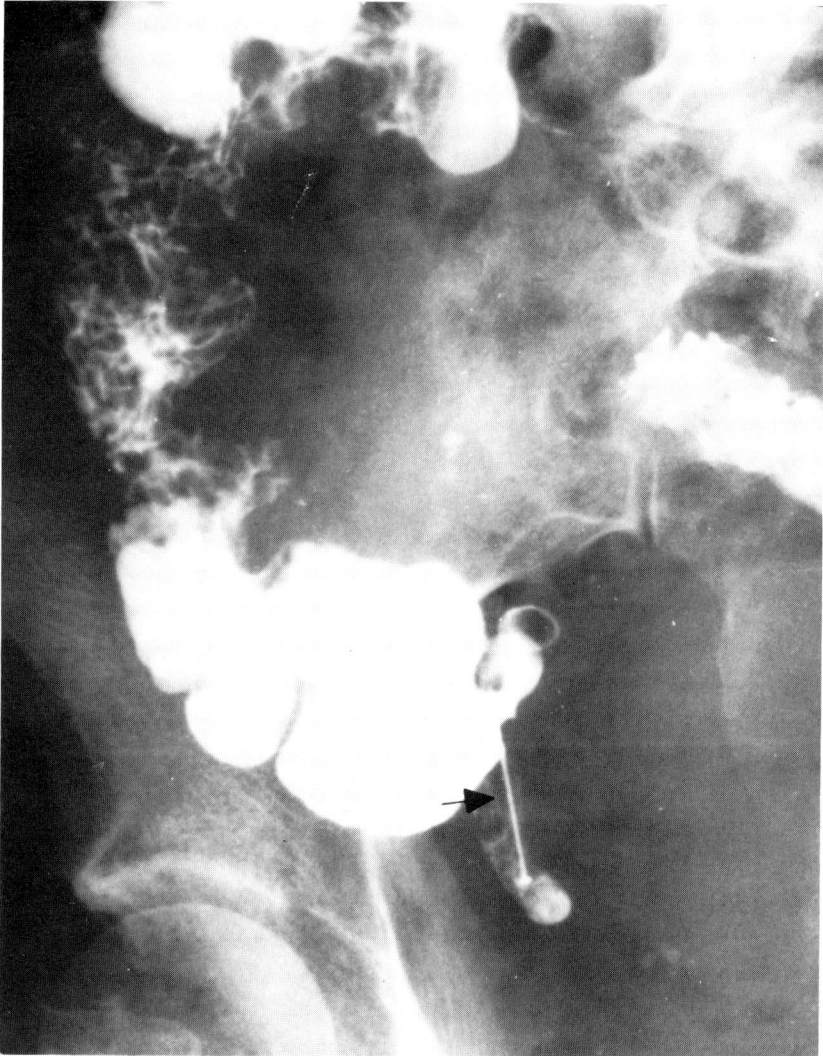

Figure 3-14. Postevacuation radiograph from a barium enema examination showing a straight pin in the appendix.

Radiologic Diagnosis of Appendicitis

The Plain Film of the Abdomen

The plain film of the abdomen is abnormal in approximately 50 per cent of patients with acute appendicitis.[10] While the radiologic findings are often nonspecific, the x-ray examination is valuable in suggesting that acute appendicitis is likely or at least that significant intra-abdominal disease is present.[11, 12, 13]

The full diagnostic value of the radiographic study is possible only when the examination is interpreted with complete knowledge of the clinical findings. The interpretation of subtle, sometimes normal variations must be flavored by the clinical data and amalgamated into an opinion based on mature judgment. Proper radiographic technique is essential. The examination should be an acute abdominal series which consists of three radiographs: a supine projection, an upright projection which includes the pelvis and an upright chest posterior-anterior projection. The pelvis must be included on the upright abdominal radiograph so that air-fluid levels in the pelvis can be identified. The diaphragm need not be included on the upright projection because this area can be studied for free air on the chest examination. In addition, the chest radiograph may reveal pneumonia, which on occasion presents clinically in a manner suggesting appendicitis. Occasionally, right and left lateral decubitus projections are of value.

CALCIFIED FECALITHS OF THE APPENDIX. Calcified fecaliths, also termed appendicoliths, coproliths or appendiceal stones, are the single most important and specific sign of acute appendicitis (Figs. 3–15 to 3–18). Not only is their detection on the plain abdominal radiograph strong evidence of acute appendicitis, but their presence suggests perforation early in the course of the disease, often before it is suspected clinically. Calcified fecaliths should be distinguished from non-calcified fecal material noted in the appendix on barium studies (Fig. 3–19). In Faegenburg's series of 17 patients with appendicoliths visible on the plain abdominal radiograph, all had acute appendicitis and 12 were complicated by gangrene or perforation.[14] Felson and Bernhard reviewed 110

cases of appendicoliths and found an almost constant presence of acute appendicitis with an incidence of perforation of nearly 50 per cent.[15] It would appear that obstruction of the appendix is more complete and distention is more marked in the presence of an appendicolith, so that perforation is more frequent and occurs earlier than in the absence of such stones.

The appendiceal stones, which vary in size from 0.3 to 6.0 cm., can be detected radiographically in from 8 to 12 per cent of patients with acute appendicitis.[16] Most likely these fecaliths originate from the precipitation of calcium phosphate on inspissated fecal material which acts as a nidus. When the calcified fecalith becomes large enough to obstruct the lumen, appendicitis develops. According to Mayer and Wells, the average fecalith is composed of approximately 20 per cent organic residue, 25 per cent inorganic phosphate and 55 per cent soaps and cholesterol.[17]

Appendicoliths usually present as a laminated density in the right lower quadrant.[14] Unusual locations occur when the appendix is in an abnormal position or when the stone lies free in the peritoneal cavity following perforation of the appendix. Calcified lymph nodes, ectopic gallstones and enteroliths in a Meckel's diverticulum, cecal diverticula or regional enteritis must be differentiated. The characteristic oval shape and laminated appearance of the fecalith of the appendix are useful features. Additional radiographic projections including oblique and stereoscopic views and a barium enema examination may be of value.

Rarely, retained barium in the appendix may form an opaque concretion which is visible on the plain abdominal radiograph. A small number of such examples of "barium appendicitis" have been reported[19] (Fig. 3–20). Persistent retention of barium in the appendix, however, occurs with sufficient frequency that no significance can usually be attached to the finding. Vukmer and Trummer reported a case of acute appendicitis that occurred 12 days after a barium enema in which the barium appeared to be the cause of obstruction of the appendix.[20] They state that retention of barium may predispose to the development of acute appendicitis and recommend that patients who demonstrate prolonged reten-

(Text continued on page 93)

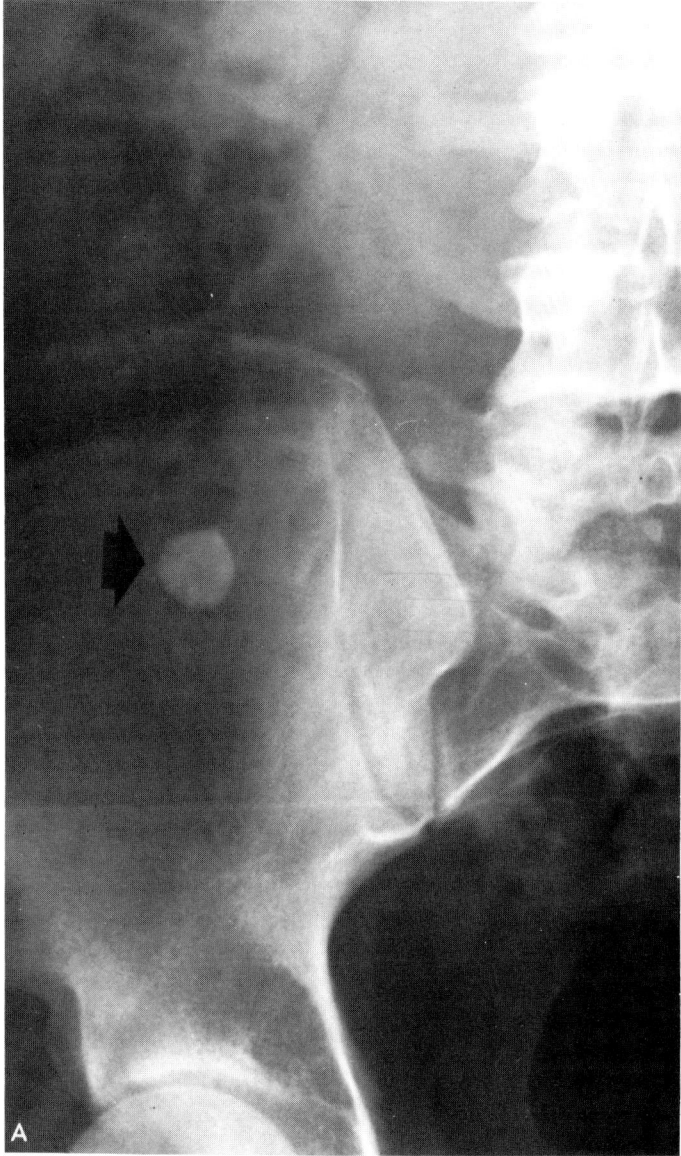

Figures 3–15 to 3–18. Plain abdominal radiographs showing an appendicolith in four patients with acute appendicitis (A) (*continued on opposite page*)

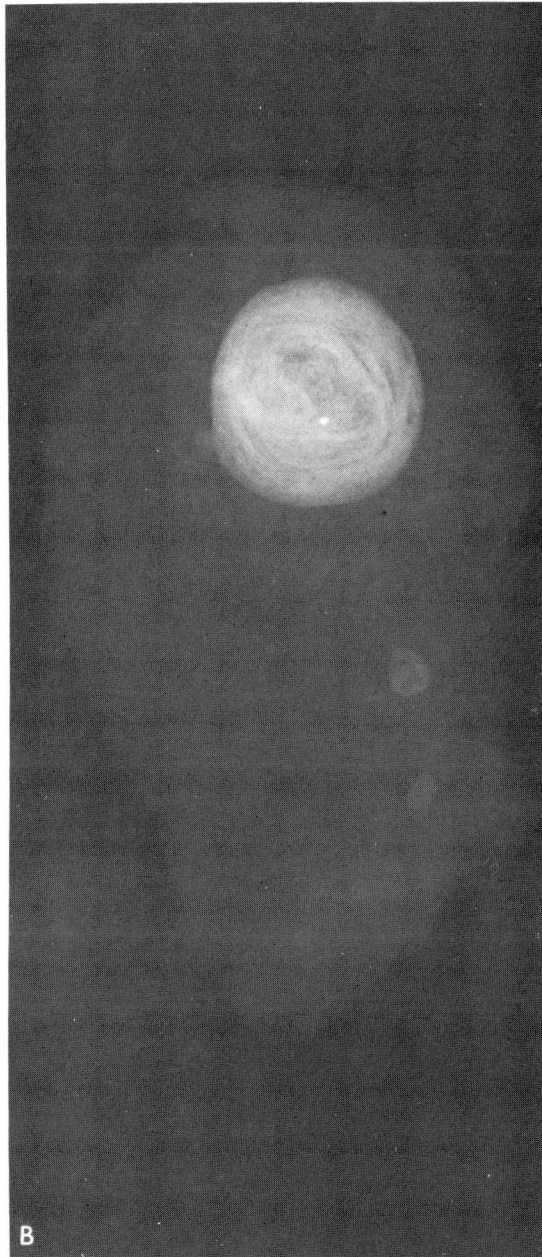

Figure 3-15. (*Continued*).   with a radiograph or photograph of the surgical specimen after appendectomy (B). The appendix was perforated in each case. The laminated appearance of the appendicolith is typical.

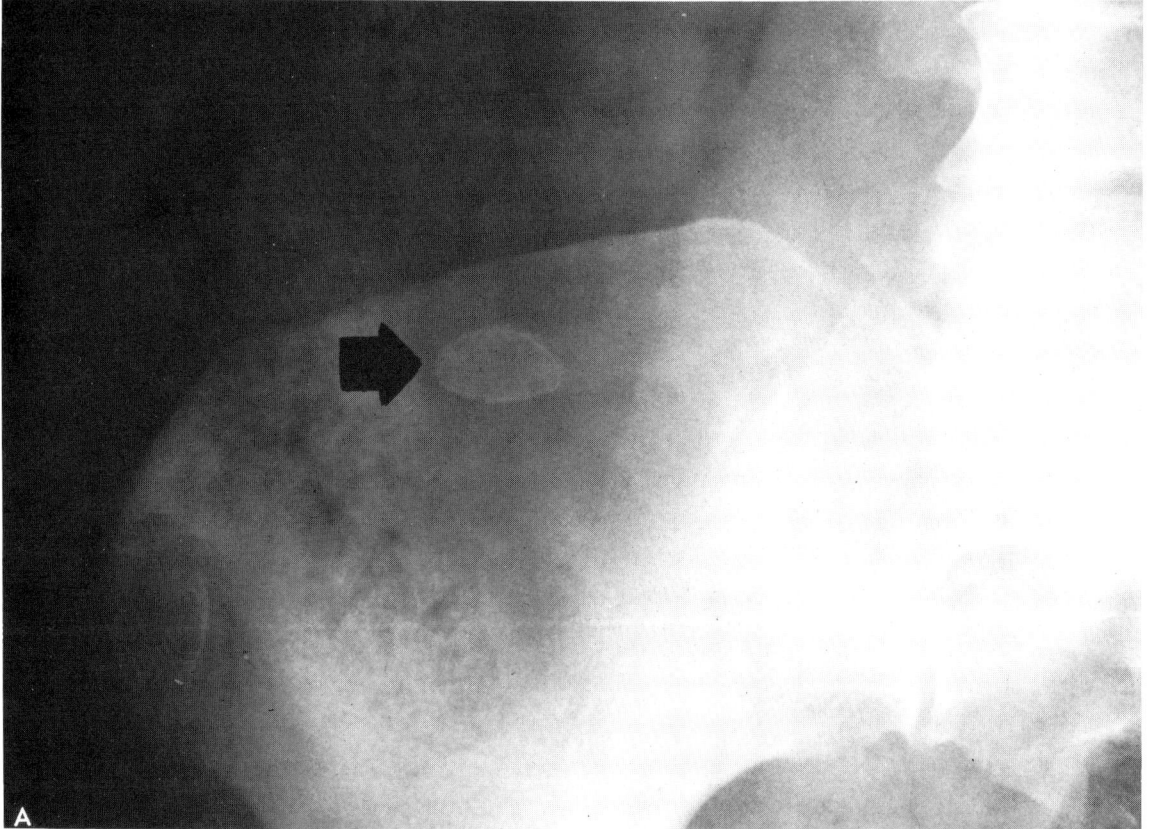

Figure 3–16. *(See legend on pp. 80 and 81.)*

Figure 3–16. (*Continued*)

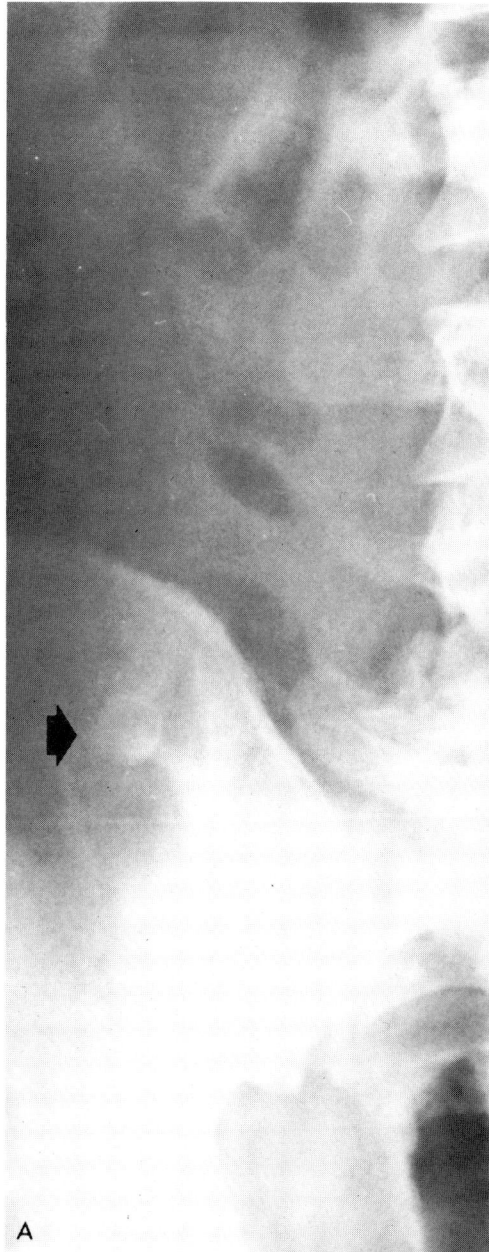

Figure 3-17. *(See legend on pp. 80 and 81.)*

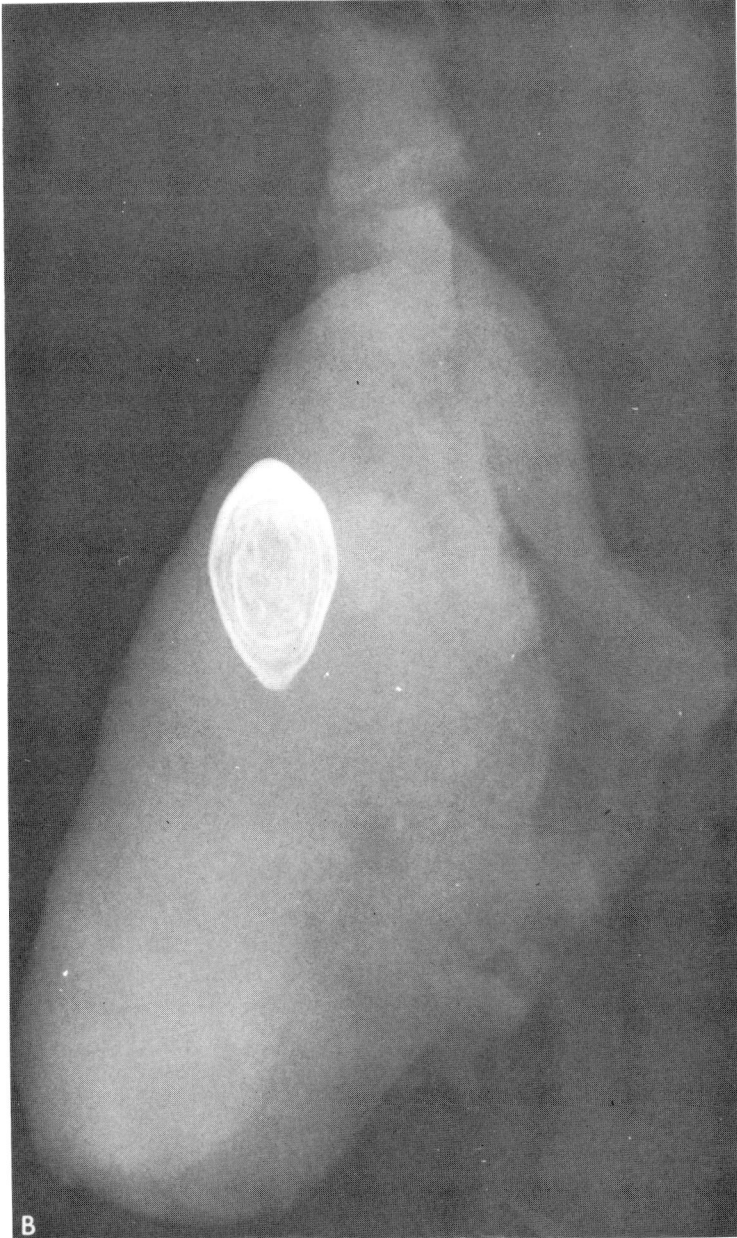

Figure 3–17. (*Continued*)

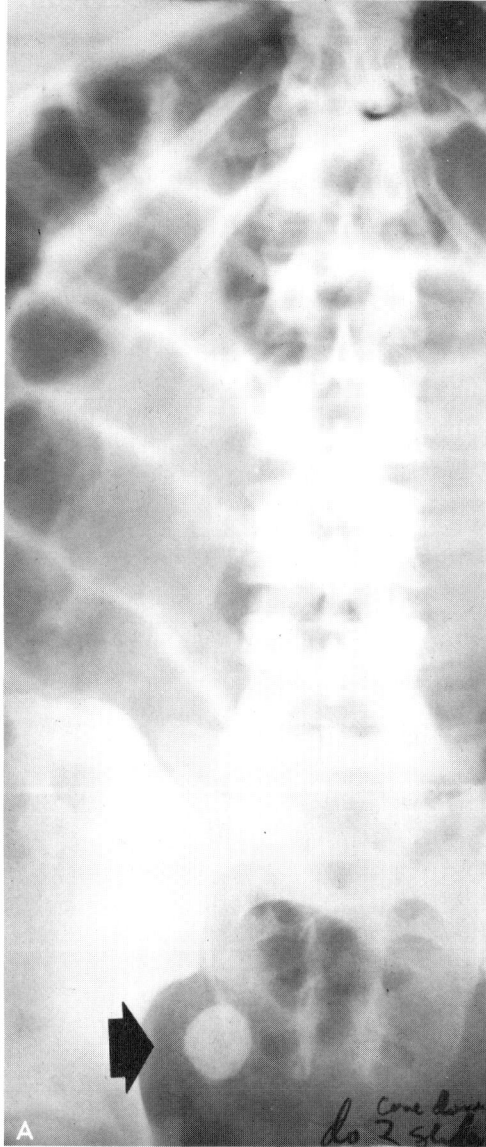

Figure 3–18. *(See legend on pp. 80 and 81.)* (From Beneventano, T. C., et al.: The roentgen aspects of some appendiceal abnormalities. Amer. J. Roent., 96:344, 1966. Courtesy of Charles C Thomas, Publisher.)

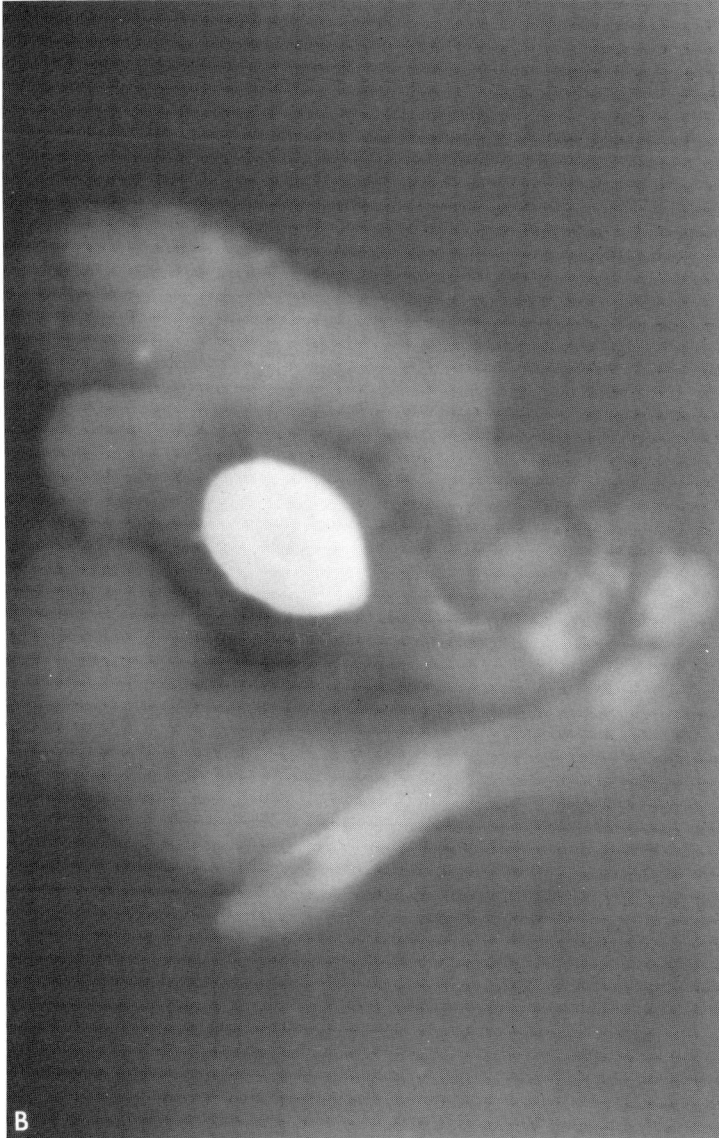

Figure 3–18. (Continued)

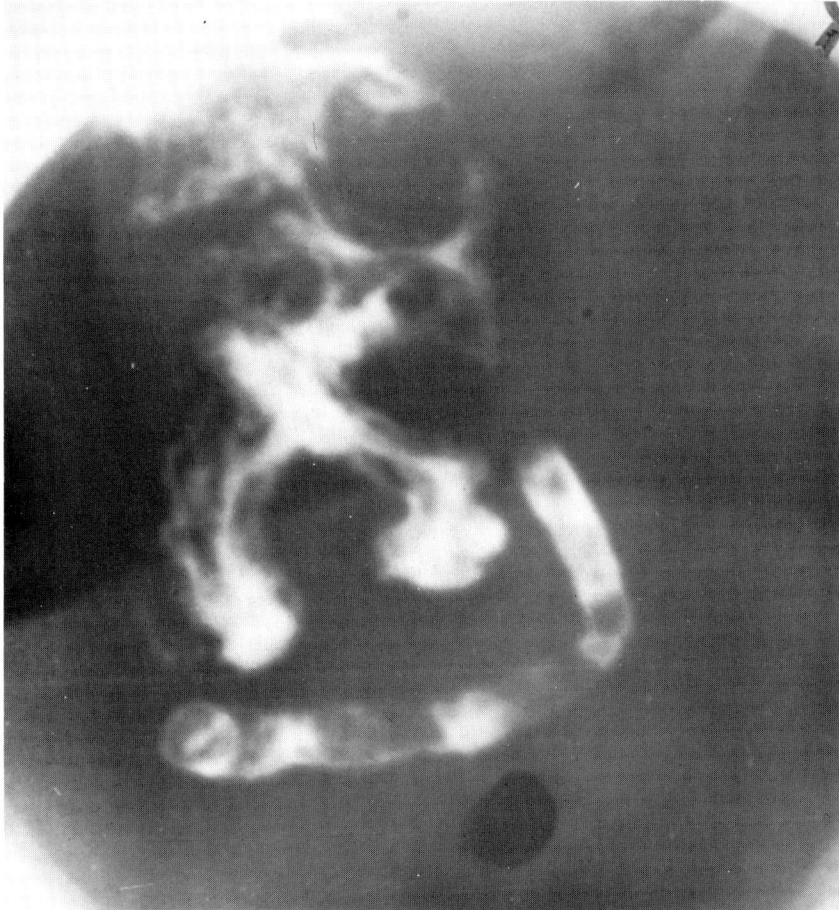

Figure 3–19. Barium in the appendix discloses multiple radiolucencies characteristic of fecaliths. These noncalcified collections must be distinguished from calcified stones which are associated with obstruction.

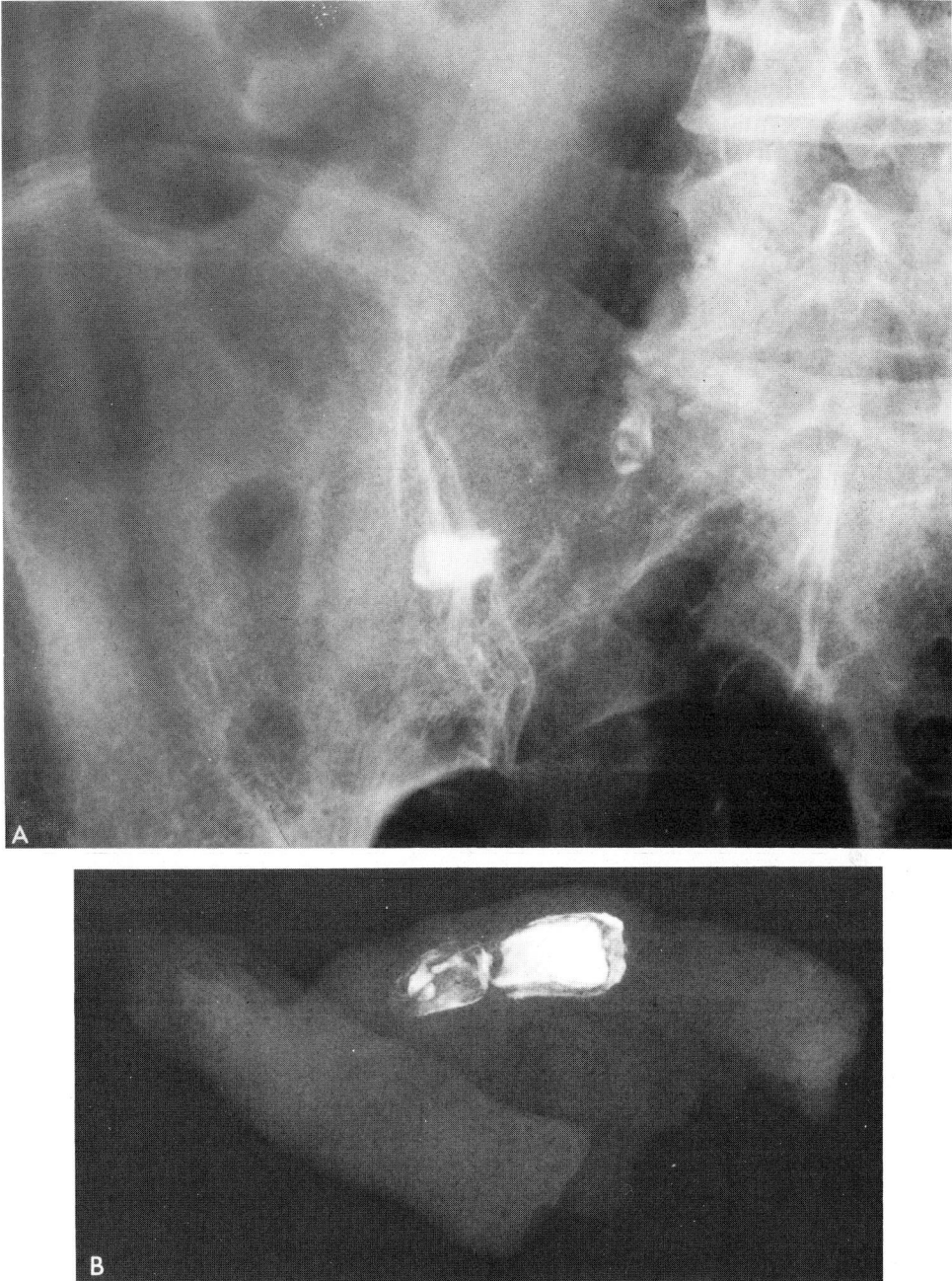

Figure 3-20. Plain abdominal radiograph (*A*) showing barium in the appendix in a patient who had had a barium enema six months earlier. The barolith produced obstruction of the appendix and acute appendicitis. (*B*) Radiograph of the surgical specimen after appendectomy.

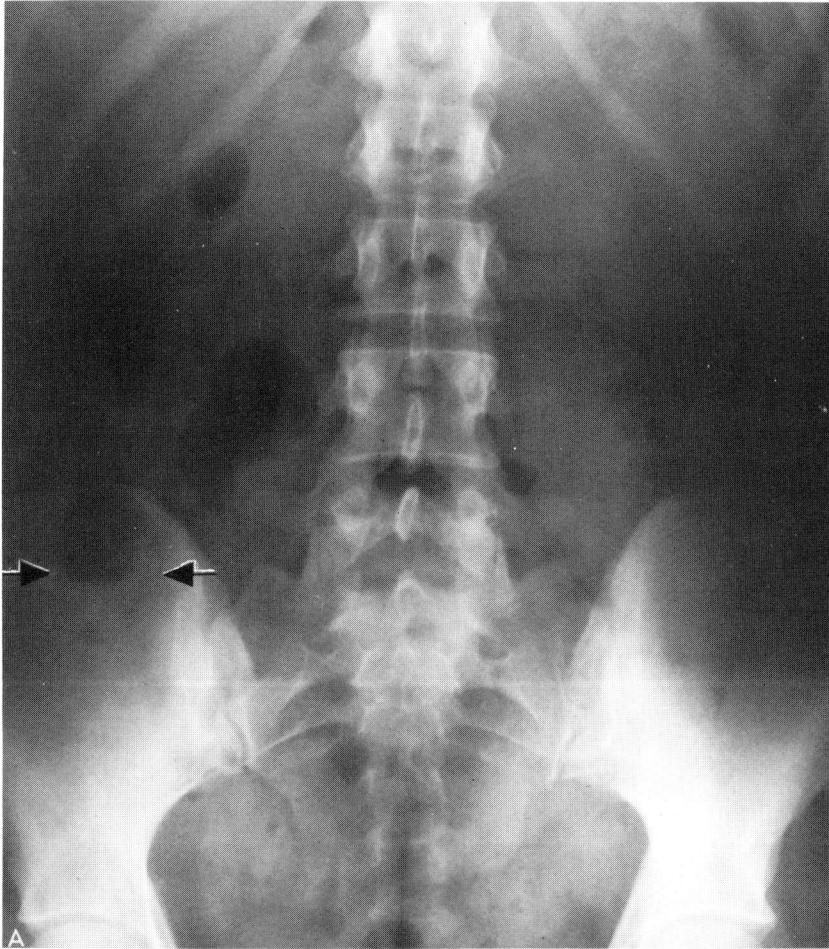

Figure 3–21. Upright plain abdominal radiograph (A) in a patient with acute appendicitis showing a short air-fluid level in the cecum (arrows). This is a frequent finding in cases of early appendicitis. *(Legend continued on opposite page.)*

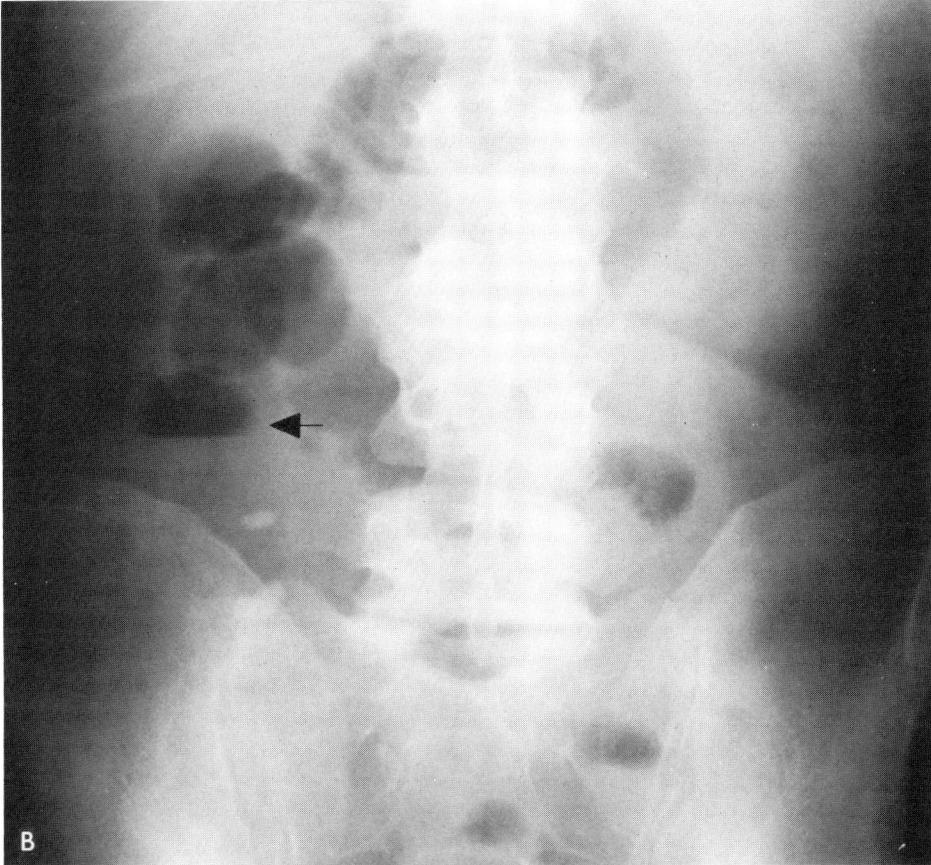

Figure 3–21. (*Continued*) Upright plain abdominal radiograph (*B*) in another patient with acute appendicitis. Sentinel ileus is evident (arrows). The calcifications are in mesenteric lymph nodes.

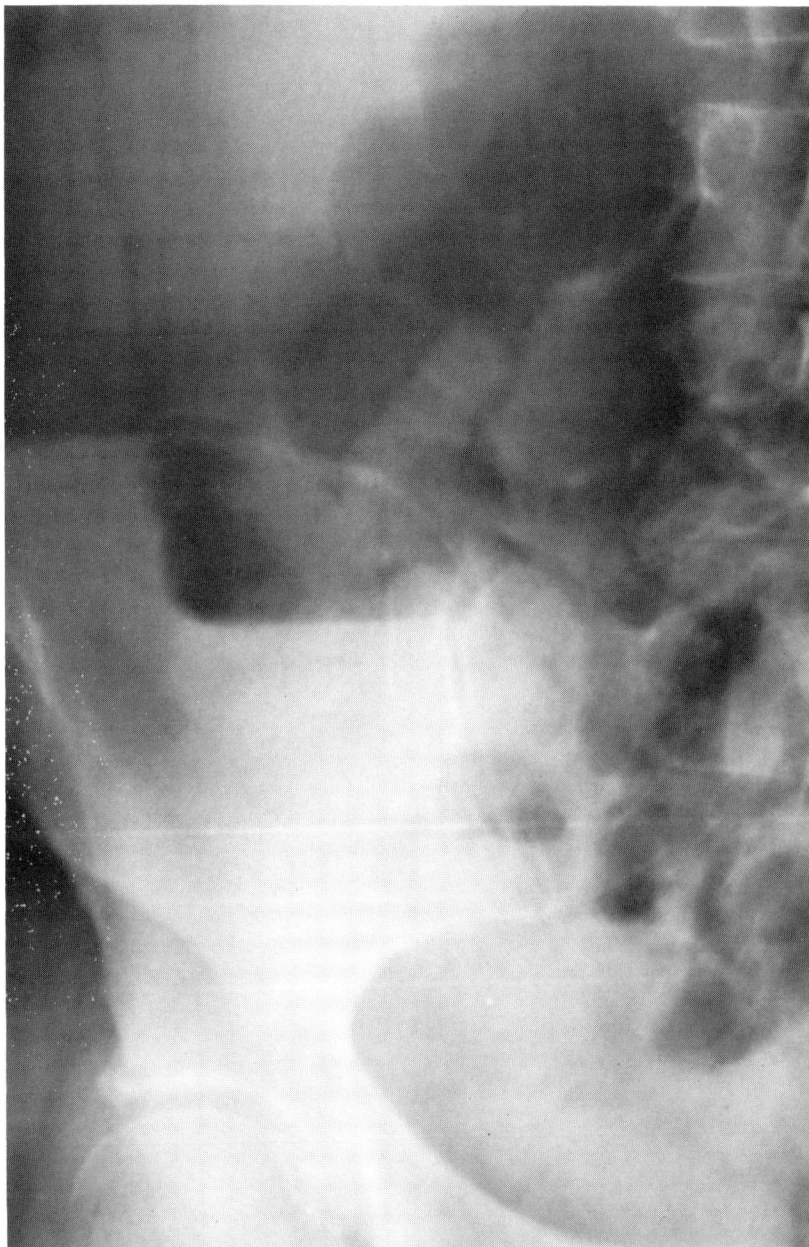

Figure 3–22. Upright plain abdominal radiograph showing an air-fluid level in the cecum. Fifty per cent of patients with acute appendicitis present with this finding (compared to 18 per cent of normal patients).

tion of barium in the appendix be informed of the finding, so that they will seek medical attention without delay at the onset of symptoms suggestive of appendicitis.

CHANGES IN THE CECUM. Because of the proximity of the cecum to the inflamed appendix, paralytic ileus of the cecum (typhlitis) is often one of the initial findings on the plain film examination of patients with appendicitis.[10, 11, 12, 13] Fluid collection in the cecum produces a poorly defined density in the right lower quadrant. On upright or decubitus radiographs the fluid layers in the bottom of the dilated cecum creating an air-fluid level. In Graham and Johnson's series of 100 cases of verified appendicitis a fluid level in the cecum was present in nearly 50 per cent, while in 200 normal plain films, 18 per cent had a fluid level in the right lower quadrant[13] (Figs. 3–21 and 3–22). Air-fluid levels, then, may be a normal finding in the right lower quadrant, particularly in patients who have had an enema or have recently taken a cathartic. Consequently, the finding must be interpreted in light of the clinical circumstances.

CHANGES IN THE TERMINAL ILEUM. Similar changes occur in the terminal ileum with the accumulation of air and fluid, indicating stasis due to adjacent inflammation in the appendix. When air-fluid levels in the ileum are associated with dilatation of the wall of the ileum, the finding is particularly significant (Fig. 3–23). Graham and Johnson found air-fluid levels in the terminal ileum in 51 per cent of their cases of acute appendicitis compared to 5 per cent in the normal.[13] Gammill and Nice reported that only 14 of 300 normal patients had an air-fluid level in the small bowel and in only 3 of the 14 was the small bowel dilated.[21] Air-fluid levels confined to the terminal ileum represent a sentinel ileus, indicating adjacent inflammatory changes, just as segmental dilatation of the small bowel in the upper abdomen is associated with acute pancreatitis or acute cholecystitis.[22] In these circumstances acute peritoneal irritation produces abnormal transit through the involved segment, leading to stasis and dilatation.

GENERALIZED DILATATION OF THE SMALL INTESTINE. Generalized dilatation of the small intestine without proportional distention of the colon occurs in acute appendicitis due either to a mechanical small

bowel obstruction or to paralytic ileus confined to the small bowel. Ten per cent of patients with acute appendicitis present with small bowel obstruction due to adhesions and twisting of the terminal ileum caused by the inflammatory changes involving the appendix[23] (Fig. 3–24). In cases of perforated appendix, a generalized paralytic ileus is the usual plain film finding, with equal dilatation of both the large and small bowel (Fig. 3–25). However, on occasion paralytic ileus may be confined to the small intestine, perhaps because of extension of the inflammatory process into the root of the small bowel mesentery (Figs. 3–26 and 3–27). In these cases differentiation from small bowel obstruction on a radiologic basis alone may be impossible.

When mechanical obstruction of the ileum due to acute appendicitis results in a volvulus of the intestine, strangulation may occur as in any other small bowel obstruction. In these cases the plain film may show the changes of intestinal ischemia which include a fixed air- or fluid-filled loop of small bowel, air in the bowel wall, or air in the portal venous system (Fig. 3–28).

GENERALIZED DILATATION OF THE COLON. When generalized peritonitis develops following perforation of the appendix, dilatation of the entire colon with multiple air-fluid levels on upright and decubitus projections is a frequent finding. While this is usually associated with an equal degree of dilatation of the small bowel, occasionally the colon is chiefly involved (Fig. 3–29). In this situation of colon ileus, differentiation from distal colonic obstruction or any of the causes of megacolon such as scleroderma, hypothyroidism or parkinsonism may be difficult on a radiologic basis alone.

OBLITERATION OF THE FAT ALONG THE PSOAS AND OBTURATOR MUSCLES. Failure to visualize the psoas muscle on one side or the other is a frequent finding in normal patients and is usually due to slight rotation of the patient. Hence, partial or complete loss of psoas visualization must be viewed with circumspection. When the lower half of the psoas margin is obliterated, one can suspect distortion of the retroperitoneal fat along the psoas muscle due to the retroperitoneal accumulation of pus, blood or urine (Fig. 3–30).

Similarly, when inflammation involves

(Text continued on page 102)

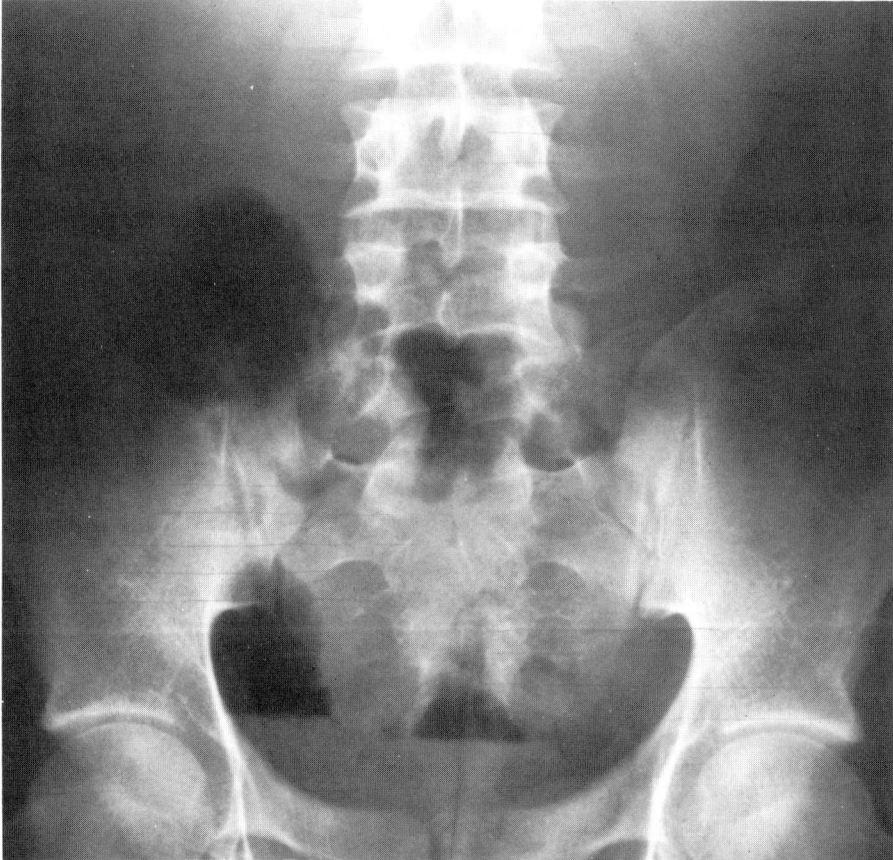

Figure 3-23. Upright plain abdominal radiograph showing multiple air-fluid levels in the terminal ileum in a patient with acute appendicitis. Sentinel ileus of this type suggests adjacent inflammatory disease, such as appendicitis.

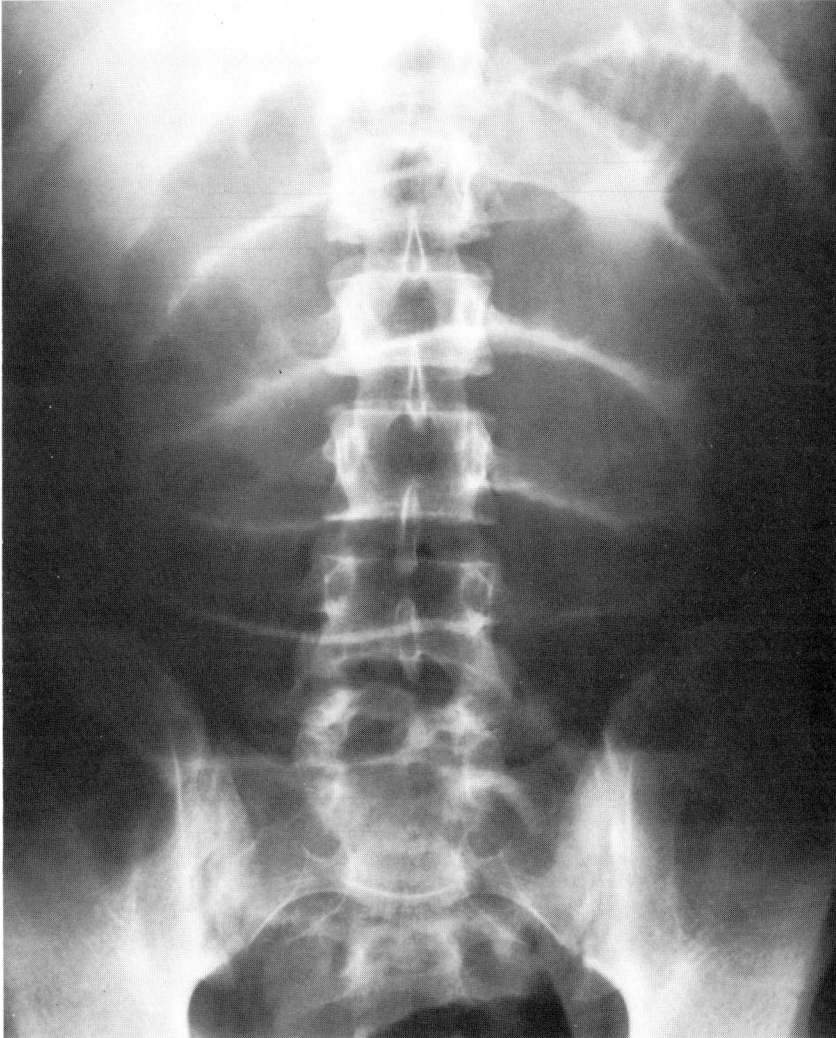

Figure 3–24. Supine plain abdominal radiograph showing multiple dilated loops of small bowel in a patient with acute appendicitis. At operation, the terminal ileum was adherent to the inflamed appendix and was twisted, producing a small bowel obstruction.

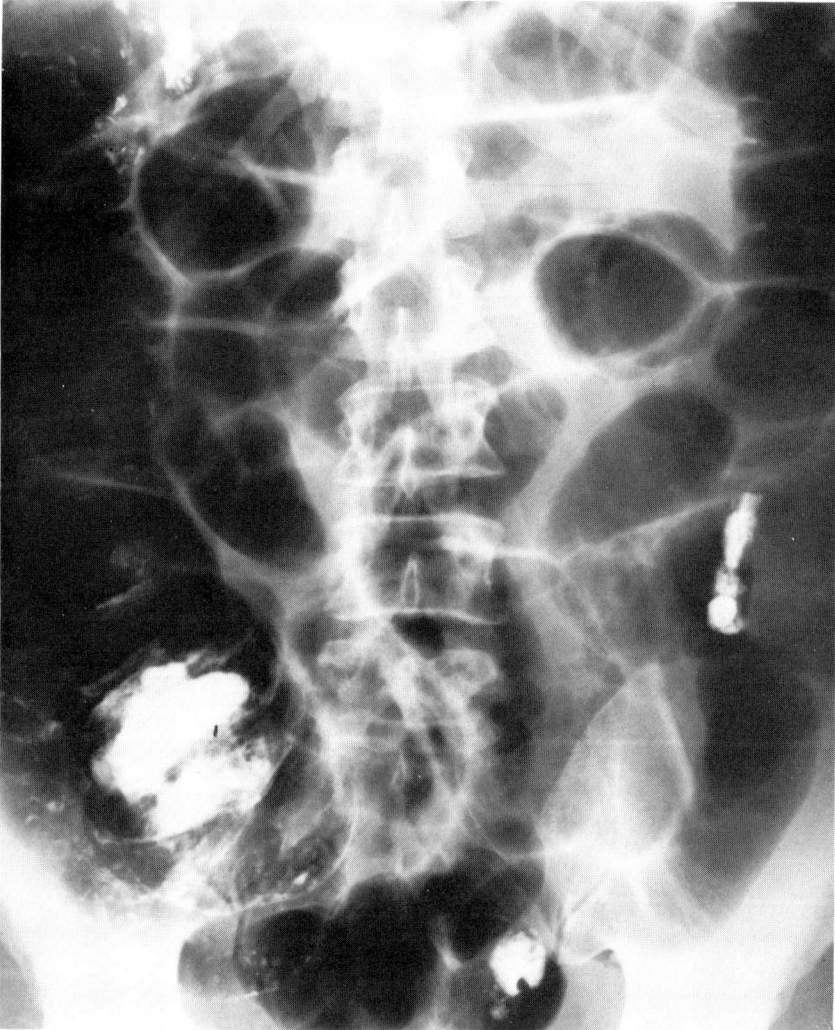

Figure 3–25. Supine plain abdominal radiograph showing a generalized paralytic ileus in a patient with appendicitis with perforation. There is proportional distention of the small and large bowel. Barium is present from an earlier examination.

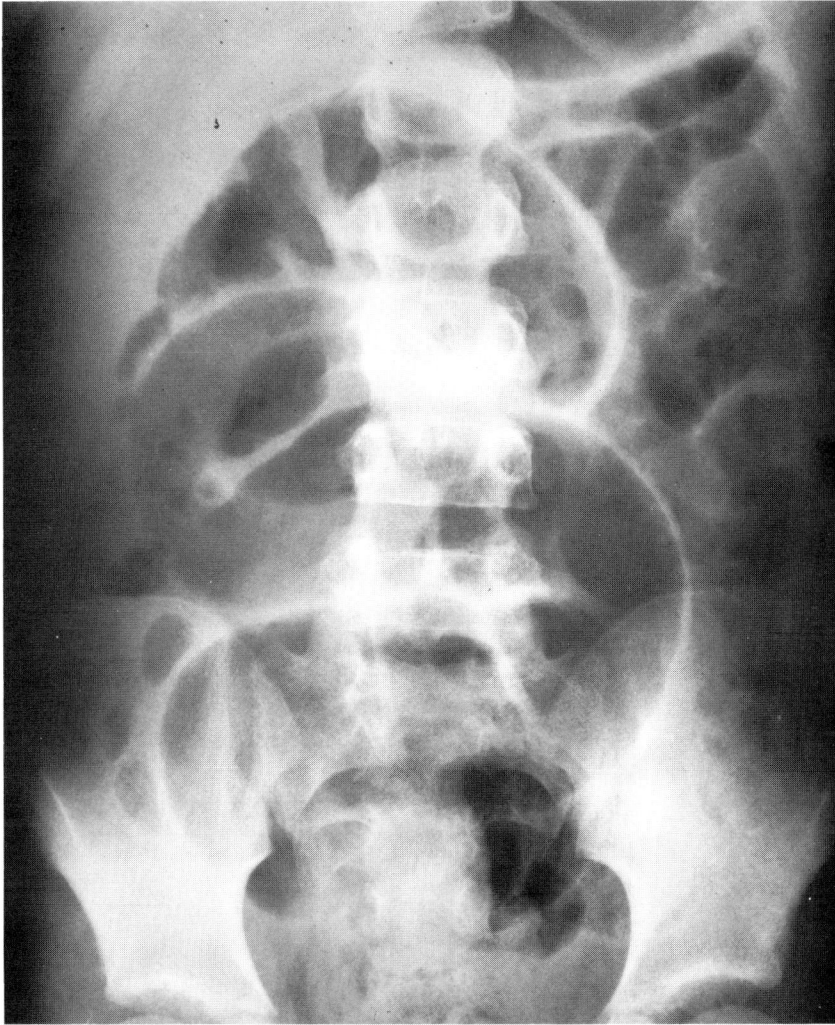

Figure 3–26. Supine plain abdominal radiograph showing miltiple loops of dilated small bowel in a patient with appendicitis with perforation. A small bowel ileus may occur without mechanical obstruction in some cases of acute appendicitis.

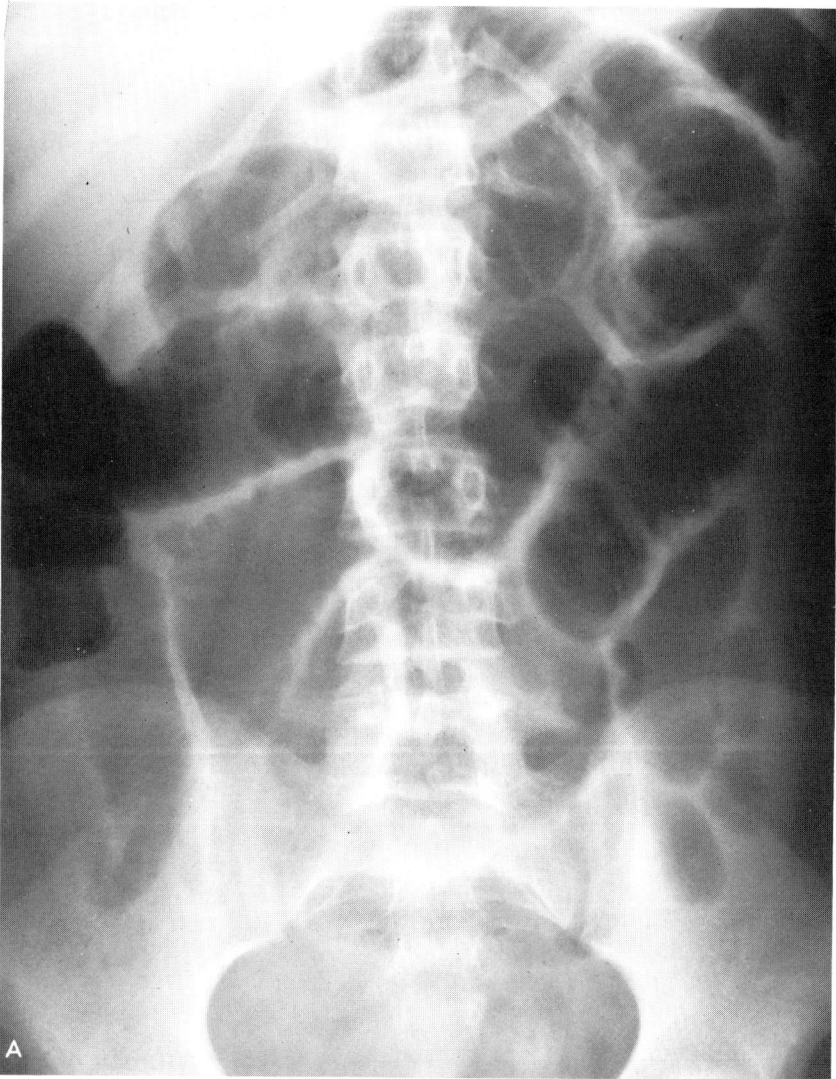

Figure 3-27. Supine (A) and upright (B) plain abdominal radiographs showing dilated loops of small bowel in a patient with acute appendicitis with perforation. Paralytic ileus involving only the small bowel may be related to extension of the inflammatory process into the small bowel mesentery.

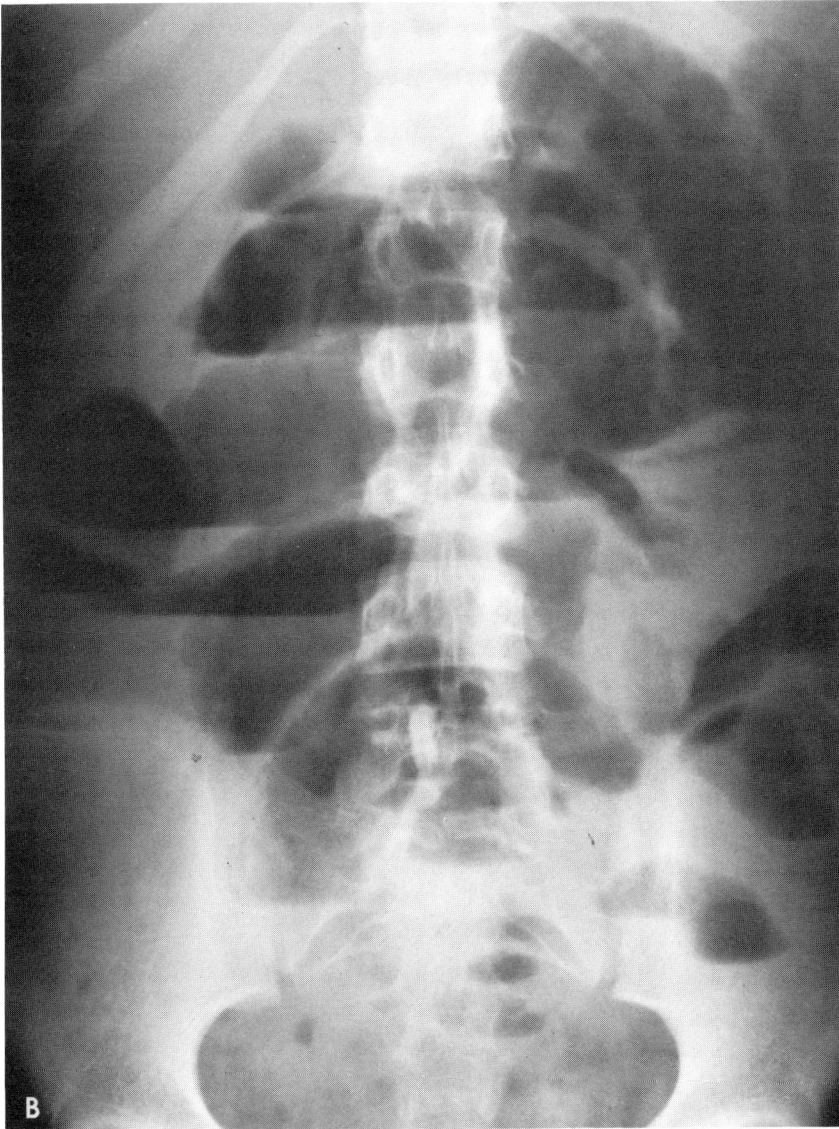

Figure 3-27. (Continued)

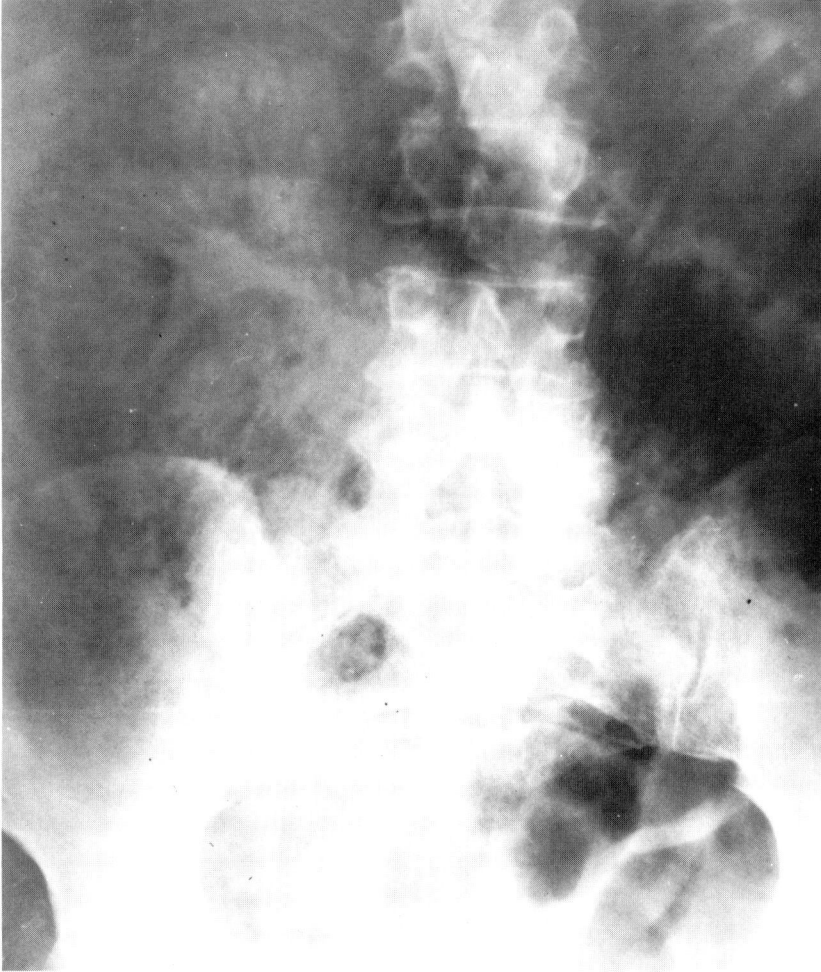

Figure 3–28. Supine plain abdominal radiograph showing air in the wall of the small bowel in an elderly patient with infarction of a long segment of the terminal ileum due to acute appendicitis. Adhesions of the ileum to the inflamed appendix resulted in a volvulus with strangulation.

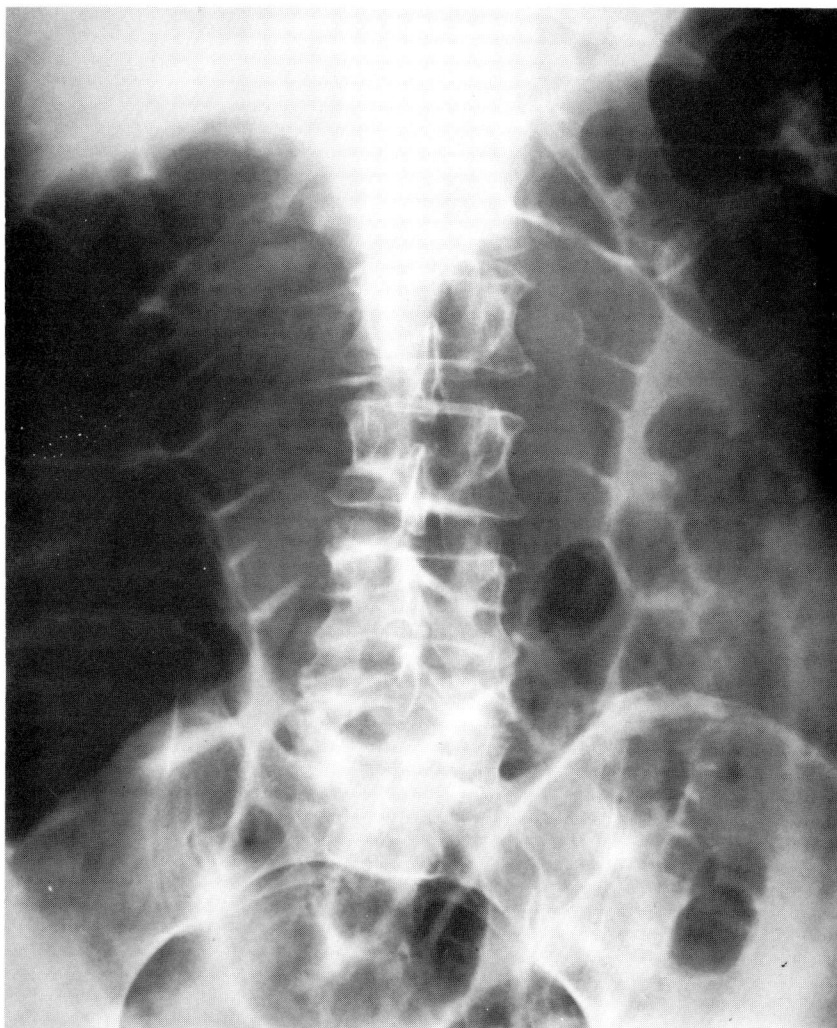

Figure 3–29. Supine plain abdominal radiograph showing disproportionate dilatation of the colon in a patient with acute appendicitis. A colon ileus of this type must be differentiated from other causes of acute megacolon.

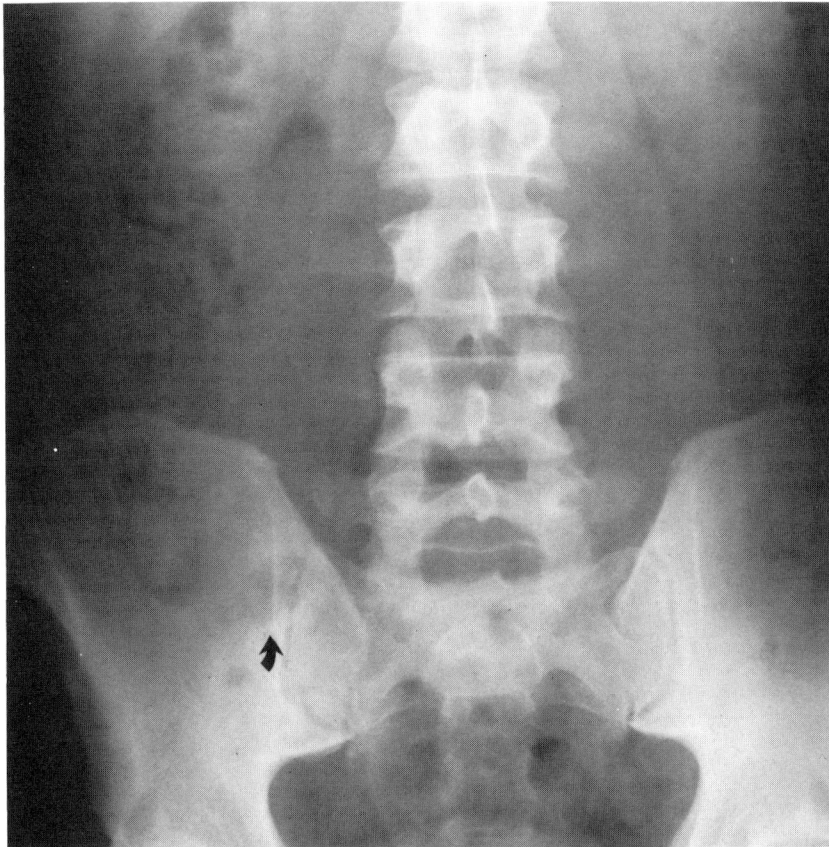

Figure 3–30. Supine plain abdominal radiograph showing obliteration of the inferior half of the right psoas margin in a patient with acute appendicitis. Loss of visualization of part of the psoas margin is more significant than failure to identify its entire length. An air collection is visible in an appendiceal abscess in the right lower quadrant (arrow).

the fat overlying the right obturator muscle in the pelvis, radiographic visualization of the muscle plane is lost (Fig. 3–31). This also occurs in other inflammatory conditions in proximity to the muscle.

FREE INTRAPERITONEAL FLUID. Frimann-Dahl has emphasized the importance of recognizing small amounts of fluid in the flank as a sign of acute appendicitis.[10] Casper detected this finding in over one-half of his patients with acute appendicitis.[24]

In normal individuals a tangential segment of the parietal peritoneum can commonly be detected on properly exposed abdominal radiographs as a fine line in the flank (Fig. 3–32A). The peritoneum is visible because of extraperitoneal fat superficial to the peritoneum and omental or pericolic fat deep to it. Thus, the peritoneum is radiopaque because it is of water density compared to the more radiolucent fat on either side. Also, in the normal right flank the ascending colon can be seen because of its content of gas and fecal material. The liver edge is visible in the right upper quadrant outlined by adjacent fat.

Pus in the right flank can be identified by loss of visualization of the peritoneum and the liver edge and by replacement of the radiolucent fat between the peritoneum and the ascending colon with water-density fluid. The ascending colon is often on a short mesentery, so that when fluid fills the lateral gutter between the parietal peritoneum and the colon, the colon is moved medially, widening the space between the preperitoneal fat and the colon (Fig. 3–32B). The radiolucent preperitoneal fat which is outside the peritoneal cavity is not obliterated, and its detection must not be taken as evidence of absence of fluid within the peritoneal space.

When large amounts of pus collect in the

(Text continued on page 106)

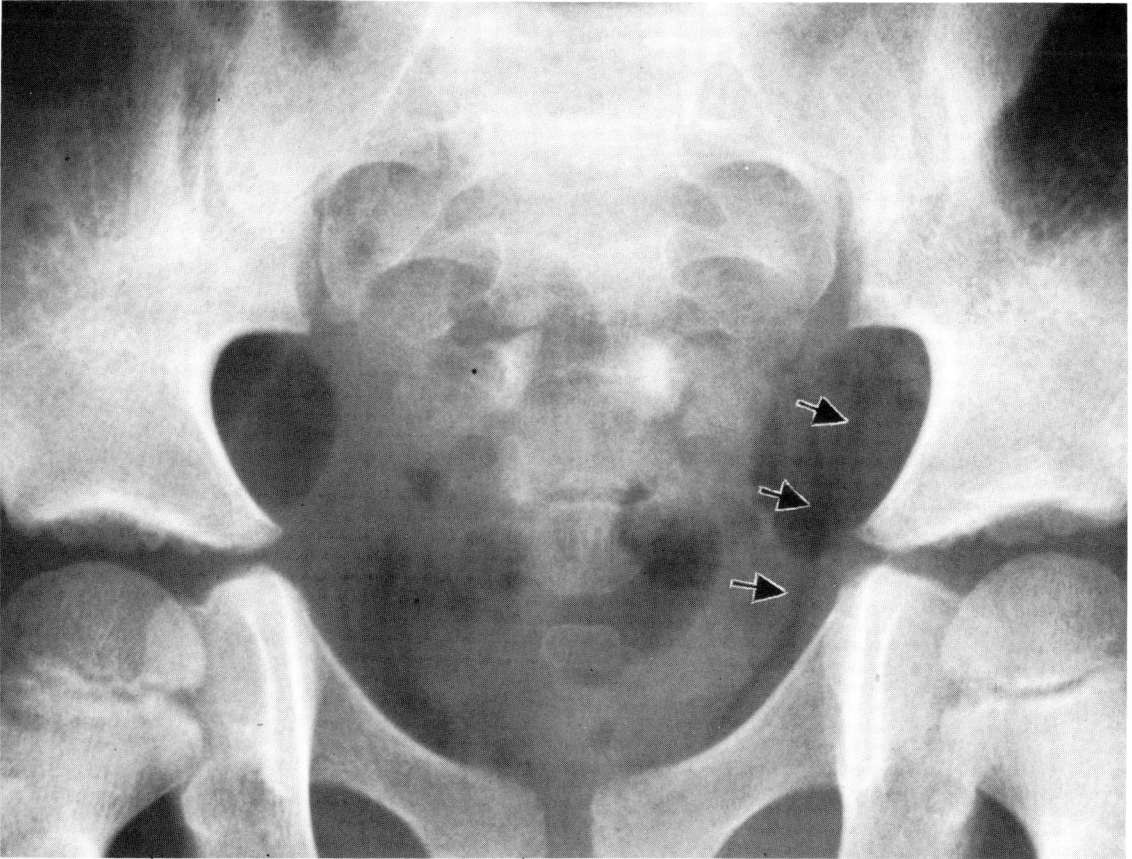

Figure 3–31. Supine plain abdominal radiograph showing the pelvis in a child with acute appendicitis. The fat stripe over the obturator muscle on the right has been lost owing to the inflammatory process. The normal obturator fat is present on the left (arrows).

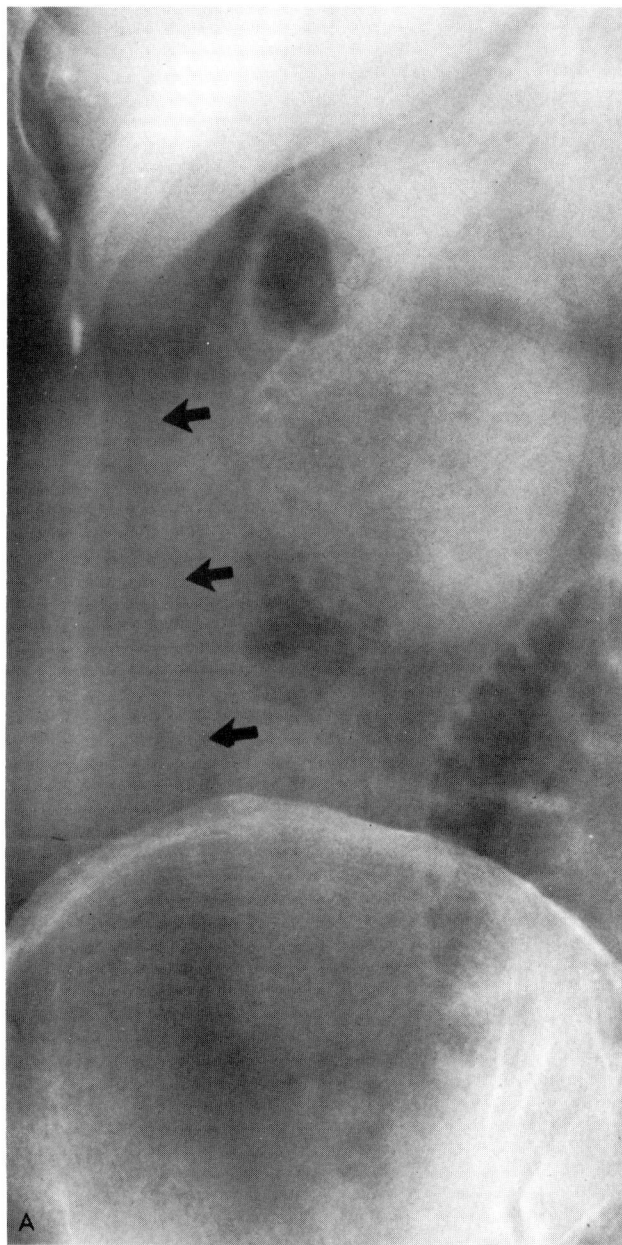

Figure 3–32. Supine plain abdominal radiograph (A) showing the right flank in a normal patient. The parietal peritoneum is visible in tangent (arrows). The normal radiolucent fat is present on either side of the peritoneum, and the liver margin can be identified. *(Legend continued on opposite page.)*

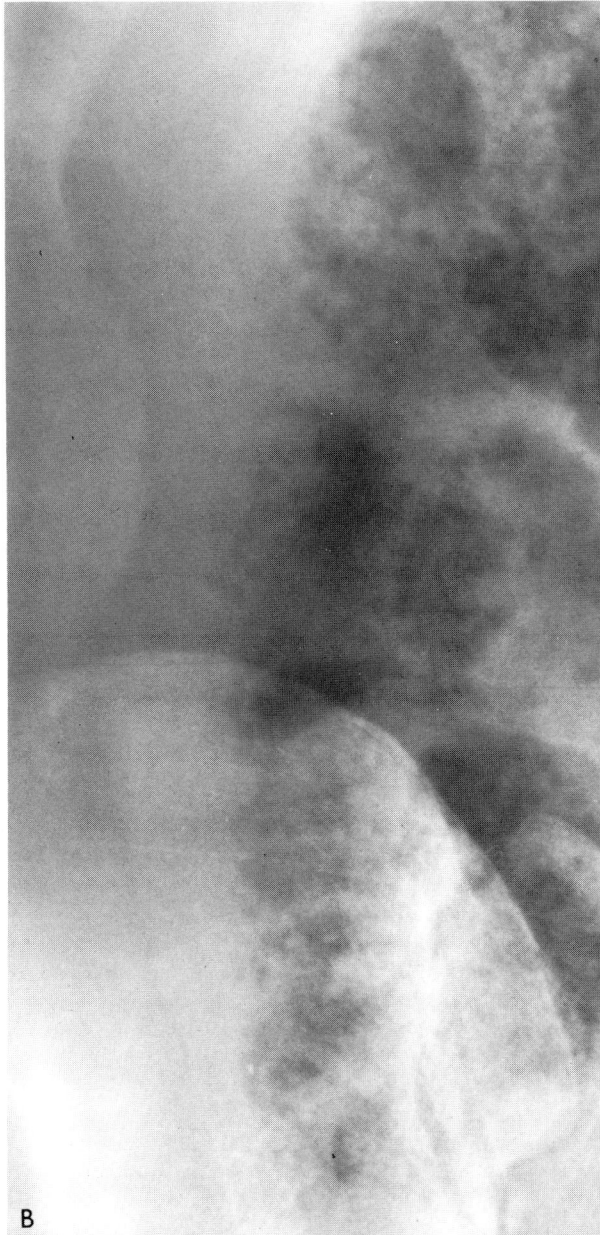

Figure 3–32. (*Continued*). Supine plain abdominal radiograph (*B*) showing the right flank in a patient with pus in the abdomen due to appendicitis with perforation. The peritoneum and liver edge can no longer be identified and the radiolucency between the colon and preperitoneal fat has been replaced with water-density pus.

peritoneal space as in cases of perforated appendicitis with advanced peritonitis, the pus may flow from the right lateral gutter into the pelvis, where it accumulates in the hollow of the sacrum on either side of the rectum in the pouch of Douglas. There the fluid can be identified as water-density masses above the bladder (the dog-ear sign) (Fig. 3–33). B mode ultrasound scans are useful in determining the presence of pus both in the flank and in the pelvis.

AIR IN THE APPENDIX. Air in the lumen of the appendix occurs both in normal patients and in patients with acute appendicitis[25] (Figs. 3–34 to 3–36). When gas is detected in an appendix which is in a normal location, it is suggestive evidence of acute appendicitis. However, gas in the lumen of an abnormally placed appendix, such as one in the right upper quadrant, is of little diagnostic significance. When the appendix is anatomically located above the cecum, air may persist in the lumen. In these cases the tubular gas pattern is usually easily distinguishable from gas in the biliary tree, the portal vein, a liver abscess or the gallbladder. When gas persists in cases in which the appendix is dependent to the cecum, appendicitis should be suspected. Gas in the appendix is more likely to be significant when the lumen is dilated or an air-fluid level is identified in the appendix on the upright radiograph (Fig. 3–37).

FREE AIR IN THE PERITONEAL SPACE. Pneumoperitoneum is a rare finding in acute appendicitis because occlusion of the appendiceal lumen is basic in the pathogenesis of the disease. Consequently, in nearly all cases there is no communication between the colonic lumen and the tip of the appendix.

However, on occasion free air is present in the peritoneal space in cases of acute perforated appendicitis and can be detected on plain abdominal radiographs. McCort reported six cases of pneumoperitoneum in 648 patients with appendicitis.[26] There was delay in diagnosis in all six from 2 to 7 days. Usually, the quantity of air is small and can be detected best on upright frontal chest radiographs where the air can be identified under the diaphragm (Fig. 3–38). Decubitus projections may be helpful (Fig. 3–39). When large amounts are present, the air may be visible between loops of bowel, so that gas inside and outside of the bowel

lumen outlines the bowel wall (Fig. 3–40). If the air collects over the liver, the falciform ligament may become visible on the abdominal radiograph. Massive pneumoperitoneum may produce an oval radiolucency on supine abdominal radiographs (Fig. 3–41). Rarely, free air may outline the lateral umbilical ligaments as they course along the lower anterior abdominal wall.

INTRA-ABDOMINAL ABSCESS. When perforation of the appendix occurs in patients with acute appendicitis, there is a marked increase in the morbidity and mortality of the disease due to bacterial contamination of the peritoneal space. The subsequent infection and associated inflammatory reaction are usually sealed off by contiguous viscera, the omentum and the involved parietal and visceral peritoneum. The infection continues to progress, but contained within an inflammatory barrier. Depending on the virulence of the bacteria and the severity of the infection, the tissue reaction varies from frank necrosis and pus to a mild phlegmonous reaction. When the infection is not localized, generalized peritonitis develops.

Perforated appendicitis is the most common cause of an intra-abdominal abscess.[27] The abscess may be intraperitoneal or retroperitoneal. It is most often located in proximity to the appendix, so that the inflammatory mass usually lies adjacent to the cecum and the terminal ileum. Depending on the position of the appendix, this may be in the right iliac fossa or in the pelvis. When the appendix is in an abnormal location or if it is unusually long, the abscess may develop anywhere in the abdomen. Abscesses have been described in the left lower quadrant, in the right flank, in the anterior abdominal wall, in the lesser sac and in the subhepatic and subdiaphragmatic spaces (Figs. 3–42 to 3–45).

When an abscess is present, plain abdominal radiographs show a poorly defined mass, often with displacement of adjacent loops of bowel. Irregular radiolucencies in the mass due to bubbles of gas produced by bacteria in the abscess often produce a mottled pattern radiographically[28] (Fig. 3–46). This may be difficult to distinguish from the stippled radiographic appearance of fecal material. The distinction can usually be made by noting the distribution of the mottled pattern. Fecal material is contained within the normal course of the right colon

(*Text continued on page 123*)

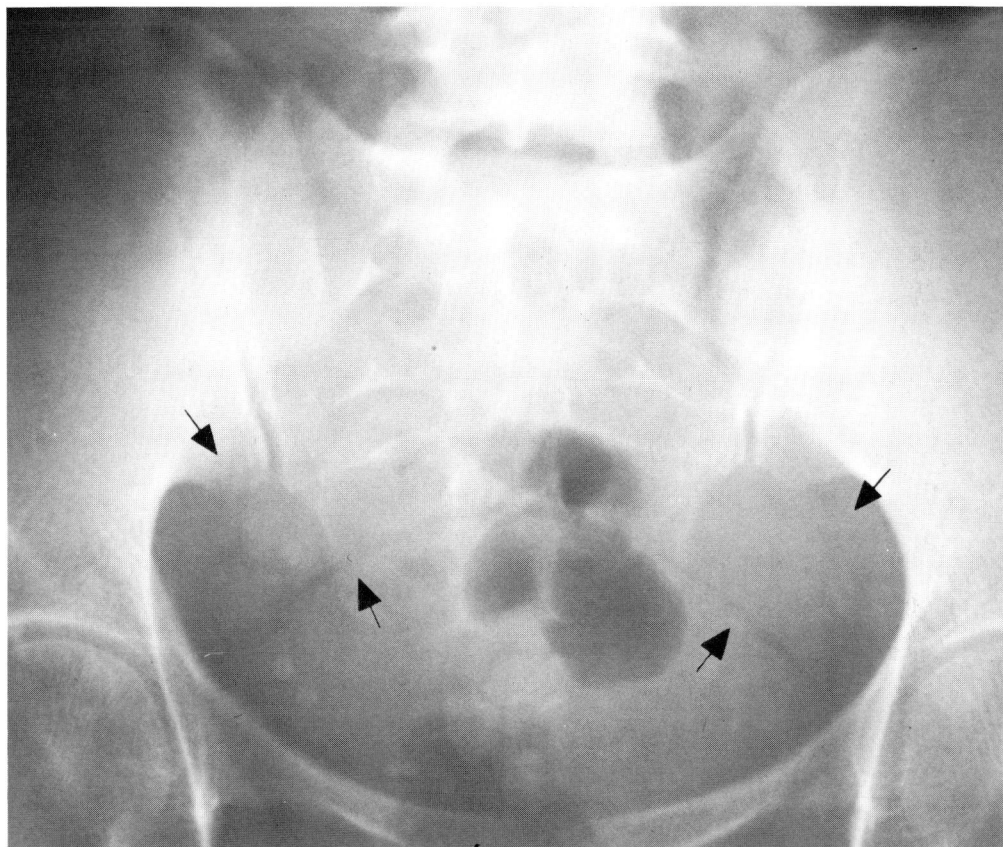

Figure 3–33. Supine plain abdominal radiograph showing the pelvis in a patient with appendicitis and pus in the abdomen. Normal extraperitoneal fat over the bladder permits differentiation of the bladder from water-density pus above and behind the bladder (the dog-ear sign) (arrows).

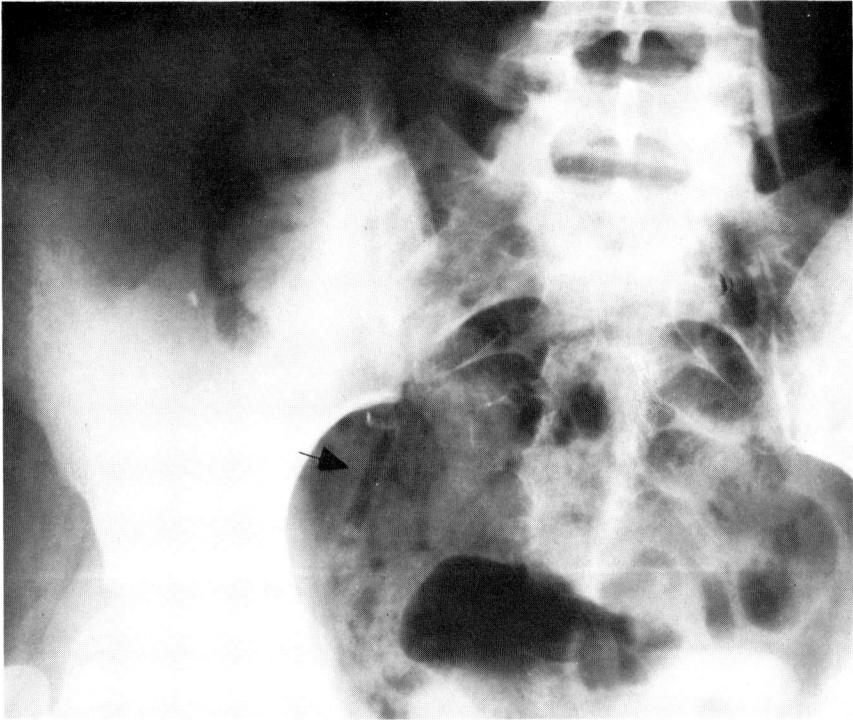

Figure 3-34. Supine plain abdominal radiograph showing gas in the appendix (arrow) in an asymptomatic patient. Gas in an appendix which is in a normal position is unusual in the absence of disease. (From Beneventano, T. C., et al.: The roentgen aspects of some appendiceal abnormalities. Amer. J. Roent., 96:344, 1966. Courtesy of Charles C Thomas, Publisher.)

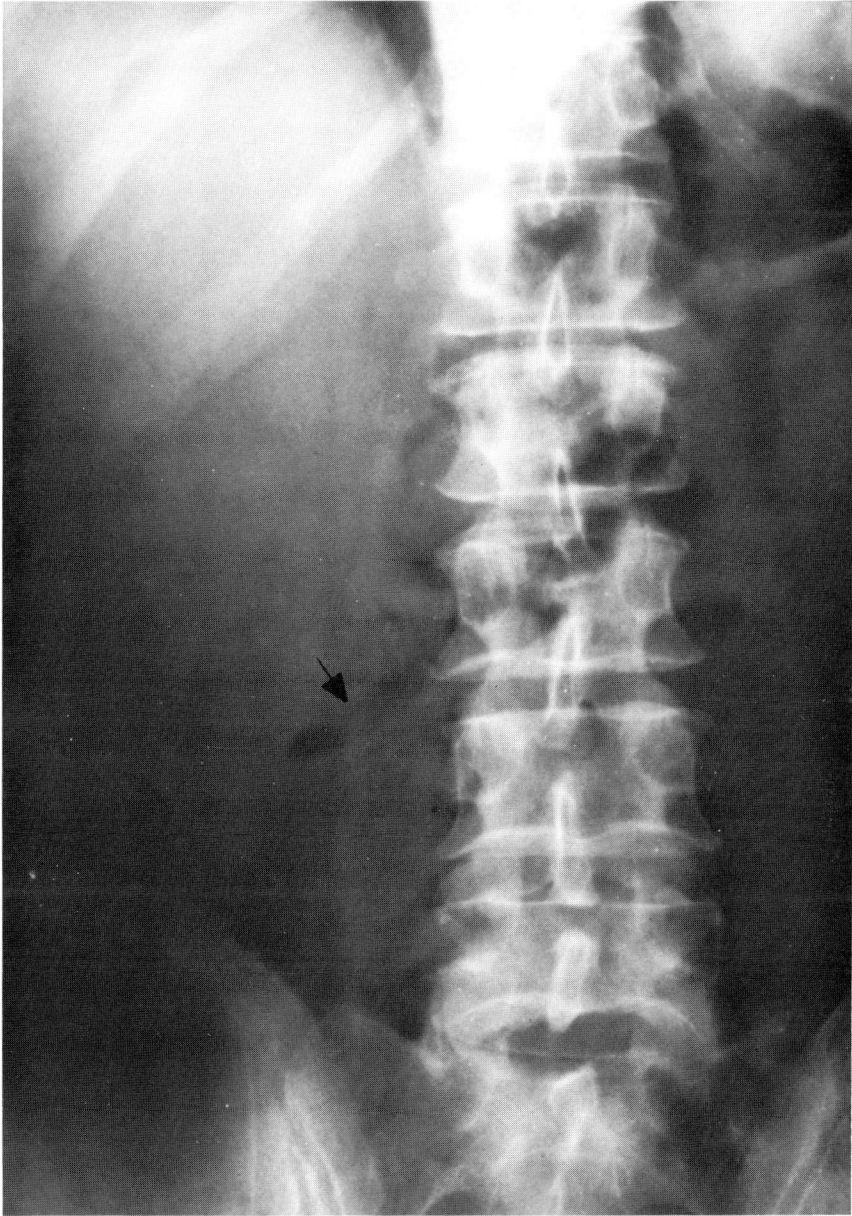

Figure 3–35. Supine plain abdominal radiograph showing gas in the appendix (arrow) in a patient with acute appendicitis. (From Beneventano, T. C., et al.: The roentgen aspects of some appendiceal abnormalities. Amer. J. Roent., 96:344, 1966. Courtesy of Charles C Thomas, Publisher.)

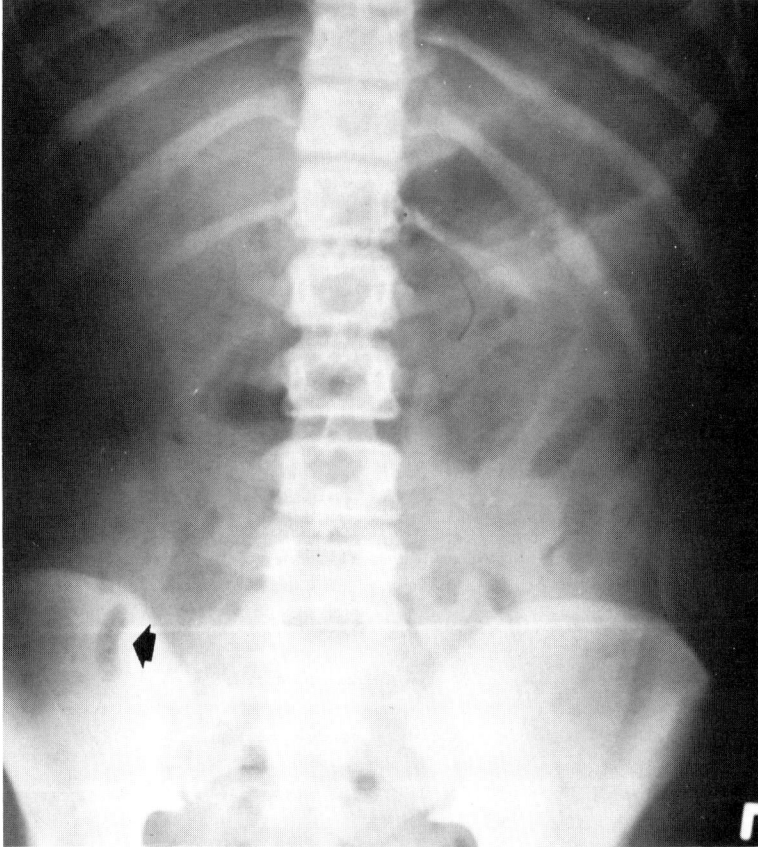

Figure 3–36. Supine plain abdominal radiograph showing air in the appendix (arrow) in a patient with acute appendicitis. Pus is visible in both flanks. (From Joffe, N.: Some uncommon roentgenologic findings associated with acute perforative appendicitis. Radiology, *110*:301, 1973.)

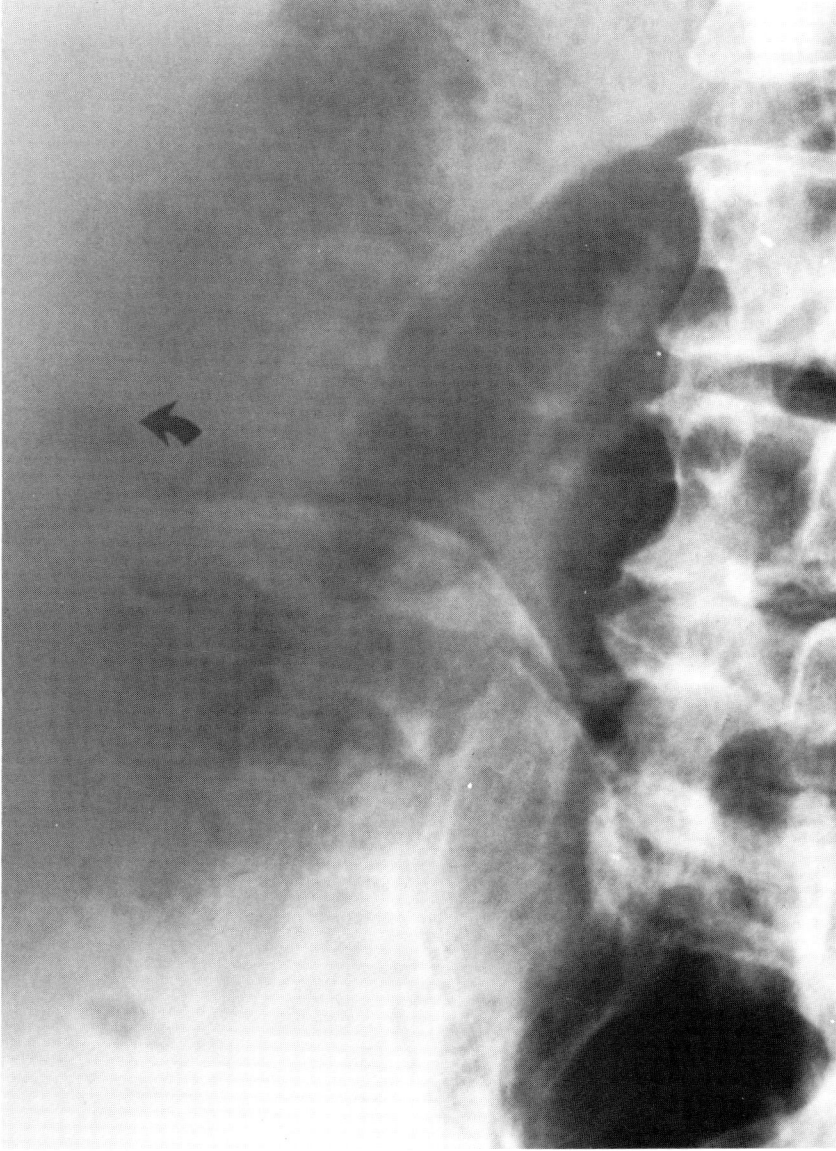

Figure 3–37. Upright plain abdominal radiograph showing an air-fluid level in the appendix (arrow) in a patient with acute appendicitis. Dilatation of the appendix and the presence of an air-fluid level are helpful in differentiation of a normal appendix containing air and an air-filled appendix due to acute appendicitis.

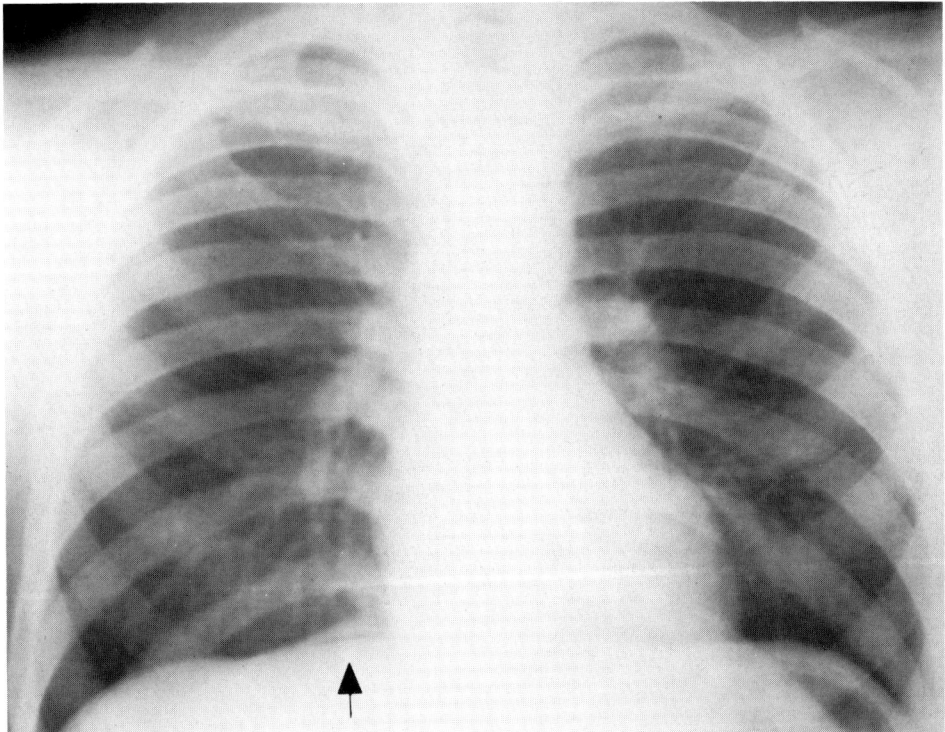

Figure 3–38. Upright posterior-anterior chest radiograph showing a small amount of free intra-peritoneal air under the right hemidiaphragm (arrow) in a patient with appendicitis with perforation. (From Beneventano, T. C., et al.: The roentgen aspects of some appendiceal abnormalities. Amer. J. Roent., 96:344, 1966. Courtesy of Charles C Thomas, Publisher.)

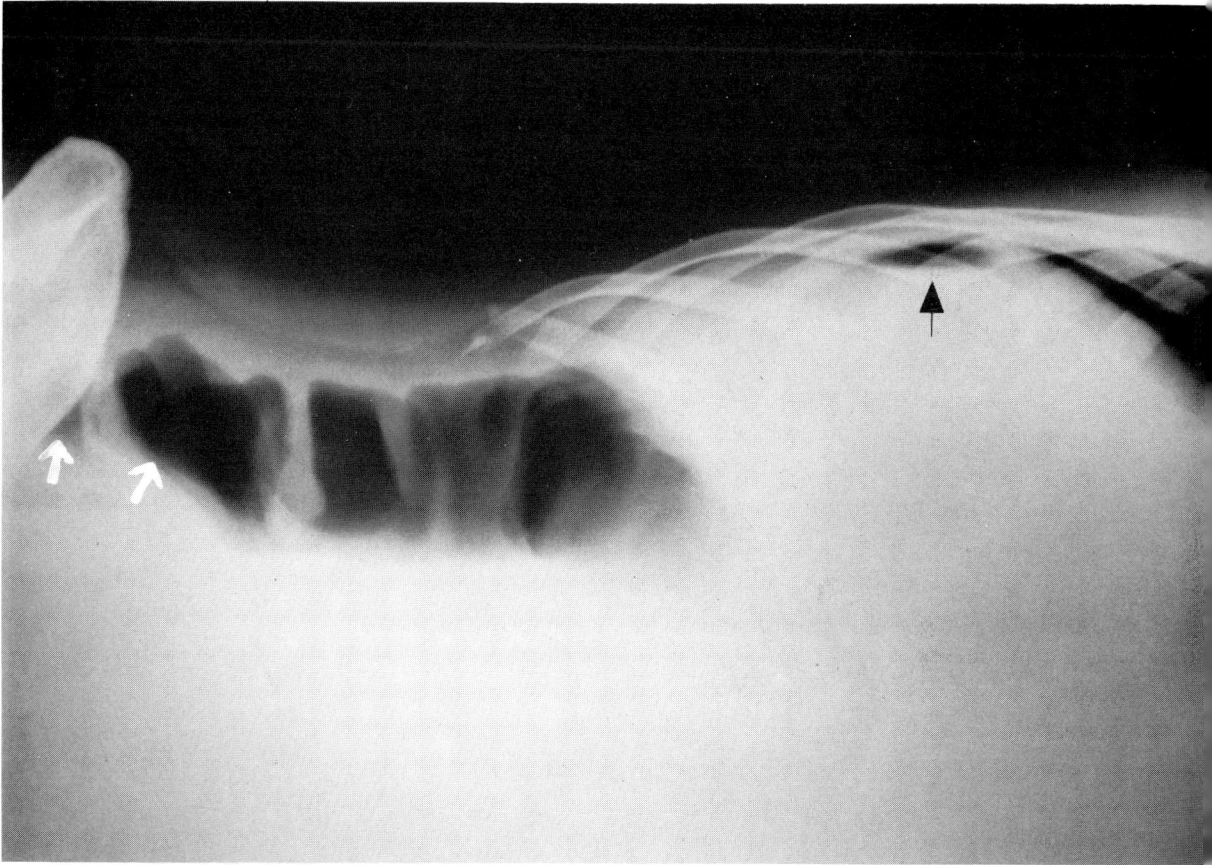

Figure 3–39. Lateral decubitus radiograph of the abdomen showing a small amount of free intraperitoneal air over the liver (dark arrow) in a patient with acute appendicitis with perforation. An abscess is visible compressing the cecum (light arrows). (From McCort, J. J.: Extra-alimentary gas in perforated appendicitis. Amer. J. Roent., 84:1087, 1960. Courtesy of Charles C Thomas, Publisher.)

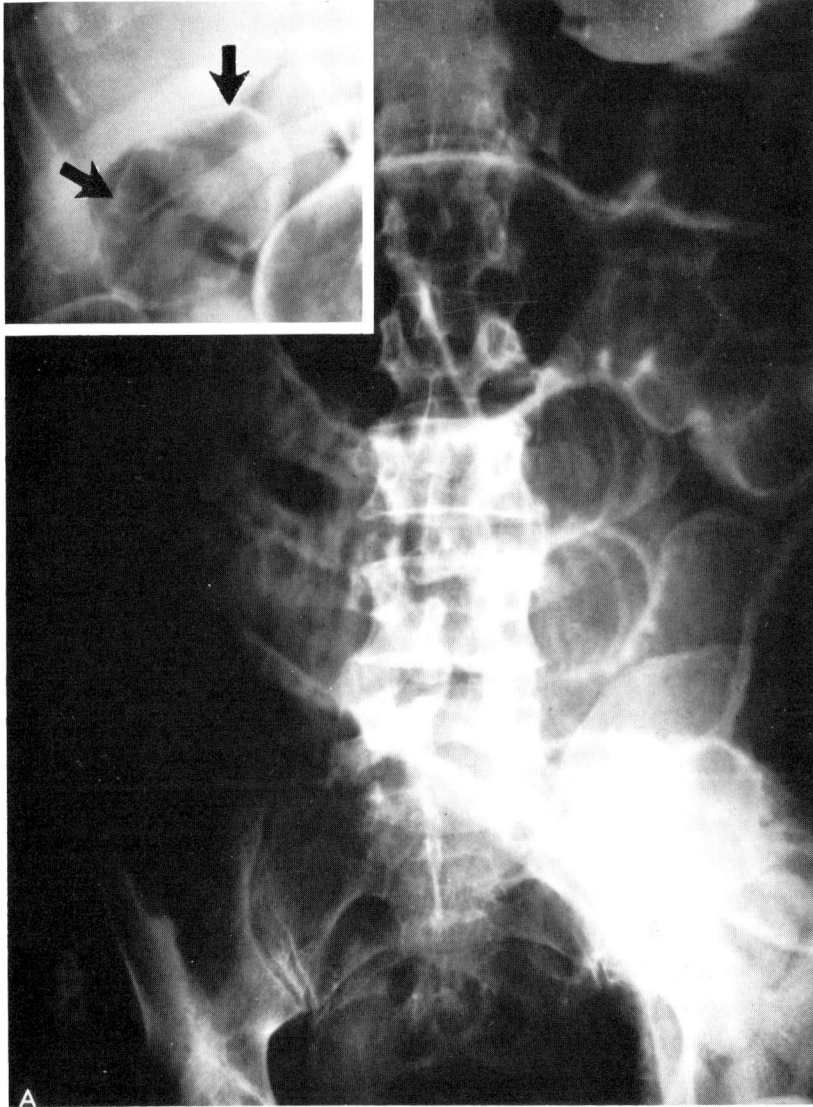

Figure 3-40. Supine plain abdominal radiograph showing free intraperitoneal air in two patients with acute appendicitis with perforation (*A* and *B*). Both sides of the bowel wall can be identified, indicating the presence of air in the peritoneal cavity (arrows) as well as in the bowel lumen. (Part *A* from McCort, J. J.: Extra-alimentary gas in perforated appendicitis. Amer. J. Roent., *84*:1087, 1960. Courtesy of Charles C Thomas, Publisher.)

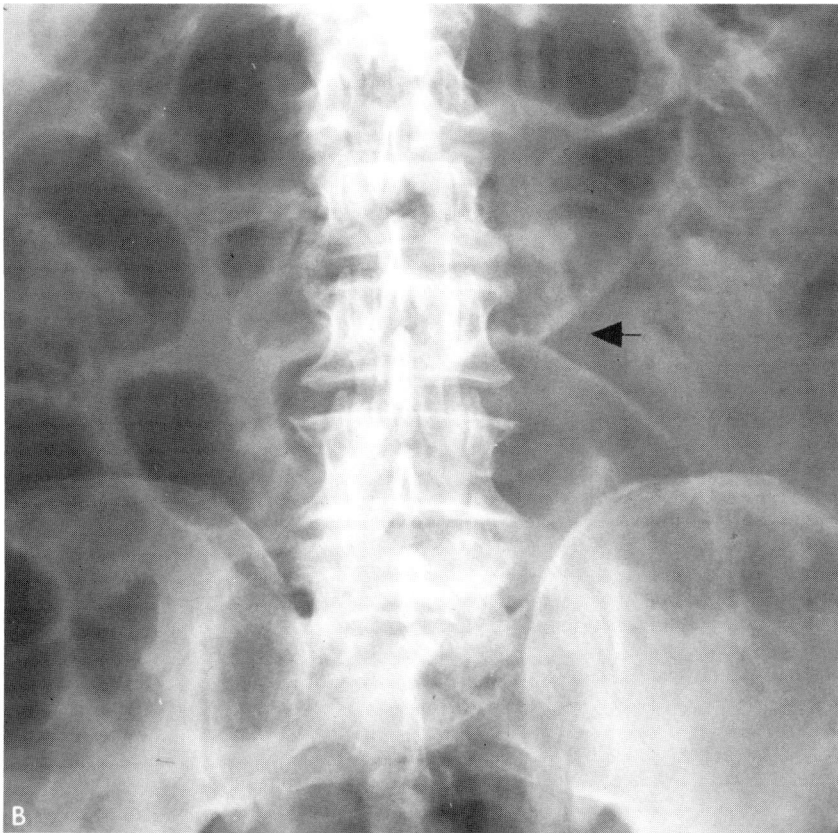

Figure 3–40. (*Continued*)

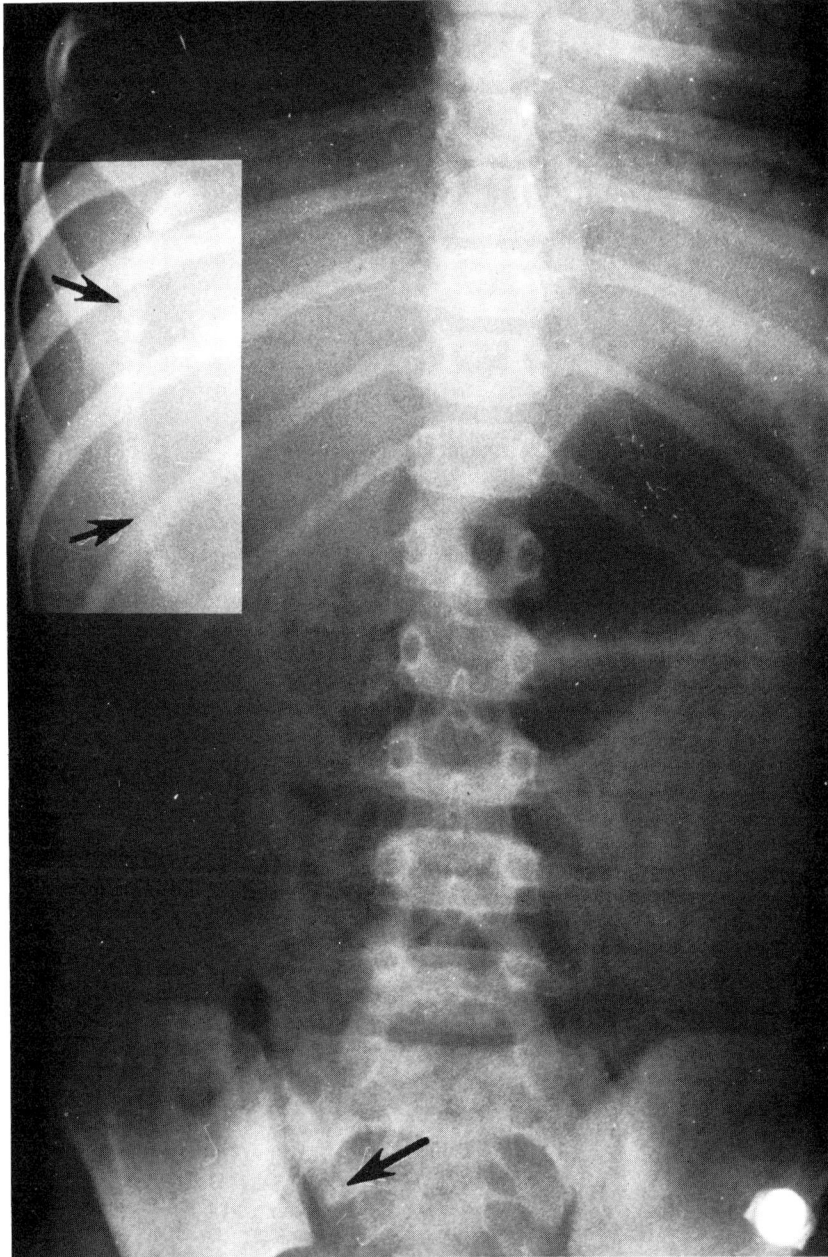

Figure 3-41. Supine plain abdominal radiograph showing free intraperitoneal air in a patient with fatal acute appendicitis with perforation. There is massive pneumoperitoneum with an oval radiolucency in the midabdomen due to air collected anteriorly. The intraperitoneal air and pus form a distinct margin (small arrows) over the liver. An appendicolith is present in the pelvis (large arrow). (From McCort, J. J.: Extra-alimentary gas in perforated appendicitis. Amer. J. Roent., *84*:1087, 1960. Courtesy of Charles C Thomas, Publisher.)

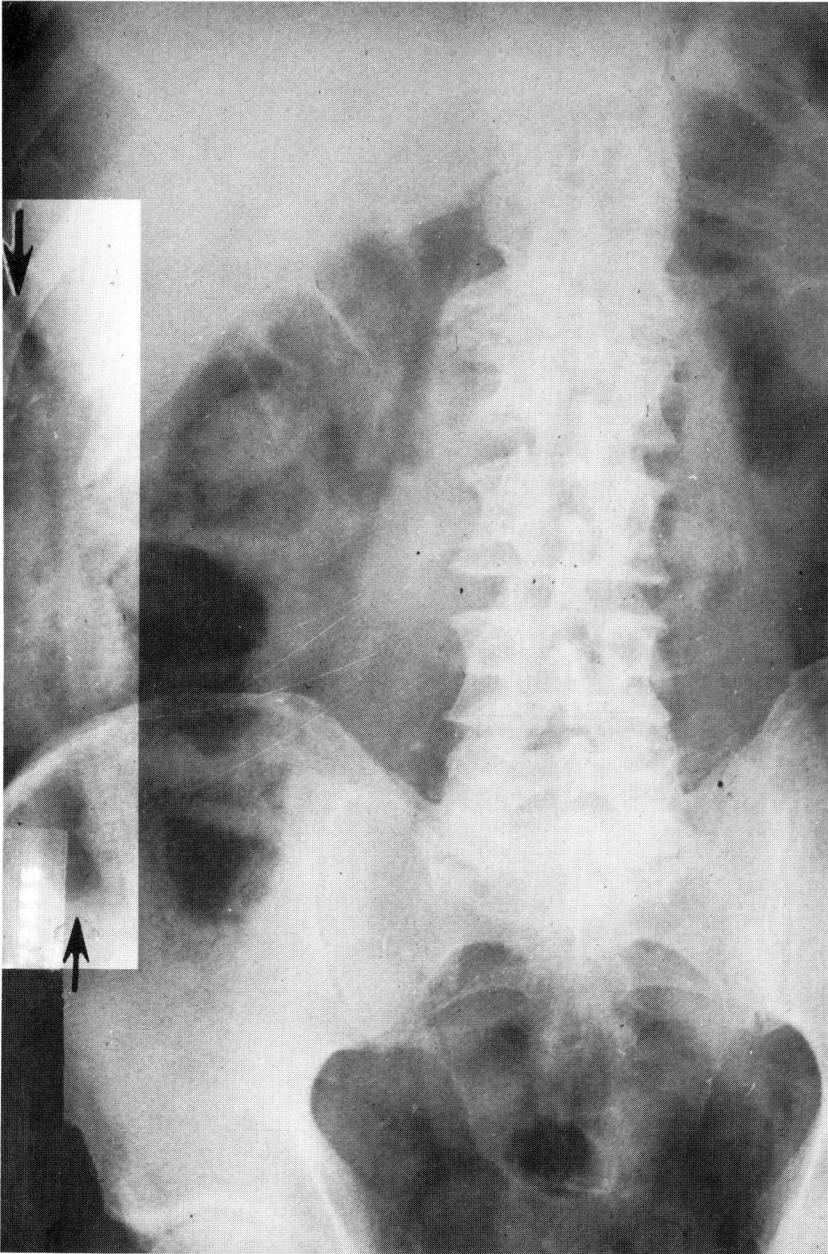

Figure 3–42. Supine plain abdominal radiograph showing gas in a flank abscess due to appendicitis with perforation (arrows). (From McCort, J. J.: Extra-alimentary gas in perforated appendicitis. Amer. J. Roent., *84*:1087, 1960. Courtesy of Charles C Thomas, Publisher.)

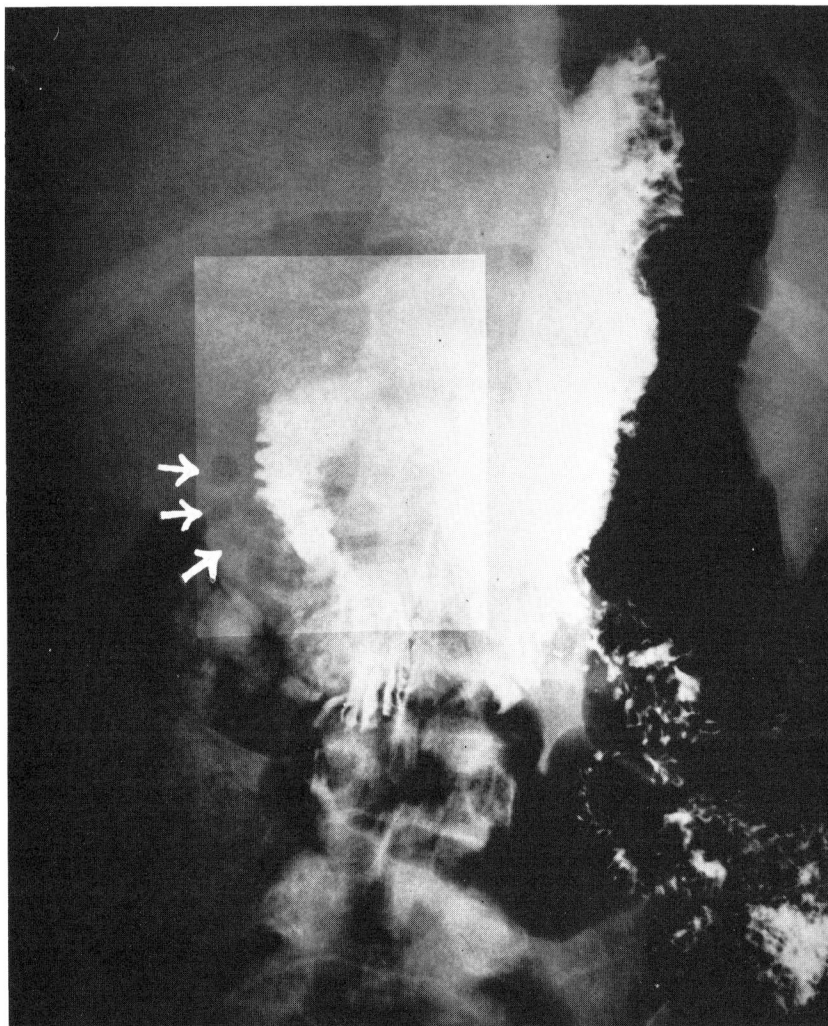

Figure 3-43. Supine radiograph from an upper gastrointestinal examination showing gas in a subhepatic abscess secondary to appendicitis with perforation (arrows). (From McCort, J. J.: Extra-alimentary gas in perforated appendicitis. Amer. J. Roent., *84*:1087, 1960. Courtesy of Charles C Thomas, Publisher.)

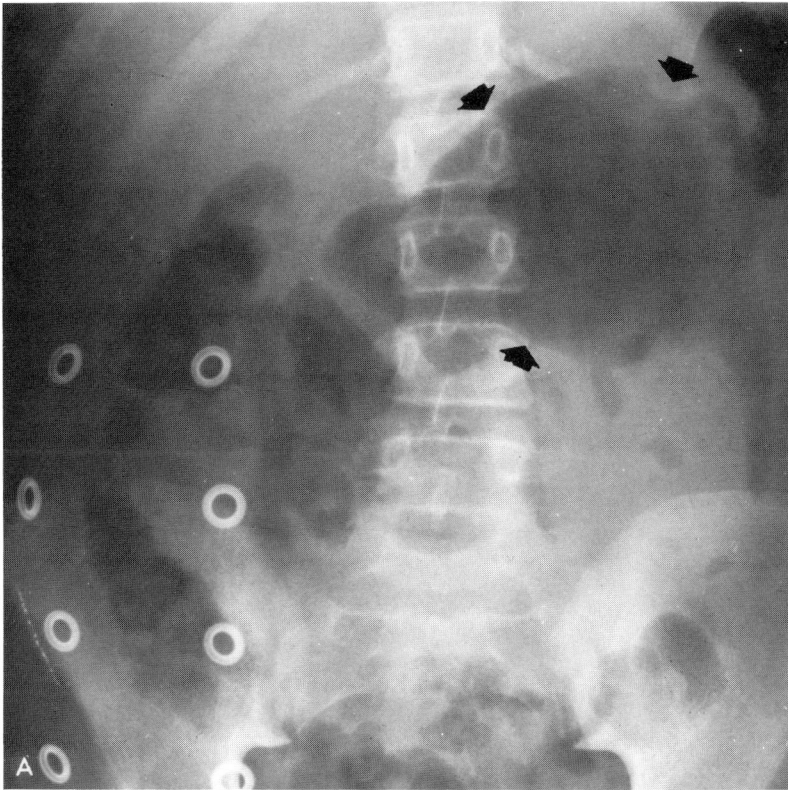

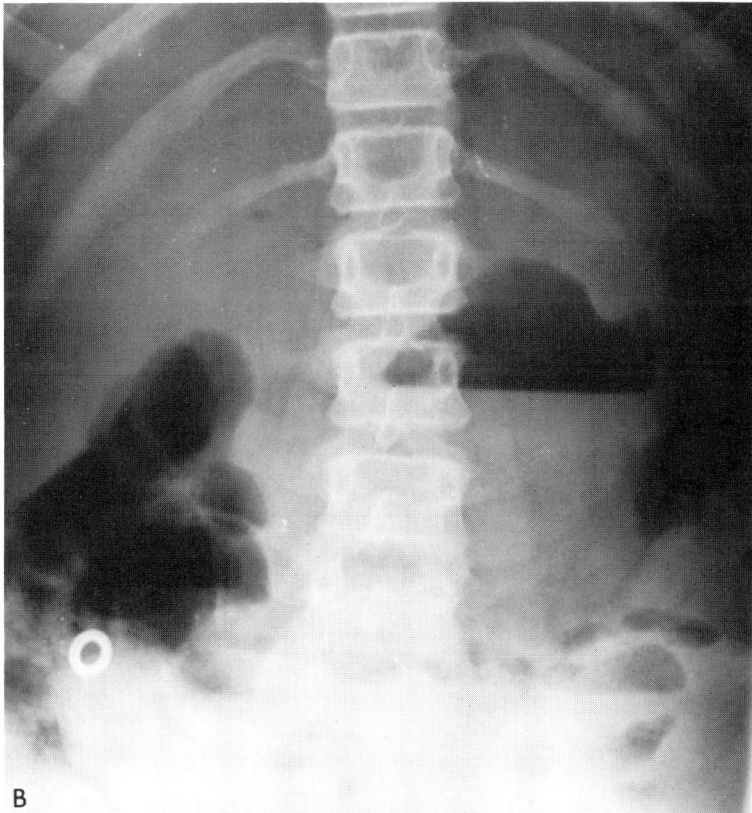

Figure 3–44. Supine (A) and erect (B) plain abdominal radiographs showing a large gas and fluid collection due to lesser sac abscess secondary to appendicitis with perforation (arrows). (From Joffe, N.: Some uncommon roentgenologic findings associated with acute perforative appendicitis. Radiology, *110*:301, 1973.)

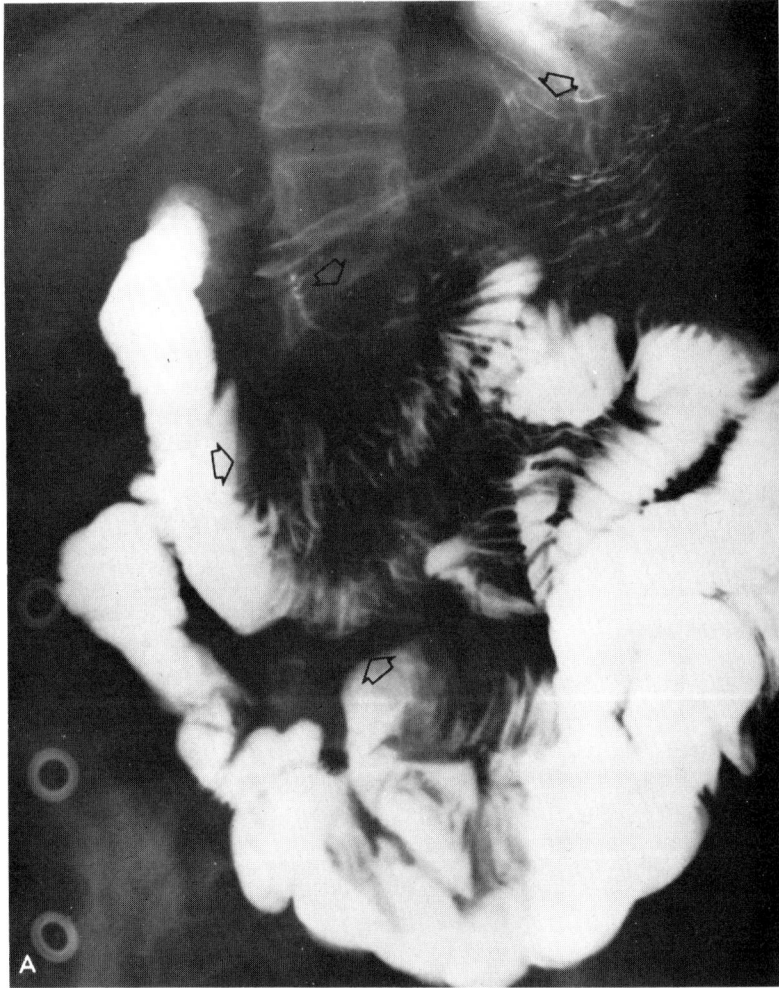

Figure 3–45. Supine (A) and erect (B) radiographs from an upper gastrointestinal and small bowel examination showing a large gas and fluid collection in the lesser sac due to an abscess from appendicitis with perforation. (From Joffe, N.: Some uncommon roentgenologic findings associated with acute perforative appendicitis. Radiology, *110*:301, 1973.)

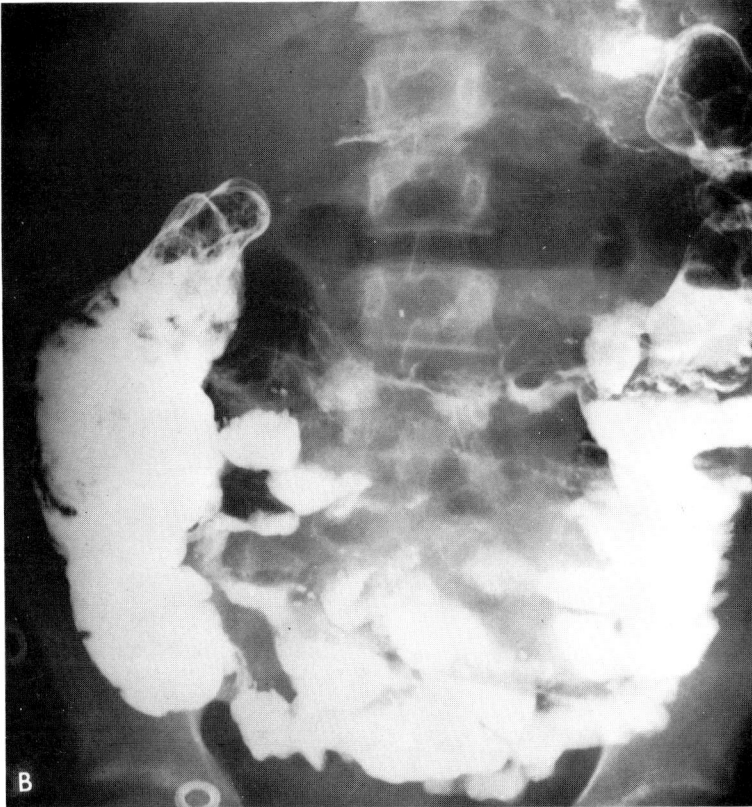

Figure 3–45. (*Continued*)

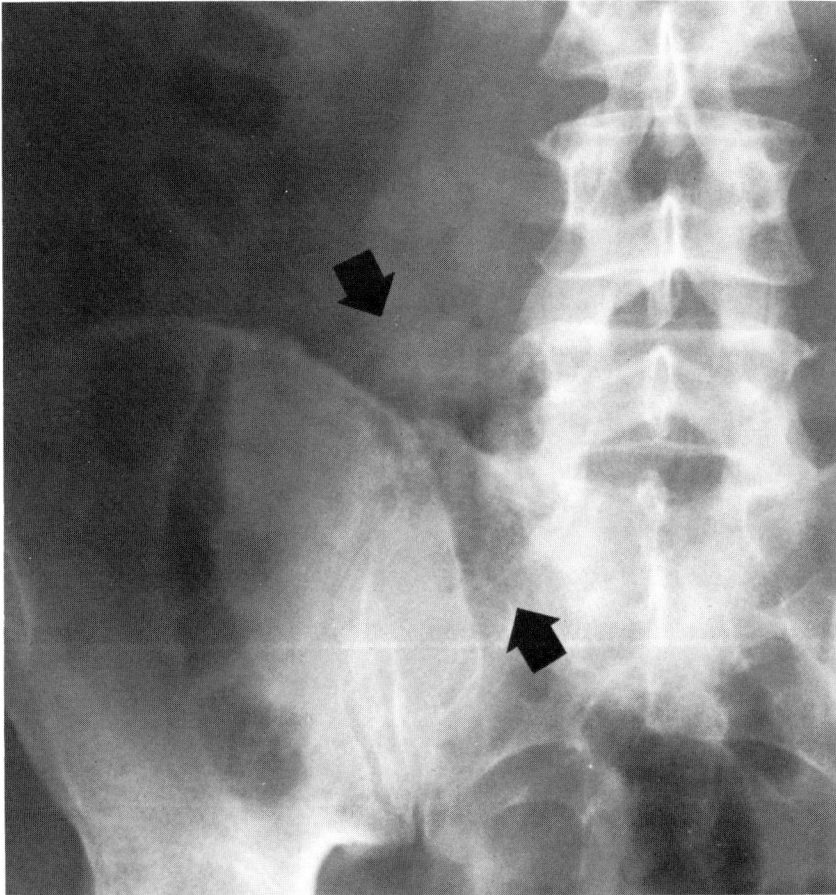

Figure 3–46. Supine plain abdominal radiograph showing stippled gas in an appendiceal abscess (arrows).

and there are often feces elsewhere in the colon. The colon is usually empty of feces in the presence of peritonitis.

In some patients the abdominal abscess forms a unilocular mass containing air and pus (Figs. 3–47 to 3–51). In some of these cases, upright and decubitus radiographs reveal a single fluid level. Such fluid levels must be differentiated from air and fluid collections in the small bowel and colon. Barium enema examination may sometimes be necessary to prove that the mass lies outside of the colon. Air-fluid levels in an abscess may be distinguished from those in the stomach by the nature of the fluid interface. Fluid levels in the stomach often have an irregular or fuzzy surface, whereas those in an abscess invariably are sharp.[29]

Barium Enema Examination

A barium enema examination should be performed on every patient suspected of having acute appendicitis in whom the clinical, laboratory and plain film radiographic findings do not permit a definite diagnosis. This is particularly true for patients older than 50 years of age in whom a number of diseases about the cecum may simulate appendicitis. Since obstruction of the appendiceal lumen is part of the pathogenesis of the disease, there is little danger of producing a perforation. Soter performed 800 barium enema examinations on patients suspected of having acute appendicitis with no complications, even in patients with a gangrenous appendix.[11] Schey reported a similar experience in 25 children.[30] The examination should be accepted as a safe, simple, prompt and reliable procedure for the diagnosis of acute appendicitis.

The enema examination should be performed in the routine fashion using barium. Judicious compression spot radiographs of the cecum and terminal ileum should be made in the customary manner. No preparation is necessary, since fecal material rarely interferes with adequate examination of the cecum in these circumstances. Air contrast studies following evacuation of the barium are frequently helpful in completely distending the cecum. The examination should be done expeditiously with a minimum of discomfort, just as for other patients.

Determination of the position of the cecum by the barium enema study is helpful in correlating the clinical findings. In female patients detection of the cecum low in the pelvis adjacent to the urinary bladder may explain why the patient's symptoms suggest pelvic inflammatory disease or bladder infection. When malrotation of the colon is present, the barium enema examination may explain clinical findings in the right upper quadrant or on the left side of the abdomen.

The barium enema examination serves to diagnose or to exclude other diseases that may simulate appendicitis such as regional enteritis, diverticulitis of the cecum, ileocecal tuberculosis and carcinoma of the cecum.

Although Dietz warns that it may be difficult to be certain that the entire appendix is filled with barium (25 per cent of appendices exceed 9 cm. in length),[31] Soter considers an appendiceal lumen filled with barium to be a reliable sign of a normal appendix. Nonfilling of the appendix is not of diagnostic significance unless it is associated with abnormalities of the cecum and terminal ileum. Schey found that the appendix fills in 92 per cent of normal children, so that failure of the appendix to fill in children with symptoms possibly due to appendicitis should be viewed with suspicion.[30]

Appendicitis on barium enema examination may be suggested by nonfilling of the appendix with a local impression on the cecum or terminal ileum.[11, 30, 32] In some cases there may be partial filling of the appendix with a sharp cutoff or with an associated mass impressing the cecum. The extent of the findings depends on the severity of the disease, which may vary from an uncomplicated inflamed appendix to a large appendiceal abscess, and on the location of the appendix in relation to the cecum and terminal ileum.

Schey considers failure of barium to enter the appendix in association with a cecal tip defect to be virtually pathognomonic of appendicitis[30] (Fig. 3–52). This has been noted by others who have suggested that edema from the inflammation creates the cecal impression.[32] Partial filling of the appendix may be suspected if a very short appendix is found. Any distortion in shape or caliber of the partially filled appendix with or without a cecal impression should be considered abnormal.

(*Text continued on page 134*)

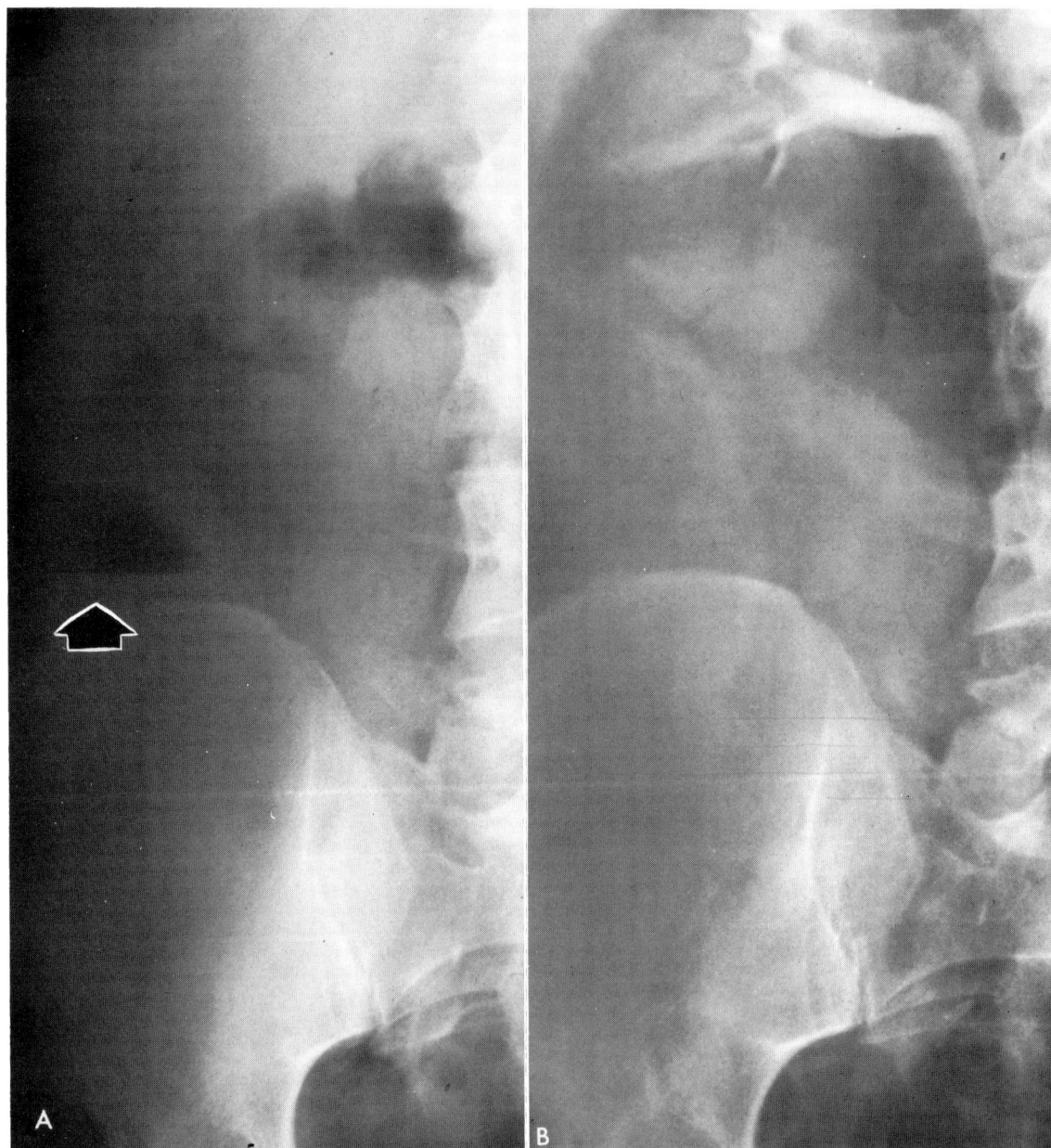

Figure 3–47. Upright plain abdominal radiograph (*A*) shows air fluid level in the right lower quadrant (arrow) in a patient with a large appendiceal abscess. The abscess is visible on a supine radiograph from an intravenous pyelogram (*B*). Marked medial displacement of the ureter indicates the mass is retroperitoneal. The barium enema examination shows that the cecum and ascending colon are displaced by the abscess (*C*).

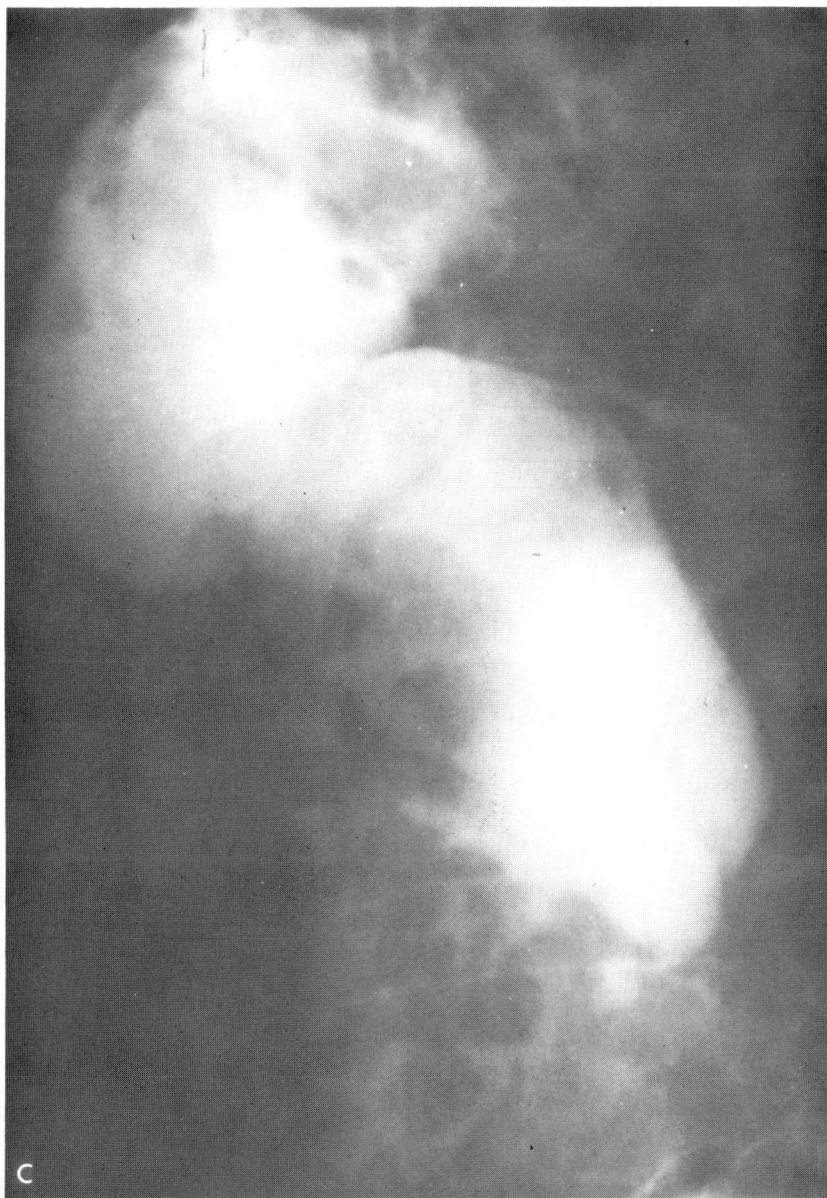

Figure 3–47. (*Continued*)

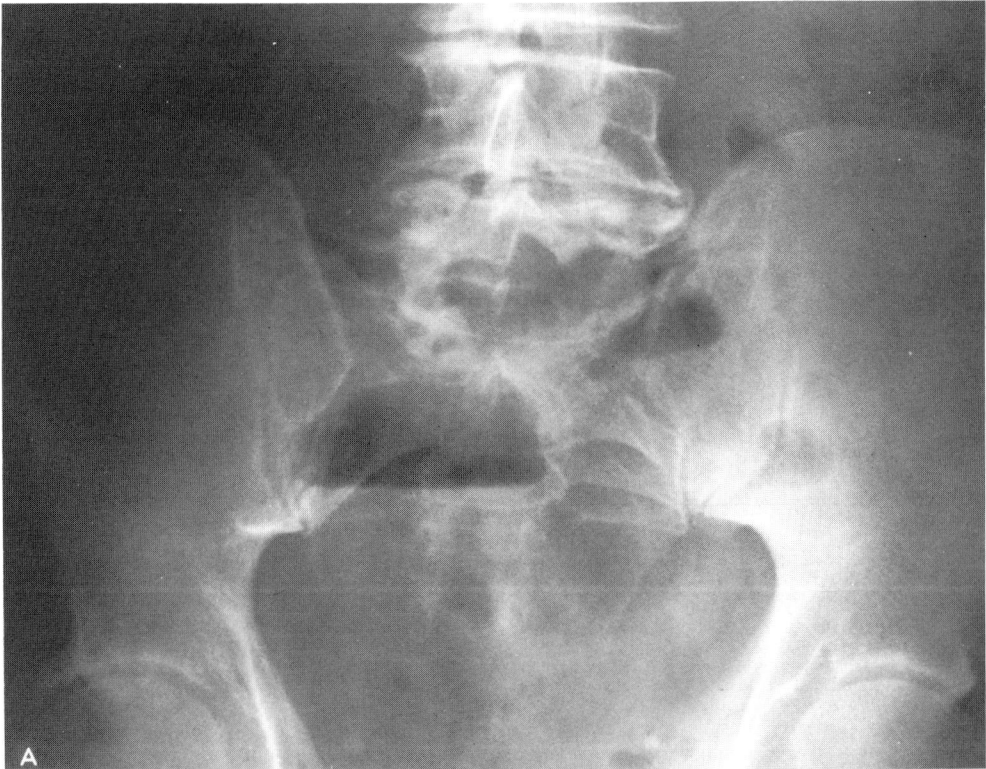

Figure 3–48. Upright plain abdominal radiograph (A) and barium enema examination (B) in a patient with a large appendiceal abscess. There is a long air-fluid level on the upright projection and marked deformity of the cecum on the barium enema study.

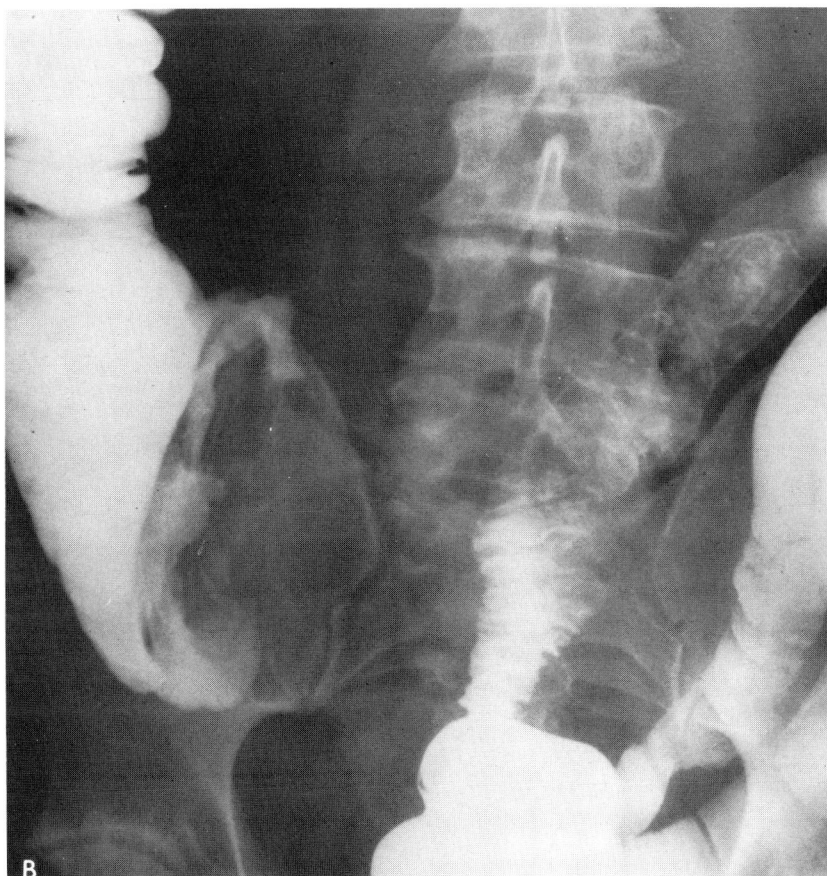

Figure 3–48. (*Continued*)

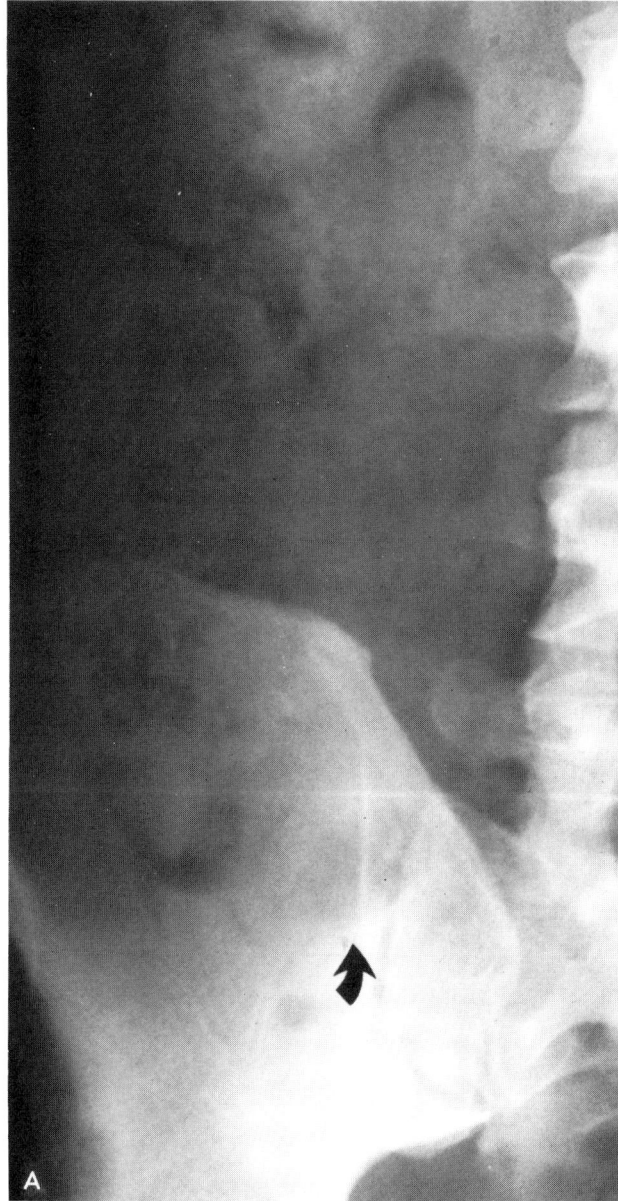

Figure 3–49. Supine plain abdominal radiograph (A) showing gas in an appendiceal abscess (arrow). *(Legend continued on opposite page.)*

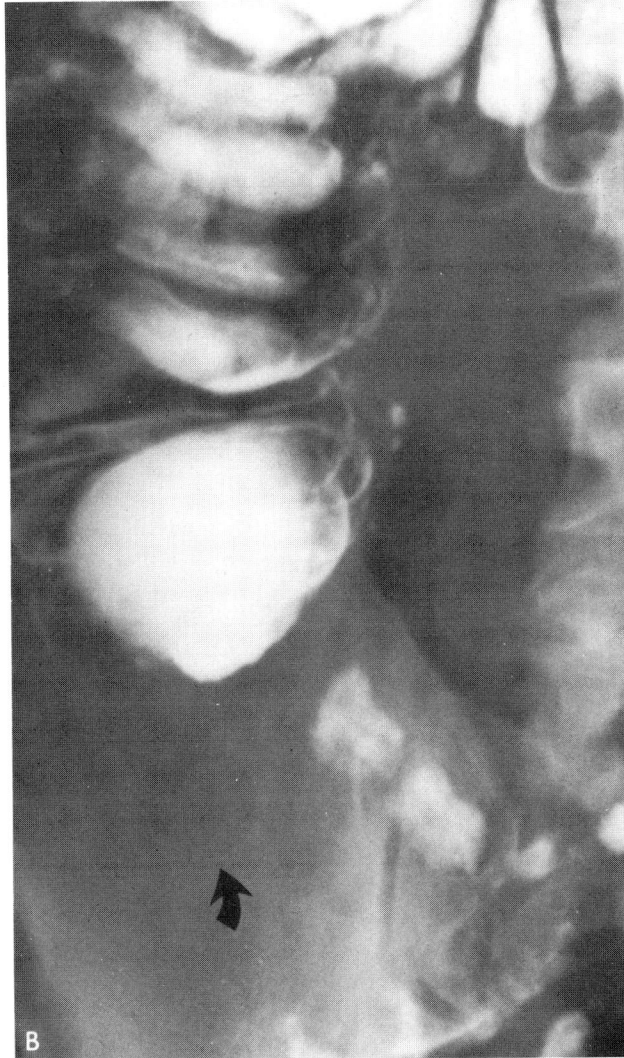

Figure 3-49. (*Continued*). The gas collection is visible adjacent to the cecum (arrow) on the barium enema examination (*B*).

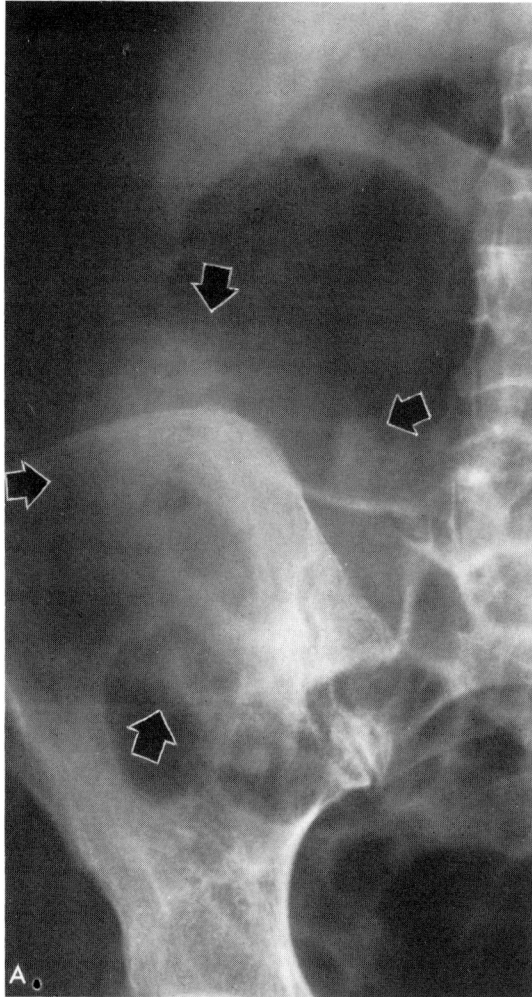

Figure 3–50. Supine plain abdominal radiograph showing a large mass in the right lower quadrant due to an appendiceal abscess (A) (arrows). *(Legend continued on opposite page.)*

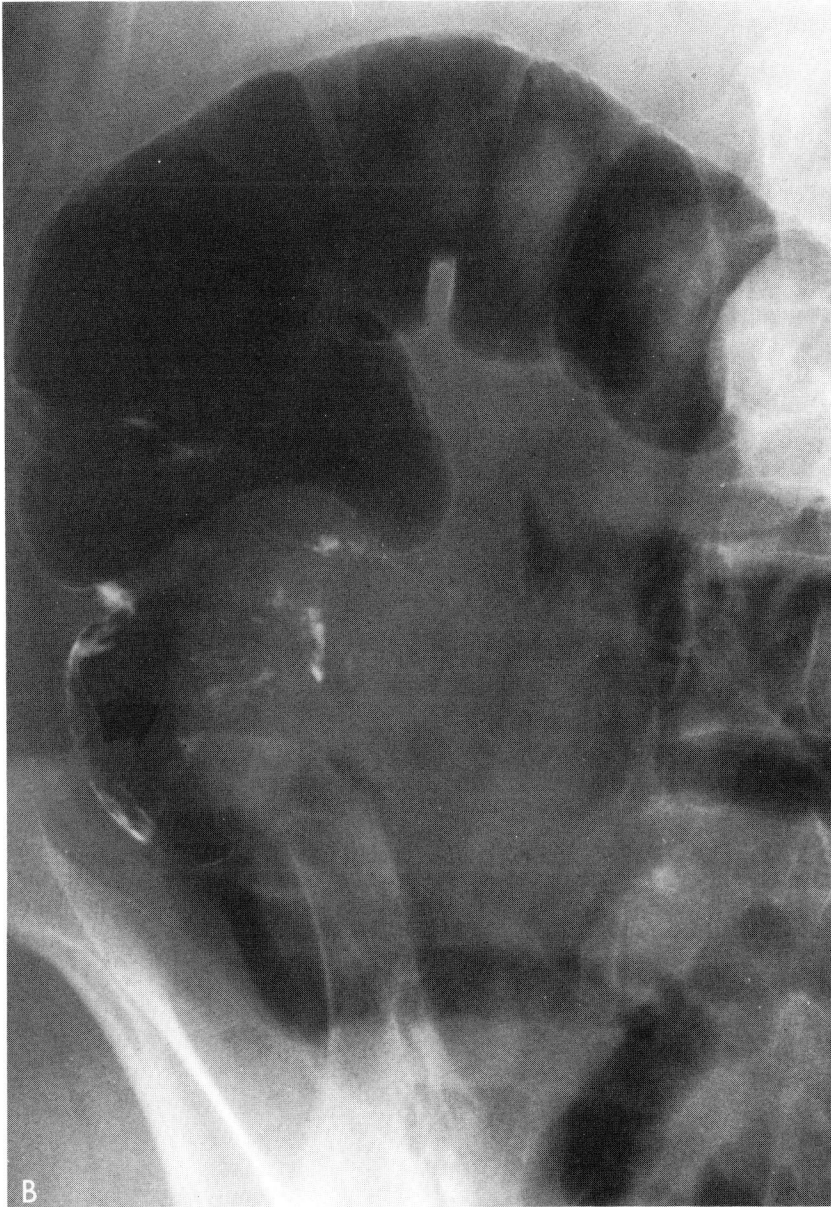

Figure 3–50. (*Continued*).   The abscess indents the cecum on the air contrast colon examination (*B*) (arrow).

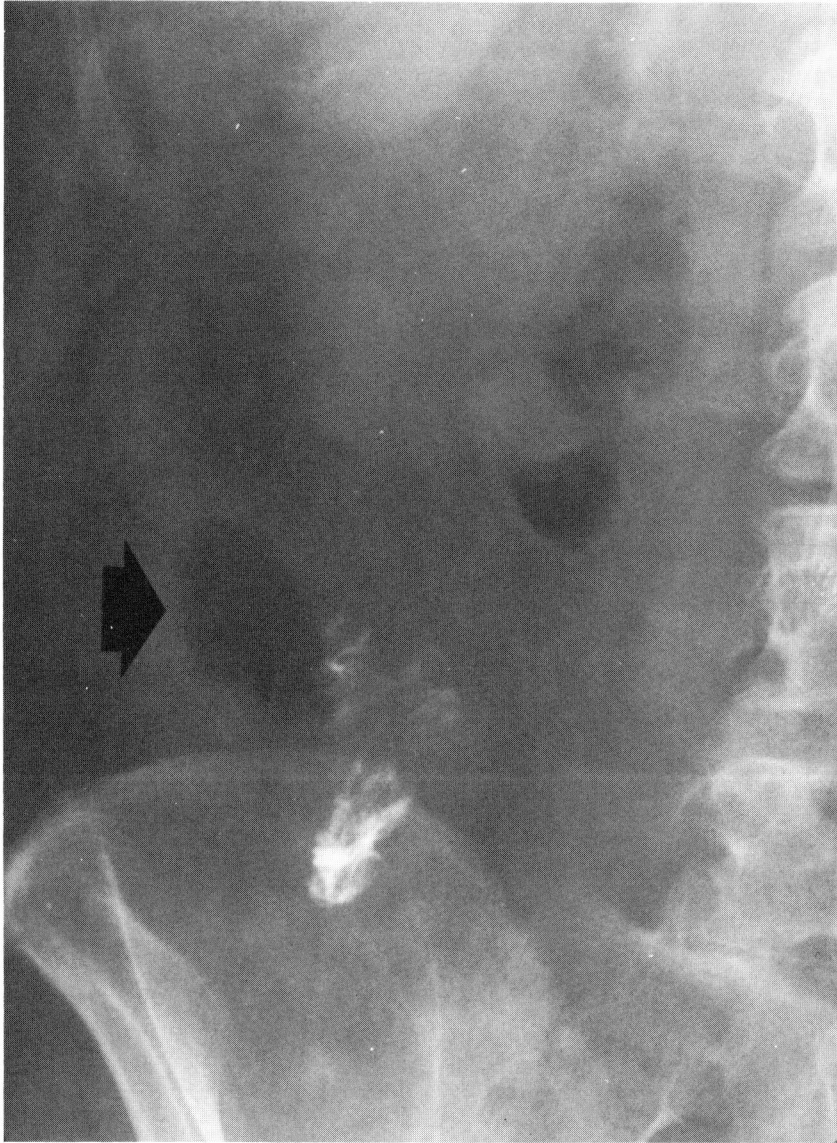

Figure 3–51. Supine plain abdominal radiograph showing a large unilocular collection of gas in an appendiceal abscess (arrow). There is a small amount of barium in the ascending colon from a recent examination. The colon is displaced medially by the abscess.

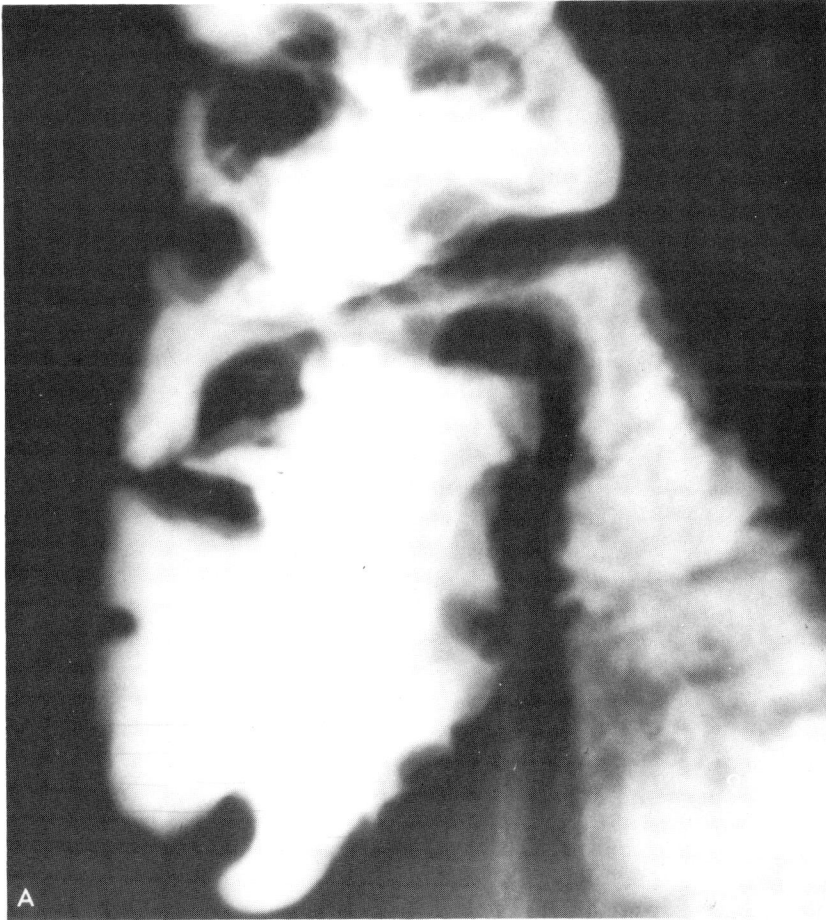

Figure 3–52. Barium enema examination in a patient with acute appendicitis. The appendix fails to fill and there is an irregularity in the base of the cecum (*A*). A spot radiograph made at fluoroscopy shows edema at the base of the appendix (*B*).

A large extrinsic compression of the base of the cecum associated with nonfilling of the appendix indicates the presence of an appendiceal abscess (Figs. 3–53 and 3–54). Displacement of the terminal ileum may also be detected (Fig. 3–55). Occasionally, when the inflammatory process extends retroperitoneally into the root of the small bowel mesentery, typical displacement of the small bowel occurs. The thickened root of the mesentery produces an oval radiolucent area to the right of the spine between the barium-filled loops of small intestine (Fig. 3–56). Huge appendiceal abscesses may displace the urinary bladder, the sigmoid colon and the rectum as well as produce compression and partial obstruction of the right ureter (Fig. 3–57). Joffe described a case of an appendiceal abscess resulting in a cecosigmoid fistula[6] (Fig. 3–58). A second patient had a sharply circumscribed smooth defect in the wall of the sigmoid colon due to a walled-off appendiceal abscess intimately adherent to and partly embedded within the wall of the colon (Fig. 3–59). When there is malrotation of the colon, appendiceal abscesses may displace structures that are ordinarily remote from the normal appendix, such as the descending colon (Fig. 3–60). Rarely, barium may enter the abscess cavity itself, indicating that the appendiceal lumen is not occluded in every case (Fig. 3–60).

Threatt and Appelman reported a case of Crohn's disease confined to the appendix presenting as acute appendicitis.[33] A mass impinged on the cecum owing to the thickened appendix which simulated an appendiceal abscess. The terminal ileum was normal.

B mode ultrasound scans and gallium[67] citrate scans of the abdomen are useful diagnostic modalities for the detection of appendiceal abscesses. These techniques are particularly valuable when the inflammatory mass is remote from the cecum and terminal ileum and fails to produce changes on the barium enema examination (Figs. 3–61 to 3–63).

Meyers and Oliphant recently reported three cases of acute inflammation of an ascending retrocecal appendix which demonstrated characteristic findings on barium enema studies.[34] In these cases abnormality in the ascending colon may be seen anywhere along its length from the cecum to the hepatic flexure, but the changes specifically involve the lateral or posterior haustral contour (Fig. 3–64). Inflammation associated with an intraperitoneal ascending retrocecal appendix occurs in the right paracolic gutter and involves the lateral haustral row of the ascending colon. Conversely, inflammation associated with an extraperitoneal ascending retrocecal appendix affects primarily the posterior haustral row, based on the normal plane of peritoneal reflection over the ascending colon. In all three of Meyers and Oliphant's cases the appendix opacified with barium and the inflammatory process, originating in the tip, did not prevent contrast material from filling the lumen. In their cases the appendix itself showed definite abnormalities, including mass displacement, sinus tracts and opacification of the abscess cavity.

MUCOCELE

Mucocele of the appendix is an uncommon condition in which there is distention of the appendix with sterile mucus associated with obstruction of the appendiceal lumen. On gross study, the appendix is distended, the lumen is filled with yellow mucus and the wall is fibrotic and frequently calcified. The incidence of the disease has been reported to be 0.15 per cent in over 30,000 autopsies.[35] The etiology of the mucus distention is generally attributed to obstruction of the appendiceal lumen with excessive mucus secretion, although some authors contend that the lesion is a mucinous cystadenoma.[36] The obstruction may be caused by a fecalith, a foreign body, a carcinoid, endometriosis, adhesions or volvulus. Grodinsky and Rubnitz produced mucoceles experimentally in rabbits by ligating the proximal appendix without occluding the blood supply after cleansing but not sterilizing the appendiceal lumen.[37]

Most mucoceles are detected incidentally during radiographic examination of the abdomen, at surgery or on postmortem examination. A smaller percentage of cases present with symptoms due to secondary infection or intussusception of the lesion.

Radiographically, the mucocele presents as a sharply outlined, globular, smooth, broad-based filling defect which invaginates into the cecum and is associated with

(Text continued on page 151)

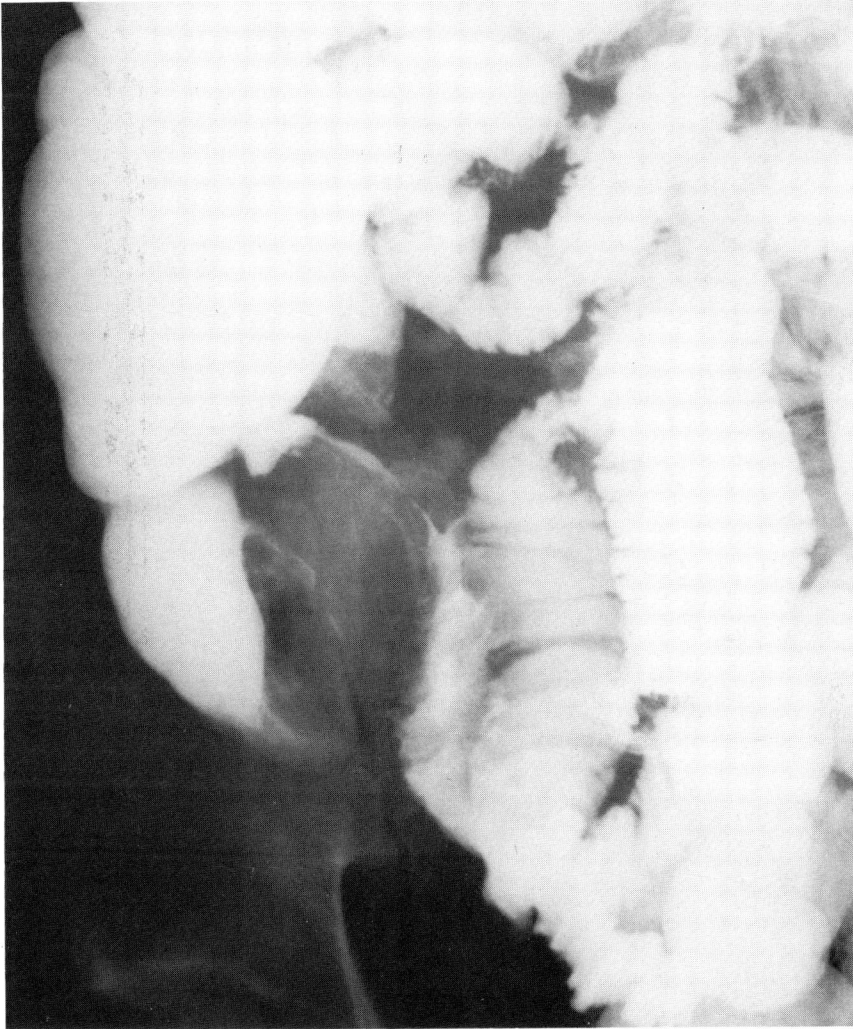

Figure 3–53. Radiograph from a small bowel examination showing extrinsic displacement at the cecal tip due to an appendiceal abscess.

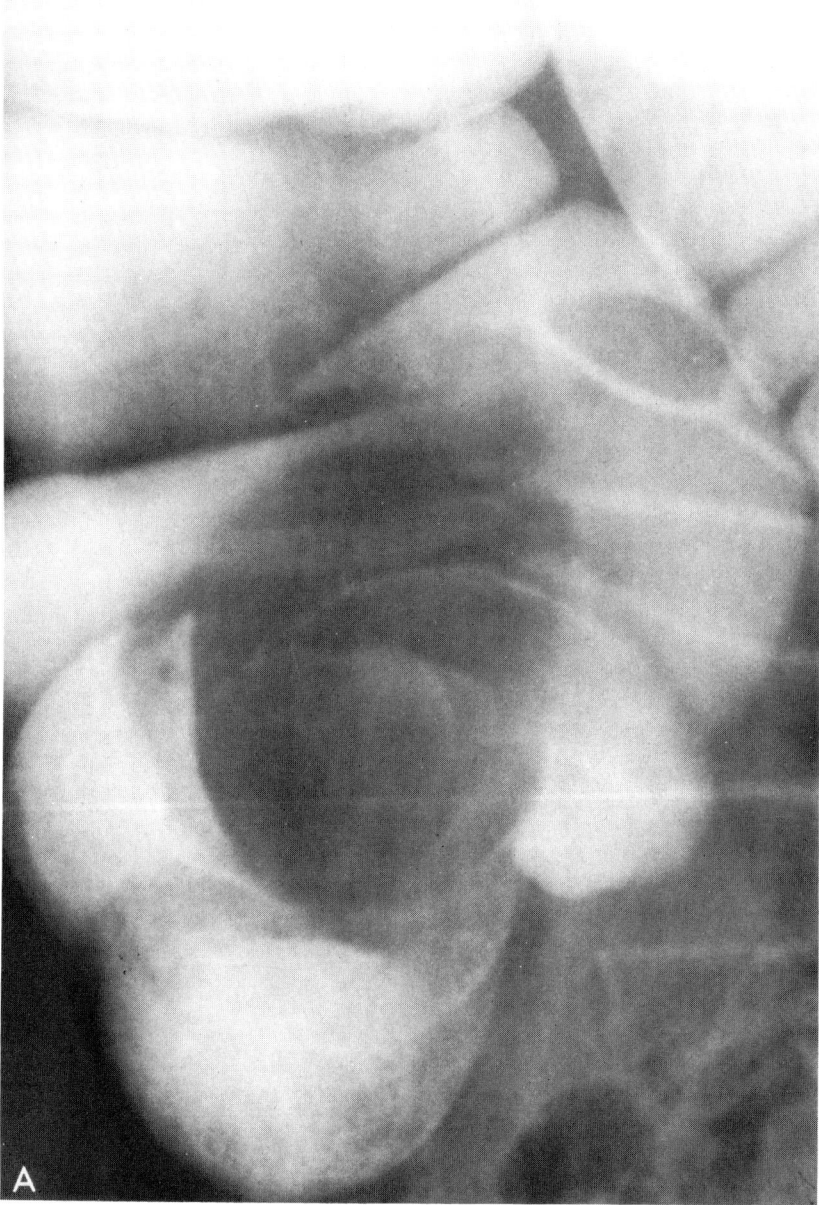

Figure 3–54. Spot radiographs of the cecum made during a barium enema examination (*A* and *B*) showing an extrinsic mass displacing the cecum owing to an appendiceal abscess.

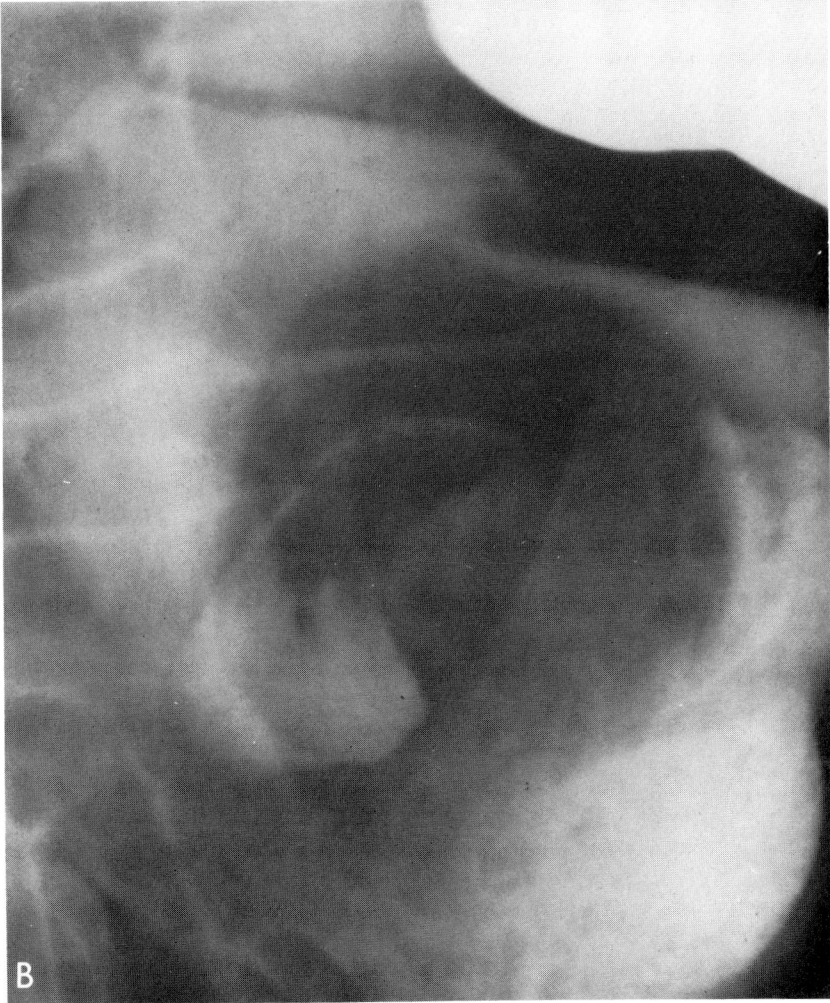

Figure 3–54. (*Continued*)

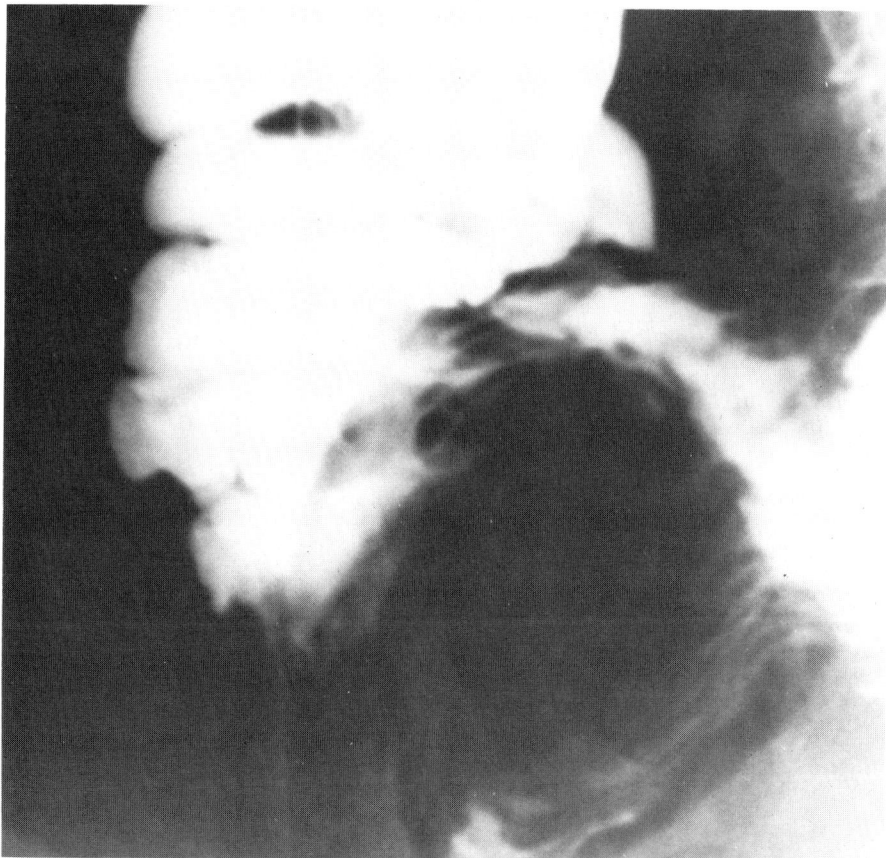

Figure 3-55. Radiograph from a small bowel examination showing marked displacement of the terminal ileum and cecum caused by an appendiceal abscess.

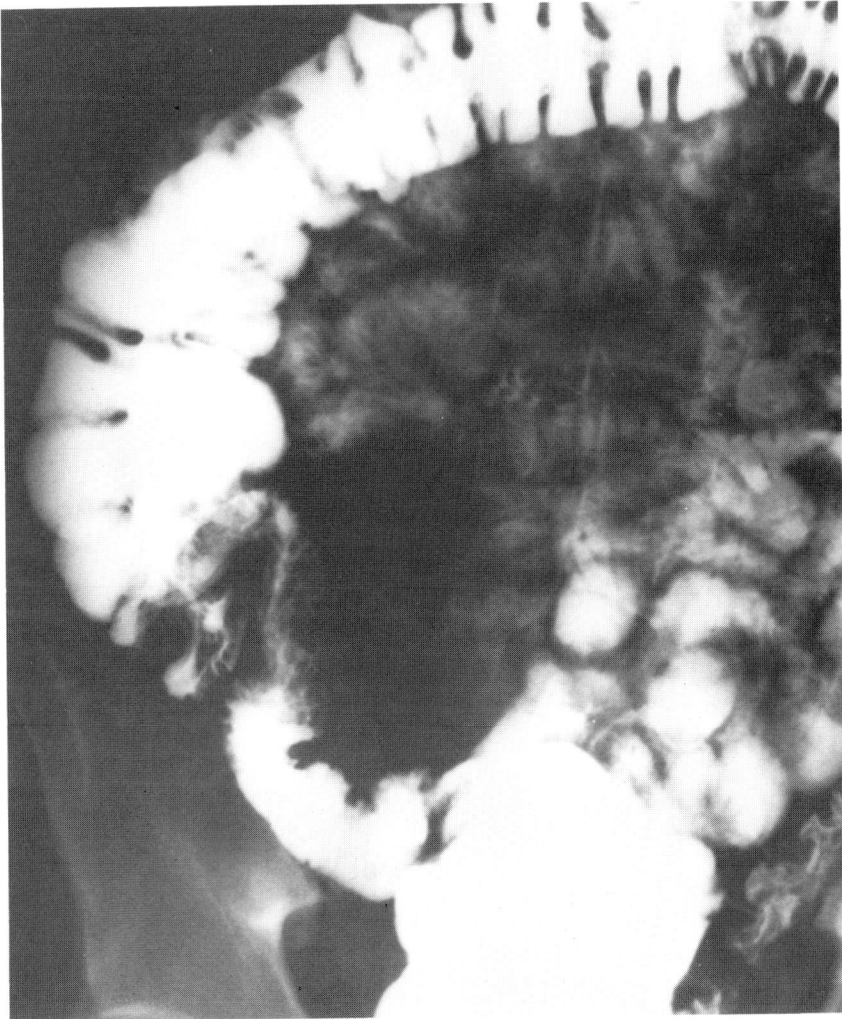

Figure 3-56. Radiograph from a small bowel examination showing displacement of the small bowel around a large appendiceal abscess in the root of the mesentery.

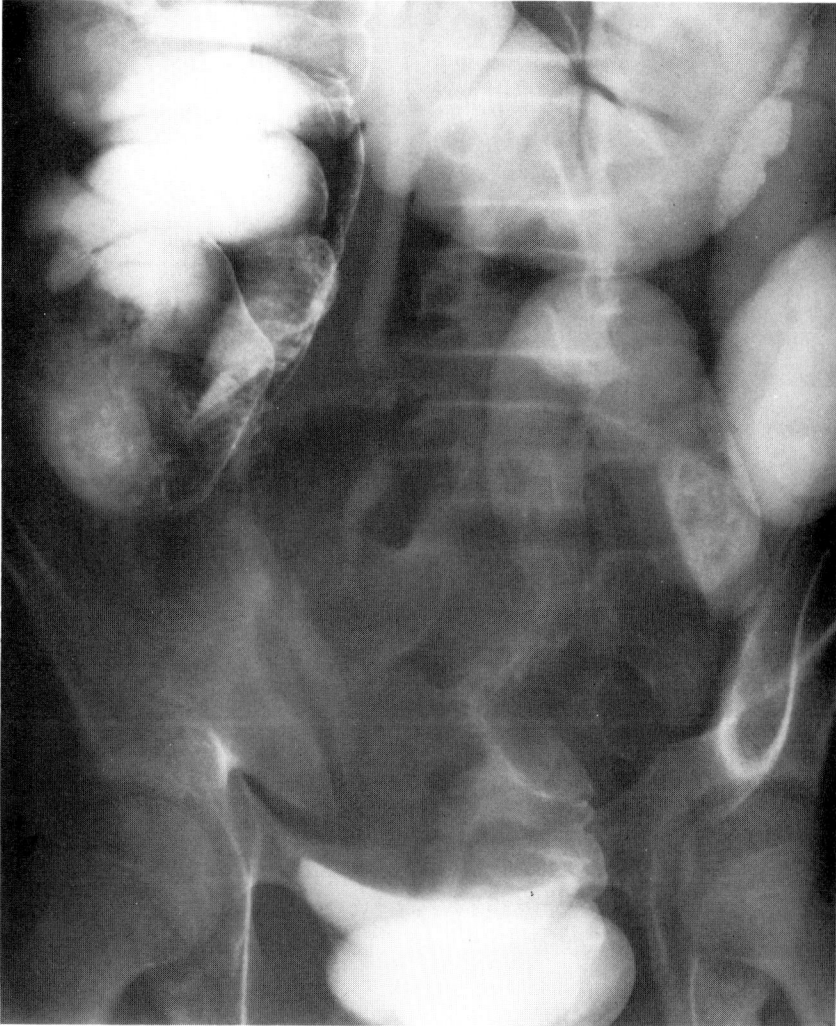

Figure 3–57. Barium enema examination performed after an intravenous pyelogram showing displacement of the cecum and the rectosigmoid by a huge appendiceal abscess which extends into the pelvis. There is partial ureteral obstruction and compression of the urinary bladder.

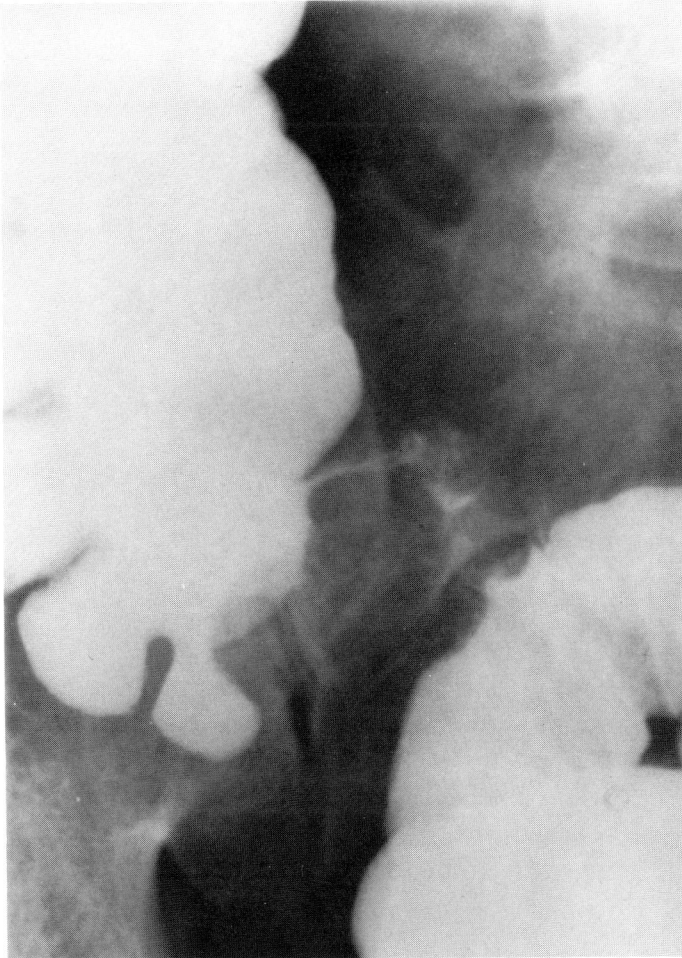

Figure 3–58. Barium enema examination showing indentation on the medial aspect of the cecum and a cecosigmoid fistula with no filling of the appendix. At operation there was a cecosigmoid fistula and a small perforation in the cecum at the base of a still-attached portion of the appendix. The distal fragment of the appendix was firmly embedded in the wall of the sigmoid colon, with its tip presenting at the site of the fistulous opening. (From Joffe, N.: Some uncommon roentgenologic findings associated with acute perforative appendicitis. Radiology, *110*:301, 1973.)

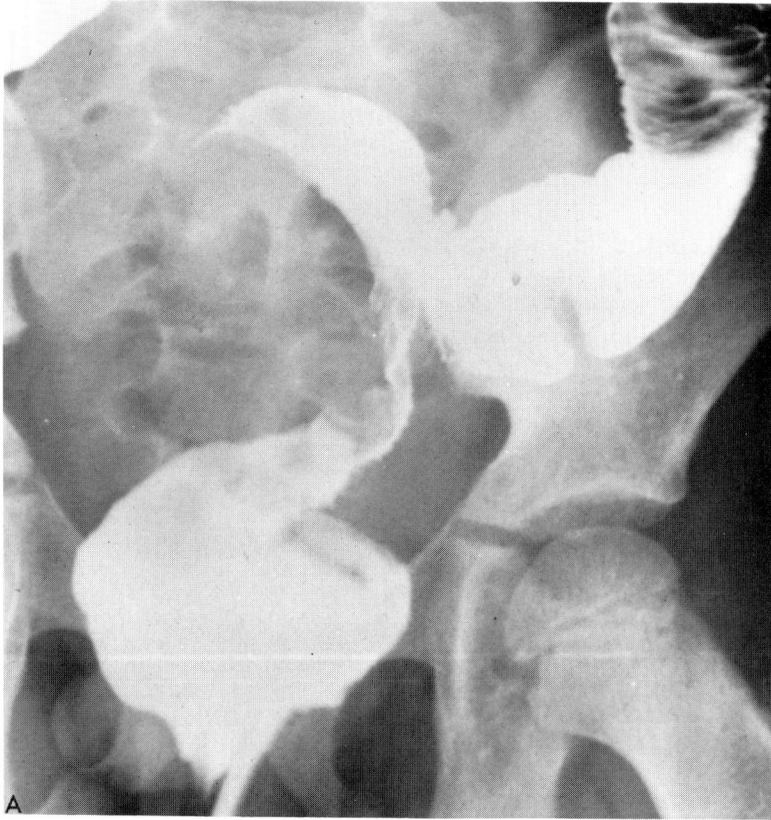

Figure 3–59. Barium enema examination showing a large circumscribed defect involving the distal sigmoid colon (*A*). Oblique projections (*B* and *C*) show acute angles at the margins of the defect and stretching of overlying mucosa. Surgical exploration revealed a perforated appendix with a walled-off abscess intimately adherent to and partly embedded within the wall of the distal sigmoid colon. (From Joffe, N.: Some uncommon roentgenologic findings associated with acute perforative appendicitis. Radiology, *110*:301, 1973.)

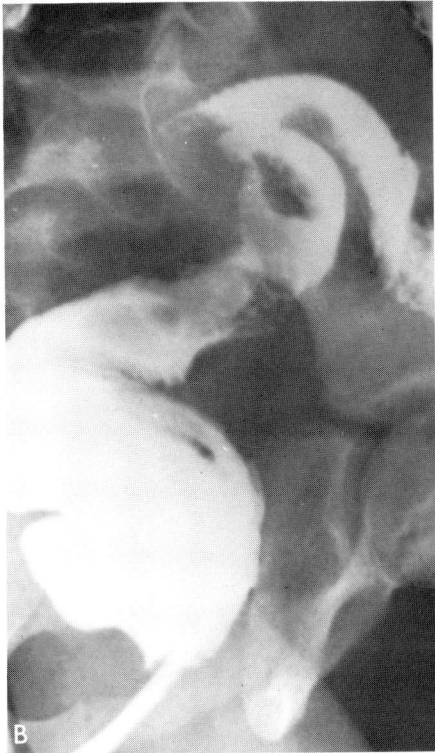

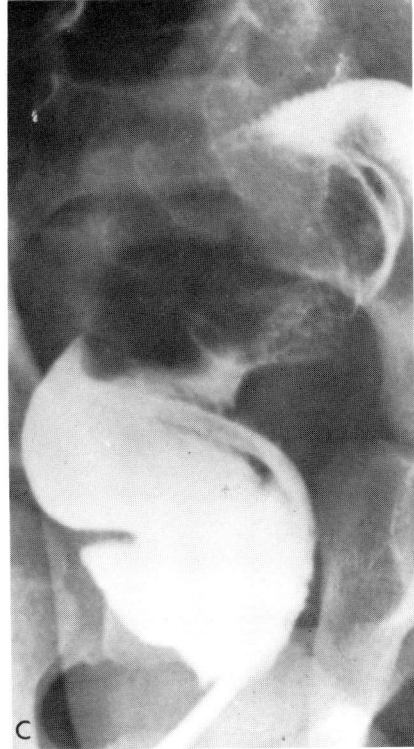

Figure 3–59. (*Continued*)

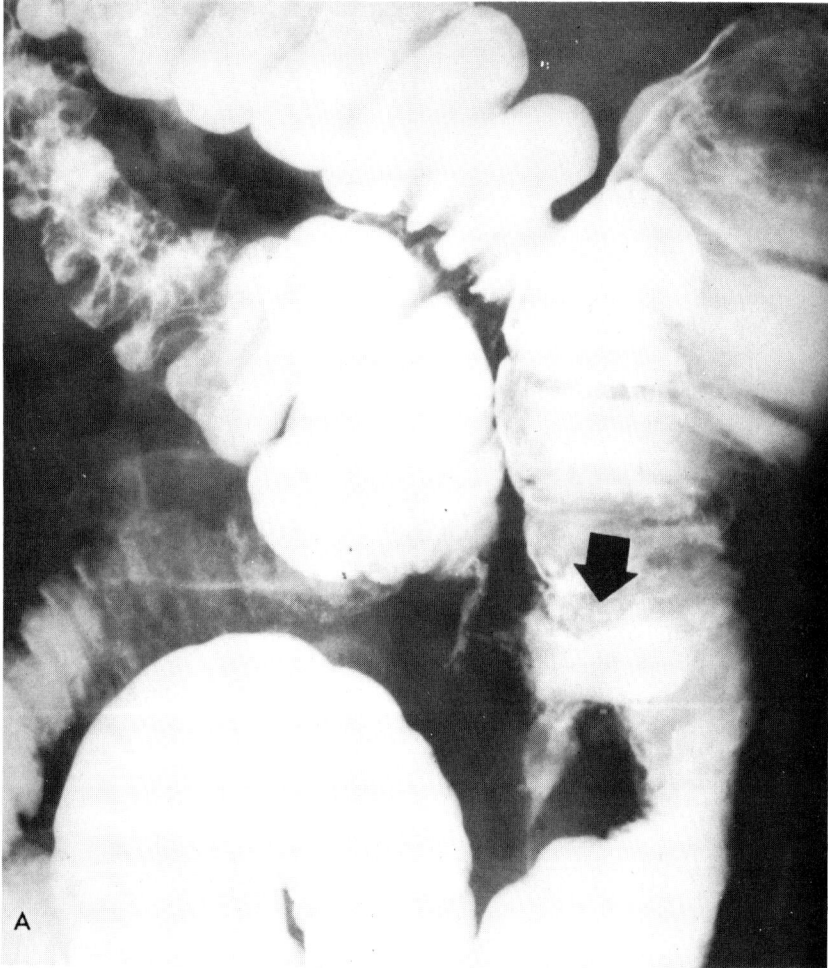

Figure 3–60. Barium enema examination showing malrotation of the colon. (A) Barium is seen entering an abscess cavity adjacent to and displacing the descending colon (arrow). (*Legend continued on opposite page.*)

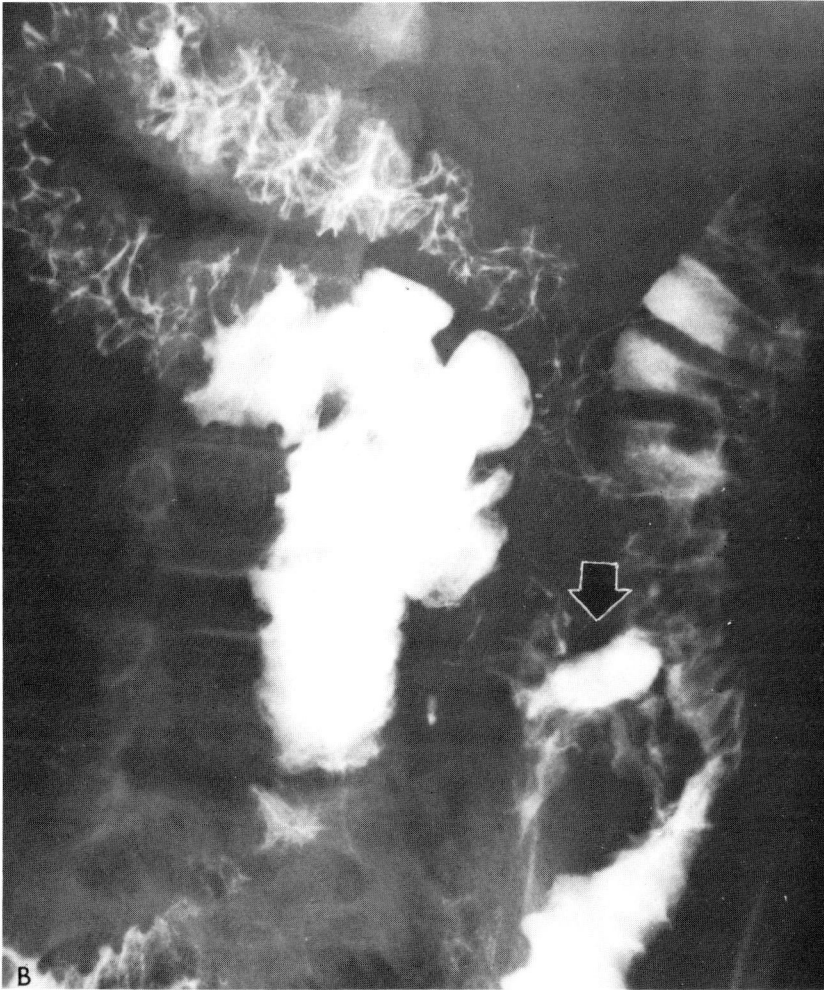

Figure 3-60. (*Continued*).   The postevacuation radiograph (*B*) shows the barium in the abscess and the displacement of the descending colon (arrow).

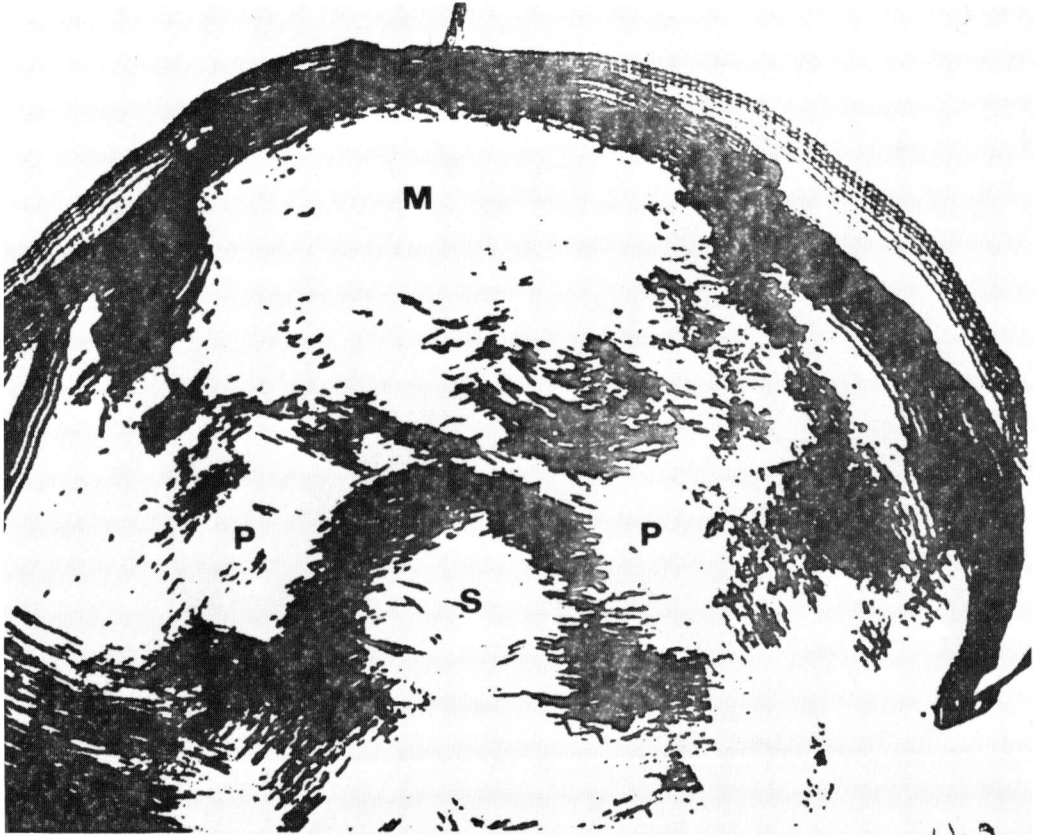

Figure 3–61. B mode ultrasound scan showing a transverse section halfway between the umbilicus and the pubis, looking downward from the head, the patient's right being the observer's right. A large relatively sonolucent mass due to an appendiceal abscess is visible (M). (S = spine; P = psoas muscle.)

Figure 3–62. B mode ultrasound scan (*A*) showing a longitudinal section to the right of the midline. The patient's head is to the observer's left. A large sonolucent mass due to an appendiceal abscess can be identified (M). Strong posterior echoes indicate that the mass is fluid filled. The bladder can be identified anteriorly (B). (A = anterior; P = posterior.)

B mode ultrasound scan (*B*) in another patient showing a longitudinal section to the right of the midline, lateral to the kidney. An appendiceal abscess (Ab) is evident lying on the psoas muscle (P). The patient's head (H) is to the observer's left. (A = anterior; F = feet; in = small intestine; P = iliacus portion of psoas muscle; p = psoas portion of psoas muscle.) Closed arrows point to right iliac bone. Open arrows outline extent of appendiceal abscess.

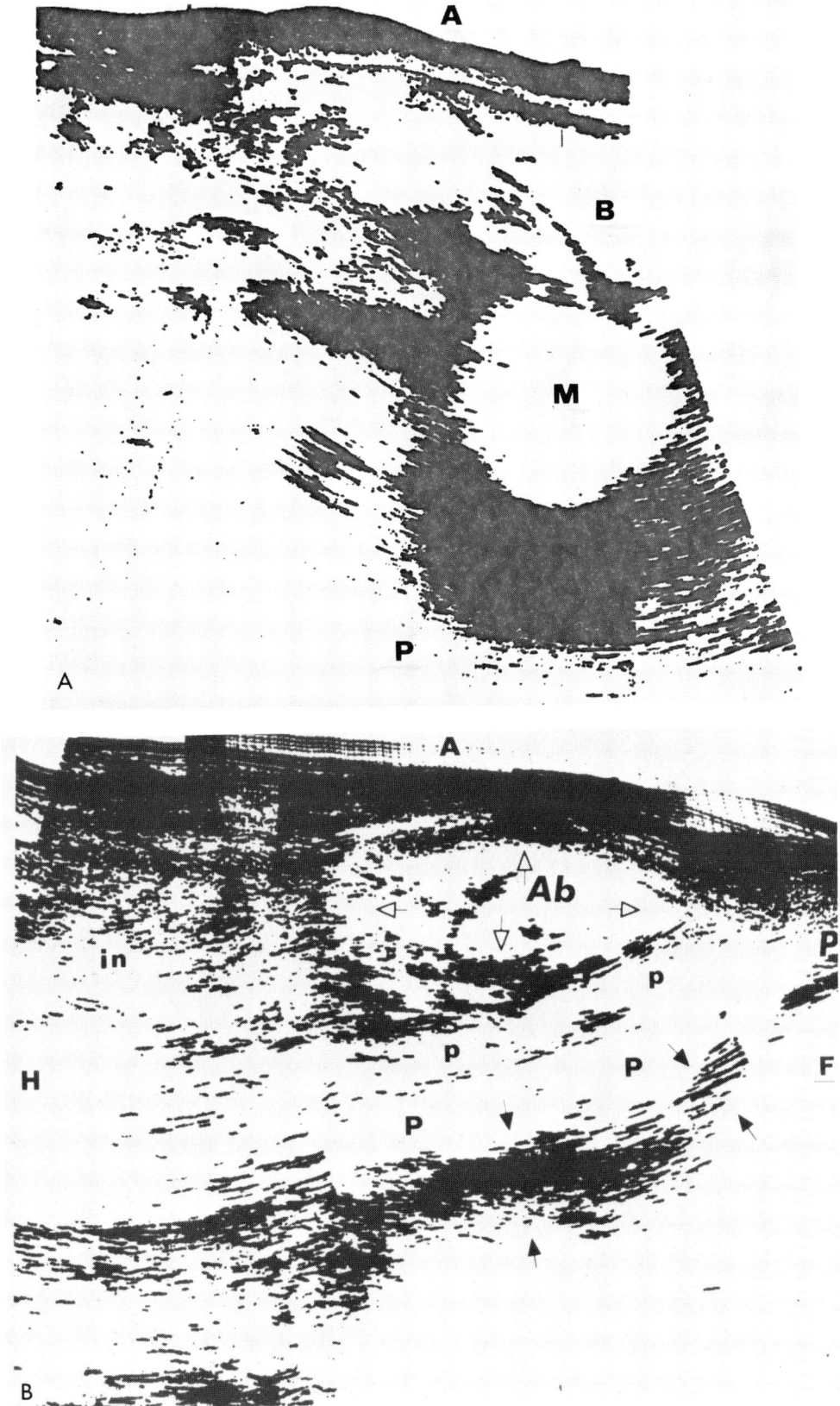

Figure 3–62. See opposite page for legend.

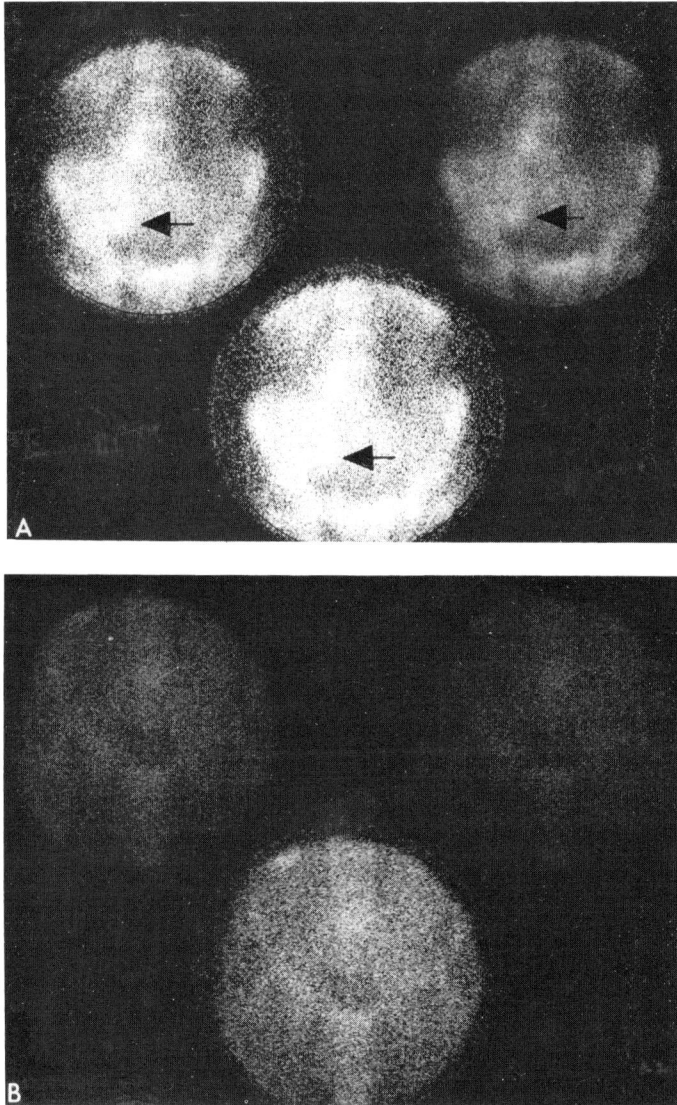

Figure 3–63. Three frontal exposures of the pelvis of gallium⁶⁷ citrate scan made with different exposure factors (*A*). Uptake by the bones of the pelvis and spine is apparent. In addition there is uptake in an appendiceal abscess in the right lower quadrant (arrows). (*B*) Normal scan.

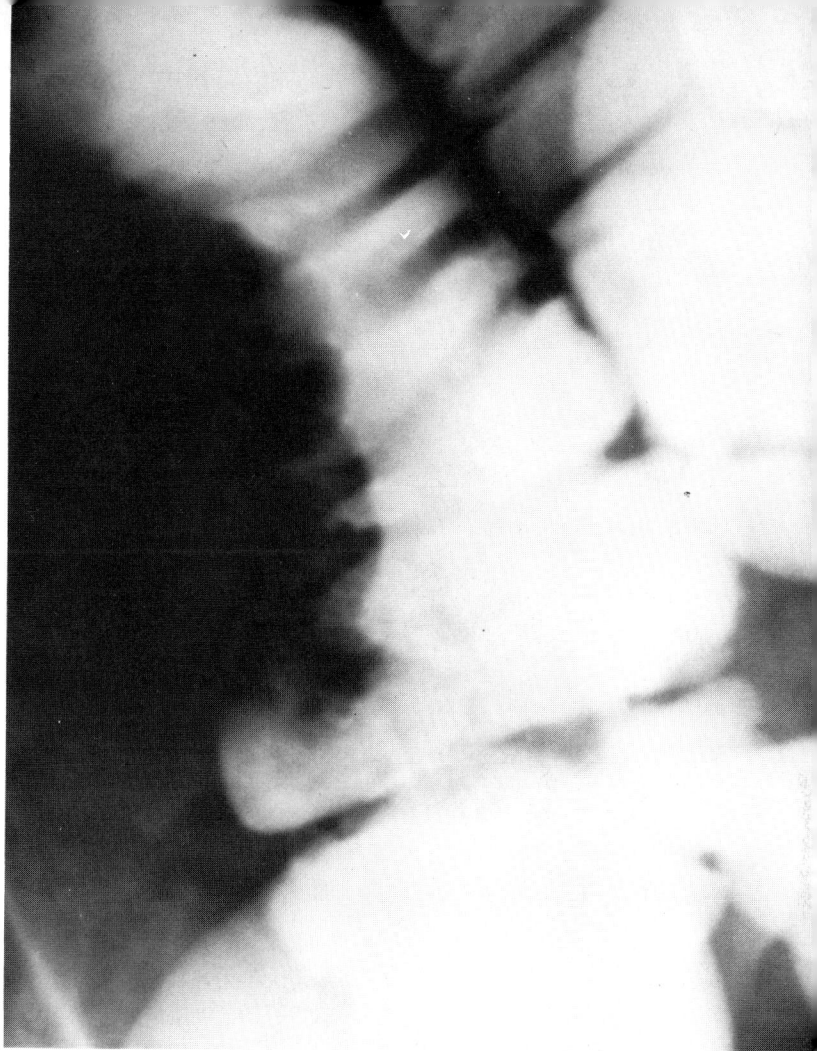

Figure 3-64. Barium enema examination showing an appendiceal abscess posterior and lateral to the ascending colon due to perforation of a retrocecal appendix.

Figure 3-65. Barium enema examination showing a smooth extrinsic impression on the cecum and terminal ileum due to a mucocele of the appendix. (From Felson, B., and Wiot, J. F.: Some interesting right lower quadrant entities. Radiol. Clin. N. Amer., 7:83, 1968.)

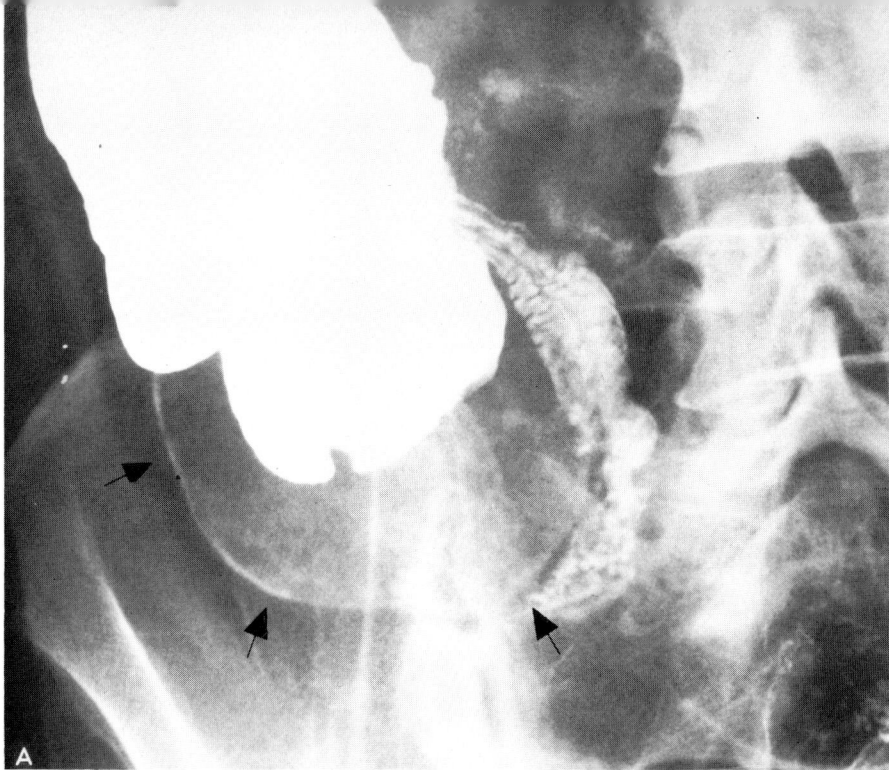

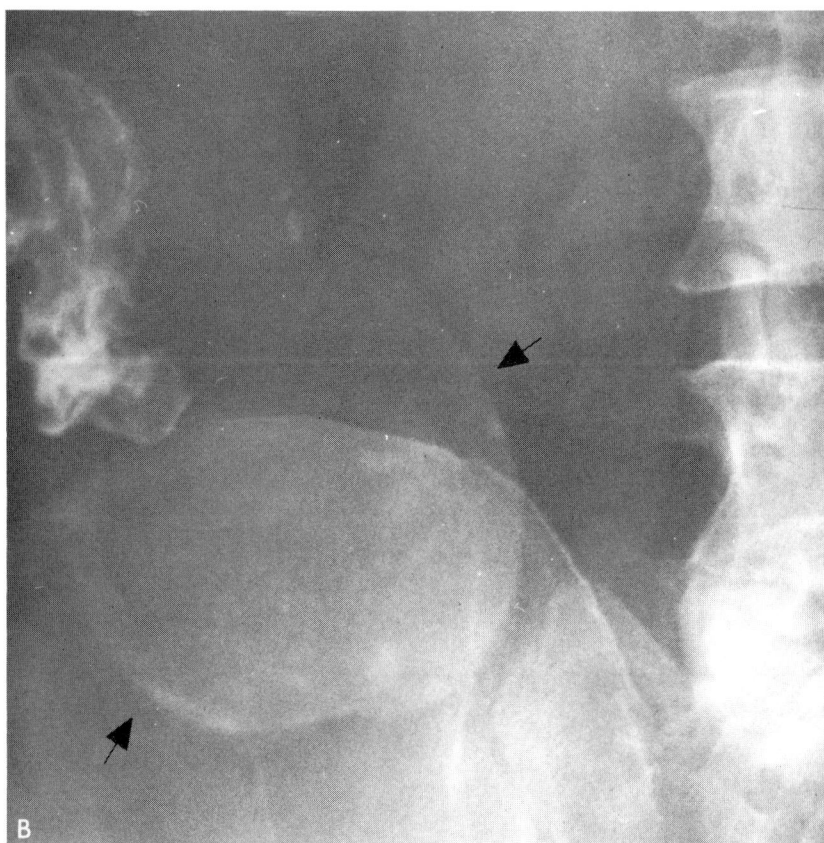

Figure 3-66. Barium enema examination (*A*) showing a large peripherally calcified mass (arrows) adjacent to the cecum displacing the terminal ileum owing to a mucocele of the appendix. The post-evacuation radiograph (*B*) shows the calcified mass (arrows).

150

nonfilling of the appendix on barium enema examination[35] (Fig. 3–65). Peripheral calcification is often present in the lesion (Fig. 3–66).

The differential diagnosis includes an inverted appendiceal stump which may produce an extrinsic impression at the cecal tip, but the defect is usually small and localized. Acute appendicitis with abscess formation is frequently associated with an extrinsic compression at the base of the cecum. This may be indistinguishable from a mucocele on radiologic examination alone, although there may be more spasm and irritability and less invagination of the cecum associated with an abscess. Differentiation between the two is usually evident clinically. Carcinoma of the cecum produces an irregular mucosal mass or infiltration of the wall of the cecum with mucosal destruction. Contrarily, mucoceles are typically smooth extramucosal lesions.

Myxoglobulosis of the Appendix

Myxoglobulosis is a term that was coined by von Hansemann in 1944 to describe a variant of mucocele of the appendix which is associated with the formation of radiopaque, calcified globules within the mucocele.[38] The globules vary from 0.1 to 1.0 cm. in size, are oval or faceted in shape and resemble tapioca or fish eggs mixed with liquid mucus (Fig. 3–67).

Felson and Wiot cite the incidence of this variant as 8 per cent, since 60 cases of myxoglobulosis have been reported compared to approximately 700 cases of appendiceal mucocele.[4] Others report an incidence of only 0.35 per cent.[39] Most authors agree that the lesion is frequently overlooked by the pathologist.

The pathogenesis of the calcified globules is uncertain. Predisposing factors that are mandatory for the formation of the mucocele include obliteration of the lumen, absence of pyogenic organisms in the lumen and continued production of mucin. Cagnetto[40] and Pohl[41] believe that cores of the globules are formed in dilated glands, and when extruded into the appendiceal lumen, they enlarge through confluence of smaller particles. Lubin and Berle described two cases in which the globules had centers consisting of granulation tissue, usually extending from the mucosa to the center of the globule.[38] In the free-floating globules the centers contained foci of granular eosinophilic debris. They suggested that initially there is an attempt to organize masses of mucin by ingrowth of granulation tissue from the wall of the appendix. These organized globules are then broken off and the centers undergo necrosis. Once detached, they serve as a nidus for circumferential deposition of mucin and calcium.

The clinical manifestations of myxoglobulosis are as minimal as those of mucoceles. The lesion is usually detected as an incidental abnormality during radiographic studies of the abdomen or at postmortem examination.

The radiologic manifestations are those of a mucocele of the appendix including a mass impressing on the cecum associated with nonfilling of the appendix. In addition, however, the calcified spherules are visible radiographically[4] (Fig. 3–68). According to Felson and Wiot the globules are usually annular and nonlaminated and may be faceted.[4] Rarely, they are solidly calcified (Fig. 3–69). The calculi are fairly uniform in size and shift within the mucocele, depending on the position of the patient. In the upright projection, the globules may layer within the mucocele.

Differential diagnosis includes appendicoliths, phleboliths, calcified lymph nodes and enteroliths. However, the combination of multiple annular calcification and a cecal defect without filling of the appendix is pathognomonic of myxoglobulosis.

Pseudomyxoma Peritonei

Pseudomyxoma peritonei is a complication of rupture of a mucocele of the appendix or ovary and is characterized by epithelial implants on the peritoneal surface, with massive accumulation of gelatinous ascites.[42, 43] The peritoneal lesions that result consist of thick mucoid material with focal hemorrhage. Careful histologic examination discloses epithelial lining within the implants. The pathogenesis of the disease is uncertain. Rupture of the mucocele may spill epithelial cells into the peritoneal cavity which are capable of implanting and growing on the peritoneum. The mucin extruded from the appendix may produce a

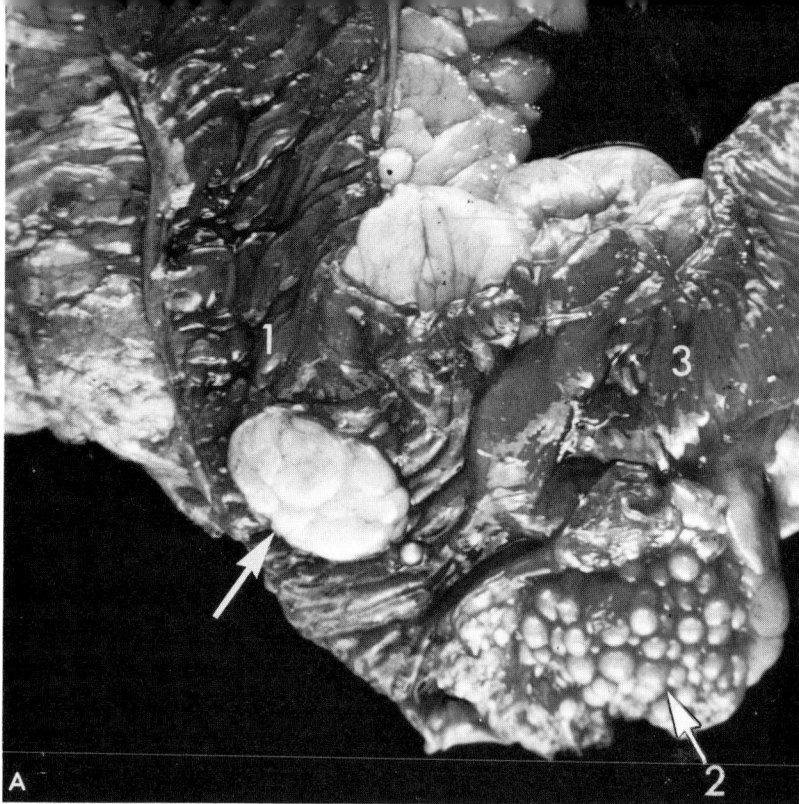

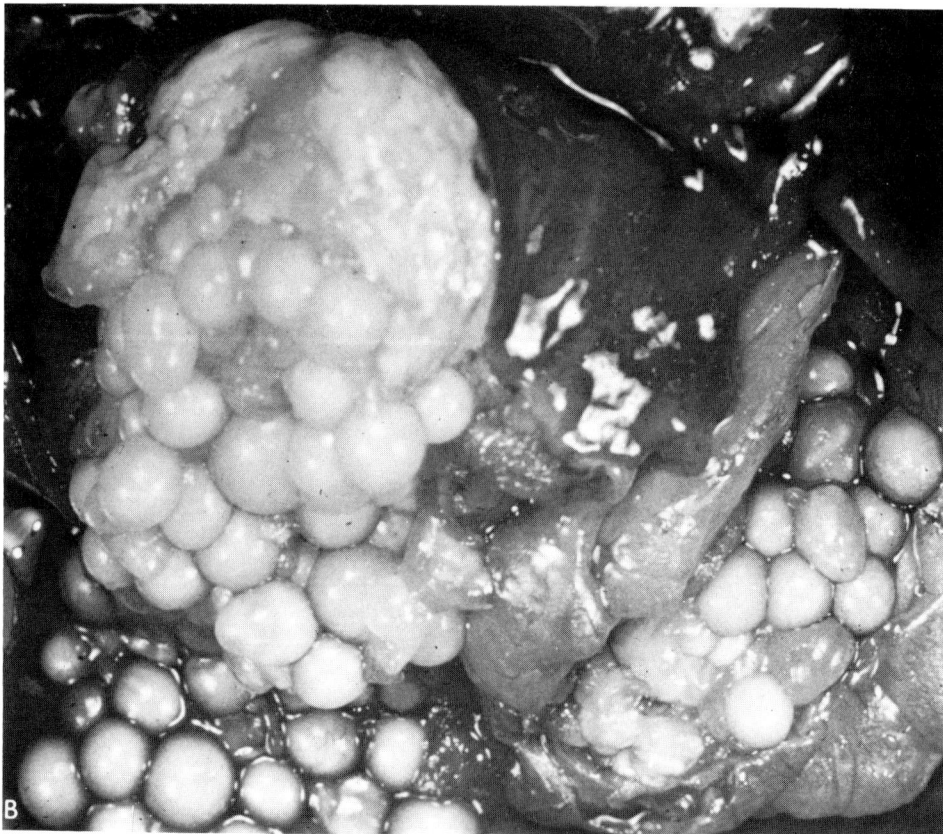

Figure 3-67. Gross appearance of myxoglobulosis of the appendix. (*A*) Cecum (1), mucocele (2) and terminal ileum (3) have been laid open. A cecal lipoma is present (arrow). (*B*) Close-up of opened mucocele showing many globules of fairly uniform size. The cecal lipoma is seen above them. (*C*) The globules (arrows) and mucus are seen in another opened mucocele. (From Felson, B., and Wiot, J. F.: Some interesting right lower quadrant entities. Radiol. Clin. N. Amer., 7:83, 1969.)

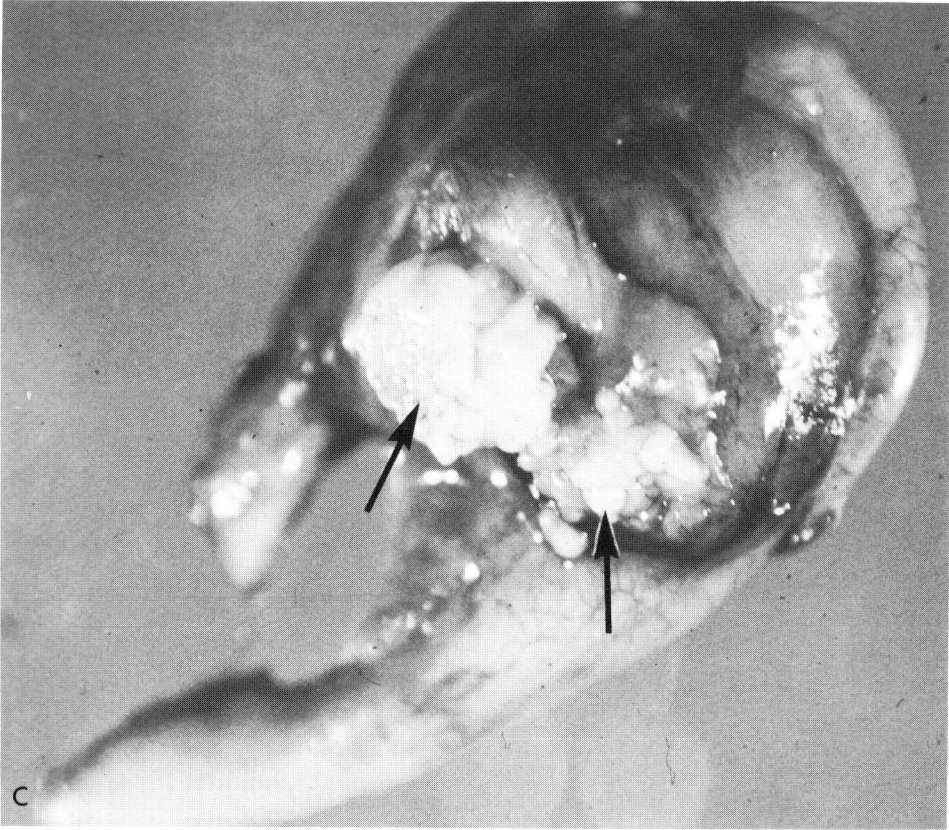

Figure 3–67. (*Continued*)

foreign body reaction with subsequent epithelialization by the peritoneum. Gibbs contends that the implants are metastatic cystadenocarcinoma from mucinous carcinoma of the appendix.[36]

The age range of the disease in one series was 31 to 80 years with an average of 61 years. Both sexes are involved equally. Symptoms are those of ascites and include abdominal distention, fullness and vague abdominal pain. Sudden sharp abdominal pain may occur at the time of rupture of the mucocele which may be associated with straining. The ascites is often massive, and recurrent small bowel obstruction is a frequent complication.

The radiologic findings are those of mucocele and ascites. Sudden decrease in size of a mucocele may indicate that rupture has occurred (Fig. 3–70). Dilatation of the small intestine occurs when small bowel obstruction develops.

POSTAPPENDECTOMY DEFORMITIES

Postoperative defects in the cecum following appendectomy may be secondary to inversion of the appendectomy stump, ligation of the stump without inversion or incomplete resection of the appendix.[44]

The inverted appendiceal stump produces a mass in the tip of the cecum at the base of the appendix on barium enema examination (Fig. 3–71). The mass varies widely in size from several mm. to 3 cm. The inverted stump is particularly prominent for several weeks after surgery owing to persistent edema and inflammation (Fig. 3–72). The surface may be smooth, lobulated or irregular, so that differentiation from mural or mucosal lesions of the cecum is often difficult, if not impossible. Air contrast enema studies are helpful in accurately

(*Text continued on page 162*)

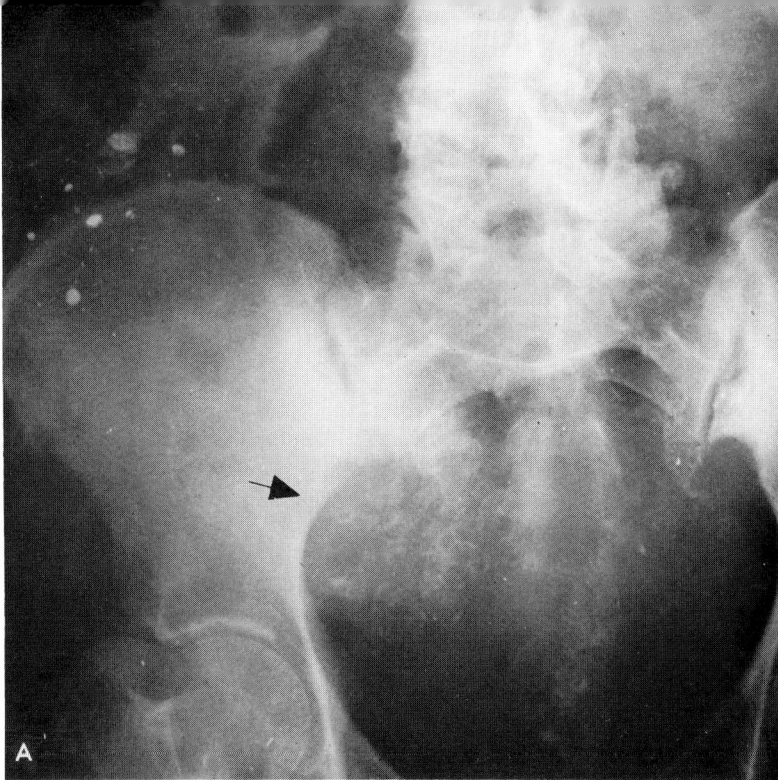

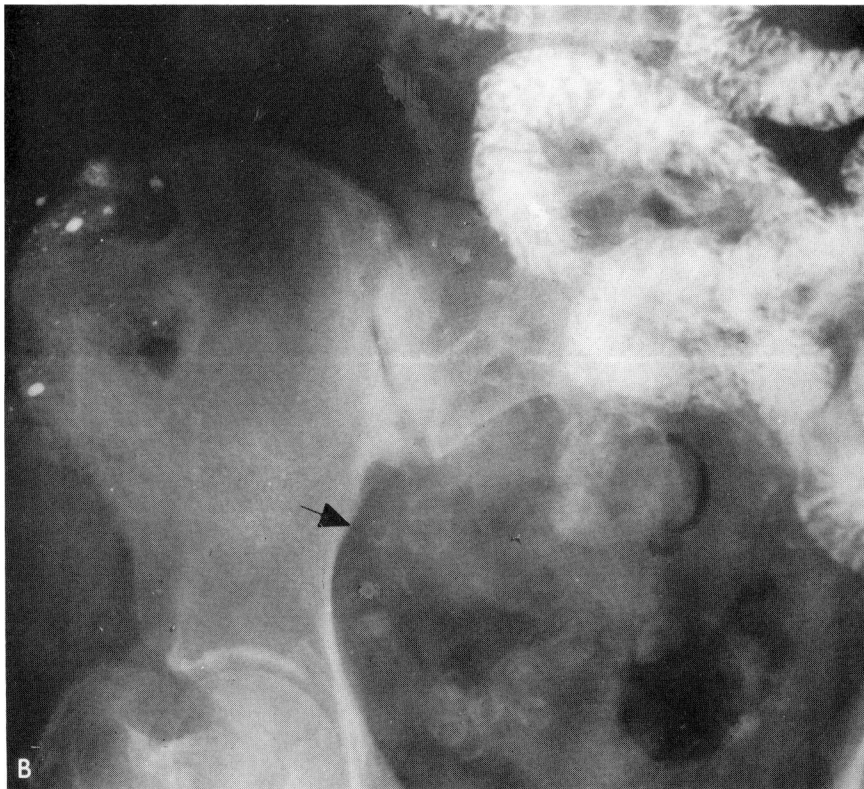

Figure 3–68. Plain abdominal radiograph (*A*) showing multiple annular faceted calcifications in a large mucocele in the right lower quadrant (arrows). Calcification is also present in the buttock. The calculi have shifted within the mass (*B*) (arrow). (From Felson, B., and Wiot, J. F.: Some interesting right lower quadrant entities. Radiol. Clin. N. Amer., 7:83, 1969.)

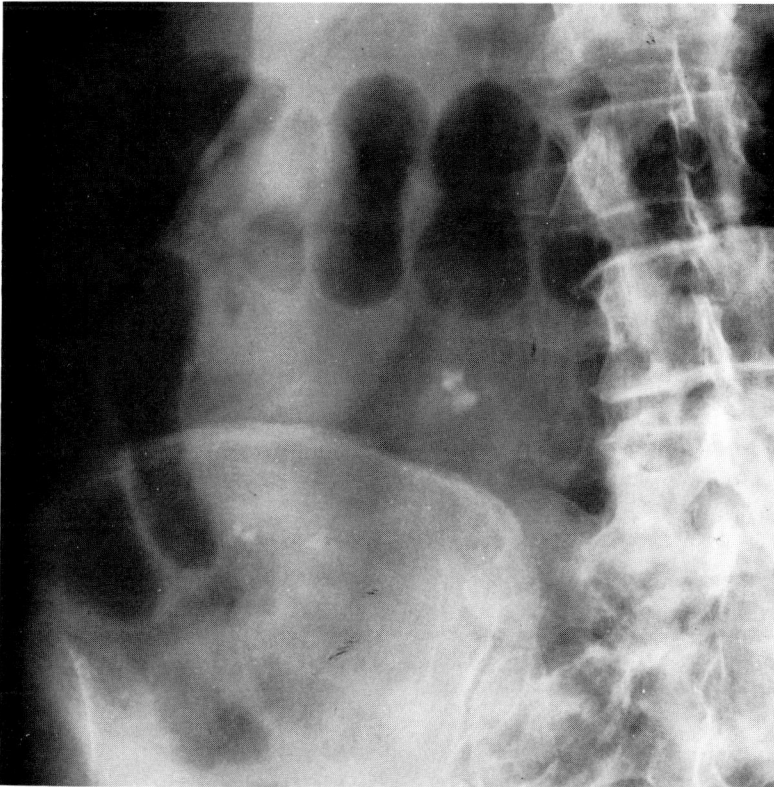

Figure 3–69. Plain abdominal radiograph showing solid calcification in globules in myxoglobulo-sis. This is a rare form of calcification. (From Felson, B., and Wiot, J. F.: Some interesting right lower quadrant entities. Radiol. Clin. N. Amer., 7:83, 1969.)

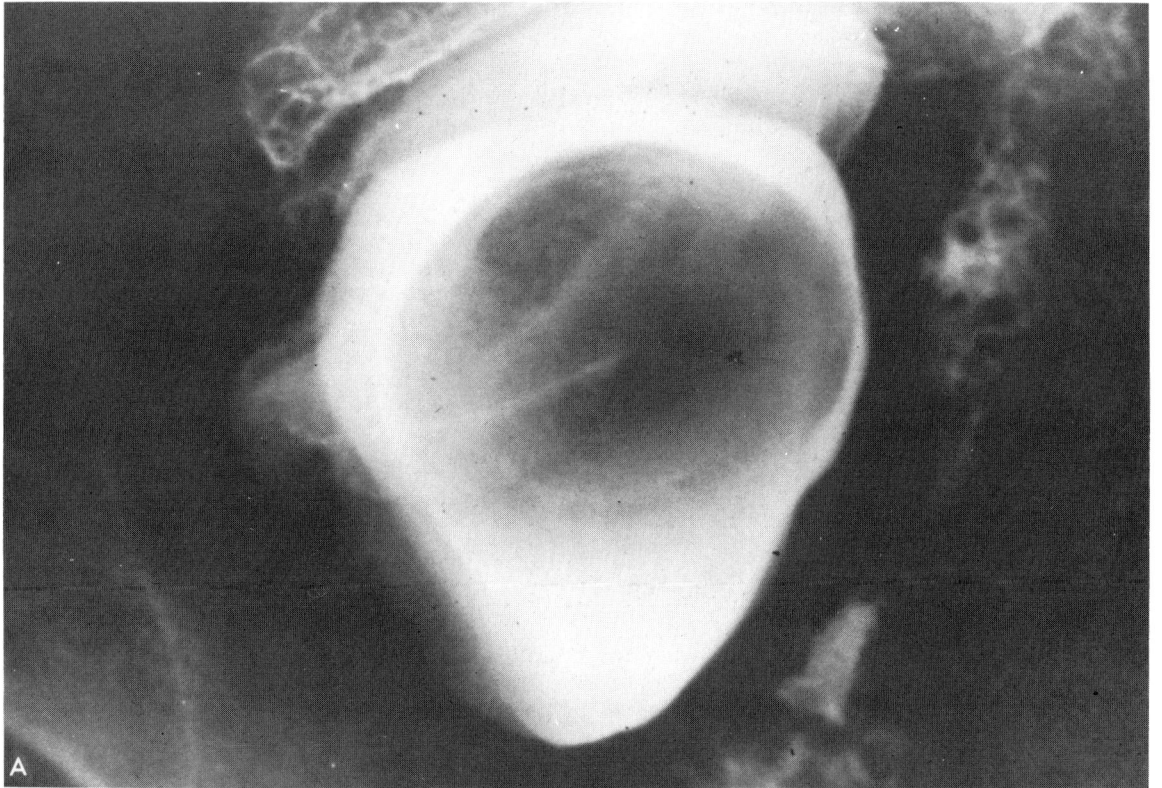

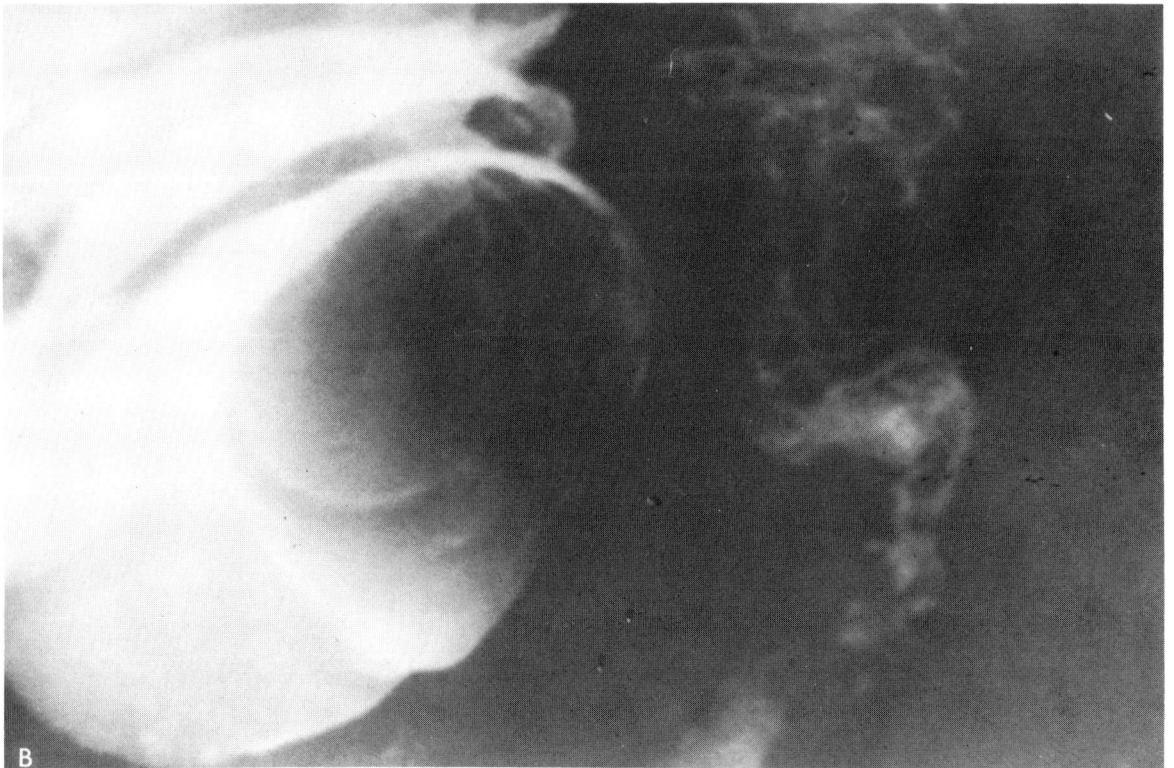

Figure 3-70. *See opposite page for legend.*

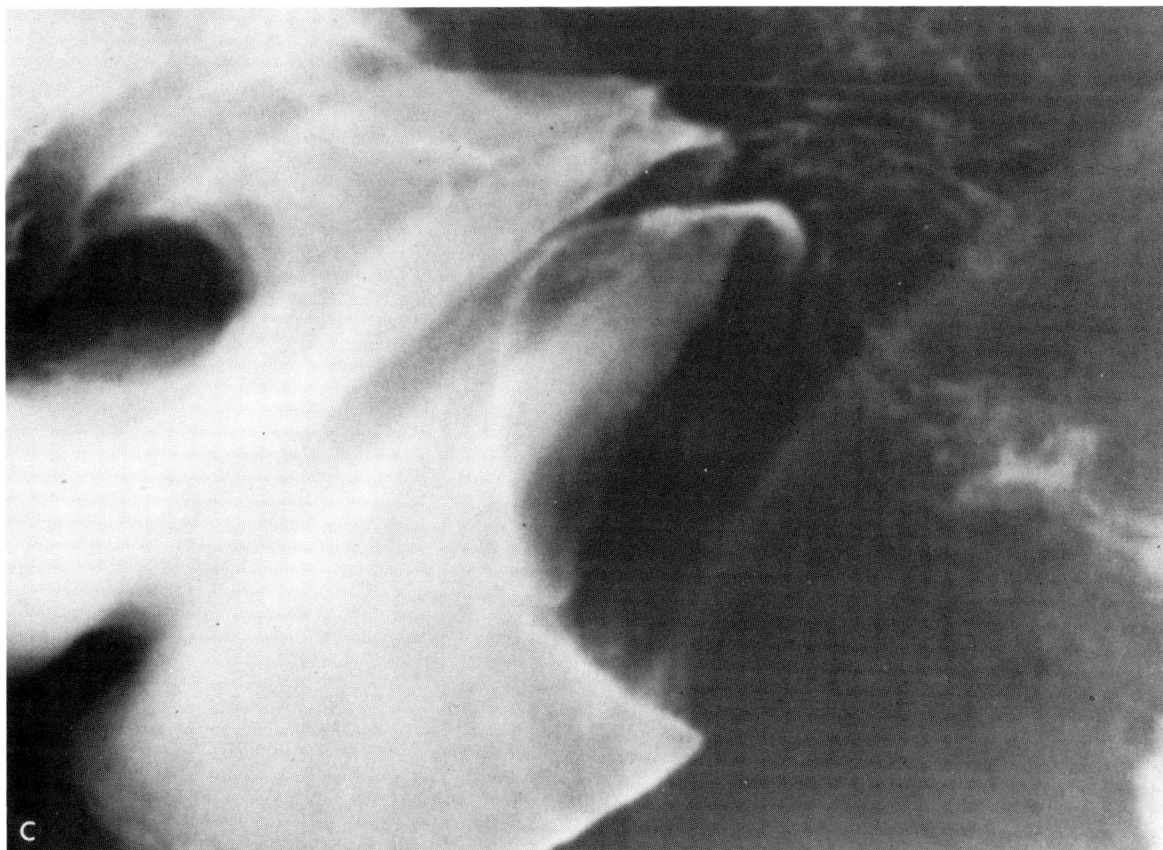

Figure 3–70. Barium enema examination (*A*, *B* and *C*) shows a smooth extramucosal mass impressing the apex of the cecum. The examination was performed because the patient complained of the sudden onset of abdominal pain while straining. The absence of fever and leukocytosis suggested that an appendiceal abscess was unlikely, and a preoperative diagnosis of pseudomyxoma peritonei from a ruptured appendiceal mucocele was made. At surgery there was a mucocele of the appendix which had ruptured, producing mucinous ascites.

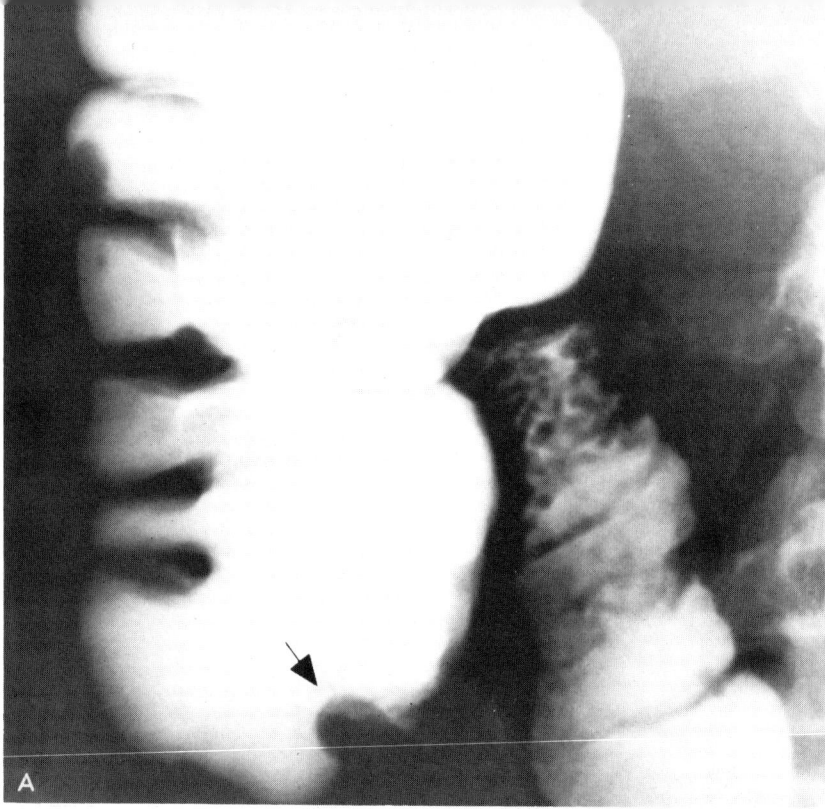

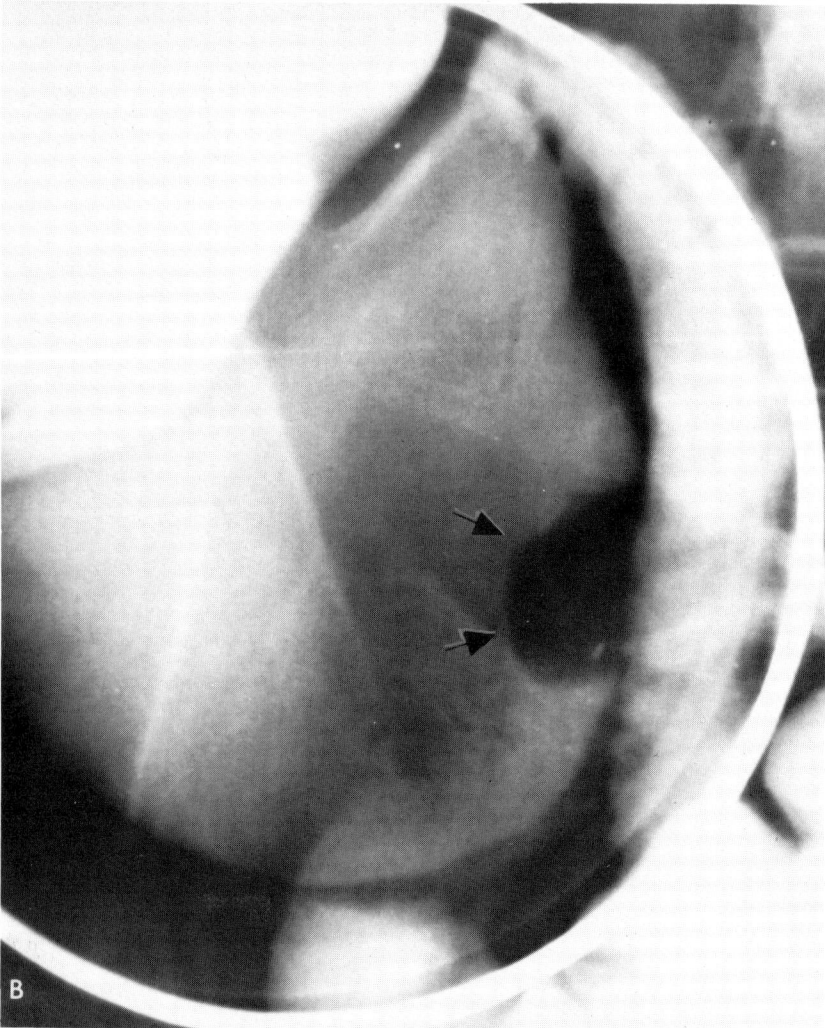

Figure 3–71. Barium enema and air contrast examinations in four different patients (*A* to *D*) showing a smooth extramucosal mass in the cecum due to an inverted appendiceal stump.

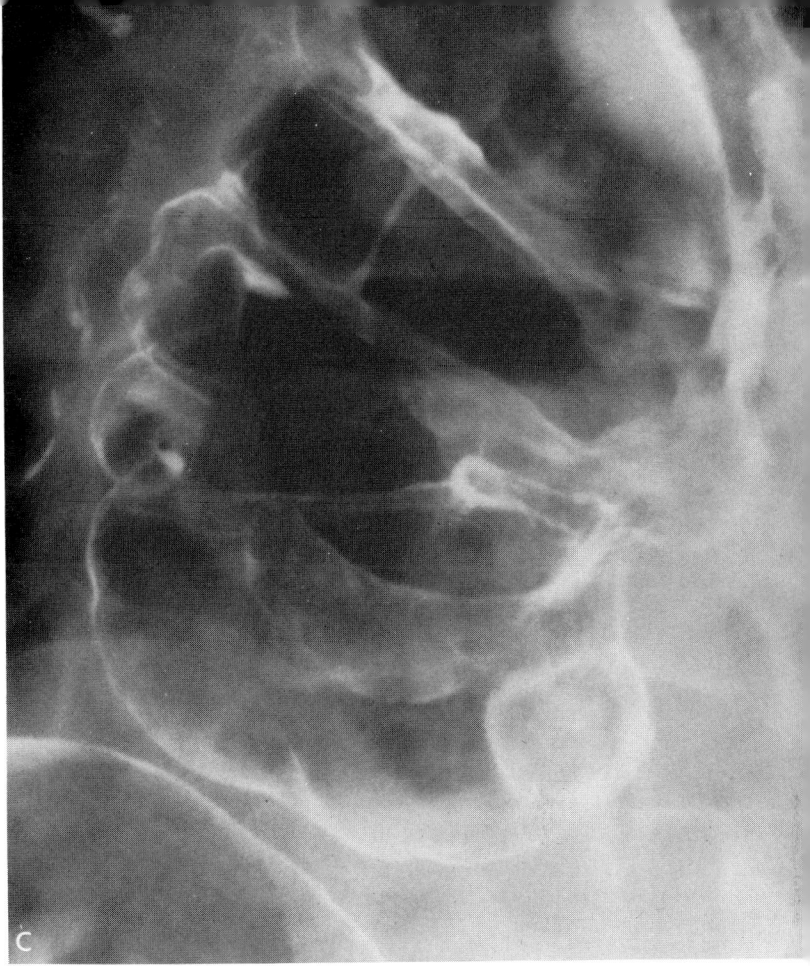

Figure 3-71. (*Continued*)

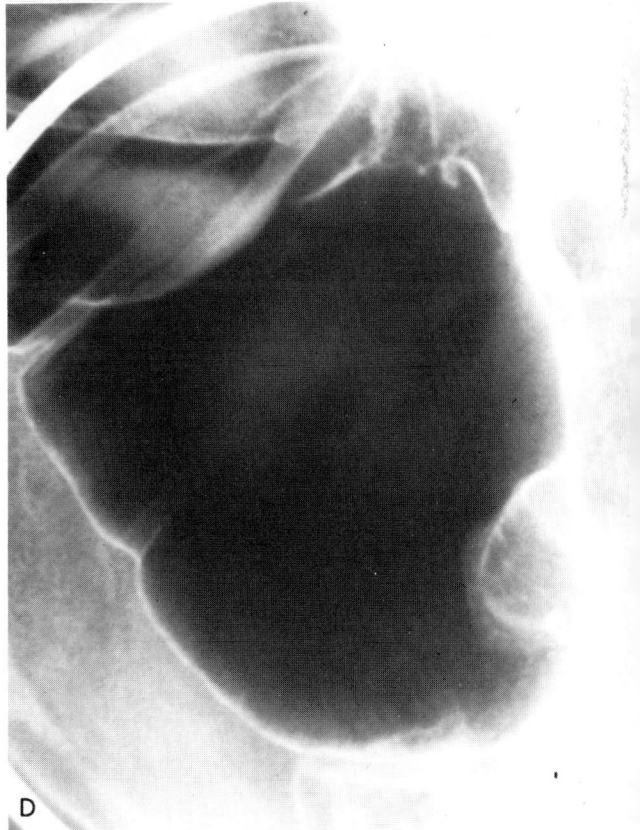

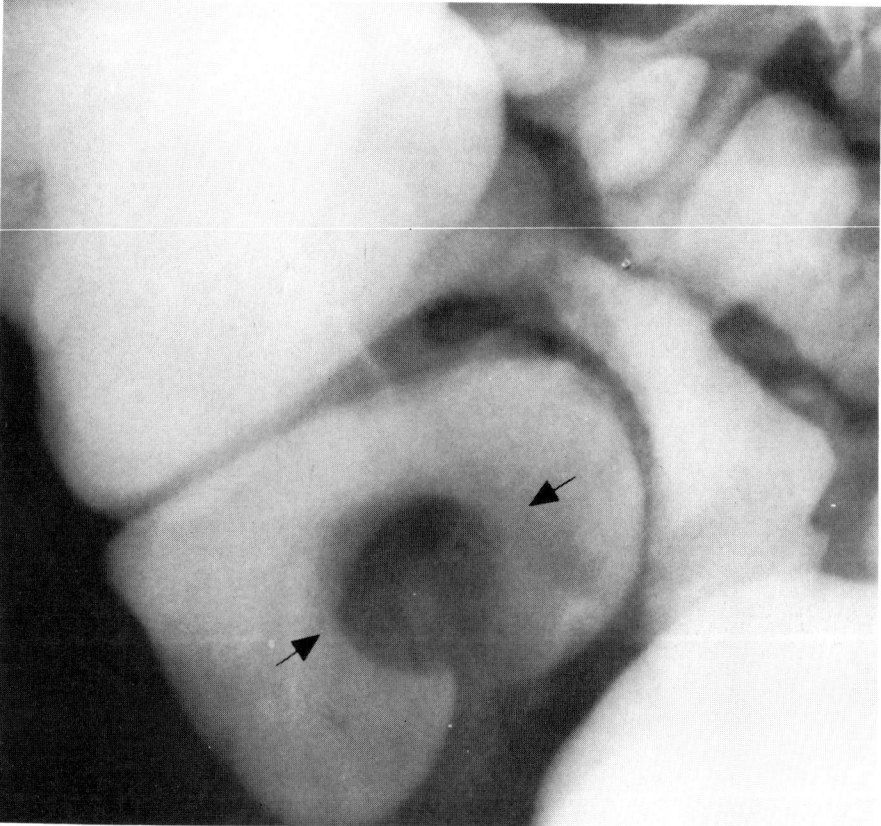

Figure 3–72. Barium enema examination showing a large inverted appendiceal stump in a patient who had had an appendectomy 3 weeks earlier. The inverted stump remains particularly prominent for several weeks after surgery owing to persistent edema and inflammation.

Figure 3-73. Barium enema (A) and air contrast (B) examination showing a large inverted appendiceal stump. When the stump is this large, the diagnosis must be established at surgery to exclude the presence of a significant lesion.

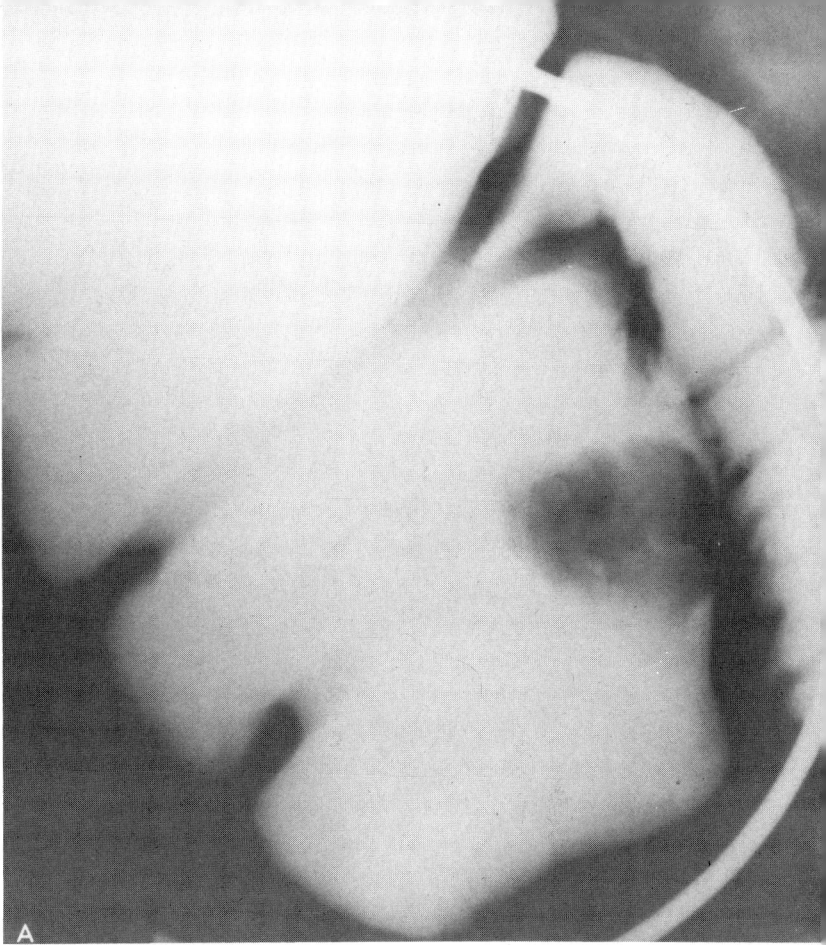

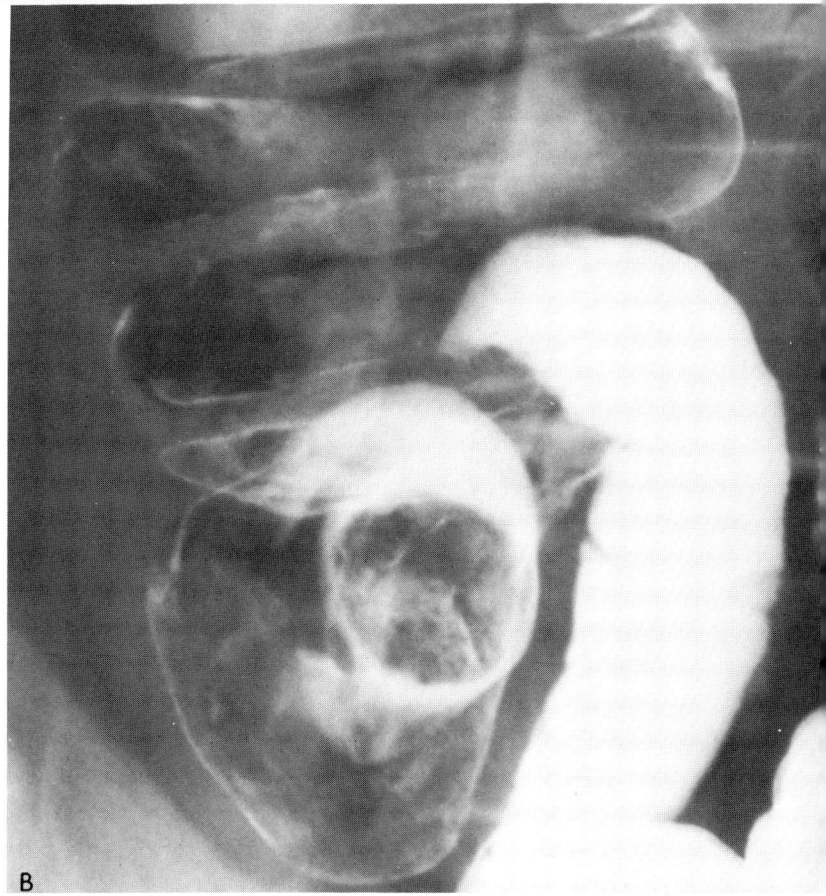

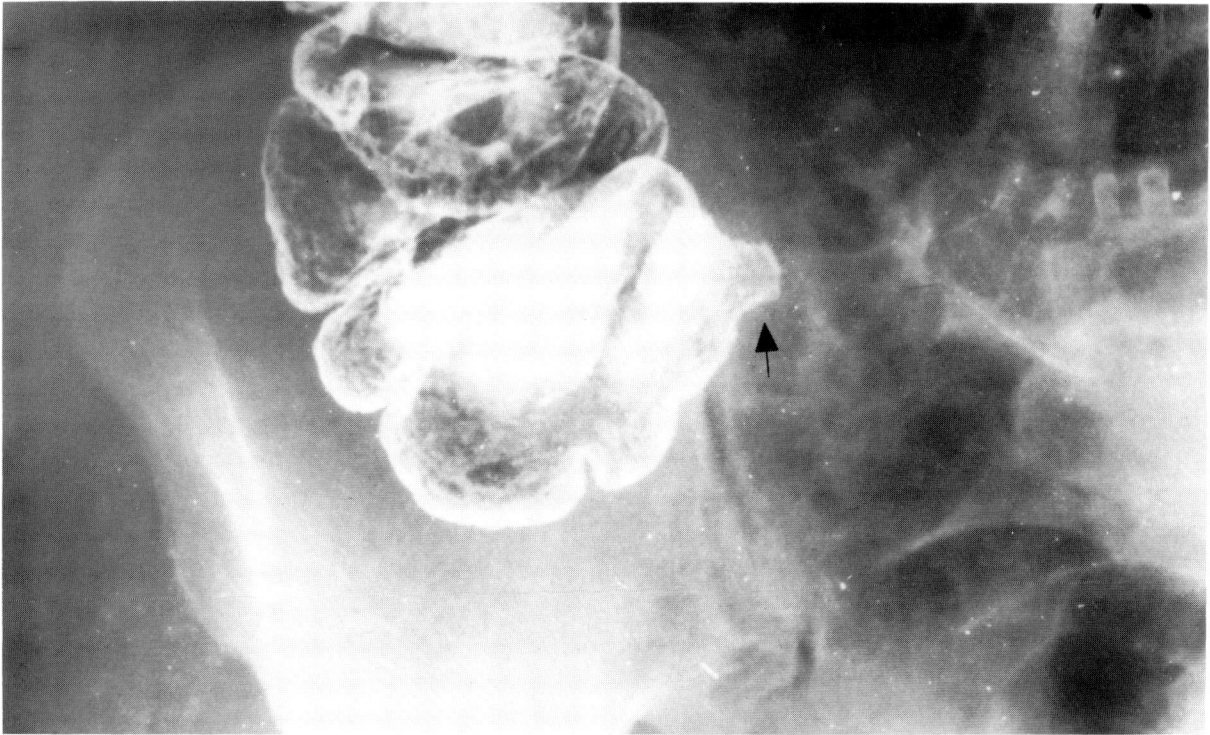

Figure 3–74. Air contrast barium enema showing an appendiceal remnant due to incomplete removal at appendectomy. (From Beneventano, T. C.: The roentgen aspects of some appendiceal abnormalities. Amer. J. Roent., 96:344, 1966. Courtesy of Charles C Thomas, Publisher.)

visualizing the radiologic features of the defect.

The importance of recognizing the defect lies in the difficulty in differentiating an inverted appendiceal stump from a significant lesion. The inverted stump rarely causes symptoms, although cases of ulceration and intussusception have been reported.

The inverted stump must be differentiated from an adenomatous polyp, a villous adenoma, adenocarcinoma, lipoma and carcinoid. If there is a history of a previous appendectomy and if the cecal defect is in the location of the appendix, it may be possible to make a radiologic diagnosis of an inverted stump with reasonable certainty. However, if the lesion is large or irregular, colonoscopy or surgery may be required for an accurate diagnosis (Fig. 3–73).

When the appendix has been incompletely removed, a short stump may fill with barium on barium enema examination (Fig. 3–74). The history of a previous appendectomy will differentiate incomplete resection of the appendix from incomplete filling due to appendicitis. Recurrent appendicitis may occur following subtotal appendectomy. In this circumstance the radiologic findings are identical to those in patients with appendicitis without previous appendectomy.

INTUSSUSCEPTION

Primary appendiceal intussusception may present as an acute surgical emergency, without time for preoperative diagnostic studies, or as a subacute recurring condition. In a third variety, there is asymptomatic intussusception of the appendix noted during a barium enema examination.[45] In these cases the intussusception is infrequent, only about 125 cases having been reported since the first description published in 1858.[45] The correct preoperative radiographic diagnosis was not made until 1970 when Hill et al. reported two patients with completely invaginated appendices, in each of whom the diagnosis was recognized on preoperative barium enema examinations.[46]

Bachman and Clemett reported four cases

characterized by barium enema demonstration of an incomplete appendiceal intussusception which reduced itself completely during the course of the same examination or was reduced at a subsequent barium enema study[45] (Figs. 3–75 and 3–76). After the reduction occurred, a normal appendix and cecum were visualized. In all cases the patients had no symptoms related to the intussusception. In several of these cases the radiographic appearance at the time of the invagination was interpreted as a cecal tumor and surgery was advised. After the intussusceptions were reduced on subsequent examination, the cecum and appendix were seen to be entirely normal in each case.

Bachman and Clemett's own experience with seven cases suggests that appendiceal intussusception occurs more often than has been previously described.[45] Fraser[47] classified the intussusception in 82 cases as follows: Intussusception of the appendix into itself (2 per cent); intussusception of the appendix into the cecum (partial, 35 per cent; complete, 10 per cent; data incomplete, 9 per cent); compound intussusception (39 per cent); and appendiceal stump inversion (5 per cent). Of those that are not compound, by far the most common variety is partial inversion of the appendix into the cecum.

Increased and abnormal peristalsis of the appendix may be a factor in the intussusception and this may be predicated on the presence of intraluminal abnormalities (foreign bodies, fecaliths, polyps and parasites) and intramural diseases (mucocele, tumor, endometriosis and lymphoid follicles).[45] The cause of the intussusception in the asymptomatic variety with a normal appendix is unknown.

The appearance of appendiceal intussusception on the barium enema study consists of an oval or round intramural filling defect arising from the medial wall of the cecum with no visualization of the appendix[45, 48, 49] (Figs. 3–77 to 3–79). A finger-like intraluminal filling defect arising from the medial wall of the cecum is virtually pathognomonic. It is important to determine whether the defect is on the same side of the cecum as the ileocecal valve, since the normal origin of the appendix is always between the apex of the cecum and the valve on the same side of the cecum as the valve.[45] Thus, if the defect is opposite the side of the ileocecal valve, it probably does not arise from the appendix.

The differential diagnosis of appendiceal intussusception includes both mucosal and intramural cecal defects due to a variety of causes such as mural and mucosal tumors of the cecum, appendiceal abscesses and inverted appendiceal stumps.[50]

TUMORS

Neoplasms of the appendix are rarely diagnosed preoperatively because they present with no specific symptom complex. In one-half of the reported cases, the patients were initially thought to have appendicitis.[51] Fifteen per cent were discovered incidentally during routine histologic examination of the resected appendix.

Tumors of the appendix are exceedingly rare. Collins reported an incidence of carcinoid (0.50 per cent), adenocarcinoma (0.08 per cent) and benign tumors (4.30 per cent).[3]

Benign tumors consist of leiomyomas, neuromas and lipomas and are rarely diagnosed by barium studies because of their small size. The diagnosis is usually made as an incidental finding in appendices removed at surgery.

The three malignant appendiceal tumors are carcinoid, "malignant mucocele" and adenocarcinoma. Carcinoids of the appendix rarely metastasize or cause the carcinoid syndrome.[52] The tumor is usually small and the diagnosis is made microscopically. While mucoceles are generally considered to be benign lesions, some authors believe that mucoceles are in fact mucinous cystadenomas and that when pseudomyxoma peritonei occurs it is caused by peritoneal metastases from mucinous cystadenocarcinoma.[36]

Adenocarcinoma of the appendix is difficult to diagnose radiographically because the appearance is simulated by a variety of other diseases in the right lower quadrant such as an inverted appendiceal stump, mucocele, appendiceal abscess and carcinoma of the cecum[53] (Figs. 3–80 and 3–81). Carcinoma of the appendix usually arises in the distal one-third of the appendix and frequently results in obstruction of the lumen and acute appendicitis. The cecum is

(*Text continued on page 176*)

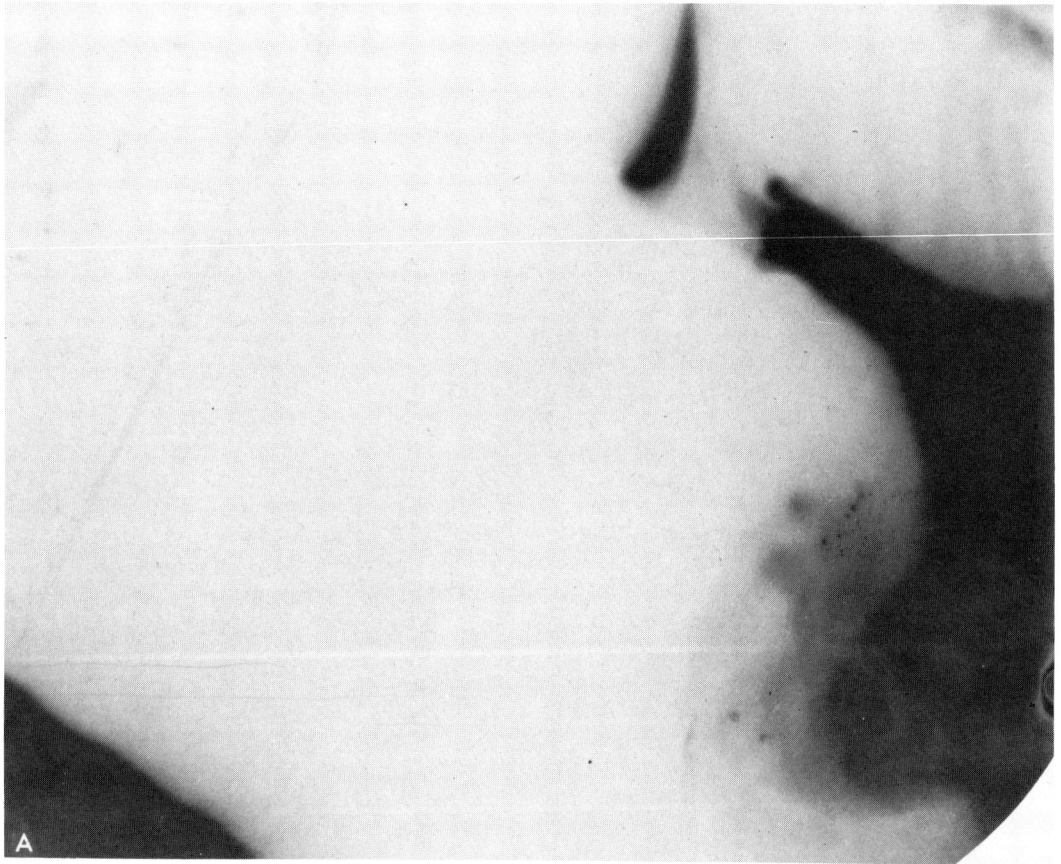

Figure 3–75. Barium enema examination (A) showing a rounded, large indentation in the medial wall of the cecum just below the ileocecal valve due to intussusception of the appendix in an asymptomatic 64-year-old female. (*Legend continued on opposite page.*)

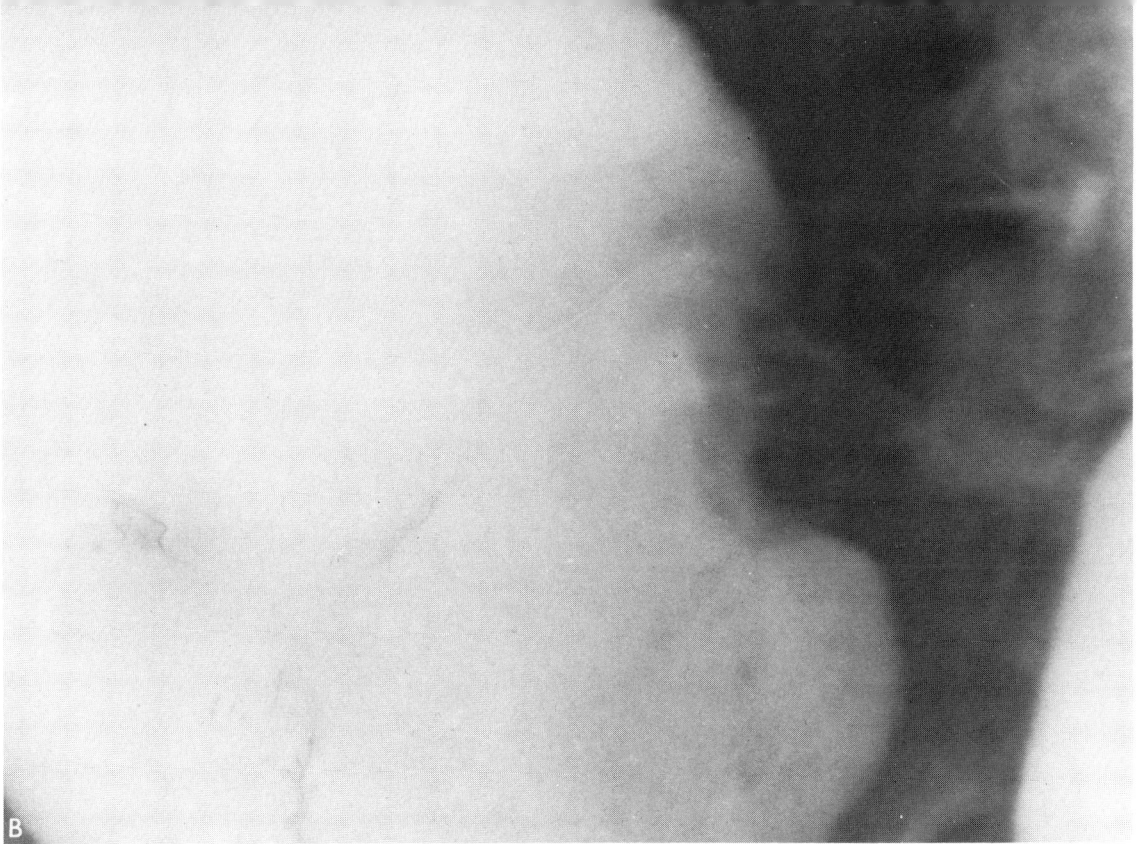

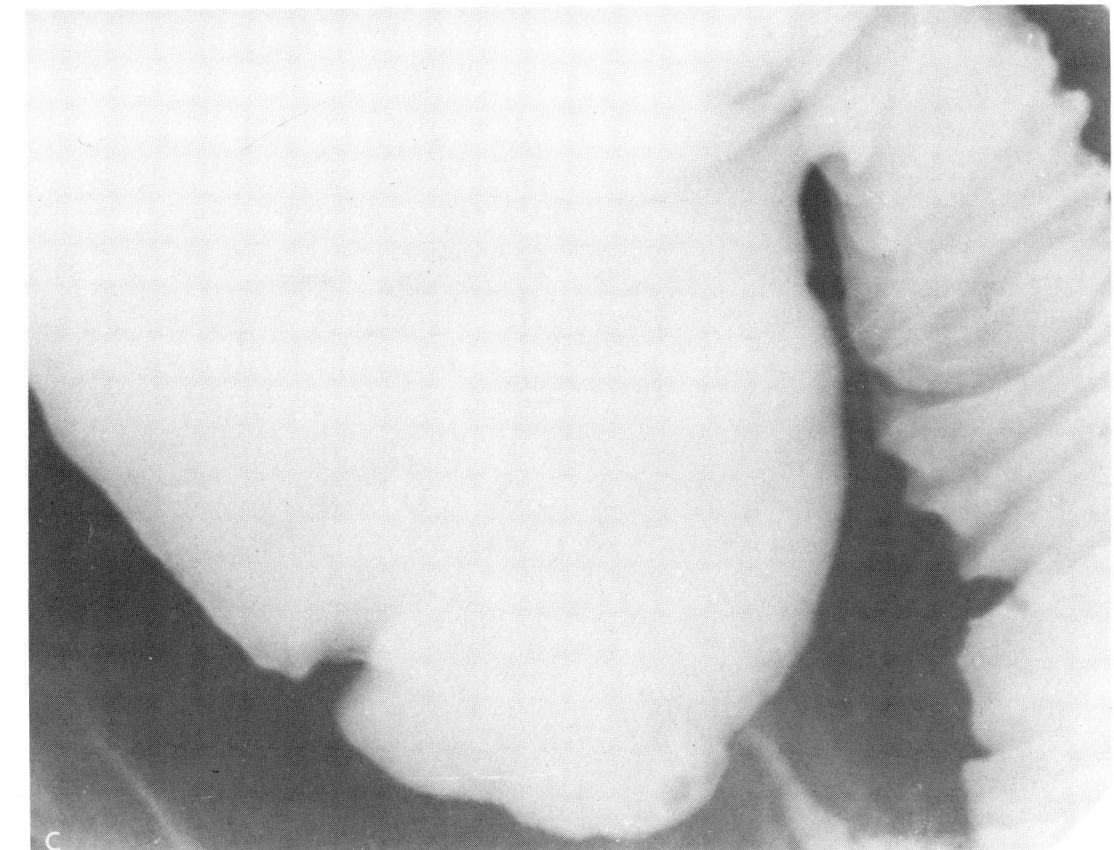

Figure 3–75. *(Continued).* *(B)* The indentation is now less pronounced; visualization of the appendix is becoming possible. *(C)* The indentation is now completely gone. The cecal wall appears normal, and a normal appendix is visualized. (From Bachman, A. L., and Clemett, A. R.: Roentgen aspects of primary appendiceal intussusception. Radiology, *101*:531, 1971.)

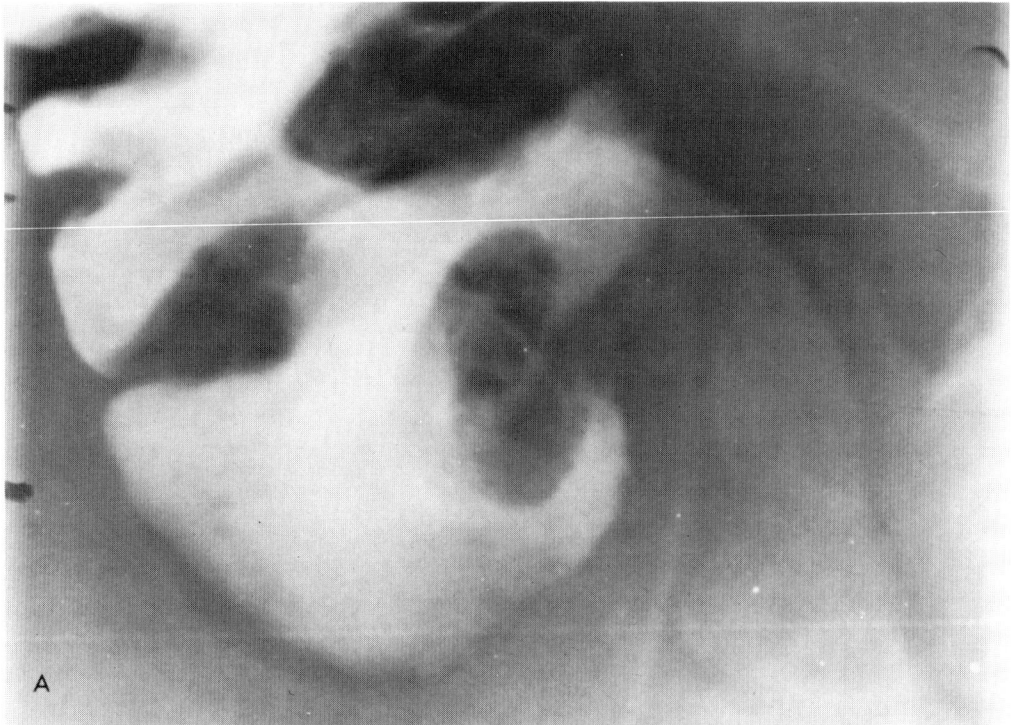

Figure 3–76. An ovoid, large smooth marginal defect is evident on barium enema examination (A) in the medial wall of the cecum below the ileocecal valve owing to asymptomatic intussusception of the appendix in a 62-year-old man. On the air-contrast examination (B) the medial cecal wall appears normal as the intussusception has been reduced in the course of the examination. The appendix arises from the region in which the defect was observed early in the examination (arrows). (From Bachman, A. L., and Clemett, A. R.: Roentgen aspects of primary appendiceal intussusception. Radiology, 101: 531, 1971.)

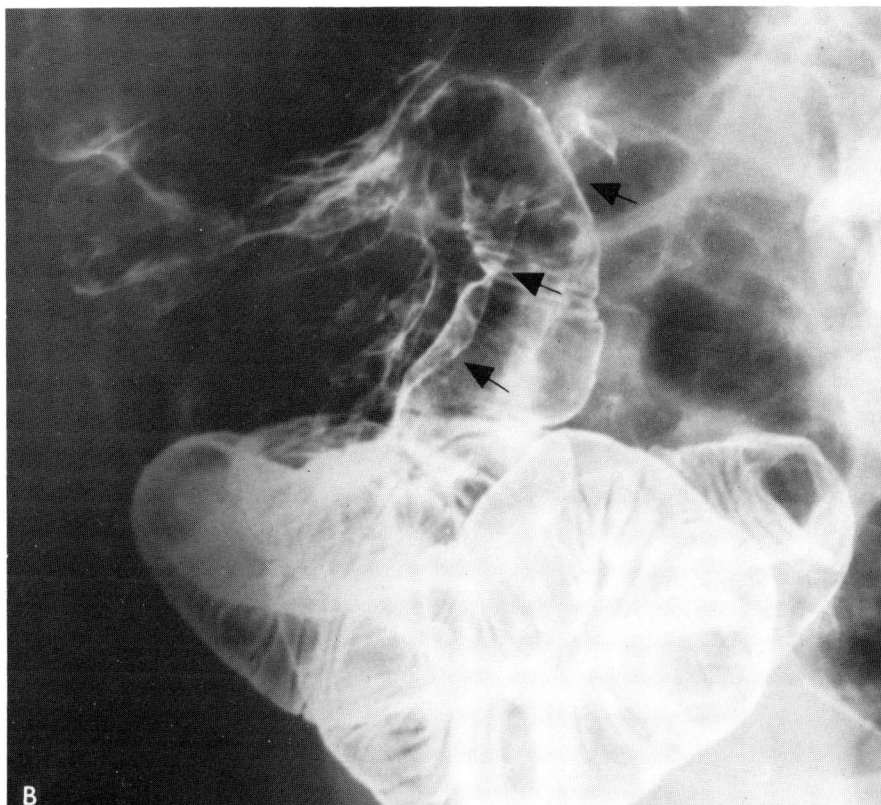

Figure 3–76. (Continued)

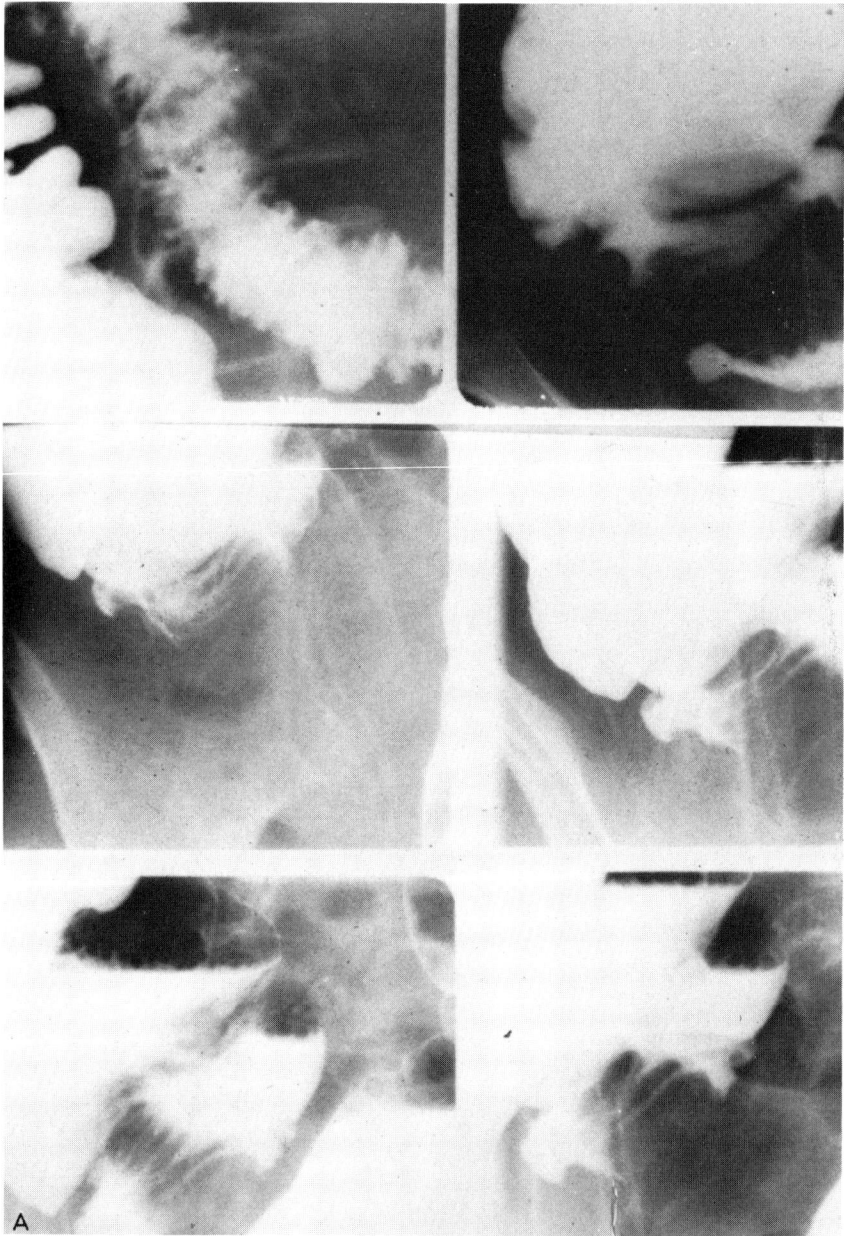

Figure 3–77. Multiple spot radiographs of the cecum made during a barium enema examination show a finger-like filling defect extending into the cecal lumen from the medial wall of the cecum near the apex (A). An air contrast study (B) shows a thick, finger-like soft-tissue density projecting into the cecum. At operation a completely invaginated and inverted appendix containing deposits of endometriosis was removed. (From Bachman, A. L., and Clemett, A. R.: Roentgen aspects of primary appendiceal intussusception. Radiology, *101*:531, 1971.)

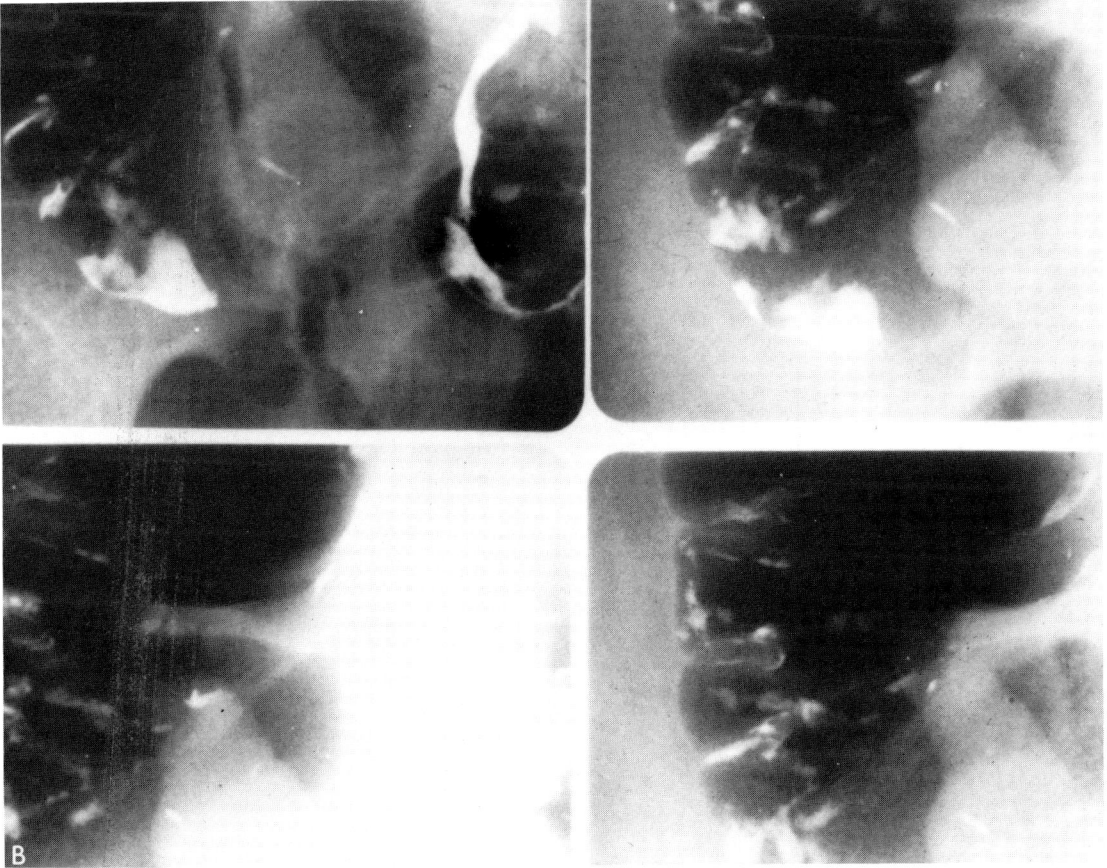

Figure 3-77. (*Continued*)

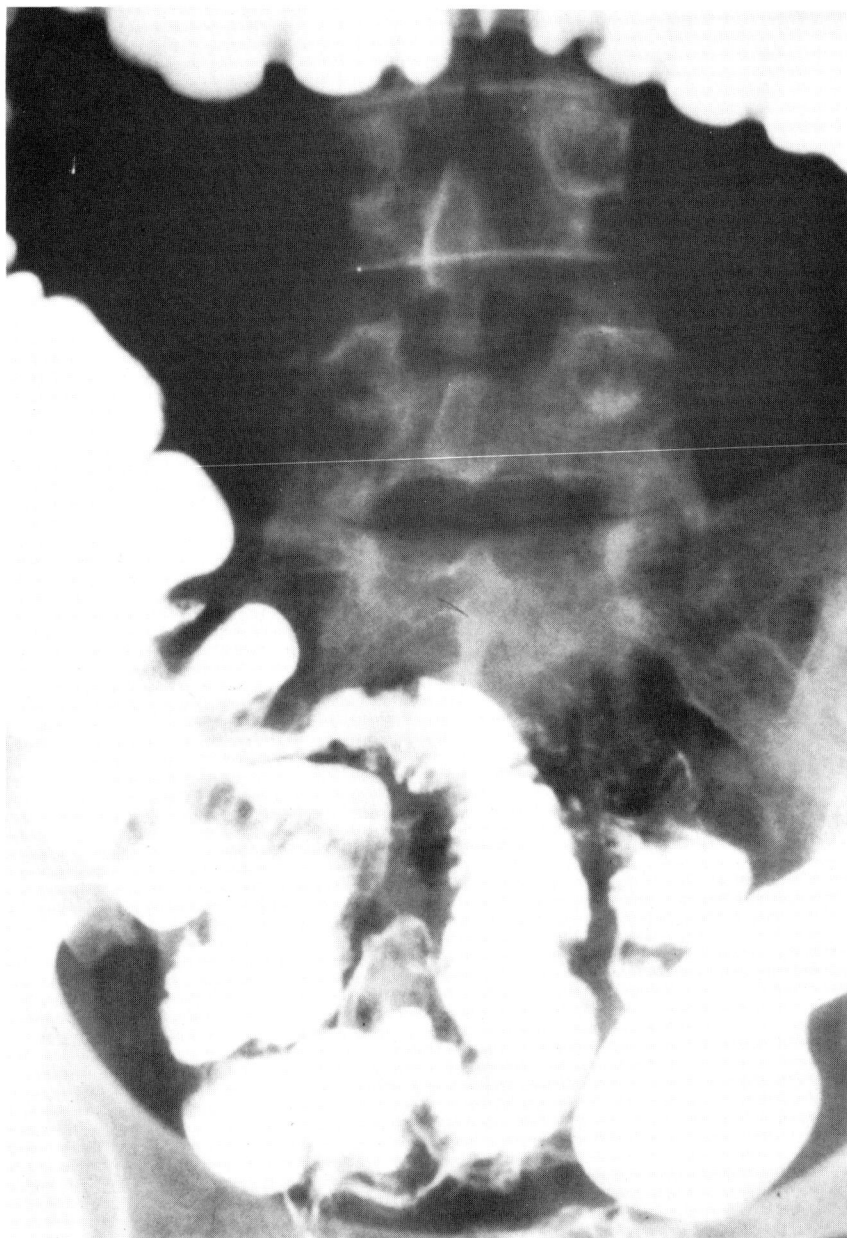

Figure 3–78. Barium enema examination showing a finger-like defect extending into the cecal lumen from the medial wall below the ileocecal valve. Surgery revealed a completely invaginated appendix projecting into the cecum. Histologically, the appendix was normal. (From Bachman, A. L., and Clemett, A. R.: Roentgen aspects of primary appendiceal intussusception. Radiology, *101*:531, 1971.)

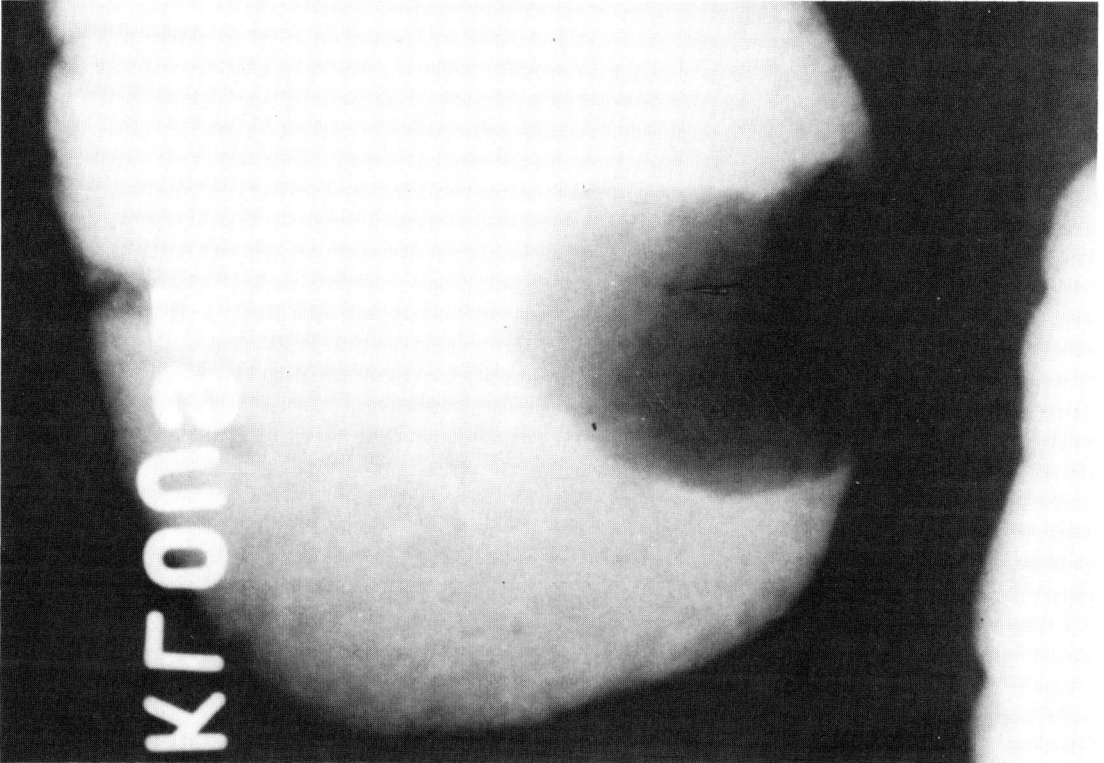

Figure 3–79. Barium enema examination showing a finger-like mass projecting into the cecum in the region of the appendix. At surgery an inverted, thickened appendix with endometrial implants was found and resected.

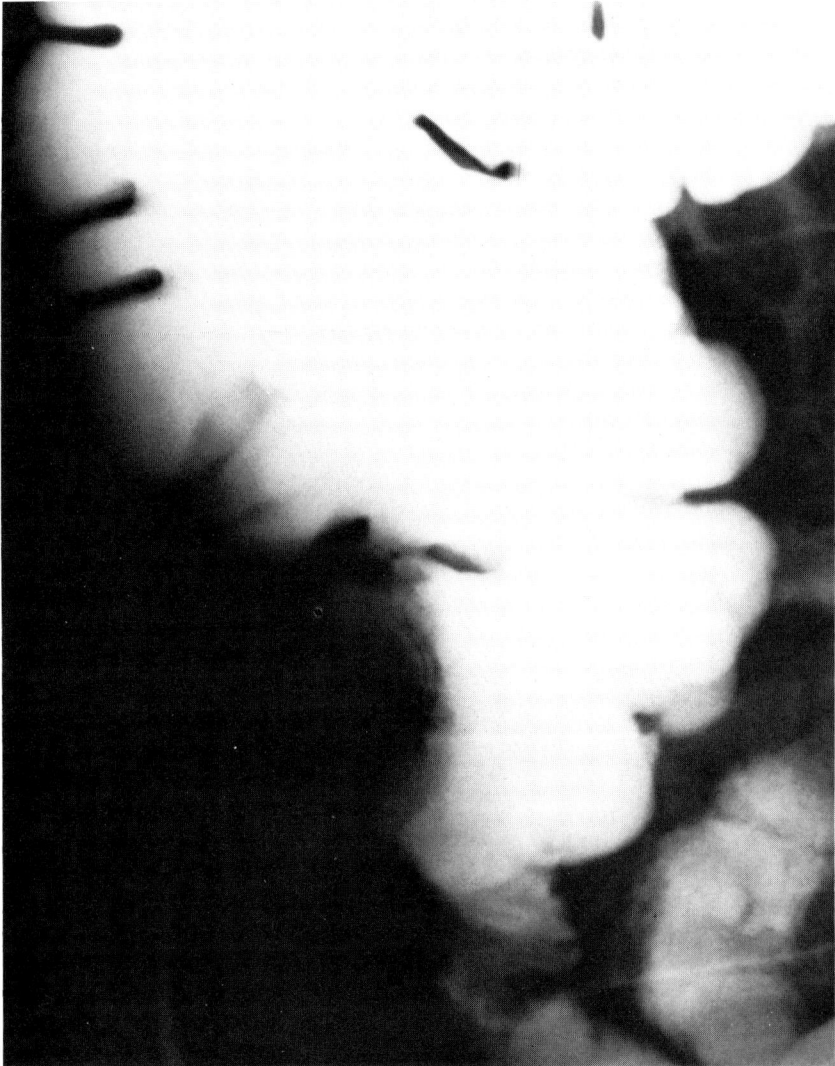

Figure 3–80. Barium enema examination showing an extrinsic mass displacing the cecum and ascending colon along the lateral and posterior aspects. At surgery, carcinoma of the appendix with extensive metastases to regional nodes and the liver was identified.

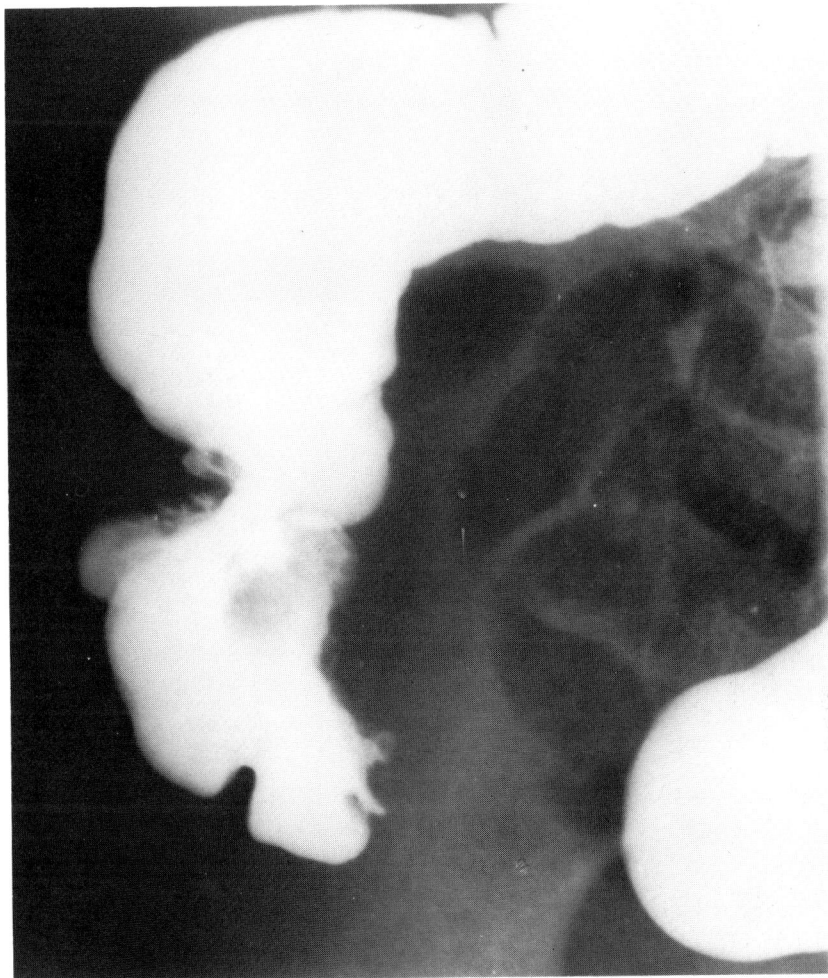

Figure 3–81.   Barium enema examination showing deformity of the cecum by a mass which proved to be an adenocarcinoma of the appendix extending into the cecum. The cecal mucosa was intact. (From Beneventano, T. C., et al.; The roentgen aspects of some appendiceal abnormalities. Amer. J. Roent., 96:344, 1966. Courtesy of Charles C Thomas, Publisher.)

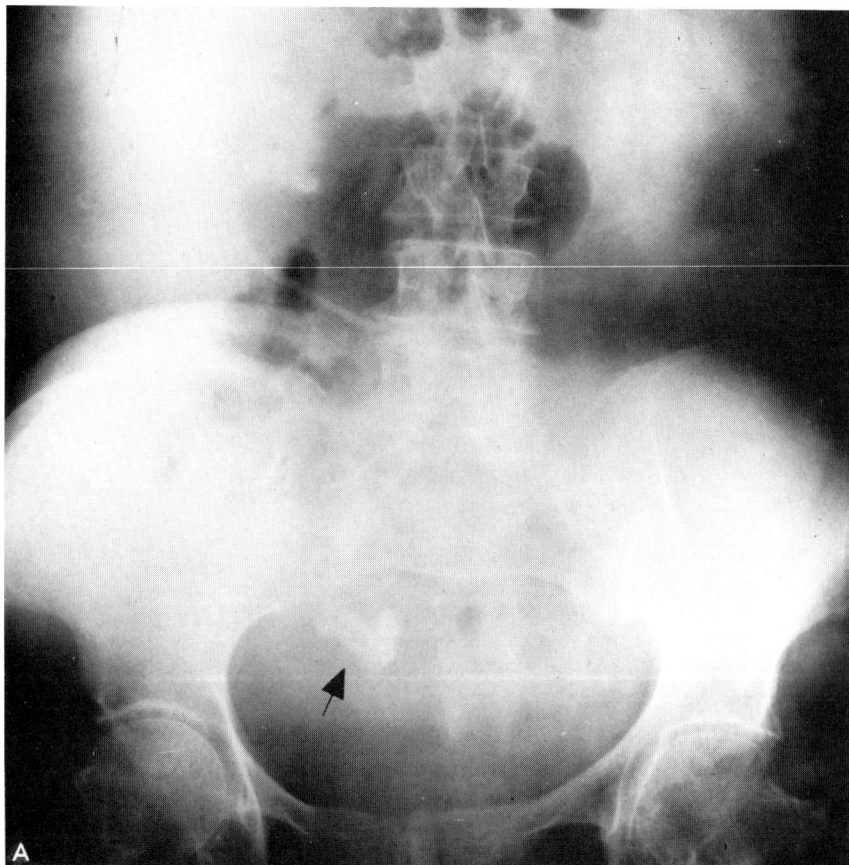

Figure 3–82.   Plain abdominal radiograph (A) showing calcification in an adenocarcinoma of the appendix. Radiograph of the appendix at postmortem examination (B) shows the calcification. (From Beneventano, T. C., et al.: The roentgen aspects of some appendiceal abnormalities. Amer. J. Roent., 96:344, 1966. Courtesy of Charles C Thomas, Publisher.)

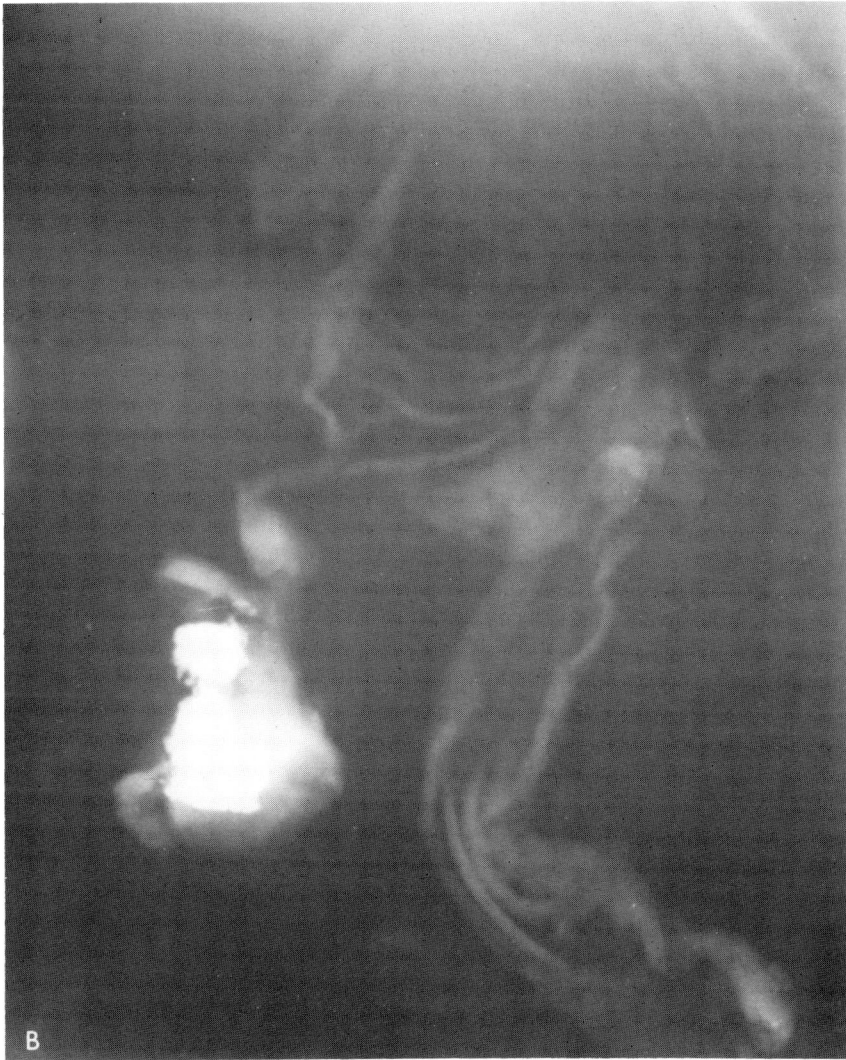

Figure 3–82. (*Continued*)

often invaded by the tumor. Calcification may sometimes be detected within the tumor on plain abdominal radiographs (Fig. 3–82).

REFERENCES

1. Beneventano, T. C., Schein, C. J., and Jacobson, H. G.: The roentgen aspects of some appendiceal abnormalities. Amer. J. Roent., 96:344, 1966.
2. Treves, F.: Anatomy of intestinal canal and peritoneum in man. Brit. Med. J., 1:470, 1885.
3. Collins, D. C.: Seventy-one thousand human appendix specimens, final report summarizing forty years study. Amer. J. Proctol., 14:365, 1963.
4. Felson, B., and Wiot, J. F.: Some interesting right lower quadrant entities, myxoglobulosis of the appendix, ileal prolapse, diverticulitis, lymphoma and endometriosis. Radiol. Clin. N. Amer., 7:83, 1969.
5. Weiner, M. E., and Jenkinson, E. L.: Diverticula of the appendix. Amer. J. Roent., 78:679, 1957.
6. Joffe, N.: Some uncommon roentgenologic findings associated with acute perforative appendicitis. Radiology, 110:301, 1974.
7. Balch, C. M., and Silver, D.: Foreign bodies in the appendix, report of eight cases and review of literature. Arch. Surg., 102:14, 1971.
8. Barnes, B. A., Behringer, G. E., Wheelock, F. C., and Wilkins, E. W.: Treatment of appendicitis at the Massachusetts General Hospital (1937–1959). J.A.M.A., 180:122, 1962.
9. Wangensteen, O. H.: Intestinal Obstructions: Physiological, Pathological and Clinical Considerations. 3rd ed. Springfield, Illinois, Charles C Thomas, 1955.
10. Frimann-Dahl, J.: Roentgen Examinations in Acute Abdominal Diseases. Springfield, Illinois, Charles C Thomas, 1960.
11. Soter, C. S.: The contribution of the radiologist to the diagnosis of acute appendicitis. Semin. Roent., 8:375, 1973.
12. Soteropoulos, C., and Gilmore, J. H.: Roentgen diagnosis of acute appendicitis. Radiology, 71:246, 1958.
13. Graham, A. D., and Johnson, H. F.: The incidence of radiographic findings in acute appendicitis compared to 200 normal abdomens. Milit. Med., 131:272, 1966.
14. Faegenburg, D.: Fecaliths of the appendix: incidence and significance. Amer. J. Roent., 89:752, 1963.
15. Felson, B., and Bernhard, M.: The roentgenologic diagnosis of appendiceal calculi. Radiology, 49:178, 1947.
16. Soter, C. S.: The contribution of the radiologist to the diagnosis of acute appendiceal disease. Med. Radiogr. Photogr., 45:2, 1969.
17. Maver, M. E., and Wells, H. G.: Composition of appendiceal concretions. Arch. Surg., 3:439, 1921.
18. Brady, B. M., and Carroll, D. S.: Significance of calcified appendiceal enterolith. Radiology, 68:648, 1957.
19. Gubler, J. A., and Kukral, A. J.: Barium appendicitis. J. Internat. Coll. Surgeons, 21:379, 1954.
20. Vukmer, G. J., and Trummer, M. J.: Barium appendicitis. Arch. Surg., 91:630, 1965.
21. Gammill, S. L., and Nice, C. M.: Air fluid levels, their occurrence in normal patients and their role in the analysis of ileus. Surgery, 71:771, 1972.
22. Young, B. R.: Significance of regional or reflex ileus in the roentgen diagnosis of cholecystitis, perforated ulcer, pancreatitis and appendiceal abscess as determined by survey examination of the acute abdomen. Amer. J. Roent., 78:581, 1957.
23. Melamed, M., Melamed, J. L., and Rabushka, S. E.: Appendicitis: "Functional" bowel obstruction associated with perforation of the appendix. Amer. J. Roent., 99:112, 1967.
24. Casper, R. B.: Fluid in the right flank as a roentgenographic sign of acute appendicitis. Amer. J. Roent., 110:352, 1970.
25. Fisher, M. S.: A roentgen sign of gangrenous appendicitis. Amer. J. Roent., 81:637, 1959.
26. McCort, J. J.: Extra-alimentary gas in perforated appendicitis. Amer. J. Roent., 84:1087, 1960.
27. Altemeier, W. A., Culbertson, W. R., Fullen, W. D., and Shook, C. D.: Intra-abdominal abscesses. Amer. J. Surg., 125:70, 1973.
28. Weens, H. S.: Gas formation in abdominal abscesses, a roentgen study. Radiology, 47:107, 1946.
29. Caruso, R. D., and Berk, R. N.: The fuzzy fluid level sign. Radiology, 98:369, 1971.
30. Schey, W. L.: Use of barium in the diagnosis of appendicitis in children. Amer. J. Roent., 118:95, 1973.
31. Dietz, W. W.: Fallacy of roentgenologically negative appendix. J.A.M.A., 208:1495, 1969.
32. Figiel, L. S., and Figiel, S. J.: Barium examination of cecum in appendicitis. Acta Radiol., 57:469, 1962.
33. Threatt, B., and Appelman, H.: Crohn's disease of the appendix presenting as acute appendicitis. Radiology, 110:313, 1974.
34. Meyers, M. A., and Oliphant, M.: Ascending retrocecal appendicitis. Radiology, 110:295, 1974.
35. Marshak, R. H., and Gerson, A.: Mucocele of the appendix. Amer. J. Dig. Dis., 5:49, 1960.
36. Gibbs, N. M.: Mucinous cystadenoma and cystadenocarcinoma of the vermiform appendix with particular reference to mucocele and pseudomyxoma peritonei. J. Clin. Path., 26:413, 1973.
37. Grodinsky, M., and Rubnitz, A. S.: Mucocele of the appendix and pseudomyxoma peritonei. A clinical review and experimental study with case report. Surg. Gynec. Obstet., 73:345, 1941.
38. Lubin, J., and Berle, E.: Myxoglobulosis of the appendix, Report of two cases. Arch. Path., 94:533, 1972.
39. Milliken, G., and Poindexter, C. A.: Mucocele of the appendix with globoid body formation. Amer. J. Path., 1:397, 1925.
40. Cagnetto, G.: Über einen eigentumlichen Befund bei Appendizitis. Virch. Arch. Path. Anat., 198:193, 1909.

41. Pohl, W.: Ein eigentümlicher Befund in der Appendix, Dtsch. Z. Chir., *126*:201, 1913.
42. Ghosh, B. C., Huvos, A. G., and Whiteley, H. W.: Pseudomyxoma peritonei. Dis. Colon Rectum, *15*:420, 1972.
43. Limber, G. K., King, R. E., and Silverberg, S. C.: Pseudomyxoma peritonei, a report of ten cases. Ann. Surg., *178*:587, 1973.
44. Freedman, E., Rabwin, M. H., and Linsman, J. F.: Roentgen simulation of polypoid neoplasms by invaginated appendiceal stumps. Amer. J. Roent., *75*:380, 1956.
45. Bachman, A. L., and Clemett, A. R.: Roentgen aspects of primary appendiceal intussusception. Radiology, *101*:531, 1971.
46. Hill, B. J., Schmidt, K. D., and Economou, S. G.: The "inside-out" appendix: a review of the literature and report of 2 cases. Radiology, *95*:613, 1970.
47. Fraser, K.: Intussusception of the appendix. Brit. J. Surg., *31*:23, 1943.
48. Toa, H., and Dunbar, J. S.: Intussusception of the appendix. J. Canad. Assoc. Radiol., *22*:33, 1971.
49. Gorske, K.: Intussusception of the proximal appendix into the colon. Radiology, *91*:791, 1968.
50. Howard, R. J., Ellis, C. M., and Delaney, J. P.: Intussusception of the appendix simulating carcinoma of the cecum. Arch. Surg., *101*:520, 1970.
51. Otto, R. E., Ghislandi, E. V., Lorenzo, G. A., and Conn, J.: Primary appendiceal adenocarcinoma. Amer. J. Surg., *120*:704, 1970.
52. Ponka, J. L.: Carcinoid tumors of the appendix, report of 35 cases. Amer. J. Surg., *126*:77, 1973.
53. Stiehm, W. D., and Seaman, W. B.: Roentgenographic aspects of primary carcinoma of the appendix. Radiology, *108*:275, 1973.

Chapter 4

INFLAMMATORY LESIONS OF THE ILEOCECAL AREA

Acute and chronic inflammatory processes affecting the terminal ileum, appendix, cecum and ascending colon make up the most common and challenging group of diseases that involve the ileocecal area. Appendicitis and its complications head the list of diseases in this category and are considered elsewhere in this text. In this chapter Crohn's disease, ulcerative colitis, tuberculosis, amebiasis, parasitic infestations, typhoid fever, evanescent colitis, angioneurotic edema, radiation enteritis, diverticulitis, urticaria, cecal ulcer and leukemic typhlitis will be discussed.

CROHN'S DISEASE INVOLVING THE SMALL BOWEL

Crohn's disease is a chronic, cicatrizing, panenteric lymphedema that involves the terminal ileum in the majority of cases but may involve any segment of the gastrointestinal tract.[1] Submucosal edema, mucosal ulceration, fissuring and thickening of the bowel wall and mesentery are characteristic pathologic findings.[2] The mucosa assumes a cobblestone pattern owing to intercommunicating crevices or fissures surrounding islands of mucosa elevated by edema and inflammation in the submucosa. Noncaseating sarcoid-like granulomas are present in about one-half to three-fourths of the patients. There has been a definite increase in the incidence of the disease in the past decade but regional enteritis is still eight times less common than ulcerative colitis.[3] Patients with regional enteritis have such a prolonged clinical course and are referred for radiologic studies so much more often than those with ulcerative colitis that the radiologist is apt to gain the impression that regional enteritis is at least as common as ulcerative colitis.

It is important for the radiologist to realize that patients with Crohn's disease, even in advanced stages, may have few or no symptoms. The classic symptoms of abdominal pain, diarrhea and weight loss may be minimal or absent. Consequently, the radiologist is in a unique position to establish the diagnosis early, if he maintains a high index of suspicion and performs a small bowel examination routinely or at the least clinical provocation as part of an upper gastrointestinal examination.

A similar situation exists in patients who present with recurrent anal fissures and fistulas. Careful attention to the terminal ileum during a barium enema examination or study of the ileum by performing a small bowel examination may establish the diagnosis of regional enteritis, even when the patient has no intestinal symptoms.

Regional enteritis involves patients of all ages ranging from infancy to advanced age. One of the most common causes of growth retardation in children and infants is Crohn's disease. A careful small bowel examination should be performed in all cases of delayed growth, even in the absence of gastrointestinal symptoms. The cause of the extreme growth retardation in these children has not been definitely established. It cannot be attributed to malnutrition or hypopituitarism in every case.[4]

The radiologic manifestations of Crohn's disease of the small intestine can conveniently be divided into the nonstenotic and stenotic phases of the disease.[1] Each acute episode of edema and inflammation of the bowel heals with fibrosis, so that over a period of years and after numerous attacks of abdominal pain, diarrhea and chills, the patient passes from the nonstenotic phase into the stenotic phase. The symptoms become more constant in the stenotic phase and consist of crampy abdominal pain and increased peristalsis, indicating partial intestinal obstruction.

The earliest radiologic manifestations of Crohn's disease of the small bowel are altered mucosal architecture and irritability of the bowel (Figs. 4–1 to 4–6). The normal, crisp, lacelike mucosal pattern of the terminal ileum becomes indistinct, coarse, nodular, thickened and shaggy. Marginal irregularities indicate superficial ulcerations. At this stage of the disease, differentiation from the normal or from lymphoid hyperplasia may be difficult on a radiologic basis alone. As the inflammation becomes more severe, typical mucosal changes occur. The normal mucosal pattern is completely replaced by a shaggy, mottled, nodular appearance, with increased irritability associated with tenderness on palpation. Involved areas more proximal in the bowel may be identified, indicating that the disease is segmental with "skip" areas (Fig. 4–7). The entire circumference of the bowel may not be affected, so that a portion of the bowel wall remains distensible and produces typical sacculations. Exudate and abnormal fluid in the lumen flocculate and dilute the barium, producing a foamy, gray radiographic pattern. If the disease is severe, fistulas to other small bowel loops, the cecum, the sigmoid and the bladder or into the mesentery may be identified (Figs. 4–8 and 4–9). The cecum is often involved with the disease, although it is sometimes deformed only indirectly because of its proximity to the inflamed ileum (Fig. 4–10).

In the nonstenotic phase of the disease, the bowel lumen remains normally distensible and the bowel wall is not excessively thickened, so that the ileum often has a normal radiographic appearance when it is completely full of barium. It is essential, therefore, that appropriate compression spot radiographs be made to demonstrate the mucosa with as much detail as possible; otherwise the diagnosis will be overlooked. In some patients the appropriate projections can best be obtained during a barium enema examination, while in others this is easier on small bowel studies. When the cecum and ileum are low in the pelvis where compression is difficult, it may be helpful to distend the bladder by giving the patient several glasses of water to drink. In these cases, the full bladder may elevate the ileum and permit compression, so that appropriate compression radiographs can be obtained.

As each acute inflammatory episode heals with scarring, the intestinal lumen narrows and the patient gradually enters the stenotic phase of the disease (Figs. 4–11 to 4–16). Areas of uninvolved bowel remain distensible and produce characteristic skip areas and sacculations. The mucosa becomes atrophic in appearance. Rigidity and thickening of the bowel become pronounced until the involved intestine becomes a fixed, rigid segment, widely separated from adjacent structures by its thickened wall. The thickened root of the mesentery displaces the small intestine, creating an oval radiolucency in the plane of the mesentery which simulates an intra-abdominal abscess. Proximal dilatation may occur, but it is seldom marked, even in the presence of pronounced narrowing of the intestinal lumen. Enteroliths and foreign bodies may be identified in the intestine proximal to the obstruction.

The linear extent of the disease in the small bowel rarely if ever increases in the course of the illness prior to surgery. Usually the same length of bowel is involved late in the stenotic phase as was diseased in the prestenotic stage. Extension into the colon is not uncommon, however, and patients with disease limited to the terminal ileum initially may subsequently develop extensive granulomatous colitis (Fig. 4–17).

Recurrent ileitis after resection of the ileum and ileocolostomy has the same radiologic appearance as the initial disease and often runs the same course (Figs. 4–18 and 4–19). The width of the anastomosis should be carefully evaluated because a narrow anastomosis in the postoperative period occurs more frequently in patients with regional enteritis compared to those with ileocolic anastomosis done for other diseases (Fig. 4–20). A narrow anastomosis may produce obstructive symptoms that can be confused initially with adhesions and later with recurrent ileitis.[5] A water enema in these patients often relieves the symptoms by washing obstructed food particles through the anastomosis. In most cases after surgery dilatation of the small bowel adjacent to the anastomosis occurs and should not be considered evidence of recurrent disease.

The complications of regional enteritis
(Text continued on page 206)

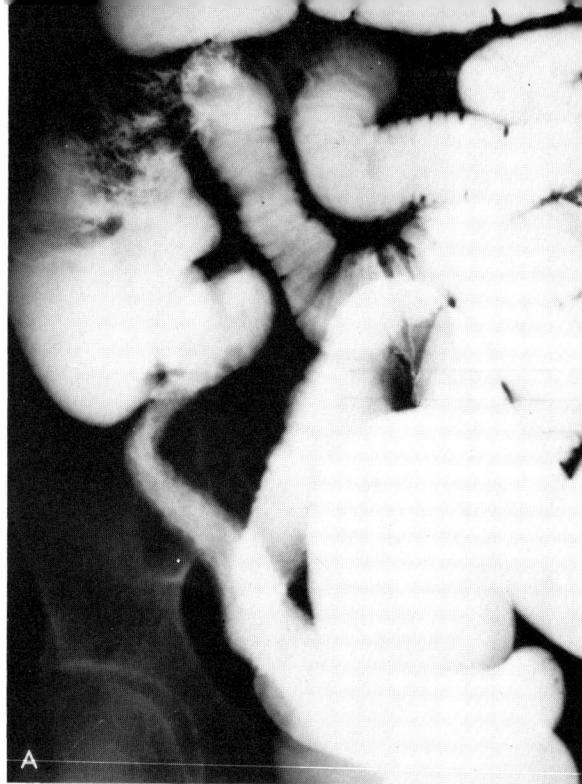

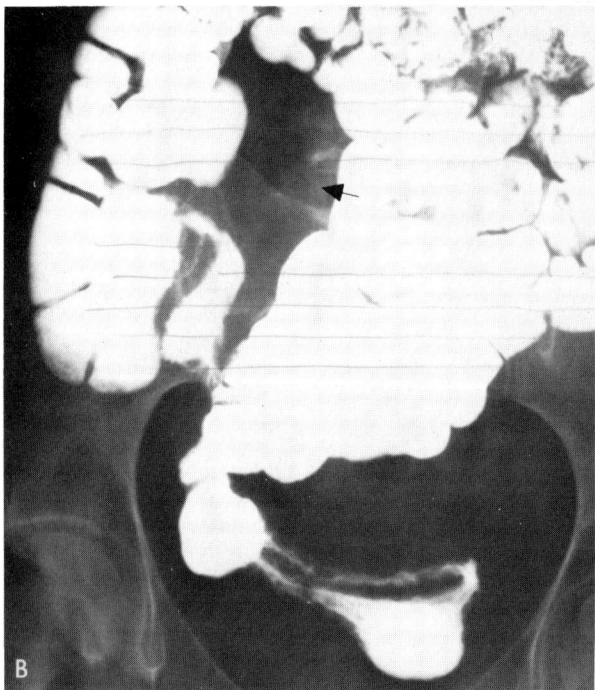

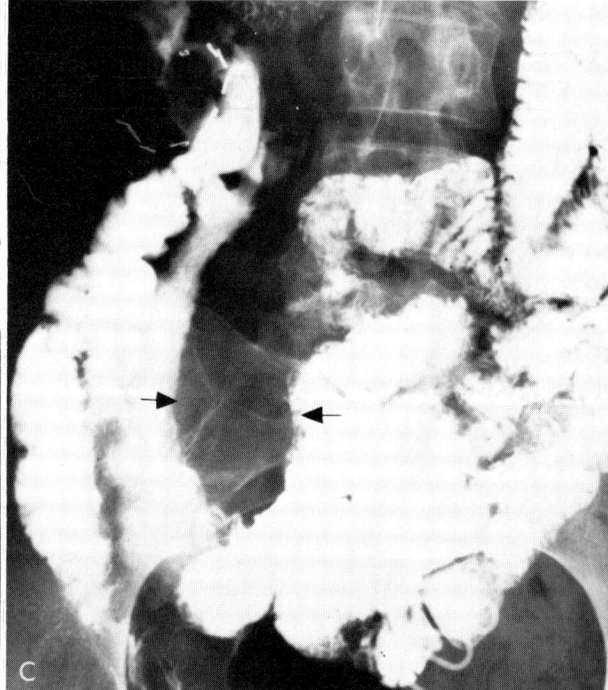

Figure 4–1. Barium enema examination (*A*) in a 19-year-old female with intermittent abdominal pain for 4 years. The examination was performed 6 months after a diagnosis of terminal regional enteritis was made at laparotomy performed because of a mistaken preoperative diagnosis of appendicitis. The shaggy coarse mucosa of the terminal ileum is typical of Crohn's disease in the nonstenotic phase.

Small bowel examination (*B*) in the same patient done 4 years later. The disease has not extended in length but has progressed to the stenotic stage. Irregular involvement of the circumference of the ileum has produced characteristic pseudodiverticula. There is thickening of the root of the mesentery as shown by displacement of the intestine (arrow).

Small bowel examination (*C*) in the same patient done one year after an end-to-side ileotransverse colostomy was performed because of chronic obstruction. The metallic clips in the right upper quadrant are on the blind end of the transverse colon. The mucosal pattern of the "new" terminal ileum resembles the findings on the original examination (*A*) and indicates recurrent ileitis. The thickened root of the mesentery is still evident (arrows). A loop of normal small bowel is adherent to the right lateral aspect of the diseased segment.

180

Figure 4–2. Barium enema examination (*A*) in a 24-year-old man who saw his physician because of an anal fistula. He had no gastrointestinal symptoms. The mucosal pattern of the terminal ileum indicates the presence of edema, nodularity and ulceration, which are typical of regional enteritis. Barium enema examination (*B*) in the same patient 7 years later. The disease has not extended in length and has not progressed to the stenotic phase. Marked nodularity and smudging of the barium indicates persistent inflammatory changes of Crohn's disease in the terminal ileum. Progression to obstruction is not invariable.

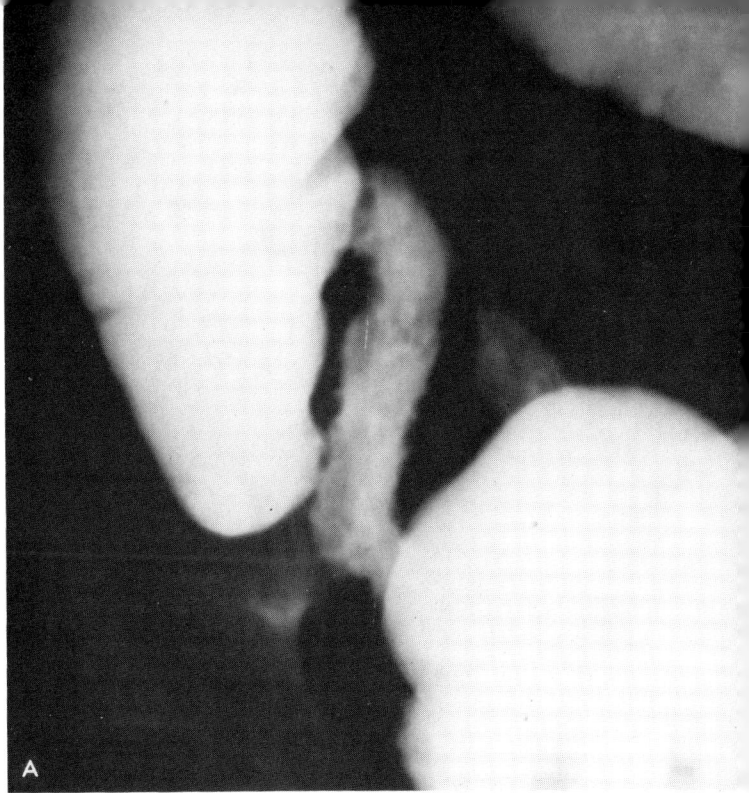

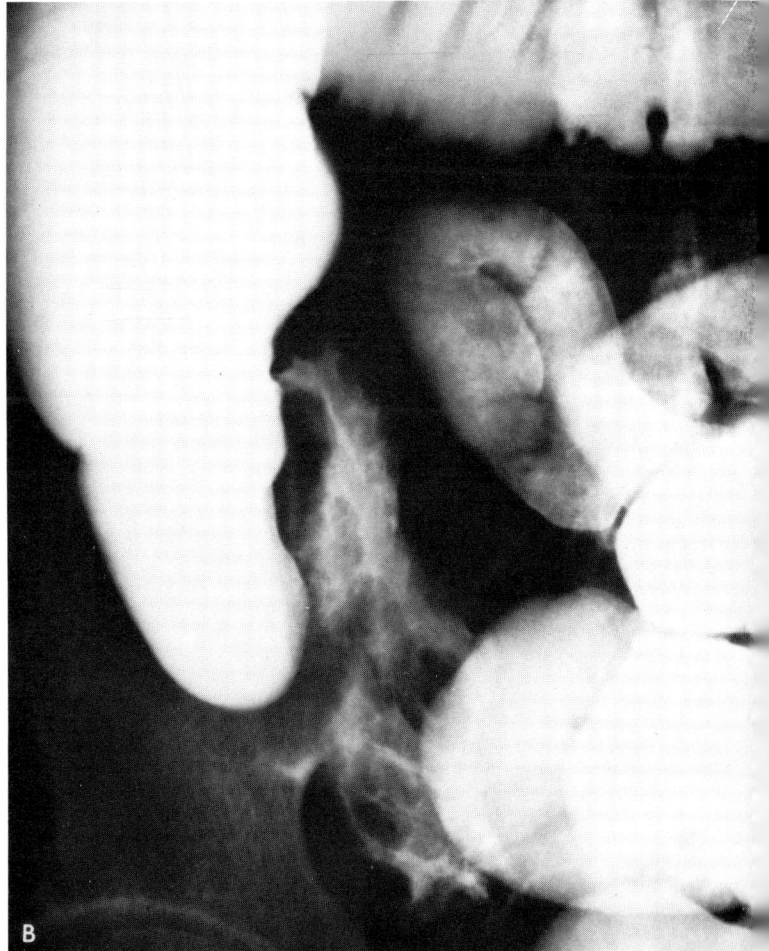

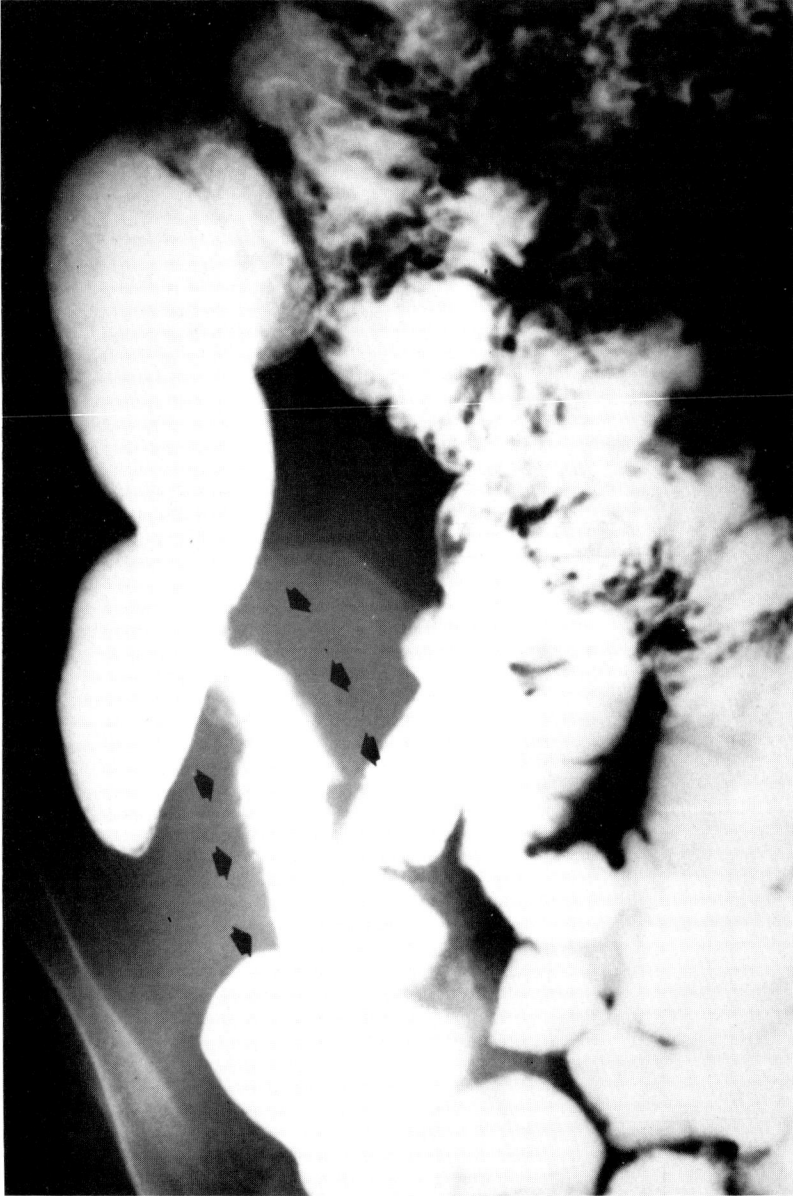

Figure 4-3. Small bowel examination showing regional enteritis of the terminal ileum (arrows) and the cecum in the nonstenotic phase of the disease. Marginal irregularities in the terminal ileum indicate ulceration and edema. The root of the mesentery is thickened and the cecum is deformed.

Figure 4-4. Small bowel examination showing regional enteritis involving the terminal ileum and cecum. A persistent area of stenosis is present (arrows). The characteristic cobblestone pattern of regional enteritis is evident in A. The thickened root of the mesentery is visible in B. The cecum is contracted and edematous.

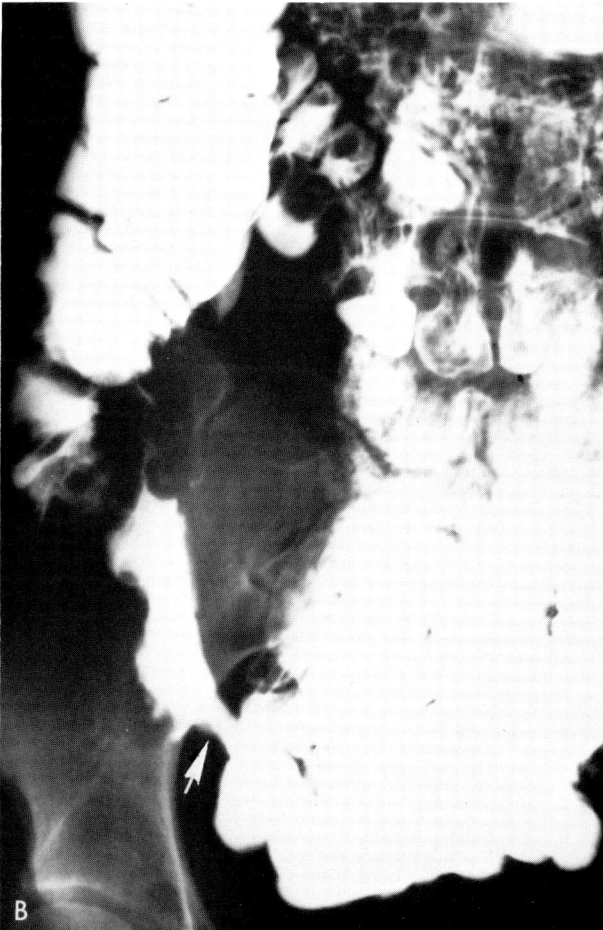

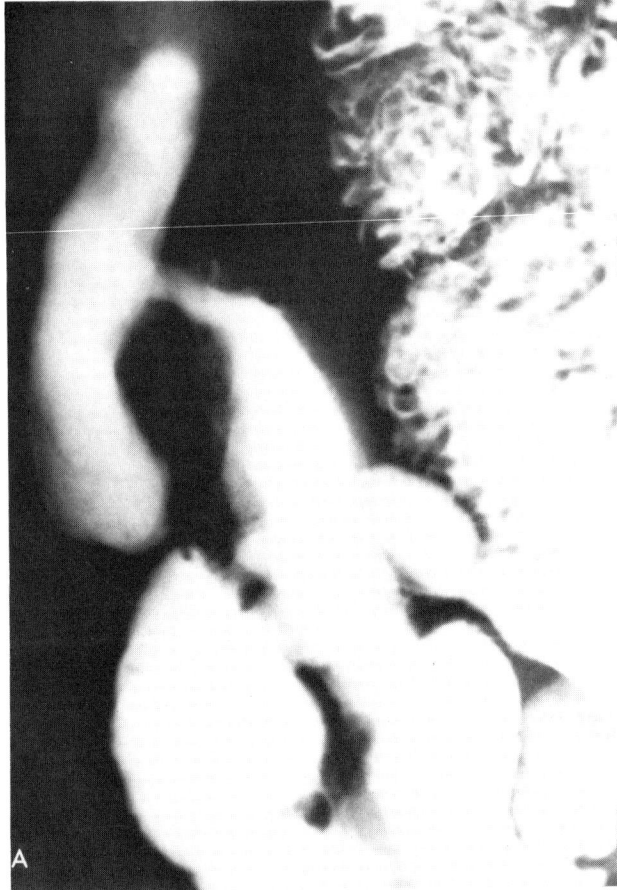

Figure 4–5. Small bowel examination showing regional enteritis of the terminal ileum and cecum. (A) Distended view of the ileum and cecum showing smudging and marginal irregularities. (B) Mucosal view showing the coarse, shaggy, thickened mucosa of the terminal ileum.

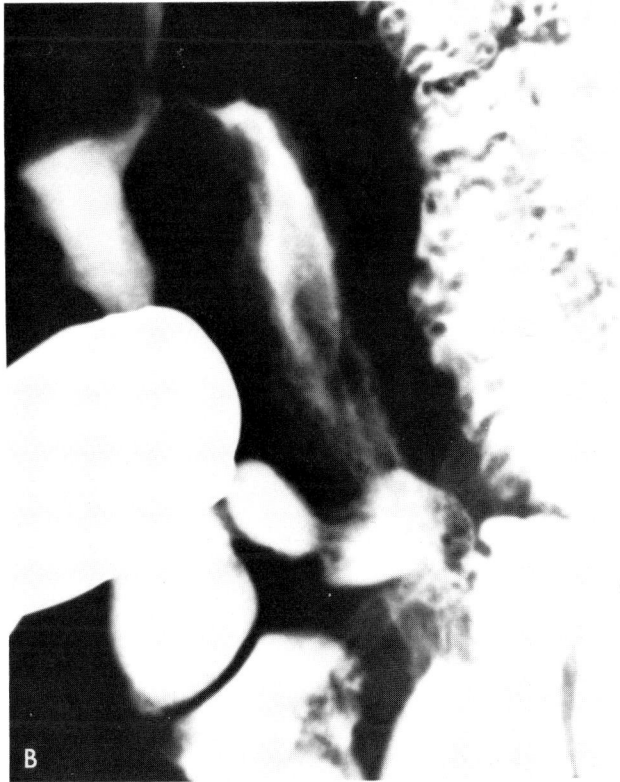

Figure 4–5. (*Continued*)

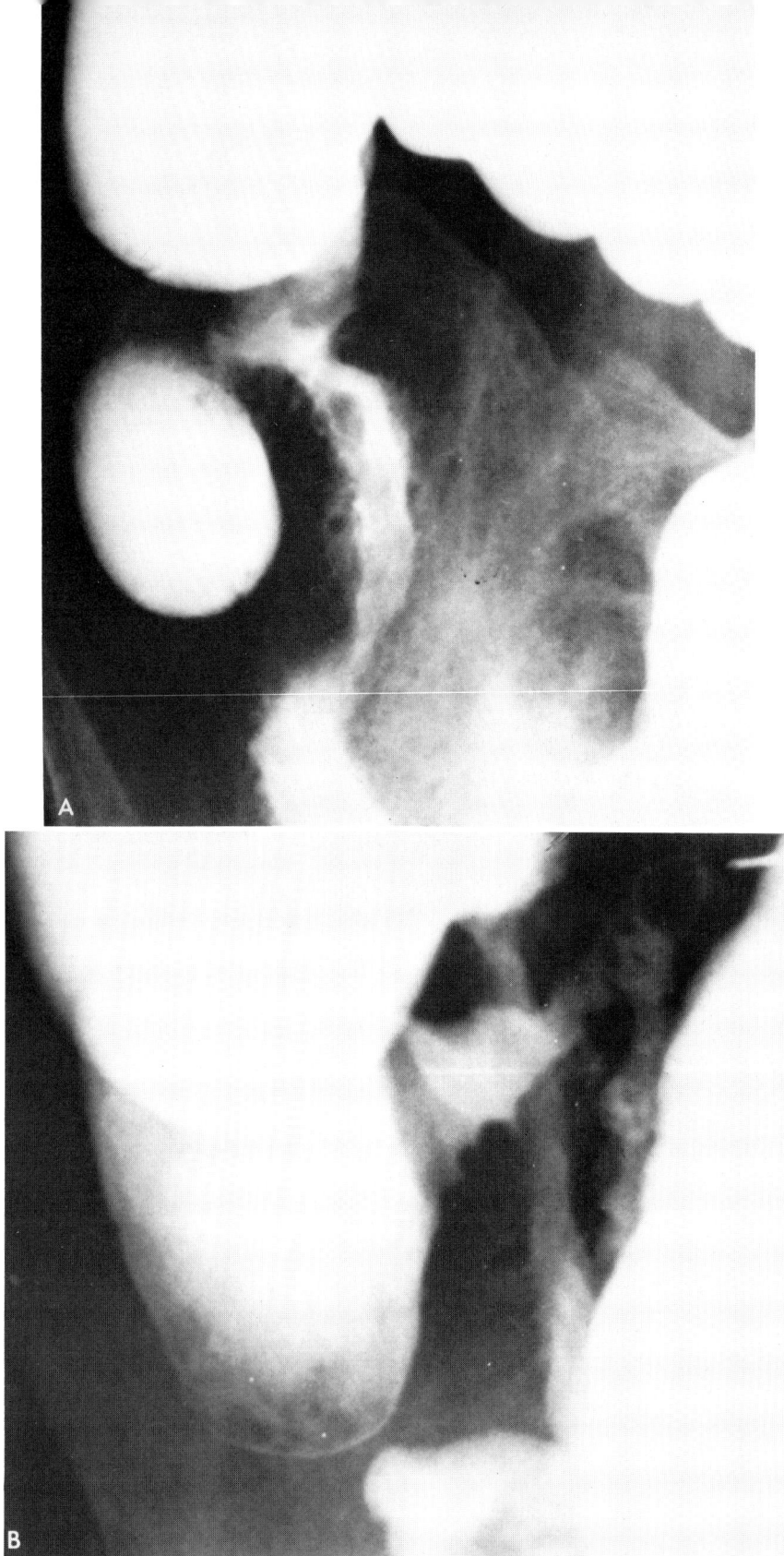

Figure 4-6. (*A* and *B*) Barium enema examination showing Crohn's disease involving the terminal ileum, cecum and appendix. The terminal ileum is nodular, irregular and narrowed. Similar changes in the appendix are visible in *B*. The cecum is mildly deformed. (*C*) Small bowel examination in the same patient showing the ileocecal involvement. The patient was an 18-year-old boy with the physical appearance of a child 10 years old. Symptoms had been present for approximately 6 years. Except for diarrhea, there were few abdominal symptoms. The patient's main complaint was that the police frequently stopped him when he was driving his car, accusing him of being too young to have a driver's license. When regional enteritis occurs before puberty, growth retardation is often profound.

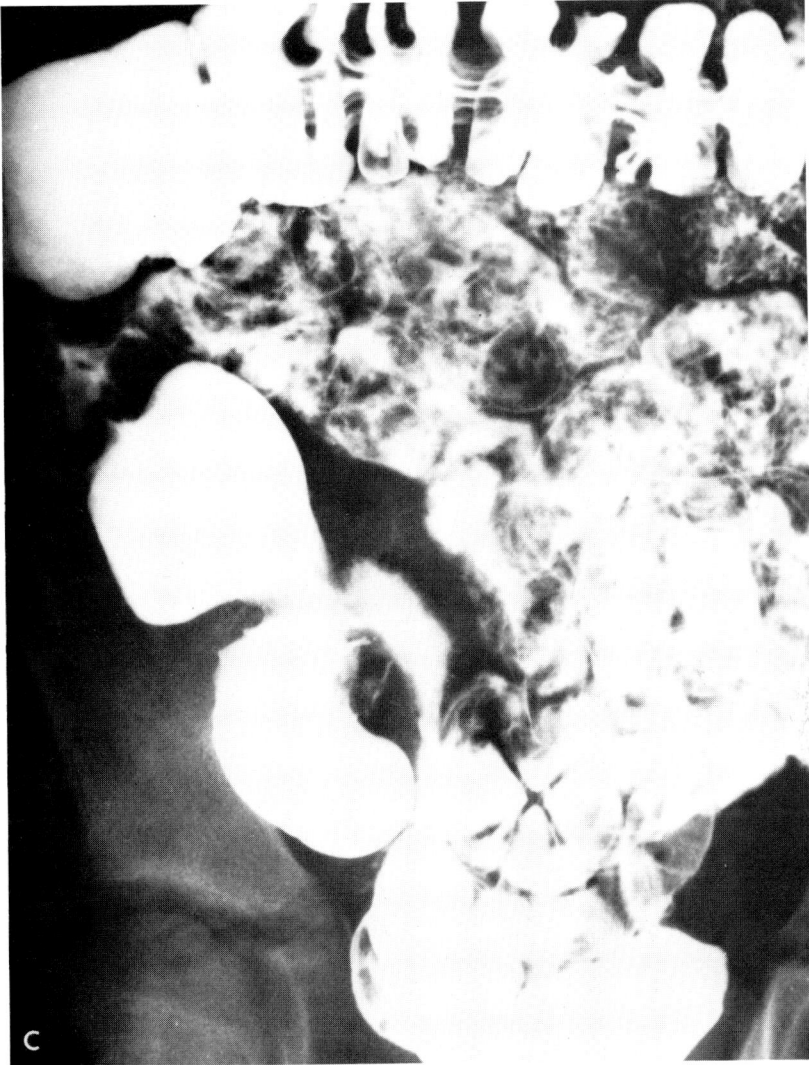

Figure 4–6. (Continued)

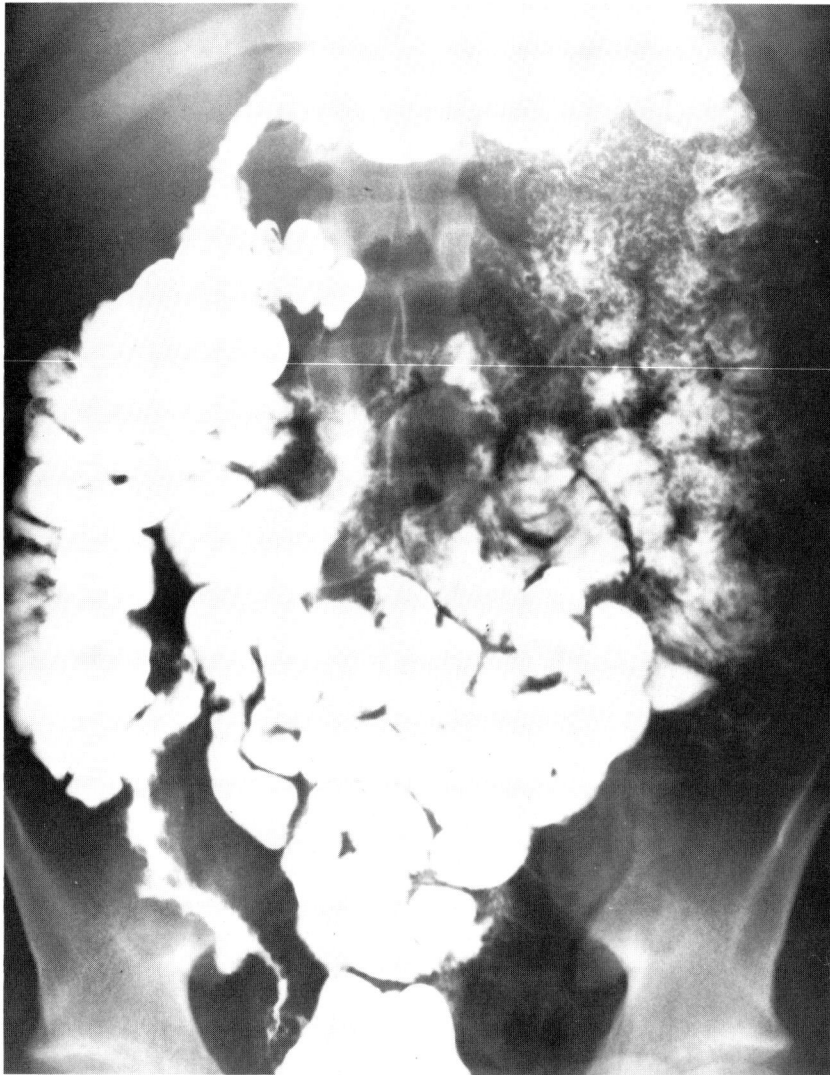

Figure 4-7. Small bowel examination showing Crohn's disease of the terminal ileum and duodenum. Both segments are narrowed, irregular and fixed. There is marked thickening of the terminal ileum.

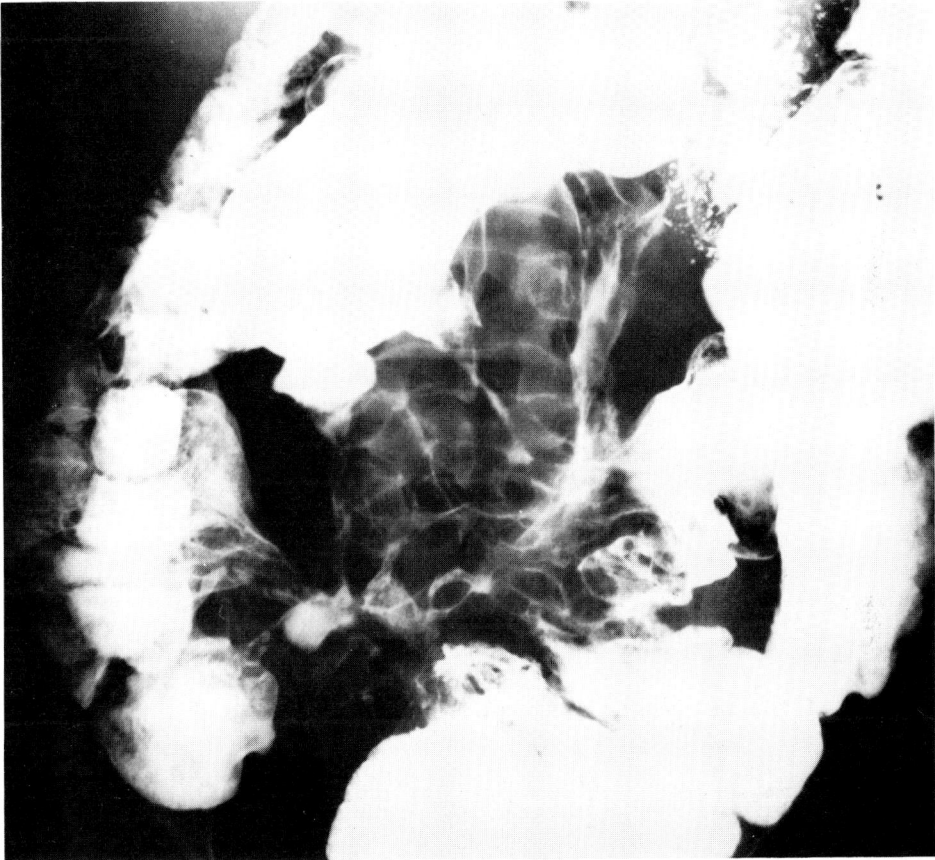

Figure 4-8. Small bowel examination showing numerous fistulas between small bowel loops and between the small bowel and colon. The patient complained only of moderate diarrhea and had no abdominal pain or weight loss.

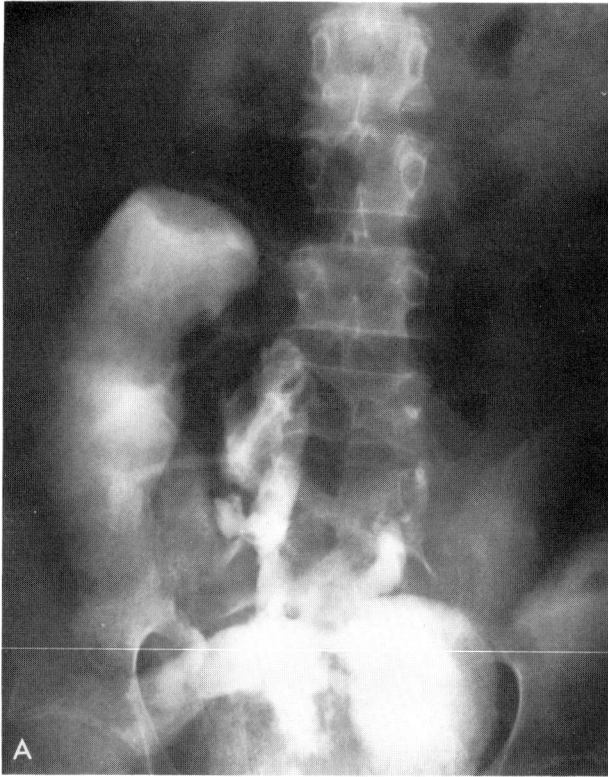

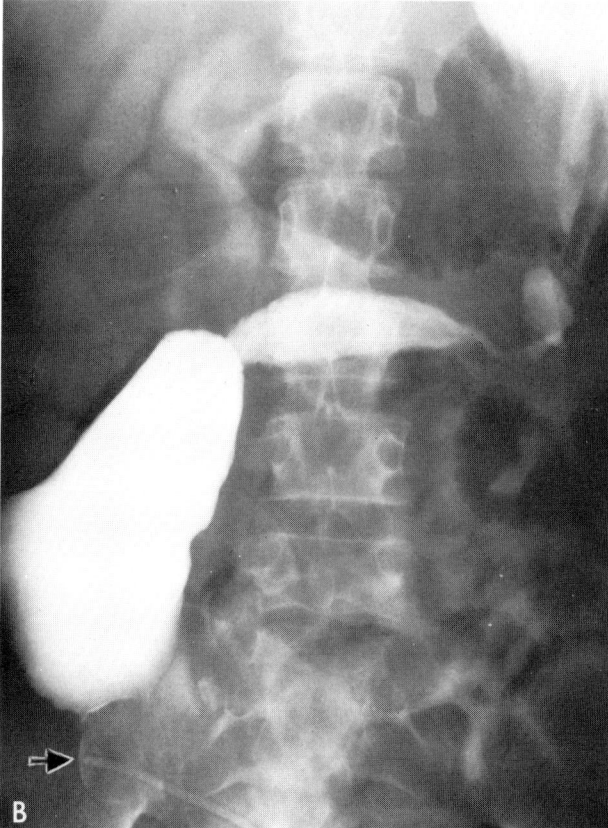

Figure 4–9. (*A, B* and *C*) Fistulograms in a cachetic 30-year-old man with a 10-year history of abdominal pain and diarrhea. He had never been seen by a physician before this admission. Perianal fistulas were so numerous and complex that it was impossible to identify the anus. Biopsy of the perianal tissue showed adenocarcinoma. (*A*) Numerous fistulas fill with barium injected from one of the perianal fistulas, demonstrating presacral, postsacral, ischiorectal and pelvic abscesses. Some barium enters the cecum and ascending colon. (*B*) A second attempt to perform a barium enema shows barium entering the ascending colon via a Foley catheter (arrow) inserted into one of the perianal fistulas. Severe granulomatous colitis of the ascending and transverse colon and a gastrocolic fistula are evident.

Figure 4-9. (Continued) (C) Lateral projection made at the same time as B showing barium in the ascending and transverse colon and the stomach. The colon is narrowed, irregular and edematous, indicating the presence of granulomatous colitis. The gastrocolic fistula is visible. Total colectomy and resection of the terminal ileum were performed with repair of the gastrocolic fistula. Symptomatic response was excellent and the patient gained 25 pounds. (D) Cystogram done 9 months later when the patient returned with uremia. Extensive tumor invasion of the bladder with ureteral obstruction is evident.

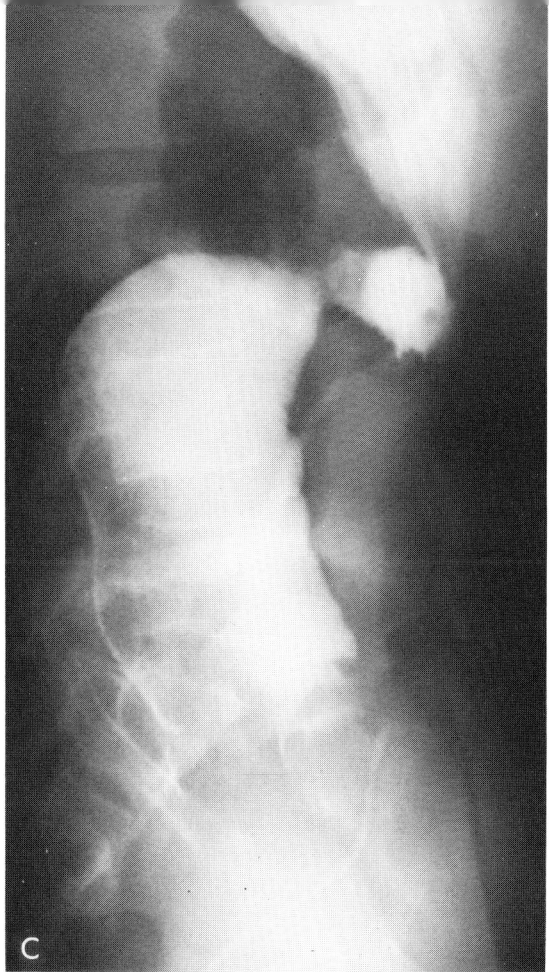

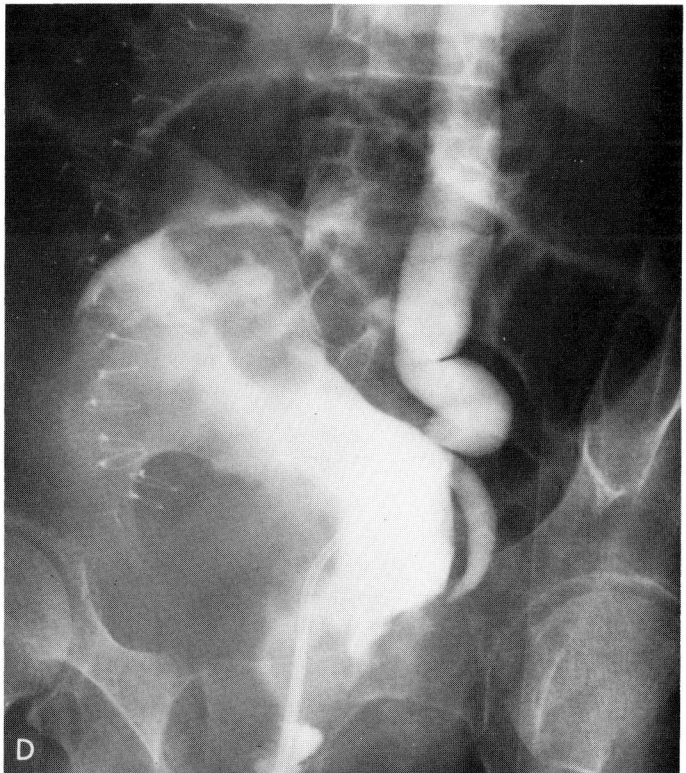

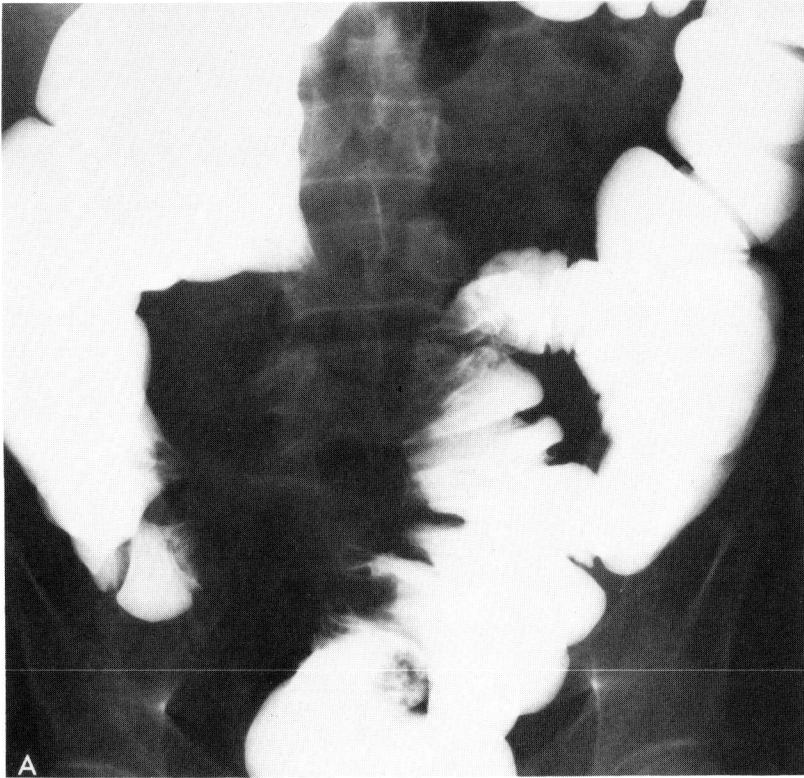

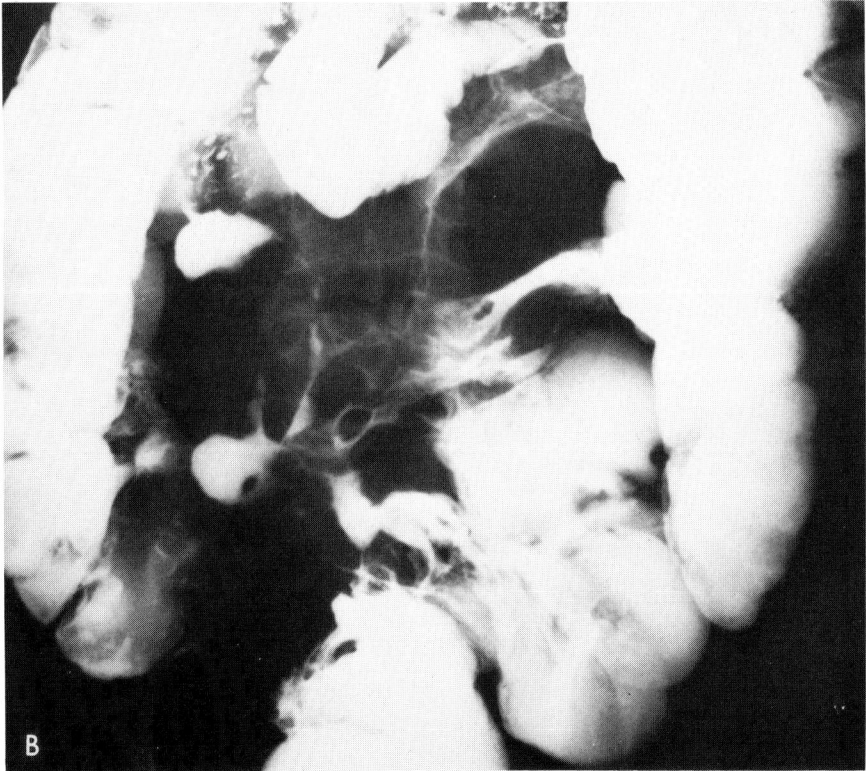

Figure 4–10. Barium enema examination (A) showing distortion of the colon at numerous points due to extensive Crohn's disease of the small bowel. Small bowel examination (B) showing Crohn's disease of the small bowel with numerous fistulas between loops of small bowel and between the small bowel and colon. The changes in the colon are secondary to the disease of the small intestine. There was no granulomatous colitis.

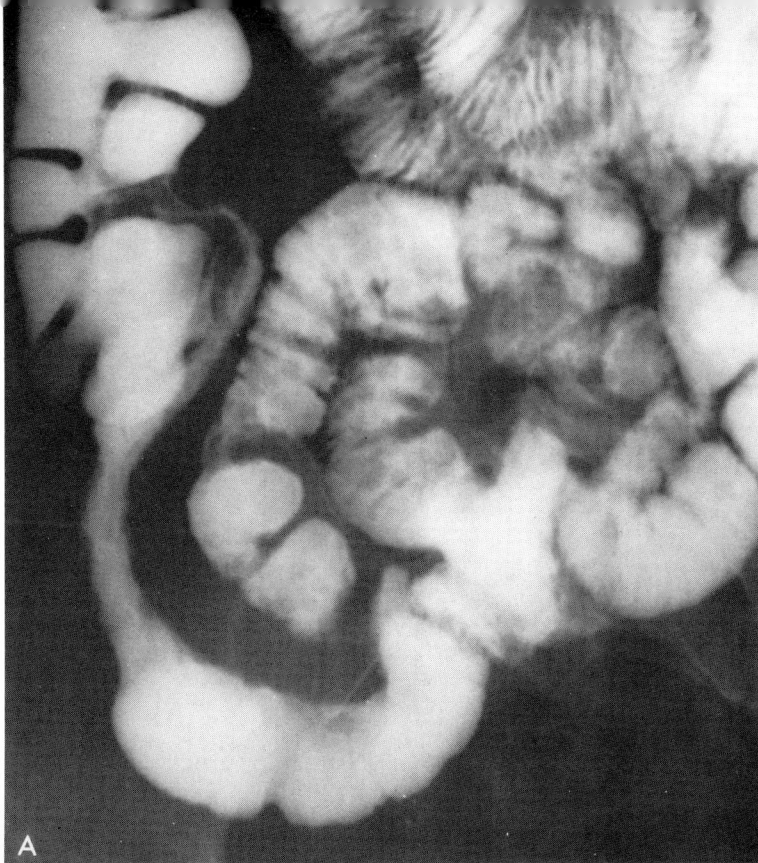

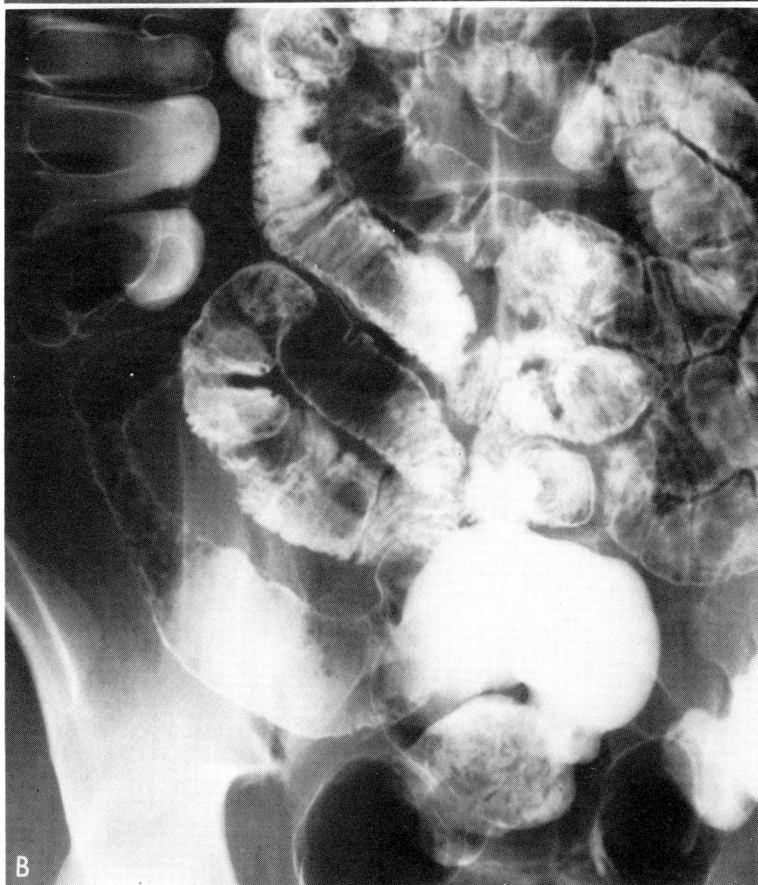

Figure 4–11. Barium enema examination (*A*) with air contrast (*B*) showing Crohn's disease of the terminal ileum. Typical thickening of the ileal wall is evident.

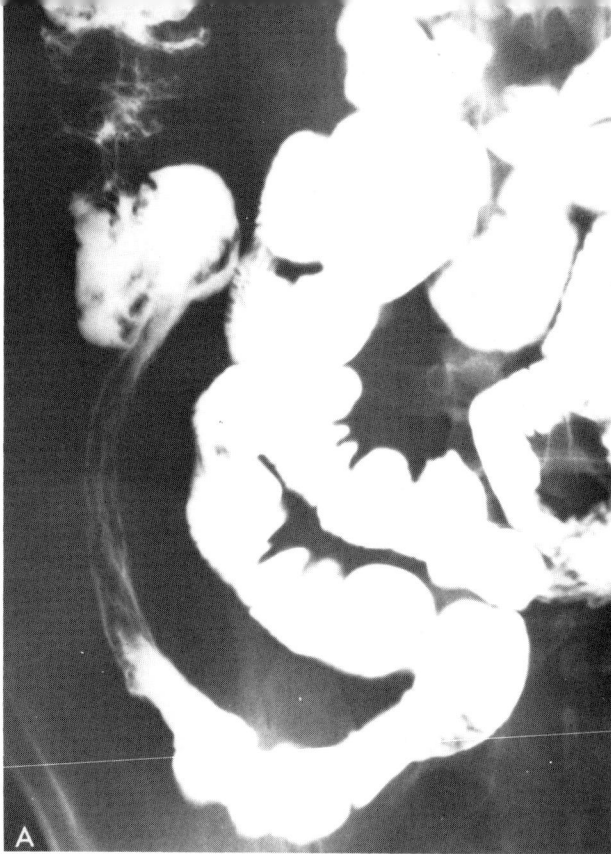

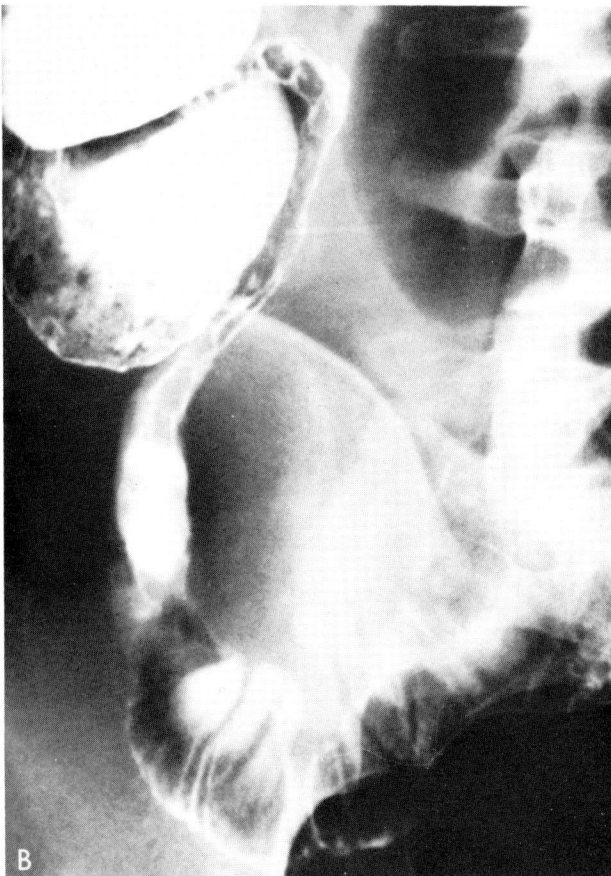

Figure 4–12. Small bowel examination (A) and barium enema (B) showing Crohn's disease of the terminal ileum. Marked thickening of the terminal ileum is apparent. There is an abrupt junction between the diseased and normal ileum.

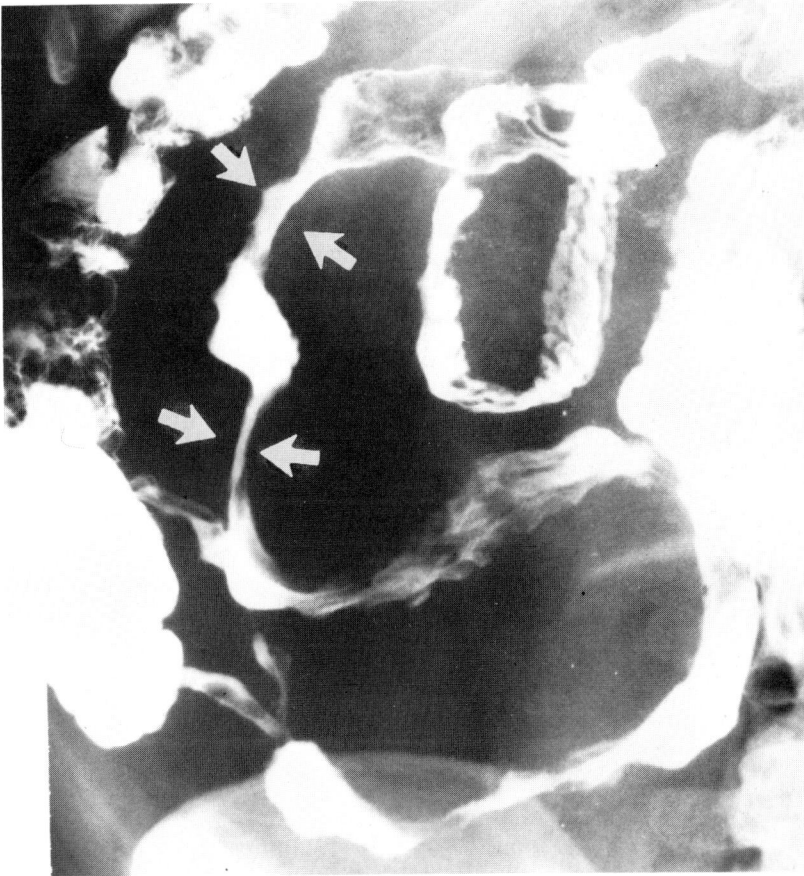

Figure 4-13. Small bowel examination showing extensive chronic Crohn's disease with marked thickening of the bowel wall and narrowing of the lumen (arrows). Wide separation of intestinal loops is typical in the advanced stenotic phase of the disease.

Figure 4–14. Small bowel examination showing extensive Crohn's disease in the stenotic stage of the disease. There is marked thickening of the bowel wall and narrowing of the lumen.

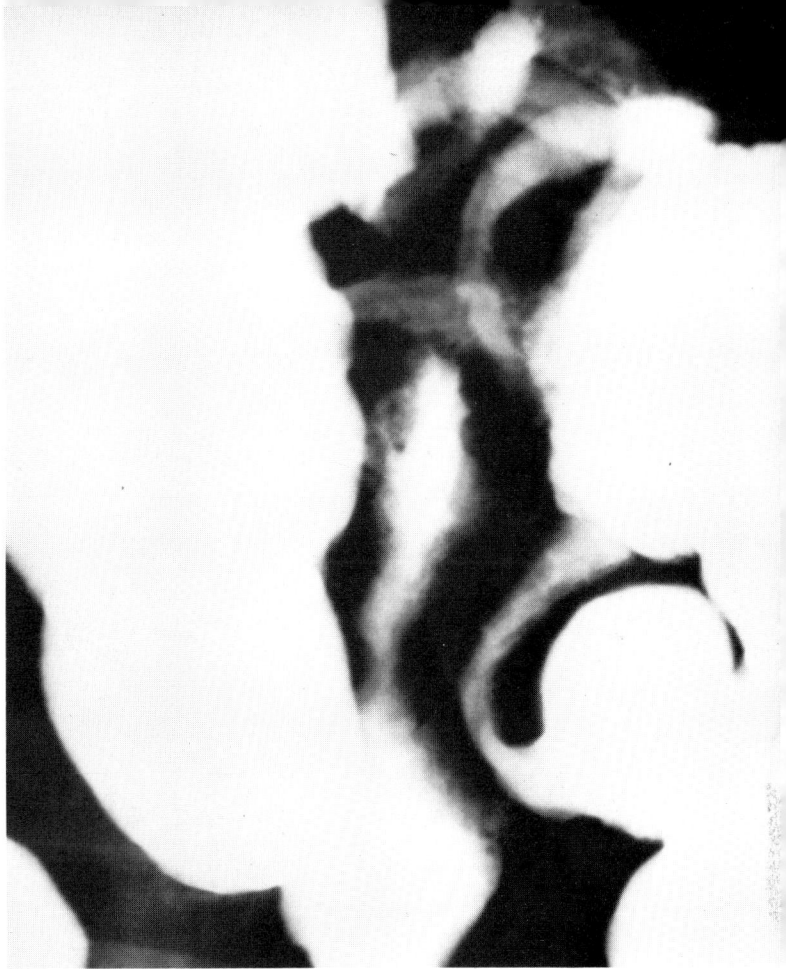

Figure 4-15. Barium enema examination showing extensive stenotic Crohn's disease of the terminal ileum with thickening of the bowel wall, narrowing of the lumen and typical mucosal changes.

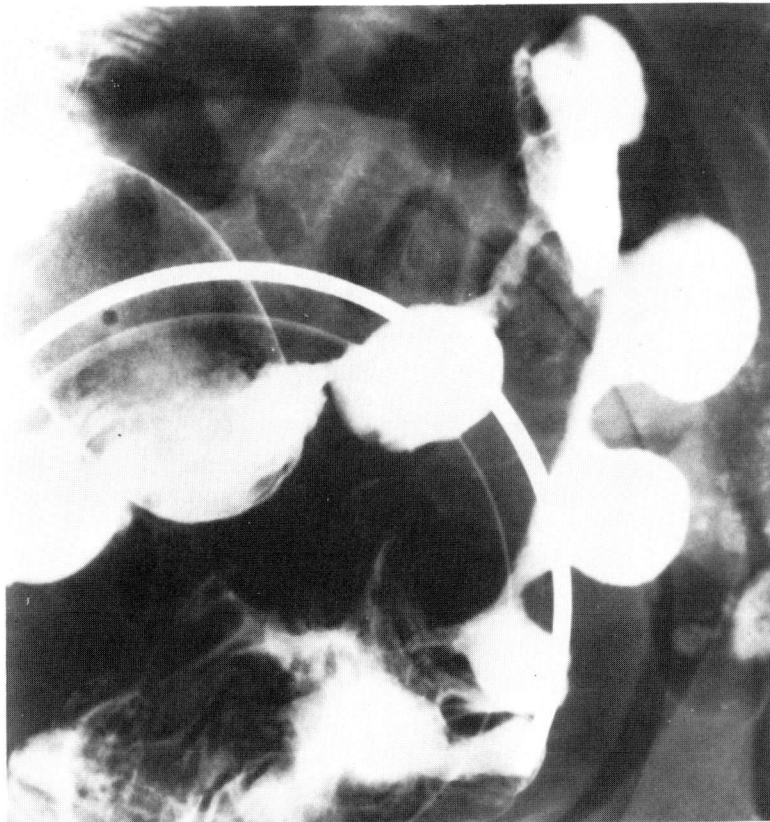

Figure 4-16. Small bowel examination showing Crohn's disease in a segment of the ileum. The disease is more marked in some areas than in others and in some segments it does not involve the entire circumference of the bowel. The relatively normal part of the bowel can distend, producing pseudodiverticula.

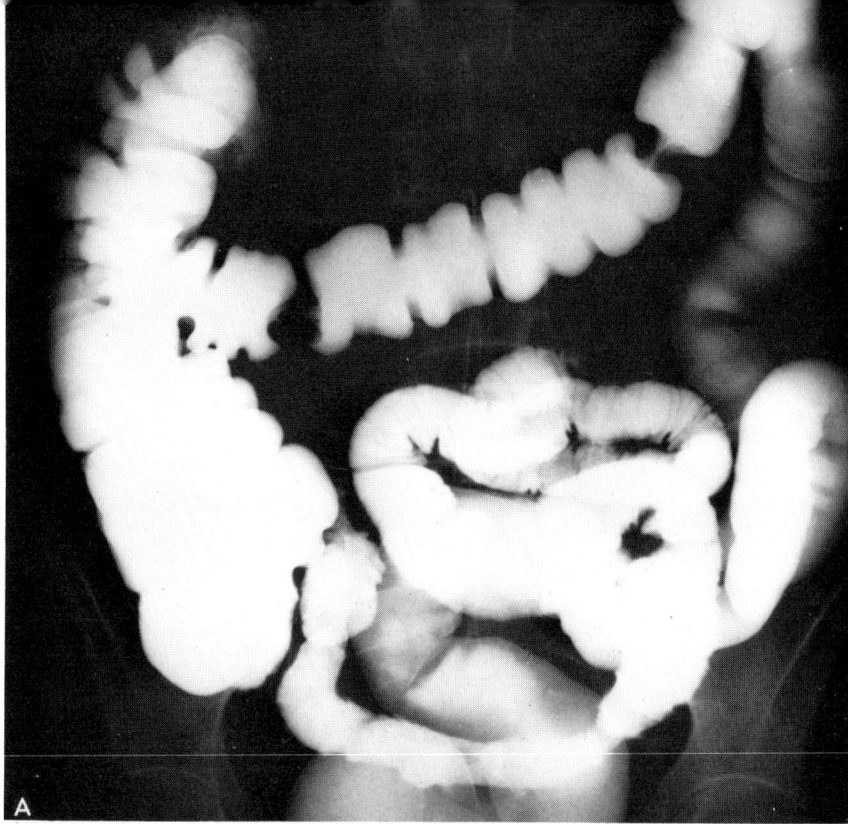

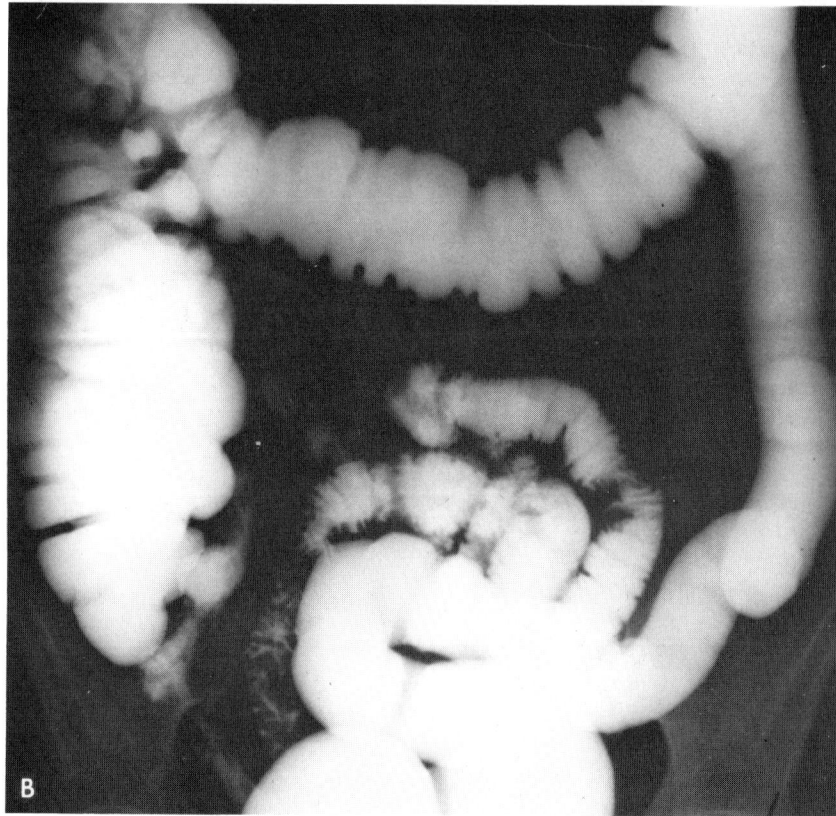

Figure 4–17.   Barium enema examination (A) showing Crohn's disease of the terminal ileum in the nonstenotic stage. The colon appears normal. Barium enema examination (B) on the same patient 4 years later showing progression of the ileal disease to the stenotic phase. The colon has remained normal. Barium enema examination (C) in the same patient 4 months after the study done in B. The haustral pattern of the colon is lost and edema is apparent throughout, with smudging and irregularity indicating extension of the disease into the colon.

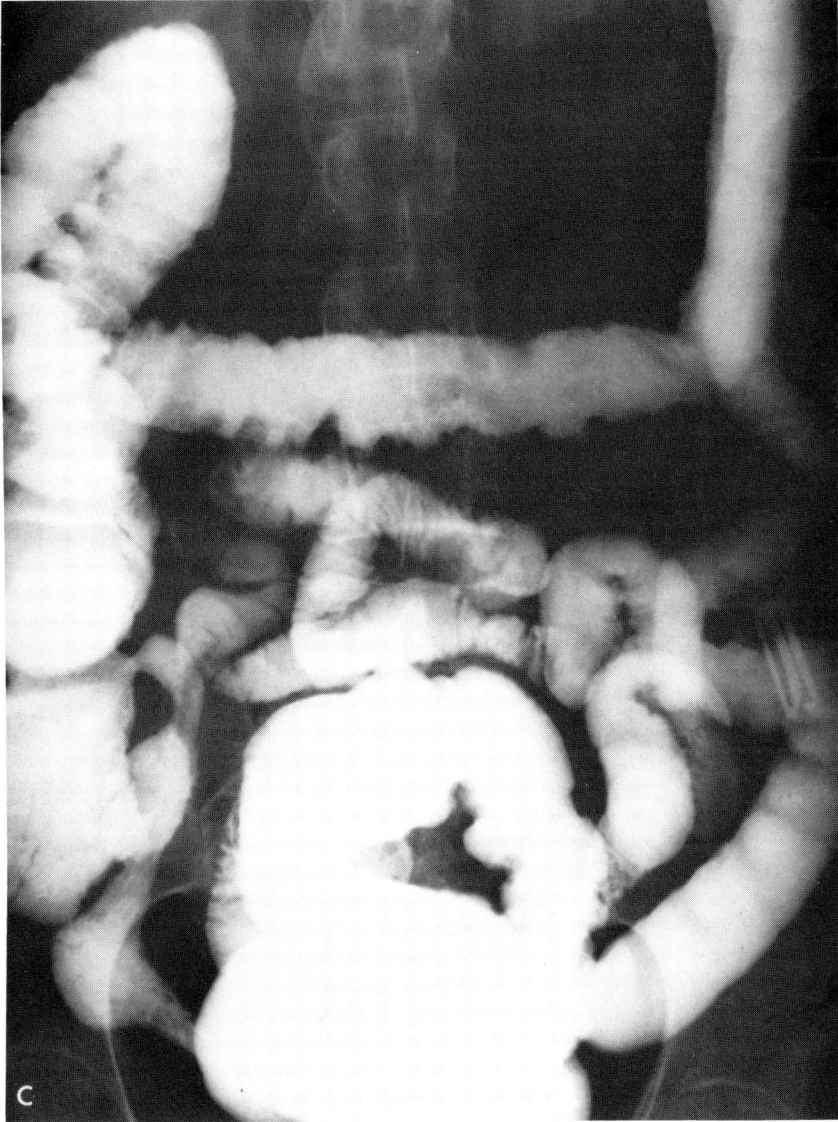

Figure 4–17. (*Continued*)

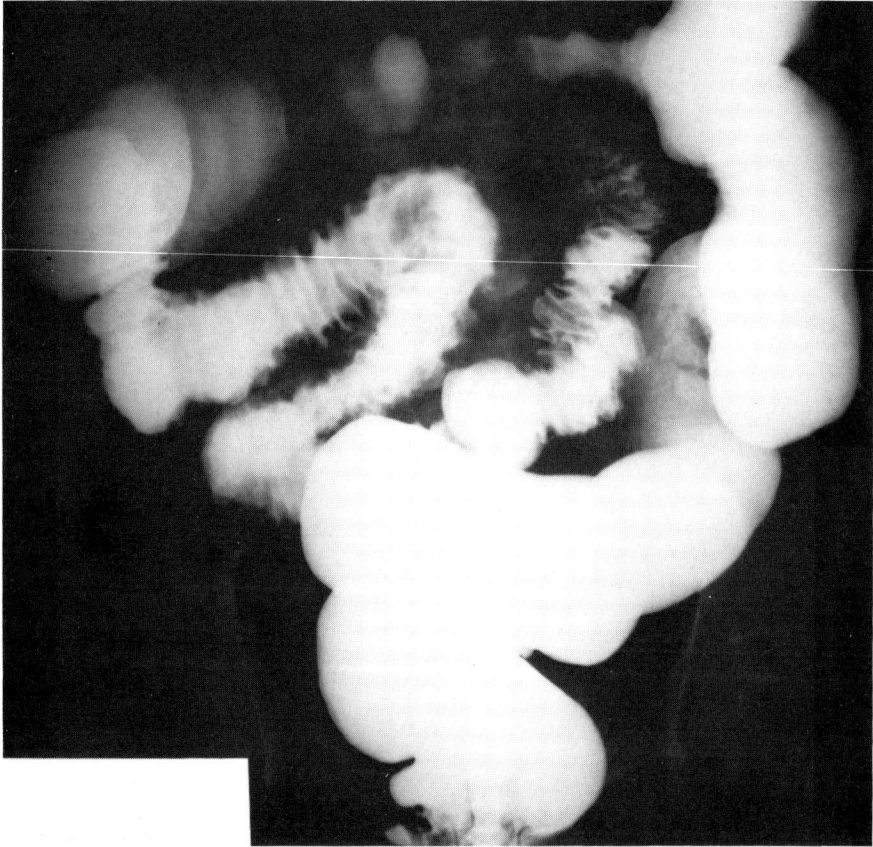

Figure 4–18. Barium enema examination showing recurrent Crohn's disease in a patient who had an end-to-end ileo-ascending colostomy with resection of the terminal ileum. Marked thickening of the mucosa of the "new" terminal ileum is apparent. The transverse colon was normal on other radiographs. There is a perianal fistula filled with barium.

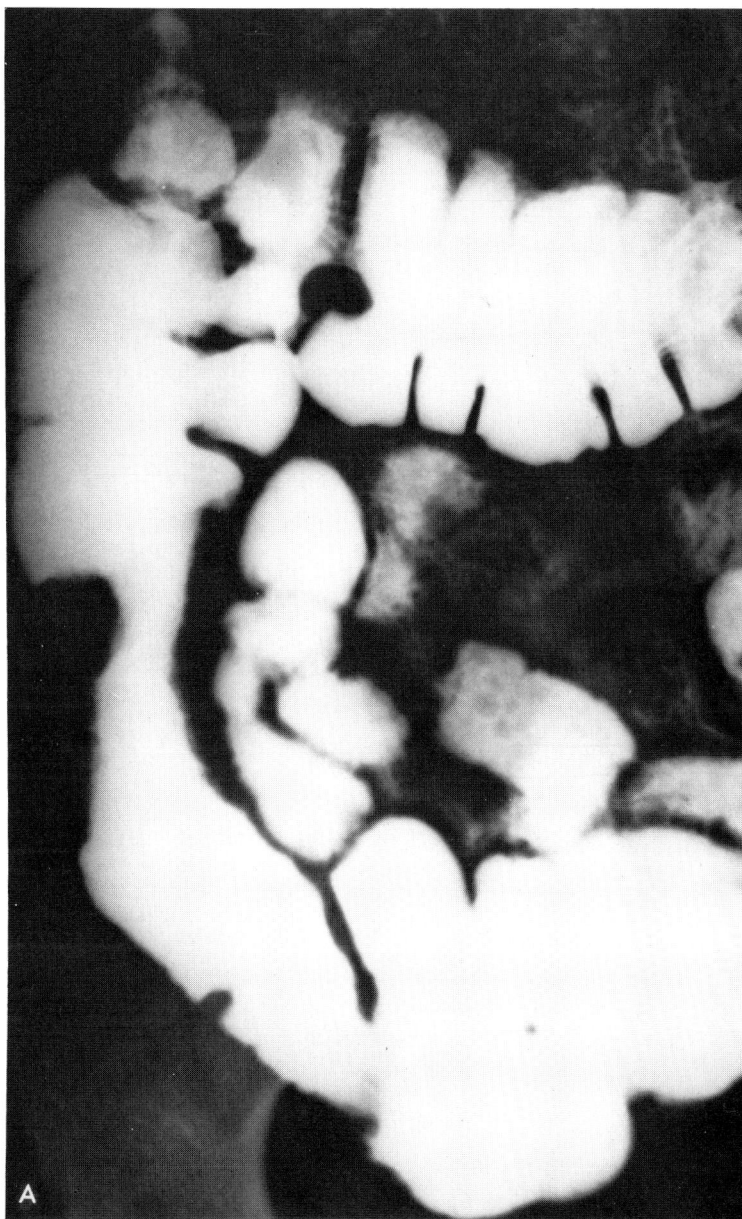

Figure 4-19. Barium enema examination showing recurrent Crohn's disease in an 18-year-old girl who had an end-to-end ileo-ascending colostomy with resection of the terminal ileum 4 years earlier. (A) Marginal irregularity indicates ulceration and edema in the "new" terminal ileum. (B) Air contrast projection shows the typical cobblestone pattern of Crohn's disease. (C) A "skip" area is present in the right side of the transverse colon. A typical "pseudodiverticulum" is present (arrow) while the opposite wall is fixed and rigid. The abnormality in the colon is not easily recognized in A.

(Figure 4-19 continued on following pages)

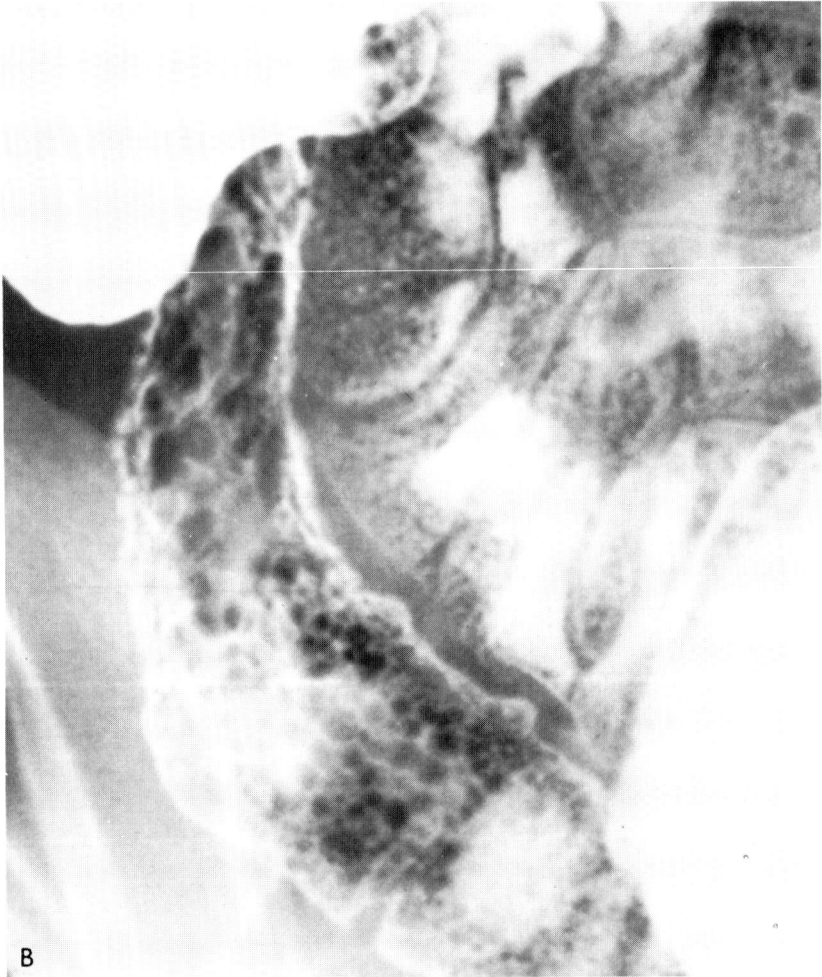

Figure 4–19. (*Continued*)

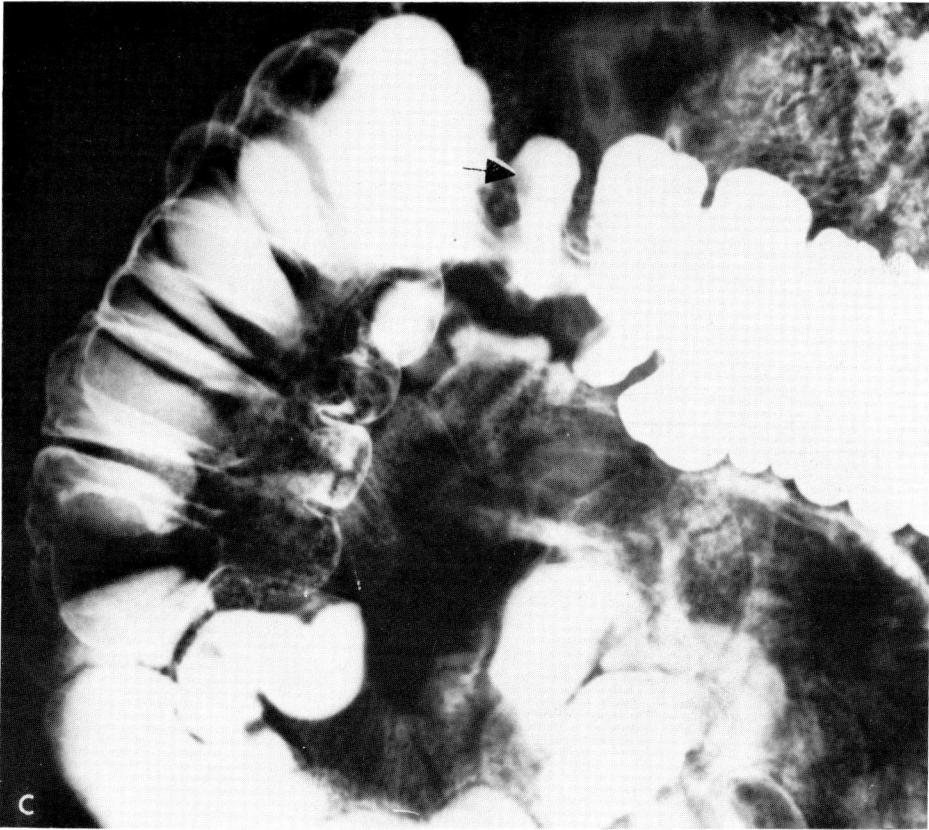

Figure 4–19. (Continued)

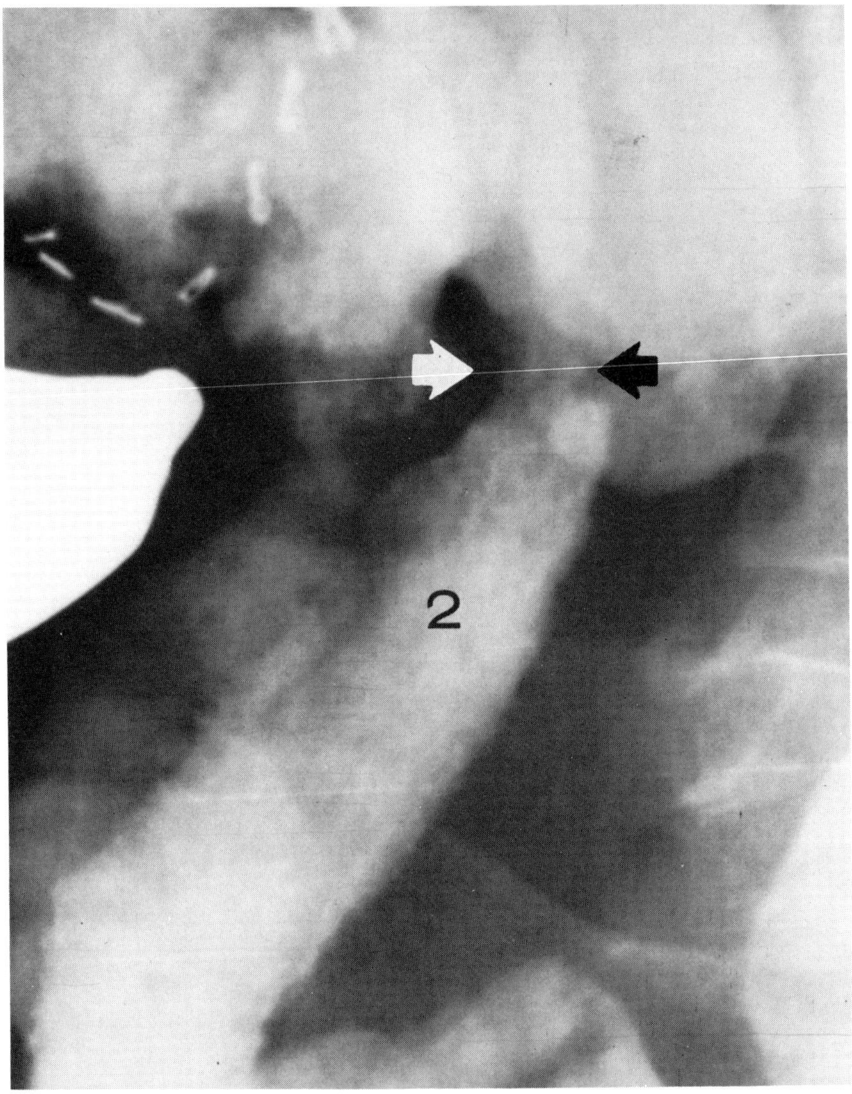

Figure 4–20. Barium enema examination in a patient who had an end-to-side ileotransverse colostomy and resection of the terminal ileum for Crohn's disease 4 years earlier. The metallic clips are on the blind end of the transverse colon. The abnormal mucosal pattern of the "new" terminal ileum (2) indicates the presence of recurrent ileitis. The anastomosis (arrows) measured only 6 mm. when surgery was performed for intermittent intestinal obstruction. (From Berk, R. N., and Chamovitz, R.: Recurrent acute small bowel obstruction due to a narrow ileocolic anastomosis. Radiology, 93:117, 1969.)

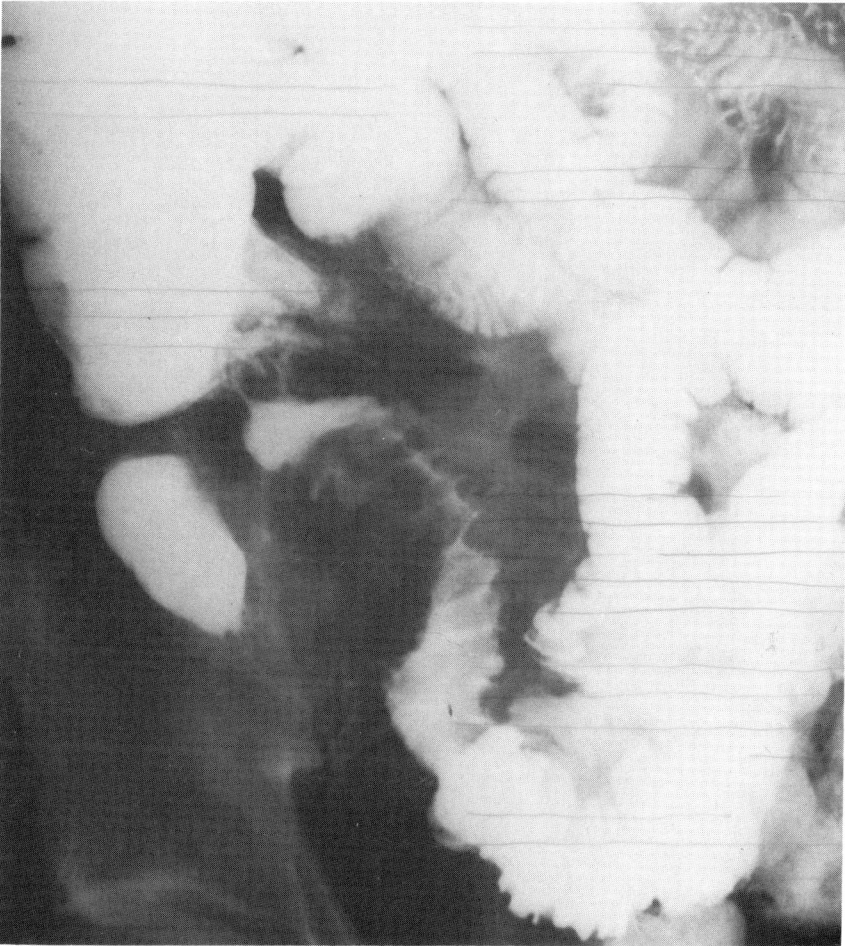

Figure 4-21. Small bowel examination in a 60-year-old man with an appendiceal abscess. The cecum is deformed and the terminal ileum is displaced around the abscess. Proximity of the ileum to the abscess has caused prominent edema of the ileal mucosa. The presence of the mass and the patient's clinical and laboratory findings suggested an appendiceal abscess rather than Crohn's disease.

that can be detected radiographically are numerous. These include intra-abdominal fistulas and abscesses, ankylosing spondylitis, ureteral obstruction (due to involvement of the ureter by retroperitoneal inflammation or by oxalate calculi), amyloidosis, peptic ulcer, pancreatitis and carcinoma. Perforation and toxic dilatation of the bowel are rare but do occur. Gallstones occur in 34 per cent of patients with regional enteritis and patients after ileal resection because of interruption of the enterohepatic circulation of bile salts.[6] The decrease in the bile salt pool leads to the development of lithogenic bile.

Crohn's disease of the small intestine must be differentiated from lymphosarcoma, which may mimic the nonstenotic phase of the disease. Thickening and blunting of the folds, irregularity of contour, ulceration and intraluminal nodules are features that are common to both diseases. The findings of inflammation (edema, irritability and exudate) are usually apparent in regional enteritis, whereas sharply defined tumor nodules may be identified in lymphosarcoma. Narrowing of the lumen is characteristic of regional enteritis but rarely occurs in lymphosarcoma, in which the diameter of the bowel is normal or increased. Bizarre ulcerations occur in lymphosarcoma.

Hodgkin's disease produces a desmoplastic reaction which may be impossible to differentiate from the chronic changes of regional enteritis. The presence of tumor nodules and a polypoid mass are features of Hodgkin's disease. Segmental infarction of the small bowel may mimic stenotic regional enteritis, but the history is acute and the cobblestone pattern of regional enteritis is absent. Carcinoma of the small intestine may produce a short stricture simulating Crohn's disease, but the absence of inflammatory changes and proximal dilatation and the presence of a sharp, overhanging margin serve to differentiate tumor from Crohn's disease. Tuberculosis involves the terminal ileum and cecum and may entirely simulate regional enteritis. In the United States, the chest radiograph usually shows active tuberculosis. The cecum is involved to a greater extent in tuberculosis than in most patients with regional enteritis. In tuberculosis, ulcers tend to be large and irregular, and the mucosal markings are coarser than in Crohn's disease. The inflammatory changes of typhoid fever in the terminal ileum may precisely simulate those of acute regional enteritis. Differentiation must depend on the clinical findings and the results of bacteriologic studies. An appendiceal abscess may produce spasm, irritability and mucosal edema of the ileum, but a mass is usually present displacing the cecum and/or ileum (Fig. 4–21). Carcinoma of the cecum may extend into the terminal ileum and produce diagnostic difficulty. This is uncommon, however, since adenocarcinoma usually fails to cross the valve, whereas lymphosarcoma does so regularly. Carcinoid tumors may involve long segments of the small bowel, producing kinking and angulation. Tumor nodules can usually be identified to help distinguish carcinoid from regional enteritis. Radiation enteritis may narrow the ileum and produce fixation and rigidity entirely indistinguishable from regional enteritis. Often this specific diagnosis cannot be made without knowledge of the history of previous radiation therapy.

CROHN'S DISEASE OF THE CECUM AND ASCENDING COLON

The radiographic features of granulomatous colitis are the same as those of Crohn's disease of the small intestine[7] (Figs. 4–22 to 4–25). These include skip lesions, deep longitudinal ulcers with transverse fissures, cobblestone mucosal pattern, eccentric involvement, fistulas, narrowing of the lumen and pseudopolypoid changes. The ileocecal valve is often thickened, and reflux into the terminal ileum may be difficult to produce during barium enema examination. Eighty per cent of patients with granulomatous colitis have changes typical of Crohn's disease in the terminal ileum.[8] Involvement of the right side of the colon often predominates, and the rectum is frequently spared. Anal fissures, internal fistulas and strictures are common. Ulcers tend to be deep and penetrating.[9]

ULCERATIVE COLITIS WITH BACKWASH ILEITIS

Ulcerative colitis is characterized by superficial involvement, mucosal edema, haustral changes and shallow ulceration in

(Text continued on page 214)

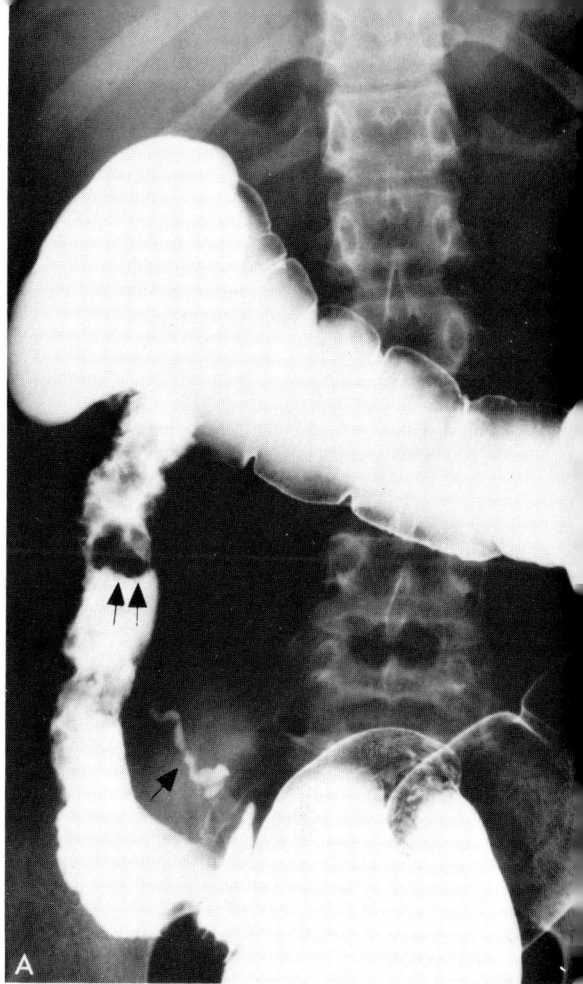

Figure 4-22. Barium enema examination (*A* and *B*) in a 16-year-old girl showing Crohn's disease of the terminal ileum, cecum and ascending colon. The ileocecal valve is edematous (double arrows). Marked cobblestone pattern is visible in both the terminal ileum and the cecum. The appendix is filled with barium (single arrow). Enlargement of the ileocecal valve usually is present in patients with Crohn's disease of the terminal ileum.

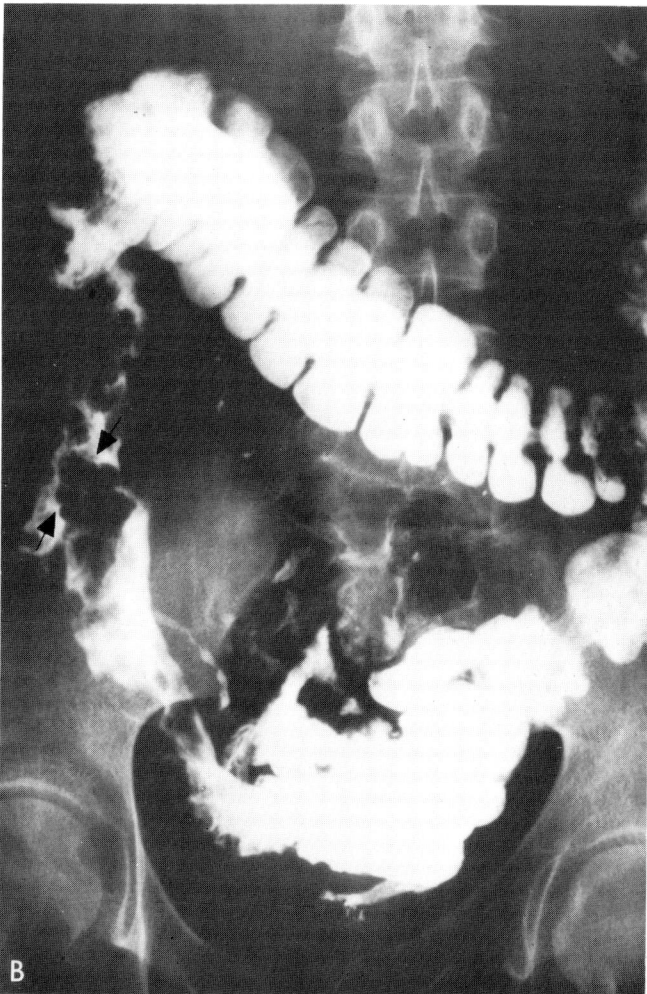

Figure 4–23. Barium enema showing deep ulcers in the ascending colon and hepatic flexure due to granulomatous colitis. The ulcers tend to coalesce along the lateral margin of the ascending colon. Deep ulcers are characteristic of granulomatous colitis.

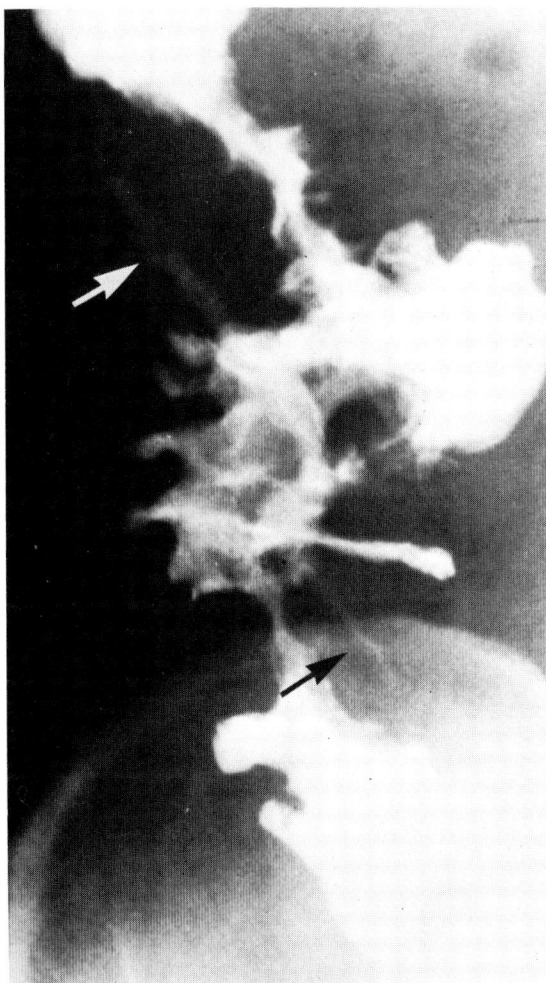

Figure 4-24. Barium enema examination showing two sinus tracts (arrows) and transverse fissures in the ascending colon due to granulomatous colitis. Long paracolic sinus tracts occur in granulomatous colitis and acute diverticulitis.

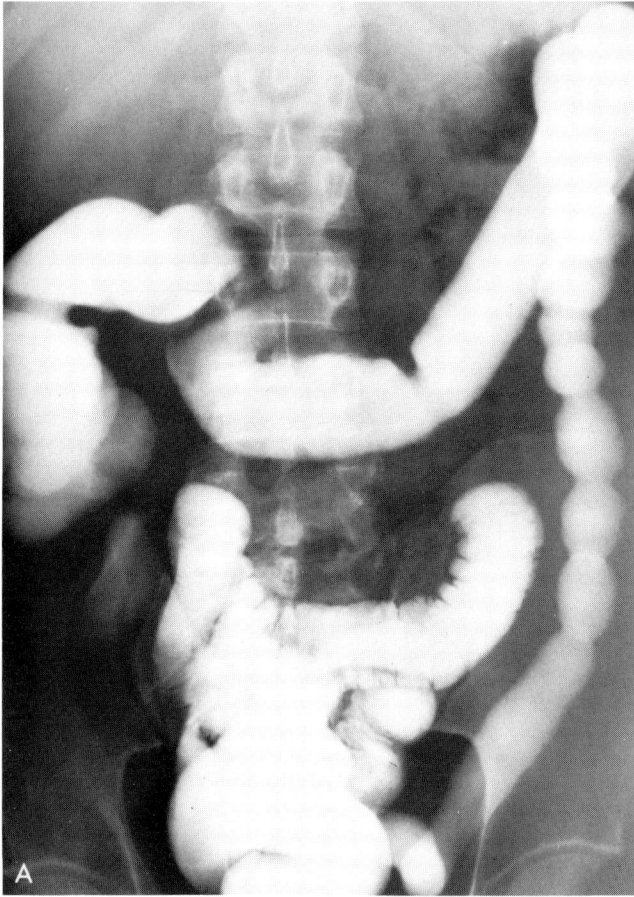

Figure 4–25. Barium enema examination in a patient with universal granulomatous colitis and terminal regional enteritis. *A* and *B* show symmetrical narrowing of the colon, loss of haustra and edema. The rectum is spared.

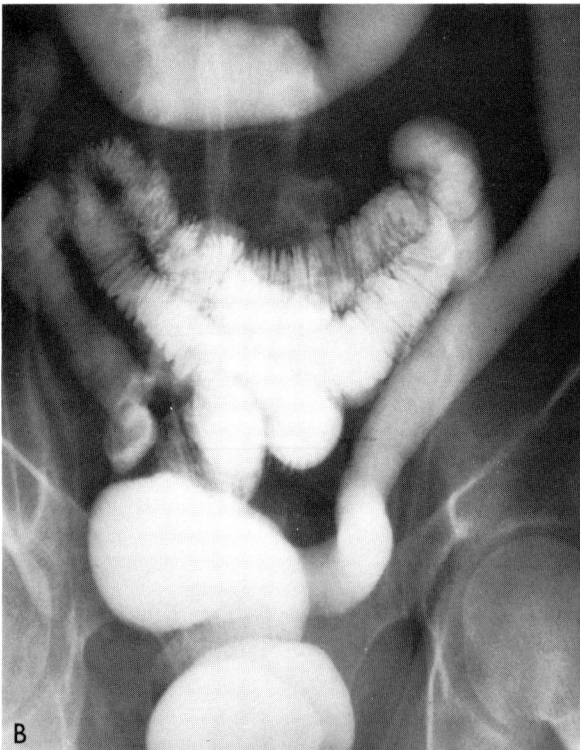

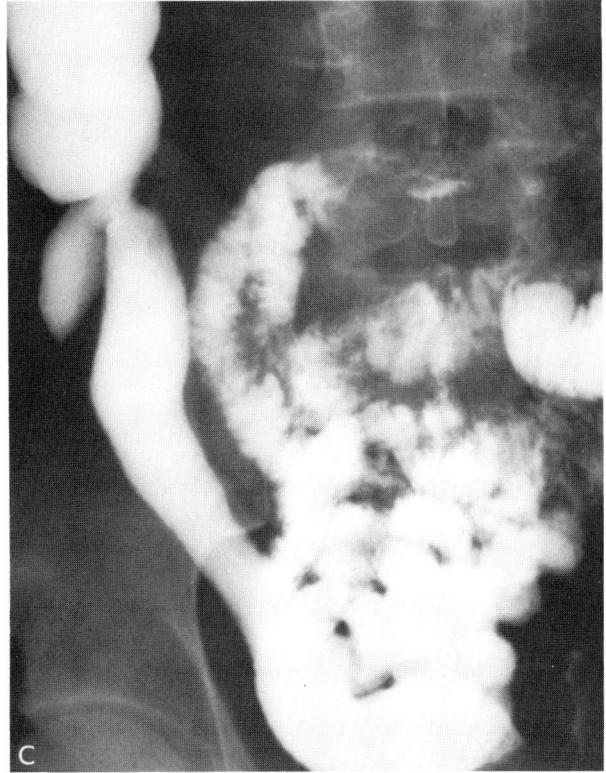

Figure 4-25. (Continued) C and D show the terminal ileum and cecum. There is marked deformity of the cecum and edema of the terminal ileum characteristic of Crohn's disease. Extensive involvement of the cecum and terminal ileum with sparing of the rectum indicates that the disease is granulomatous colitis rather than ulcerative colitis.

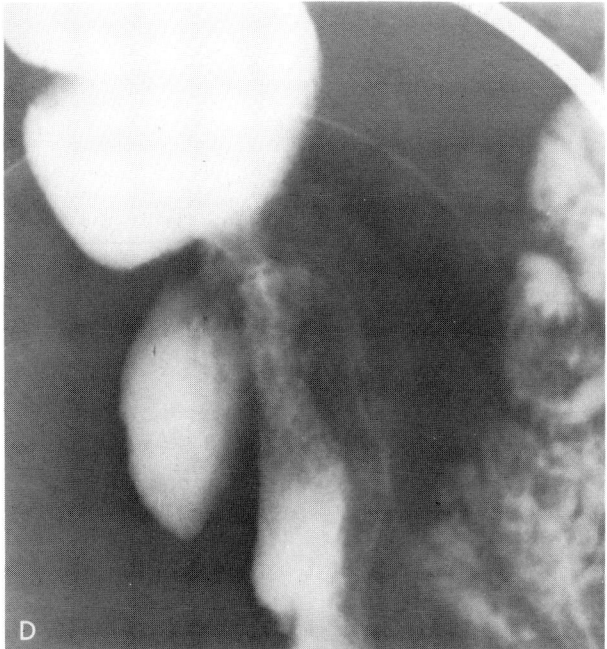

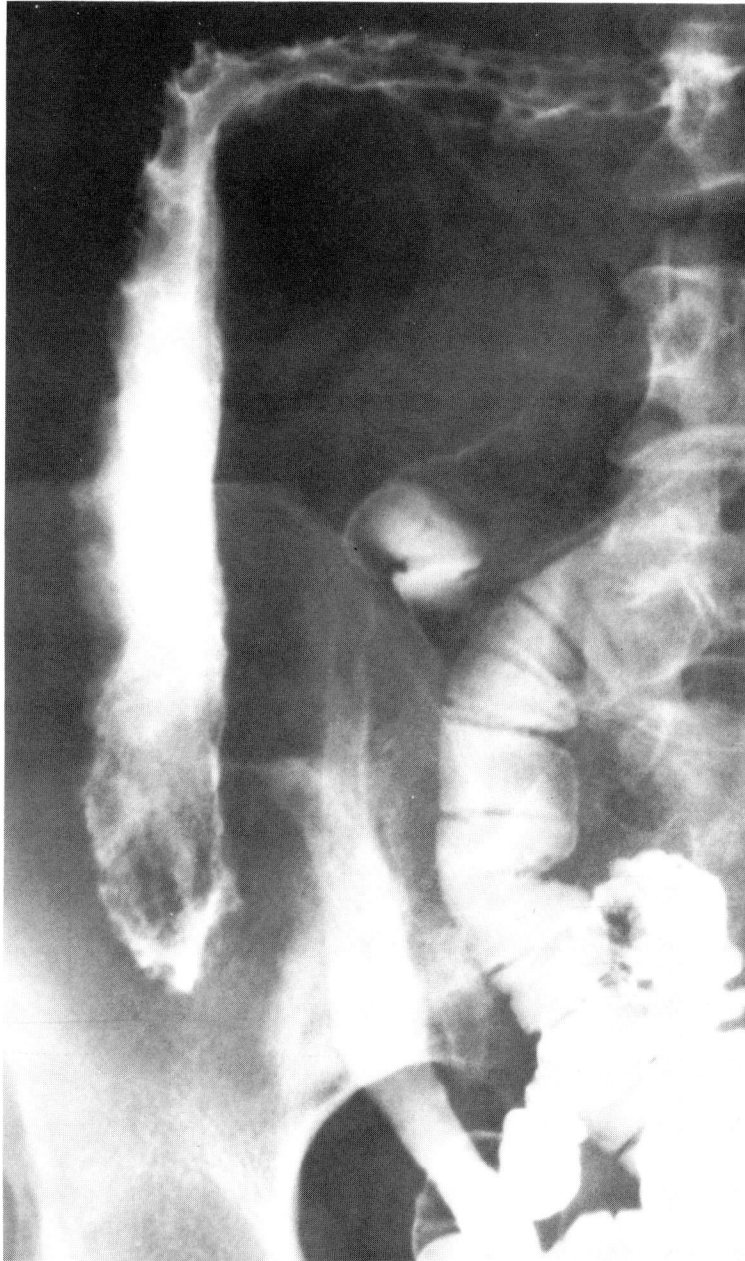

Figure 4-26. Postevacuation barium enema examination showing advanced ulcerative colitis with backwash ileitis. Marked edema, narrowing and ulceration of the colon are apparent. The entire colon was involved. Backwash ileitis occurs only in association with extensive ulcerative colitis.

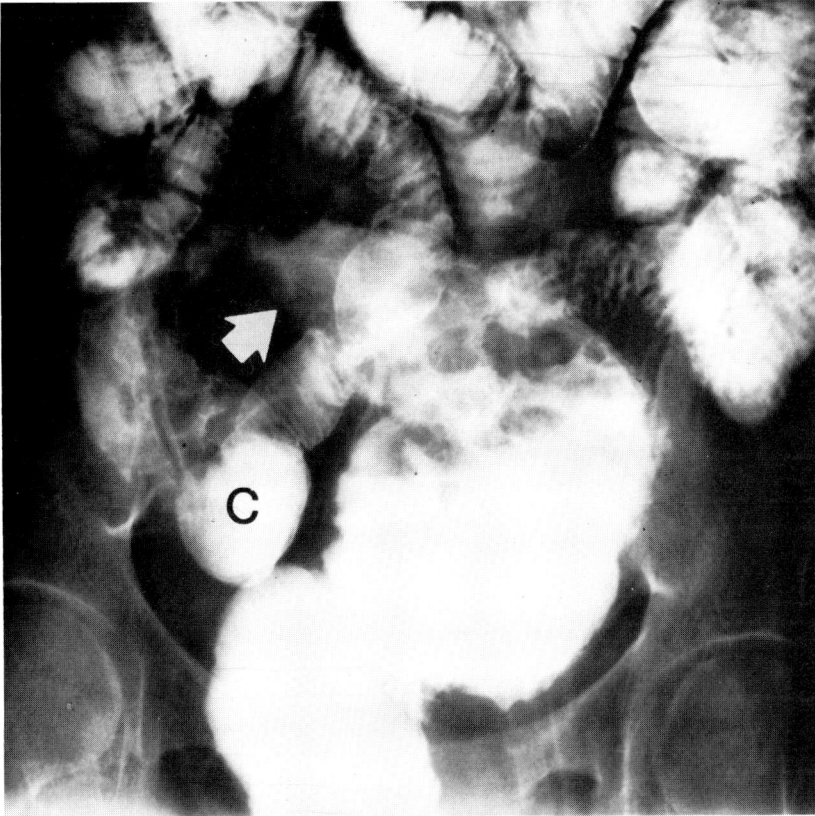

Figure 4–27.   Barium enema examination of a 37-year-old female showing remarkable shortening of the colon due to advanced ulcerative colitis. This is the first radiographic study performed on this patient and there was no previous surgery. Extensive pseudopolyposis is visible. The entire colon has shrunk into the pelvis. There is extensive barium reflux into the small bowel. (C = cecum; arrow = transverse colon.)

the majority of cases (Fig. 4–26). The disease is symmetrical and tends to extend proximally from the rectum. Marked shortening and narrowing occur in chronic cases (Fig. 4–27). Fistulas and strictures are uncommon. The ileocecal valve is often patulous and atrophic, allowing free reflux of barium into the ileum during barium enema examination.

When the right side of the colon is severely involved, the terminal ileum may be abnormal (backwash ileitis). As much as 25 cm. of ileum may be involved. The radiologic appearance may be identical to non-stenotic terminal regional enteritis.

TUBERCULOSIS OF THE ILEOCECAL AREA

Because of the marked decline in bovine tuberculosis brought about by pasteurization of milk and the development of effective treatment for pulmonary tuberculosis, the incidence of ileocecal tuberculosis in the United States has markedly decreased. In the United States tuberculous involvement of the intestine is usually secondary to the ingestion of infected sputum in patients with active pulmonary tuberculosis. This complication occurs in from 1 to 4 per cent of patients with active pulmonary lesions depending on the extent of the pulmonary disease; correspondingly, the chest x-ray shows active tuberculosis in the majority of cases in this country.[10] In countries such as India, where bovine tuberculosis is still prevalent, the disease is commonly acquired by ingestion of infected milk.

When gastrointestinal infections occur, 85 to 90 per cent of the lesions are located in the ileocecal region.[11] Both the ileum and the cecum are usually involved, with the cecal abnormality predominating. In approximately one-fifth of the cases, the cecum alone is affected. Even more rarely, the cecum is spared and the ileum is the sole site of the disease. After invasion of the mucosa, acid-fast bacilli localize in the submucosa and produce caseating granulomas. Both the ulcerative and hypertrophic forms of intestinal tuberculosis produce strictures and fibrosis.[12] In the ulcerative form irritability is marked, producing spasm of the cecum (Stierlin's sign). This is nonspecific and is seen in a number of other diseases involving the cecum. In the hypertrophic form, exuberant granulation tissue and fibrosis may produce a mass in the cecum. Complications of both forms include obstruction, abscess, fistulas, free perforation, intussusception and hemorrhage.

The radiographic manifestations of ileocecal tuberculosis, first described in 1911 by Stierlin, include edema and irritability of the terminal ileum and cecum[13] (Figs. 4–28 to 4–31). The mucosa appears coarse, smudged, nodular, irregular and ulcerated. Fibrosis shortens the ileum and cecum and distorts the ileocecal valve, often making identification of the ileocecal junction difficult. Concentric fibrosis leads to areas of localized stricture and obstruction.

Differential diagnosis from Crohn's disease, carcinoma, lymphoma and amebiasis is often difficult and sometimes impossible. Tuberculosis usually develops on both sides of the ileocecal valve and almost always involves the valve, which may become rigid and fixed in an open position. The cecum is retracted and deformed. The chest radiograph shows active tuberculosis. In Crohn's disease the cecum is more often spared or at least is less extensively involved than in tuberculosis, and the ileocecal valve is often enlarged and competent.[12] Fistulas and eccentric involvement with skip areas are characteristic of Crohn's disease. A fibrotic mass due to tuberculosis may simulate a cecal tumor. Carcinoma of the cecum rarely crosses the ileocecal valve to involve the ileum. Lymphoma, however, does so regularly, and differentiation from hyperplastic tuberculosis may be impossible. Amebiasis rarely involves the terminal ileum, but ulceration and fibrosis in the cecum and ascending colon may be identical in amebiasis and tuberculosis. Other diseases that involve the ileocecal area and produce deformity and edema of the terminal ileum and cecum such as appendiceal abscesses, metastases, diverticulitis and endometriosis may also simulate many of the radiologic changes caused by tuberculosis.

AMEBIASIS

Amebiasis is an acute and chronic inflammatory lesion of the colon caused by *Entamoeba histolytica*. It has been estimated that the incidence of the disease in the United States varies from 2 to 6 per cent,

(*Text continued on page 225*)

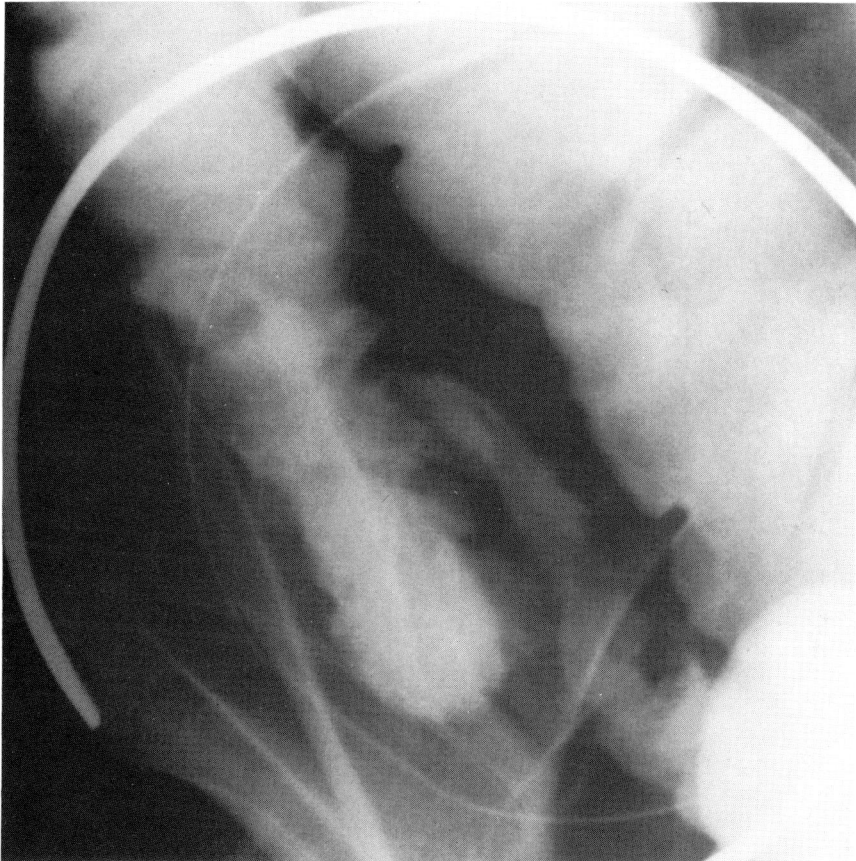

Figure 4-28. Barium enema examination showing tuberculosis involving the terminal ileum and cecum.

Figure 4-29. Barium enema examination (*A* and *B*) showing ileocecal tuberculosis. There is marked narrowing of the cecum, ascending colon and terminal ileum with dilatation of the small intestine proximal to the lesion.

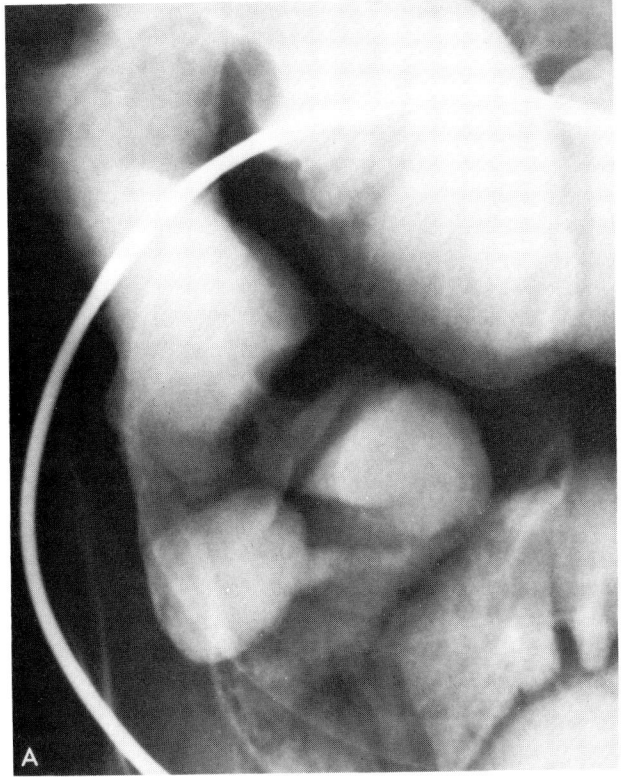

Figure 4-30. Barium enema examination (A) showing involvement of the cecum and ascending colon by tuberculosis. The terminal ileum is spared. Chest radiograph (B) showing extensive active pulmonary tuberculosis with cavitation.

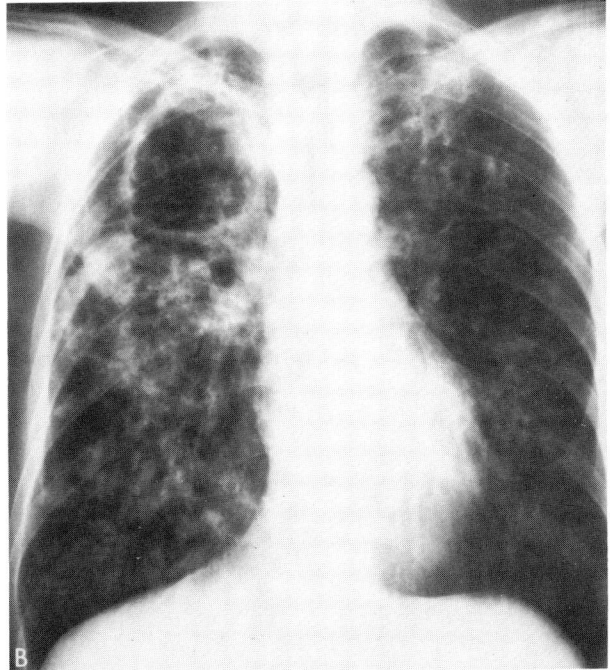

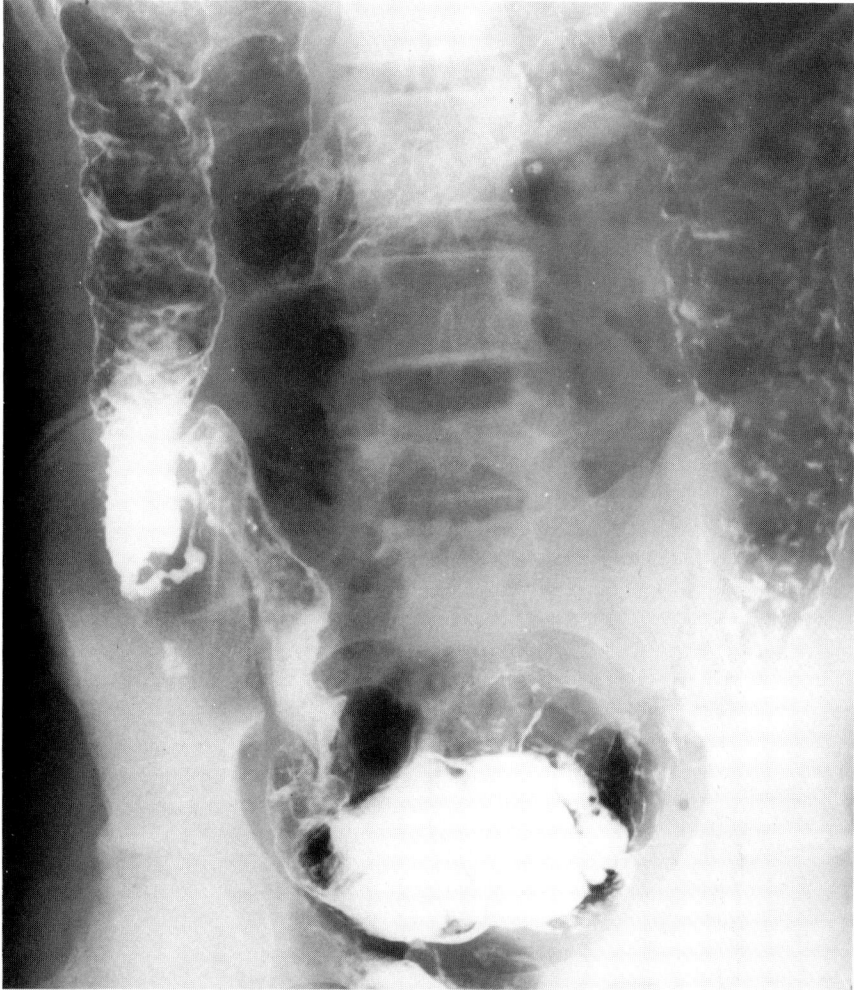

Figure 4-31. Air contrast enema examination showing tuberculosis of the ileocecal area involving the terminal ileum, cecum and appendix. (From Beneventano, T. C., Schein, C. J., and Jacobson, H. G.: The roentgen aspects of some appendiceal abnormalities. Amer. J. Roent., 96:344, 1966. Courtesy of Charles C Thomas, Publisher.)

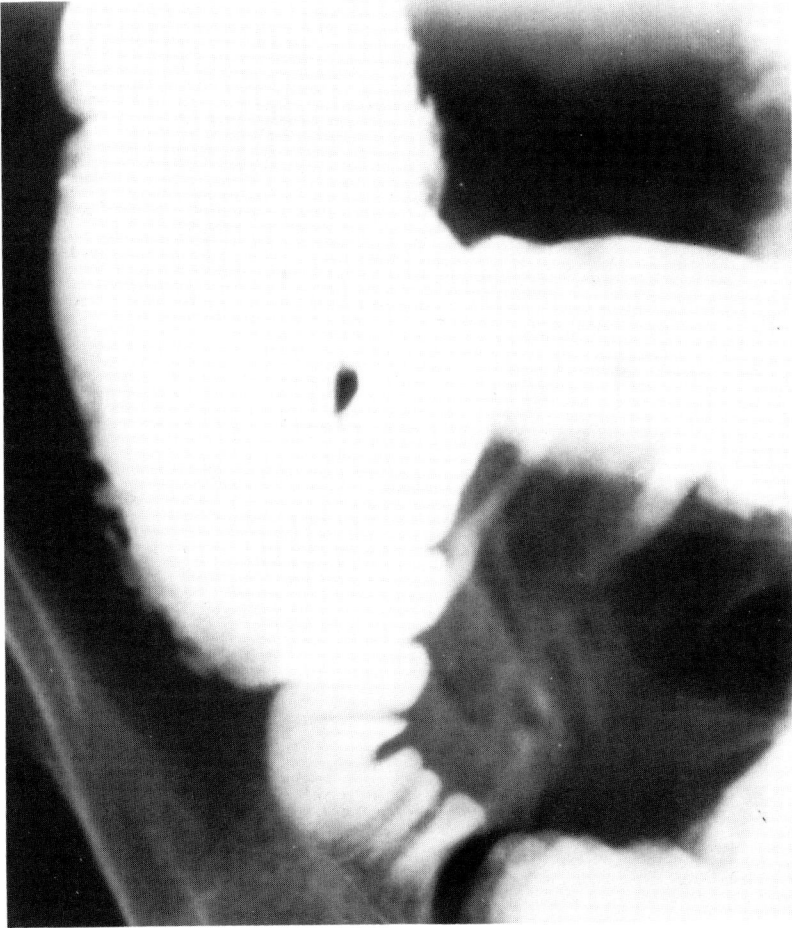

Figure 4-32. Barium enema examination showing acute amebiasis of the cecum. Deep "flask"-shaped ulcers are present.

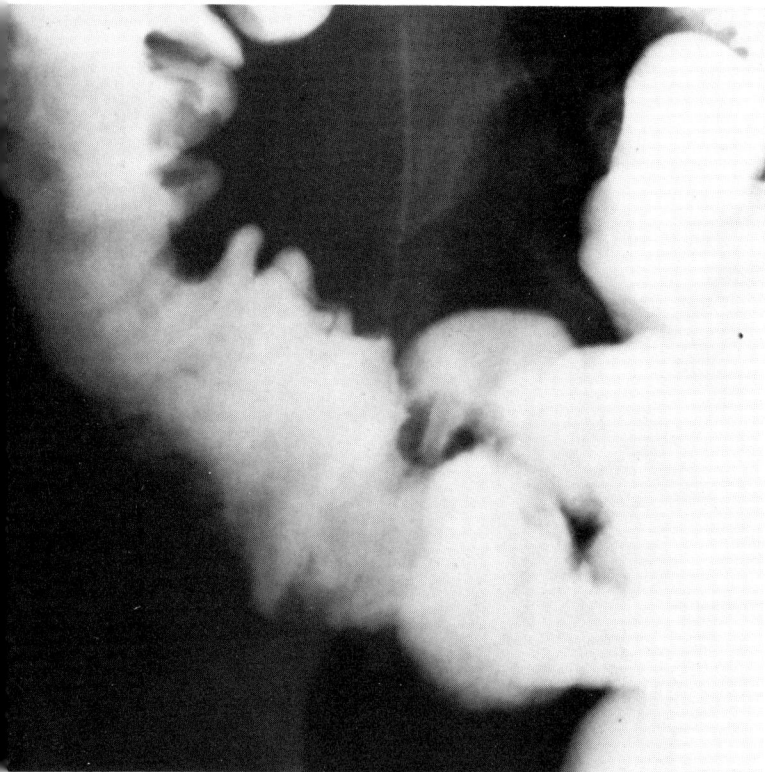

A

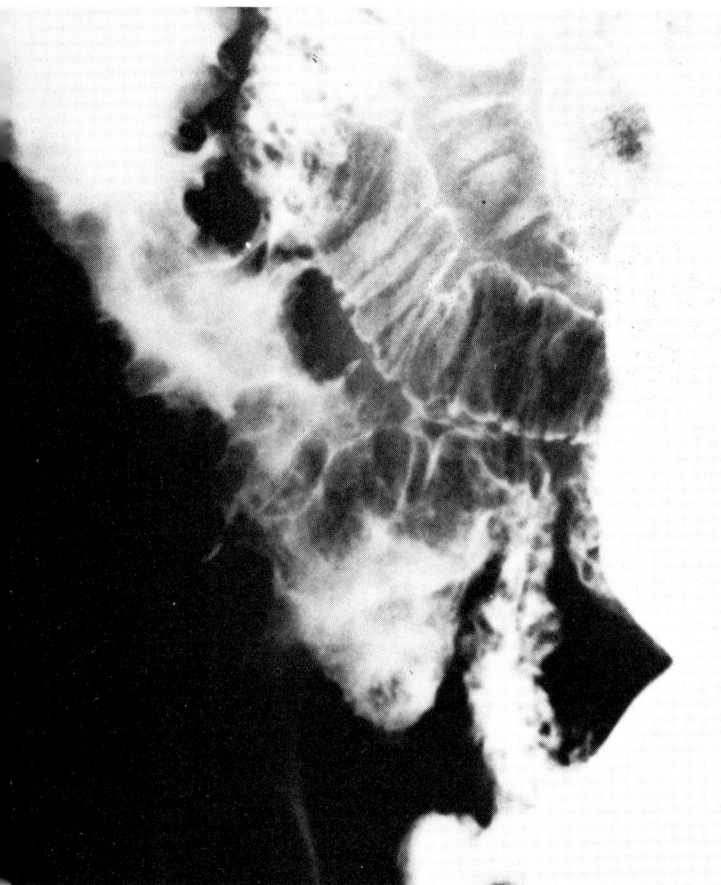

Figure 4–33. Barium enema (*A*) and postevacuation study (*B*) showing acute amebiasis. Extensive edema, spasm and ulceration are present in the cecum and ascending colon. The terminal ileum is not involved.

B

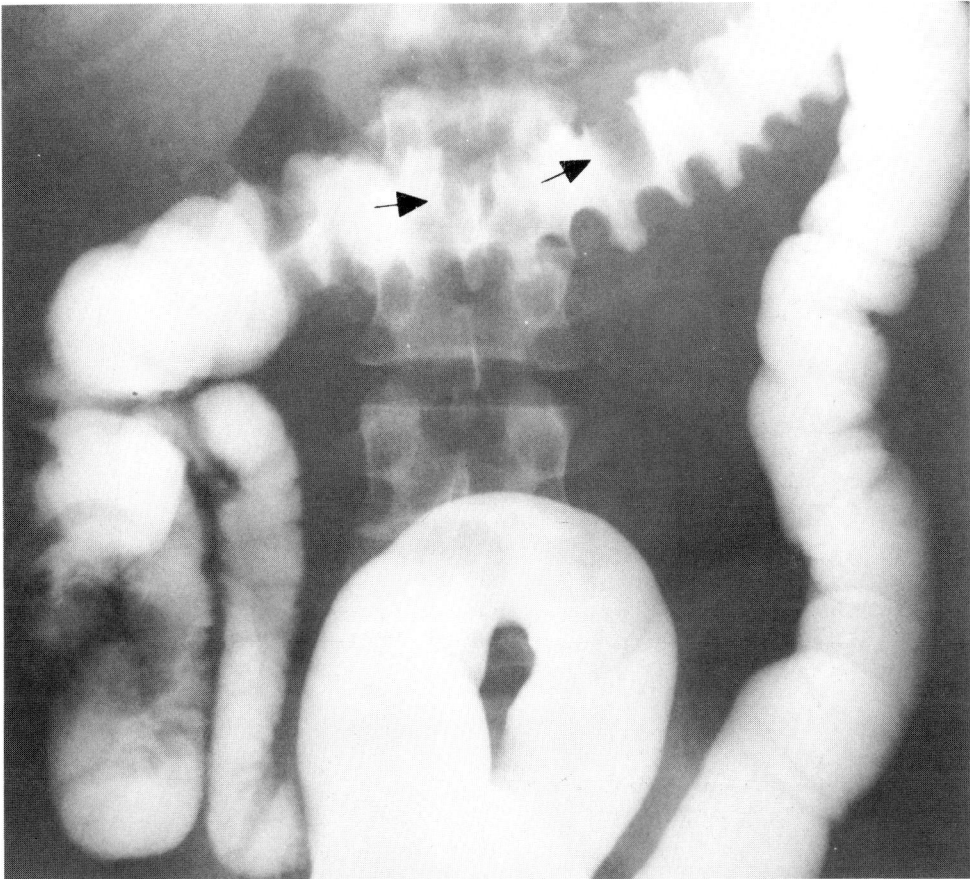

Figure 4–34.   Barium enema examination showing acute amebiasis of the cecum and ascending and transverse colon. Thumbprinting is present in the transverse colon (arrows). The cecum is deformed.

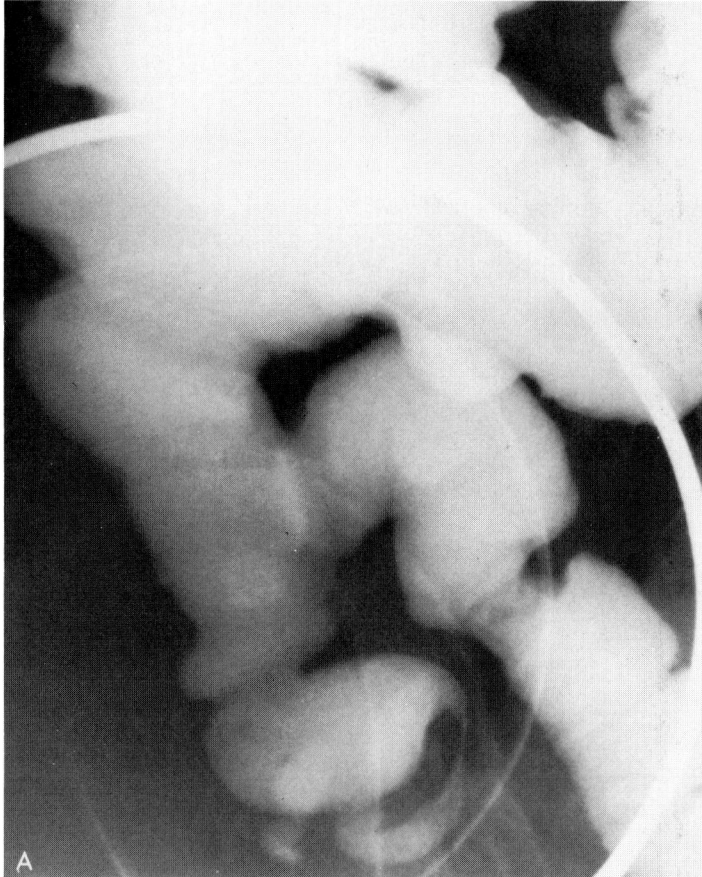

Figure 4–35. Barium enema examination (A) with air contrast study (B) showing amebiasis of the cecum. Narrowing and deformity of the cecum due to fibrosis without involvement of the terminal ileum are characteristic.

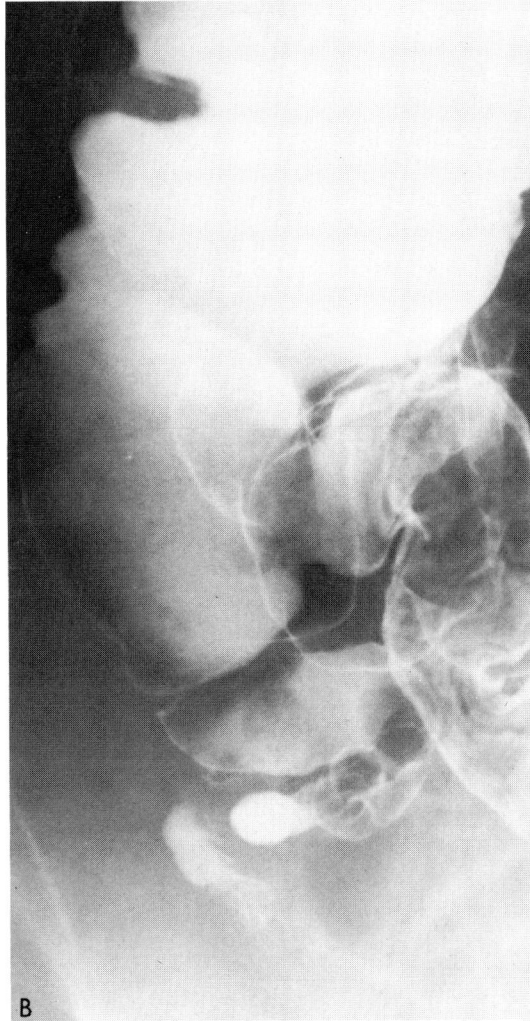

Figure 4–35. (Continued)

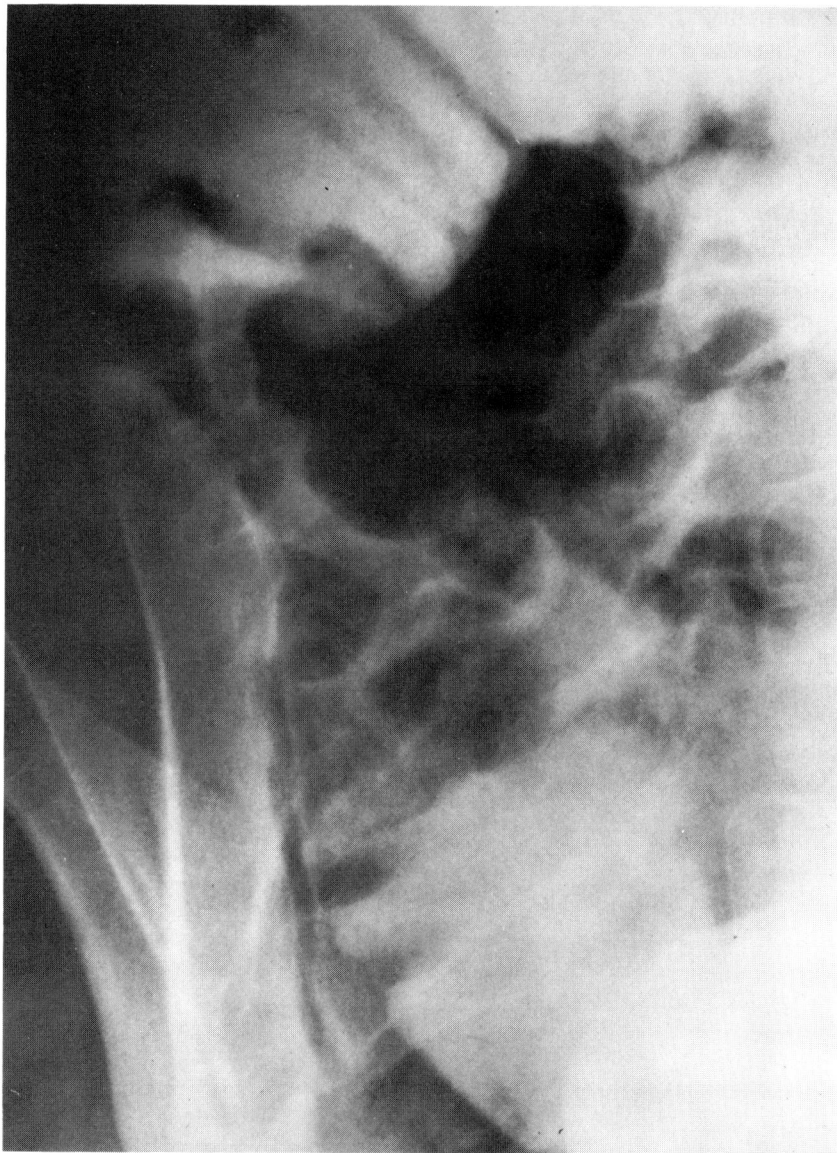

Figure 4-36. Barium enema examination showing an ameboma of the ascending colon. Localized narrowing is evident, with a mass present along the medial wall. An ameboma may simulate a polypoid or annular adenocarcinoma of the colon.

depending on the section of the country.[14] Most of the E. histolytica infections are silent or carrier infections. Less than 10 per cent of patients with proved amebiasis have acute dysentery.[15] Most symptomatic patients have an insidious course.

The infection occurs as the result of ingestion of viable cysts of E. histolytica in food and water contaminated with sewage. The walls of the cysts disintegrate in the alkaline milieu of the small intestine, forming trophozoites which invade the mucosa of the cecum and ascending colon. The cecum is involved in 90 per cent of the cases, although the rectosigmoid and transverse colon are affected in 30 per cent of the patients. Diffuse colonic involvement is not uncommon. The amebas grow in the submucosa causing small papillary areas of inflammation which lead to focal mucosal necrosis and ulceration. The ulcers are usually small and superficial, but occasionally they coalesce, extend into the submucosa and form deep, undermining, "flask"-shaped craters. The small bowel is rarely involved and then only in fulminating disease. An ameboma consisting of granulomatous and fibroplastic tissue may produce localized thickening of the bowel wall or a polypoid mass. Complications of amebiasis include liver abscess, intestinal obstruction, intussusception, toxic megacolon, hemorrhage and perforation.

Asymptomatic carriers have no radiographic abnormality. Radiographic findings occur in 70 per cent of patients with acute amebiasis and in 40 per cent of patients with chronic disease.[16] The radiographic features reflect the ulceration, inflammation and fibrosis that occur. The mucosa of the cecum becomes granular, thickened and irregular. The individual ulcers are usually too shallow to be identified radiologically, but marginal irregularity of the barium column occurs. Occasionally, ulcers coalesce producing deep craters with undermining and overhanging margins (Fig. 4–32). The inflammatory process causes spasm, edema, narrowing and deformity of the cecum (Fig. 4–33). Skip areas in other parts of the colon are common (Fig. 4–34). As healing occurs with fibrosis and scarring, there are narrowing of the lumen and shortening and contraction of the bowel, leading to the classic roentgenologic features of the cone-shaped cecum (Fig. 4–35). The ileocecal valve is usually patulous, often allowing striking reflux of barium during barium enema examination. An ameboma may produce a localized area of rigidity and deformity, intraluminal tumor mass or a combination of both[18] (Fig. 4–36).

Concentric contraction of the cecum associated with a normal terminal ileum is the typical finding in amebiasis. Eighty per cent of patients with Crohn's disease of the colon have associated disease of the terminal ileum. When the colon alone is involved, Crohn's disease is usually diffuse. Eccentric involvement, cobblestone mucosal pattern, deep ulcers and fistulas are characteristic. Carcinoma must be carefully excluded, because surgical intervention in the presence of amebiasis may be disastrous. Postoperative complications in patients with amebiasis are frequent and include fecal fistula, exacerbation of the disease, peritonitis and hemorrhage. Cecal involvement with amebiasis is more symmetrical, and irritability is more marked than with carcinoma of the cecum. Involvement of other segments of the colon may occur. Whenever the distinction between amebiasis and tumor is difficult, a course of therapy for amebiasis is indicated. The radiographic findings should improve following therapy if the changes are due to amebiasis.

Diverticulitis of the cecum usually is eccentric, produces an intramural abscess and is usually associated with other diverticula of the colon. Tuberculosis may produce marked deformity and irritability of the cecum, but the terminal ileum is usually involved. In ulcerative colitis, there is diffuse involvement of the colon, frequently with shortening and narrowing of the bowel. Localized areas of colonic disease in addition to the cecal lesion are not uncommon in tuberculosis as well as in amebiasis. Appendicitis with abscess, foreign body perforation with abscess and actinomycosis usually cause a smooth extrinsic impression on the cecum. Actinomycosis most commonly involves the appendix and is often associated with cutaneous fistulas. Ischemia or submucosal hemorrhage from leukemia, Henoch's purpura and Dicumarol therapy changes rapidly from day to day. The history is often helpful in differentiation. Thumbprinting, the classic radiographic feature in ischemia and hemorrhage, may also occur in amebiasis.

NEMATODIASIS OF THE ILEOCECAL AREA

Round worm infestations of the gastrointestinal tract include ascariasis, hookworm, strongyloidiasis, trichuriasis and anisakiasis. Of these the hookworm and strongyloidiasis involve the upper small bowel and will not be considered further. Trichuriasis and anisakiasis characteristically involve the cecum, while ascariasis may involve any portion of the small bowel, especially the jejunum.

Infestation of the intestine with the nematode *Ascaris lumbricoides* is estimated to occur in as high as 30 per cent of the world population, and in certain areas of the tropics the figure may reach 80 per cent.[18] Ninety-nine per cent of the worms are found in the small intestine, particularly the jejunum, but they may be seen anywhere from the stomach to the rectum. The ova are ingested in contaminated food, water or soil. Larvae hatch in the small bowel, migrate to the lungs via the venules and lymphatics, travel up the bronchi and trachea and are swallowed. The worm matures in the small intestine, growing to 15 to 35 cm. in length. The worm is seen as a long, smooth, cylindrical defect in the barium, and its intestinal tract may be identified as a thin string of barium (Fig. 4–37).

Trichuriasis, or whipworm, is a common nematode in warm, moist climates.[19] The ova are ingested in contaminated soil, hatch in the duodenum and develop into adult worms in the cecum, where they become embedded between villi. There is usually little reaction at the site of their attachment. Barium enema examination discloses a granular mucosal pattern with flocculation of barium. In severe cases the entire colon,

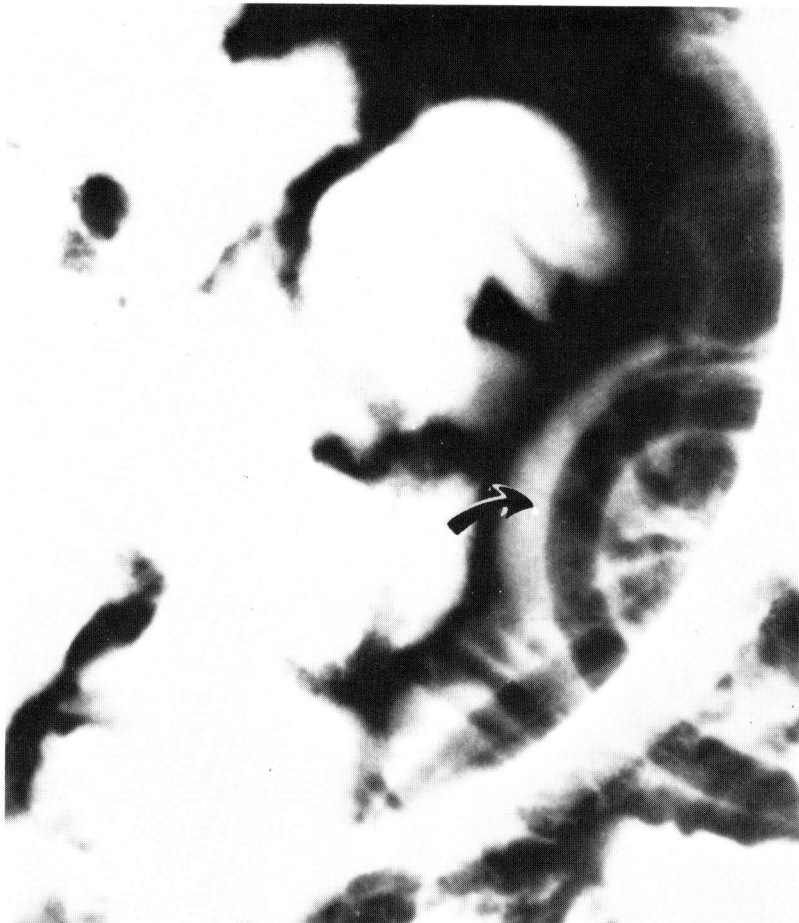

Figure 4–37. Small bowel examination showing an Ascaris in the small intestine (arrow).

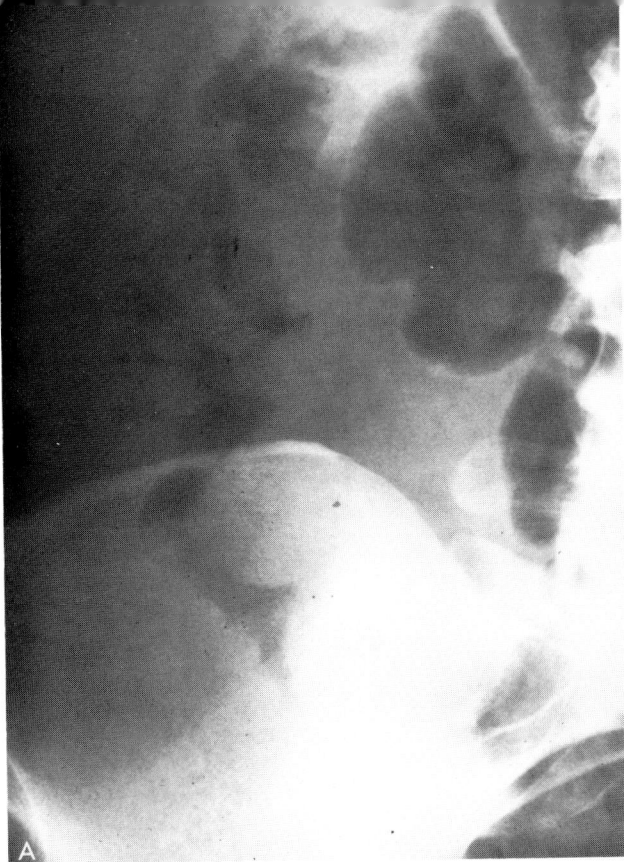

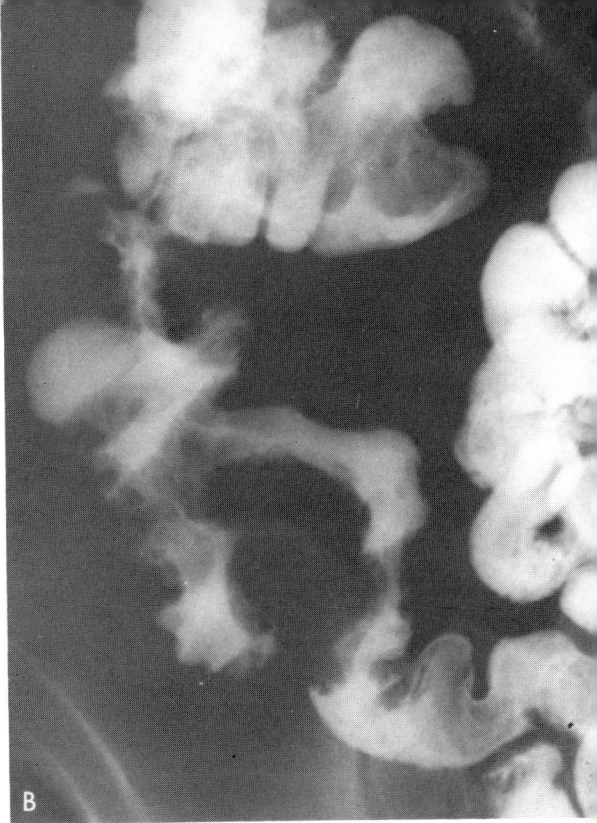

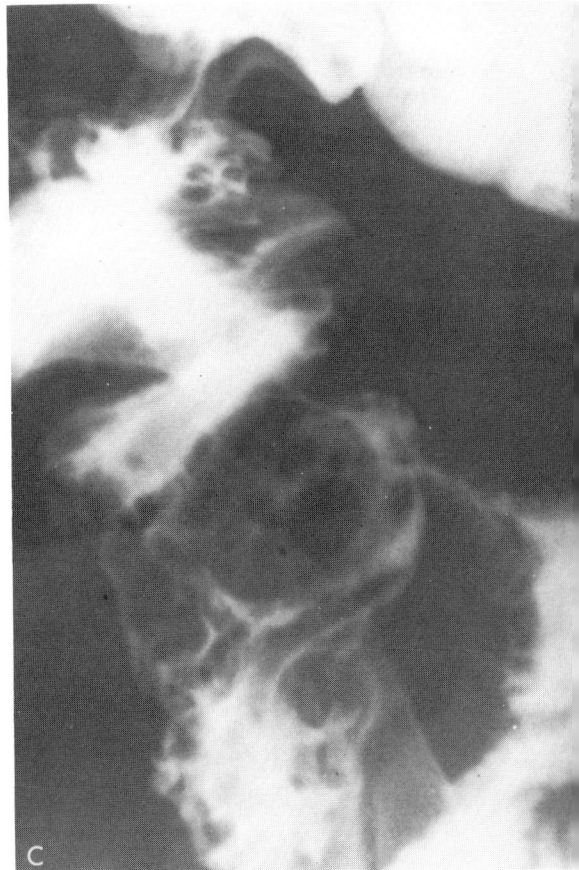

Figure 4–38. Plain film (*A*) and barium enema (*B* and *C*) showing ileocolitis due to anisakiasis. The plain abdominal radiograph shows thickening and irregularity of the cecum and ascending colon. Edema and irritability are apparent on the barium enema. (From Richman, R. H., and Lewicki, A. M.: Right ileocolitis secondary to anisakiasis. Amer. J. Roent., *119*:329, 1973. Courtesy of Charles C Thomas, Publisher.)

appendix and terminal ileum may be involved. Air contrast enema studies may have a target appearance due to the coiled portion of the worm lying in mucus. The mucosal pattern is similar to the coarse folds described in children and young adults with mucoviscidosis.

Anisakiasis is a disease produced by the nematode, anisakis. The disease should be suspected in patients with right ileocolitis who have a history of eating raw, slightly salted or vinegar-pickled fish.[20] The final hosts are marine mammals such as the whale and dolphin. The larvae hatch in sea water and are eaten by crustaceans, which in turn are ingested by sea fish such as herring, cod and mackerel. The larvae penetrate the gastrointestinal tract of the fish and develop in muscle. The cycle is completed when the marine mammals ingest the fish. When man eats raw fish, the larvae are found in the wall of the intestine in one-third of the cases and in the stomach in two-thirds. The anisakis grows in the submucosa, inciting a massive eosinophilic infiltrate and edema. Plain abdominal radiographs and barium enema studies show thickening of the bowel wall of the ascending colon or terminal ileum similar to the radiographic changes in Crohn's disease, tuberculosis, ischemia, hemorrhage and tumor (Fig. 4–38). Differential diagnosis is difficult unless the history that the patient has eaten raw, pickled or salted fish is known.

TYPHOID FEVER

Typhoid fever is an acute, often severe illness caused by *Salmonella typhosa* char-

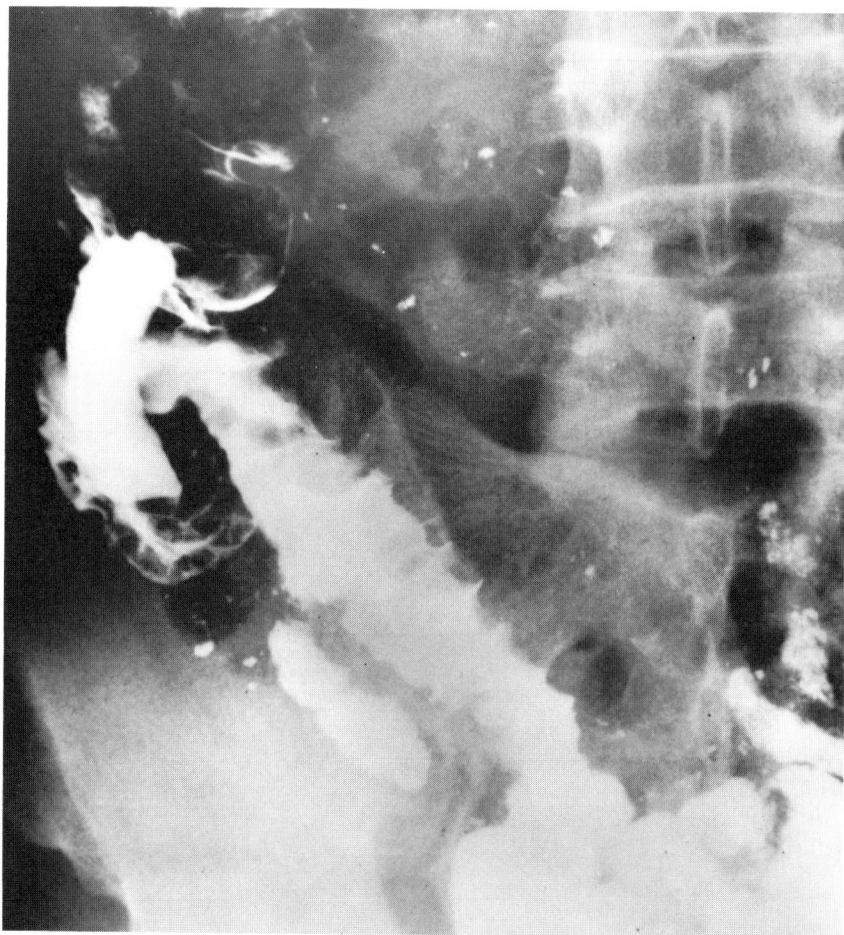

Figure 4–39. Small intestine examination showing edema and ulceration of the terminal ileum due to typhoid fever. The appearance of the ileum returned to normal promptly after antibiotic therapy. (From Francis, R. S., and Berk, R. N.: Typhoid fever. Radiology, *112*:583, 1974.)

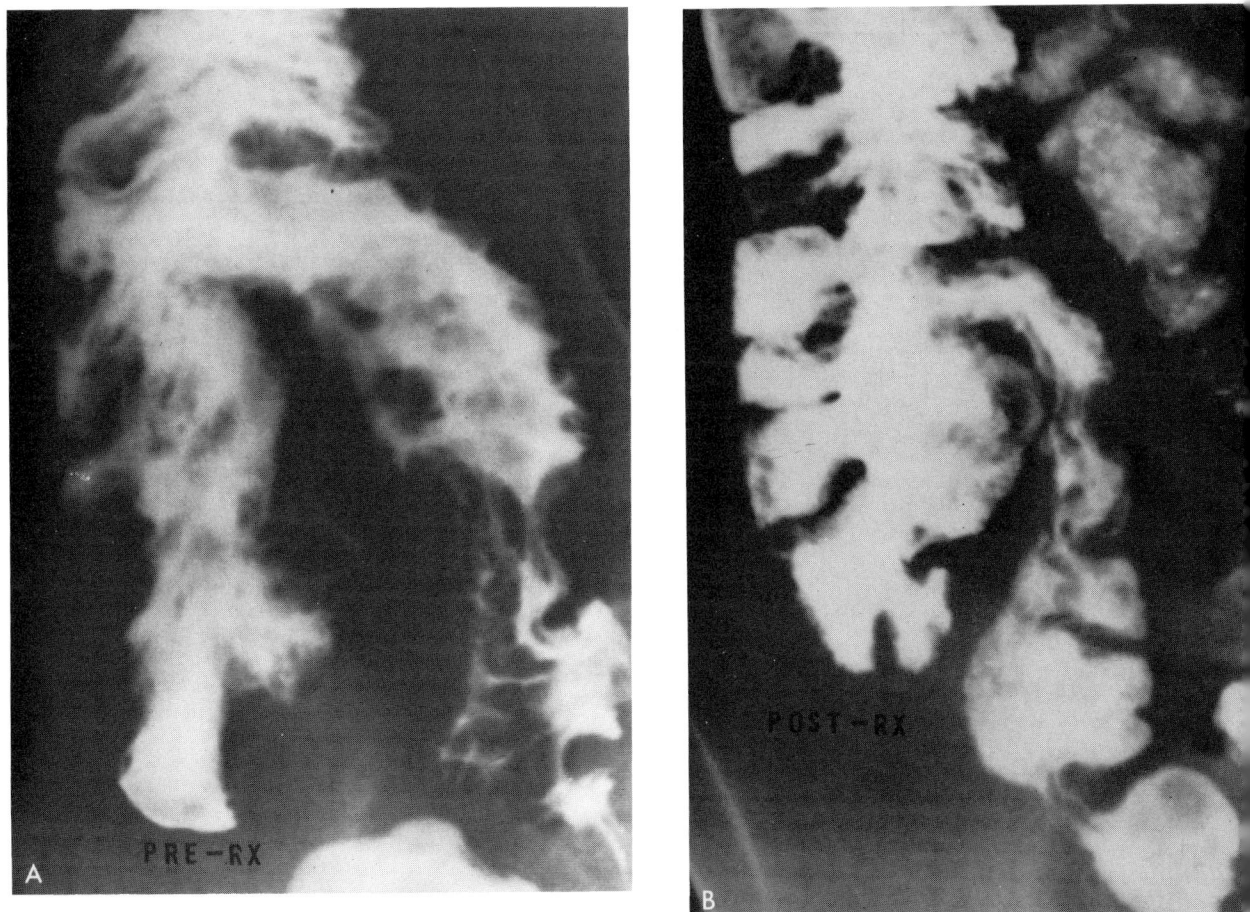

Figure 4-40. Small bowel examination (A) showing edema and nodularity of the terminal ileum and cecum due to typhoid fever. (B) The ileum and cecum have returned to normal after antibiotic therapy. (From Francis, R. S., and Berk, R. N.: Typhoid fever. Radiology, *112*:583, 1974.)

acterized by fever, malaise, anorexia, splenomegaly, a maculopapular rash and leukopenia.[21] The organism is a gram-negative, nonspore-forming bacillus which may produce an acute inflammatory response that consists of hyperplasia of the reticuloendothelial system. In uncomplicated cases the radiologic manifestations of the disease are confined to the terminal ileum and cecum.

Typhoid fever is entirely limited to man and is transmitted by bacterial contamination of food and water with human feces. Once in the gastrointestinal tract, the organisms are phagocytosed by lymphoid tissue, particularly in the Peyer's patches of the terminal ileum and the smaller lymph follicles in the cecum. There the organisms multiply and produce raised plaques which delineate the Peyer's patches. Necrosis of the overlying mucosa causes linear ulcers which are orientated parallel to the longitudinal axis of the ileum. Following the primary enteric phase of the disease, bacteremia occurs, with subsequent involvement of the reticuloendothelial elements of the liver, spleen, lymph nodes and gallbladder.

The radiologic findings in the terminal ileum in patients with typhoid fever have been described by DeBusscher[22] (Figs. 4–39 and 4–40). The author notes mucosal nodularity, spasm and flocculation of the barium. Only a minority of the cases had marginal serrations indicative of superficial ulcerations. Although healing with fibrosis and stricture may occur in some patients with typhoid fever, the ileum usually returns to normal after treatment.

Radiologic differentiation from regional

enteritis may be difficult. Typhoid rarely involves the cecum and does not produce the pseudodiverticula, fistulas and mesenteric edema that are noted in regional enteritis. Mild to moderate splenomegaly often occurs with typhoid and is absent in Crohn's disease. In tuberculosis, amebiasis and actinomycosis cecal involvement usually predominates. When the differential diagnosis is uncertain, follow-up radiologic examination after appropriate antibiotic therapy for typhoid will show resolution of the inflammatory process in the terminal ileum with return of normal radiologic features.

EVANESCENT COLITIS IN THE YOUNG ADULT

In 1971 Miller et al. reported five patients with a segmental inflammatory disease of the colon which they felt was distinct from granulomatous colitis, ischemic colitis and ulcerative colitis.[23] None of the five had gastrointestinal symptoms or pertinent medical history prior to the acute onset of symptoms of abdominal pain and diarrhea. Of the five, three had bloody bowel movements. All five were younger than 42 years of age and two were younger than 22 years of age. No patient was on any medication which has been associated with decreased coagulability states. Stool cultures and examinations were negative.

The barium enema examination showed segmental involvement of the colon, with spasm, thumbprinting and ulcerations (Fig. 4–41). In four of the five the right side of the colon was involved. Two had separate, skip areas of the disease in the colon. Symptoms disappeared rapidly within 2 weeks in all five patients. Each patient was followed from 1 to 4 years after initial hospitalization, without recurrence of gastrointestinal complaints. The follow-up barium enema studies of each patient were normal.

Miller et al. feel that segmental colitis in a young adult in which reversibility of the lesion occurs in a short time without therapy is unique.[23] Ischemic colitis may be reversible but it is rare in patients younger than 42 years of age without associated abnormalities. The clinical course and rapid reversibility are not typical of ulcerative colitis or regional enteritis. Miller et al. conclude that while evanescent colitis in

young adults may be an unusual clinical manifestation of ischemia of the colon, the disease is unique and may represent a separate entity of unknown etiology.

HEREDITARY ANGIONEUROTIC EDEMA

Hereditary angioneurotic edema is a disease characterized by circumscribed, noninflammatory edema of the skin, mucous membrane and viscera. The patients lack the inhibitor (C′ 1 esterase inhibitor) of the activated first component of complement (C′ esterase).[24]

The initial symptoms may appear at any age. Visceral manifestations may be the first or the only findings of the disease. Attacks of localized cutaneous swelling or episodes of abdominal pain occur spontaneously at intermittent intervals. Any portion of the gastrointestinal tract may be involved, but the small intestine predominates.

The radiographic features of the disease consist of thickened, irregular, smudged mucosal folds in a localized segment of the small intestine[25] (Fig. 4–42). Spiculation, thumbprinting and a "stacked coin" appearance of the mucosa may occur. The findings are present only when the patient is in a visceral crisis and resolve as the attack of abdominal pain subsides.

The differential diagnosis includes ischemia and submucosal hemorrhage from any cause. Rectal bleeding is frequent in these cases but rarely occurs in patients with angioneurotic edema. A positive family history is characteristic of angioneurotic edema. Nonstenotic regional enteritis may produce similar radiographic changes in an isolated segment of the small intestine. However, the eccentric involvement, pseudodiverticula, cobblestone pattern, fissures, ulcerations and fistulas that are characteristic of regional enteritis do not occur in angioneurotic edema. The clinical history and the rapid reversibility of the radiographic changes in angioneurotic edema are important differential features.

RADIATION ENTERITIS

Of the various parts of the small intestine, the ileum is the least sensitive to radiation injury.[26] Between 5000 and 6000 rads there

(Text continued on page 233)

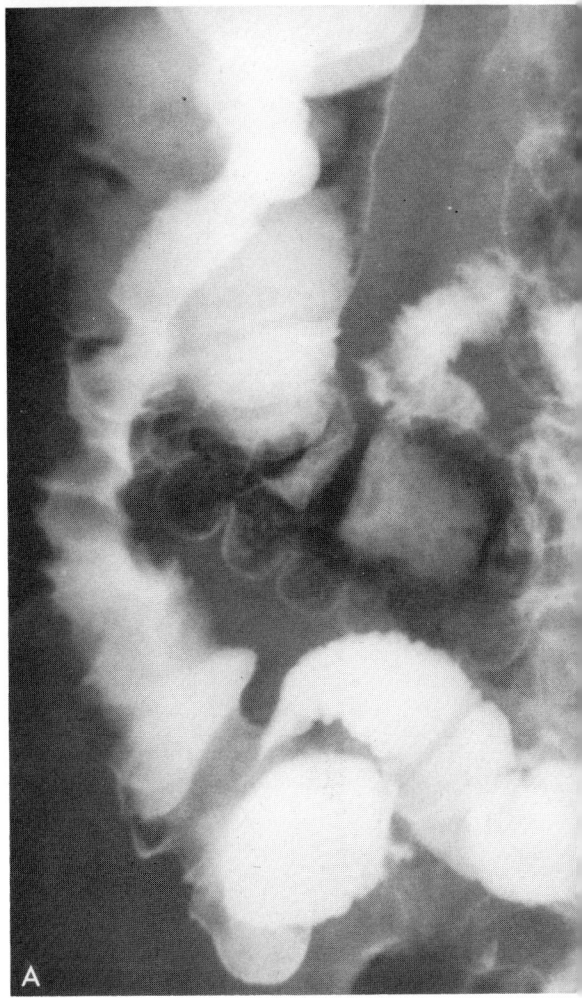

Figure 4-41. Barium enema examination (A) showing edema, irritability and mucosal thickening of the cecum and ascending colon due to evanescent colitis of young adults. The remainder of the colon was normal. The patient was a 32-year-old female with abdominal pain and diarrhea of acute onset.

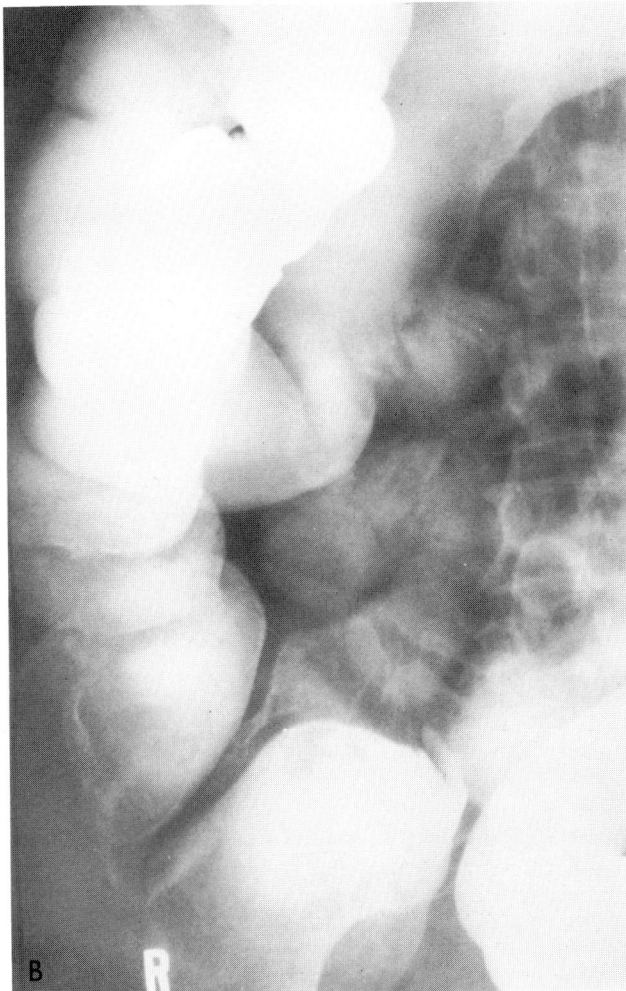

Figure 4-41. (*Continued*) Follow-up barium enema two weeks later (B) after the symptoms had subsided showing that the colon has returned to normal.

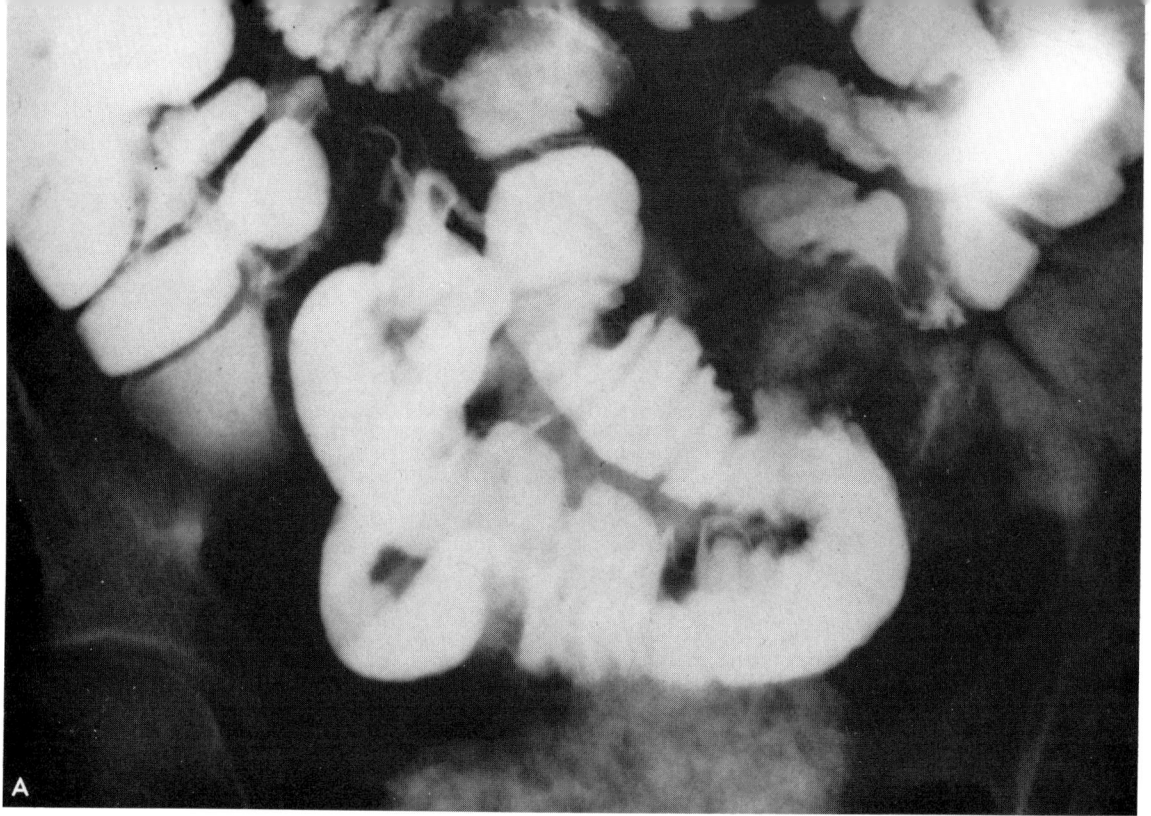

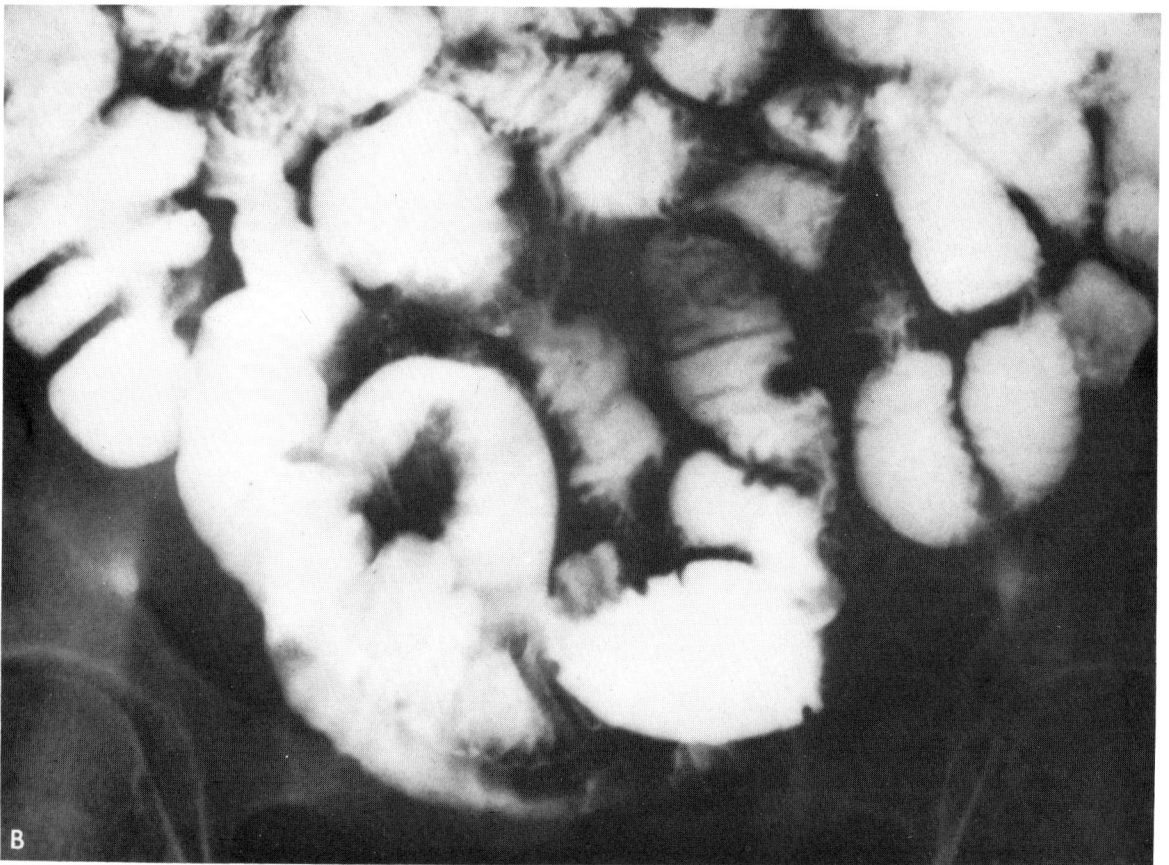

Figure 4-42. Small bowel examination in a 24-year-old woman with abdominal pain and diarrhea of abrupt onset. The patient had a history of intermittent episodes of cutaneous angioneurotic edema. Edema of the last several feet of the ileum is apparent (*A* and *B*). Repeat examination 3 weeks later when the symptoms subsided showed no abnormality of the bowel (*C*).

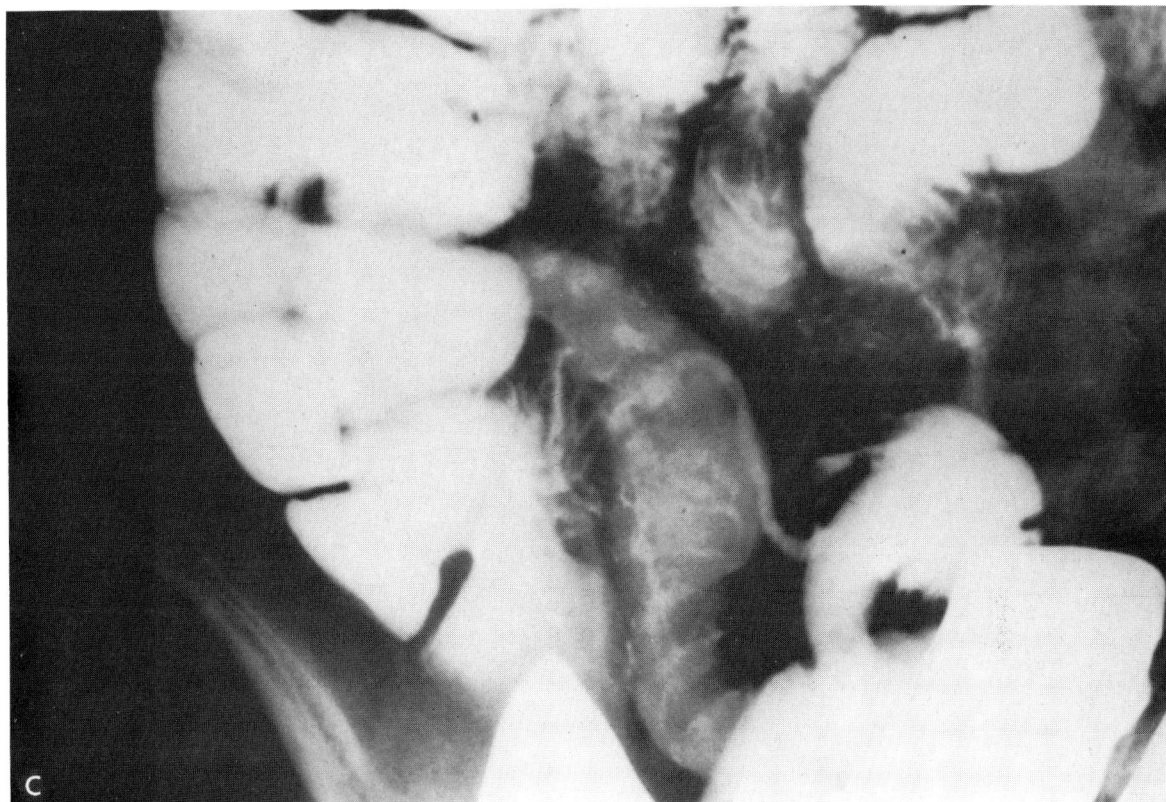

Figure 4–42. (Continued)

is a 50 per cent probability of serious radiation injury of the ileum occurring if the patient survives the malignant disease. Most radiation damage results when the small intestine is fixed in the pelvis by adhesions because the intestinal segment is not free to move out of the direct radiation beam at times during the daily radiotherapy. The incidence of radiation enteritis following pelvic radiation reported in the literature varies from 0.6 per cent to 17.0 per cent of patients.[27]

The radiation injury may be either acute or chronic. Acute damage occurs during and shortly after the course of radiotherapy. Plain abdominal radiographs may show paralytic ileus involving the affected intestinal segments. The pathologic changes of chronic radiation injury are due to endarteritis with vascular occlusion.[28] Although all layers of the bowel wall are affected, the submucosa is most severely involved. The radiologic manifestations consist of atrophy and shallow ulceration of the mucosa. The bowel wall is thickened and the lumen is narrowed (Fig. 4–43). When obstruction occurs, there is proximal dilatation. Submucosal infiltration with chronic inflammatory tissue and fibrosis causes thickening, straightening and irregularity of the folds. Nodular filling defects due to marked, irregular submucosal thickening may occur. The mesentery is often shortened and thickened, producing angulation and separation of the bowel loops. The intestine may become matted together and encased in a fibrous capsule.

Mason et al. emphasize that the early changes of radiation fibrosis are often subtle.[29] The radiographic features commonly mimic those of metastases to the small bowel, with kinking and nodularity of the involved loops. Chronic radiation enteritis may simulate stenotic regional enteritis. Differentiation usually requires knowledge of prior radiation treatment.

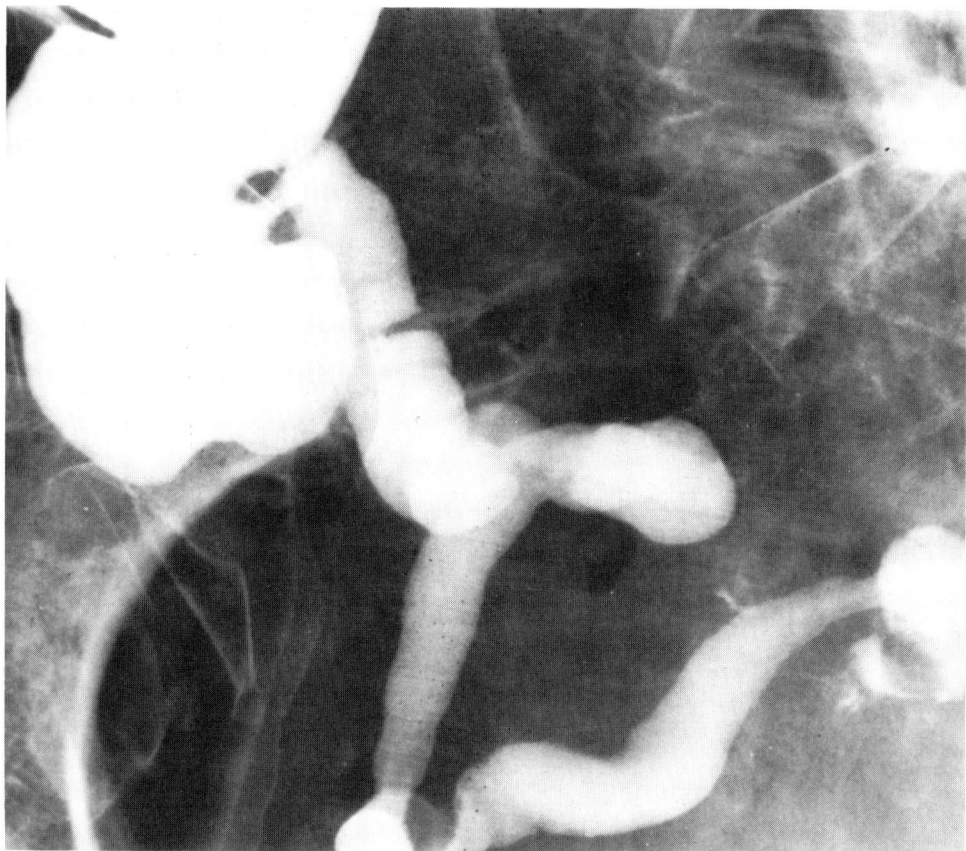

Figure 4-43. Barium enema examination showing narrowing, rigidity and atrophy of a long segment of the terminal ileum due to radiation enteritis. Six years earlier the patient received a course of radiotherapy for carcinoma of the cervix.

DIVERTICULITIS OF THE ILEOCECAL AREA

Cecal diverticulitis is a relatively uncommon disease that is usually discovered at surgery in patients operated upon with the mistaken diagnosis of acute appendicitis. Nissenbaum et al. reviewed 166 cases reported in the literature.[30] They noted that in the collected cases the patients were considerably younger and more females were affected than in cases with sigmoid diverticulitis.

Cecal diverticula are often single and generally contain all three layers of the bowel wall, rather than consisting of mucosa as is the case with sigmoid diverticula. This finding plus the fact that patients are younger suggests that the diverticula are congenital in origin rather than acquired as with sigmoid diverticula.

The clinical manifestations of cecal diverticulitis are similar to those of acute appendicitis, with generalized abdominal pain that eventually localizes in the right lower quadrant. The preoperative diagnosis is almost always acute appendicitis.

Because of the close similarities with acute appendicitis, few patients with cecal diverticulitis are given preoperative barium enema examinations. However, barium studies may disclose a localized mural abscess in the wall of the colon (Figs. 4-44 and 4-45). The lesion is usually smooth, eccentric and sharply demarcated from the adjacent colonic wall. The appendix may fill, excluding the presence of acute appendicitis. In the absence of tenderness, irritability and the presence of other diverticula, the differentiation from a polypoid carcinoma may be impossible. Correlation of the radiologic and clinical findings is mandatory.

Ileal diverticula occur on the mesenteric side of the bowel, supporting the theory that

Figure 4-44. Barium enema examination showing a mass in the ascending colon due to a diverticular abscess. The intramural mass in this case cannot be distinguished from adenocarcinoma of the colon.

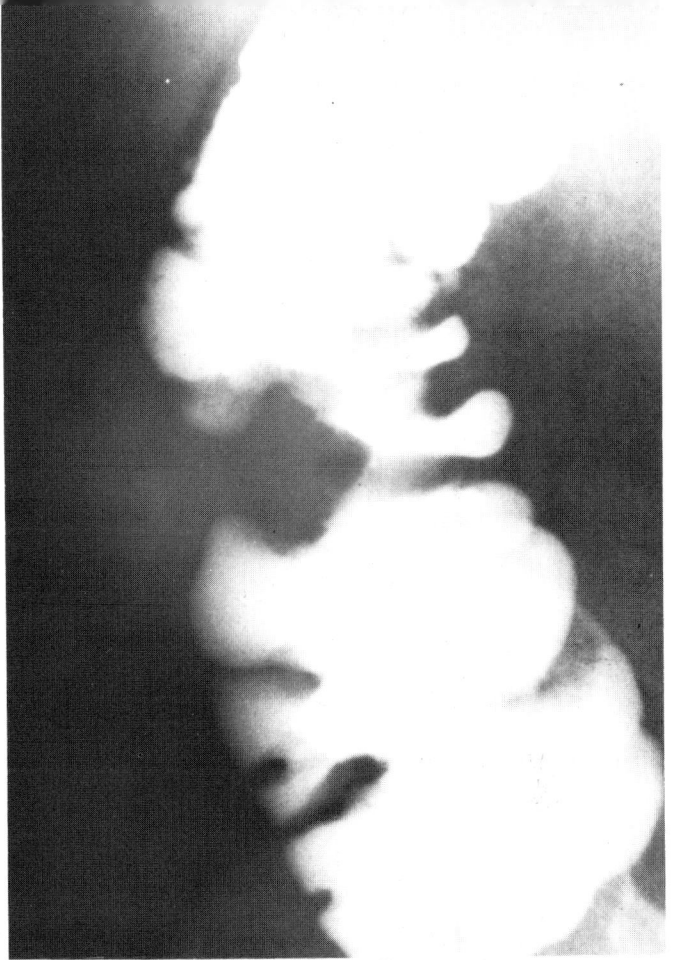

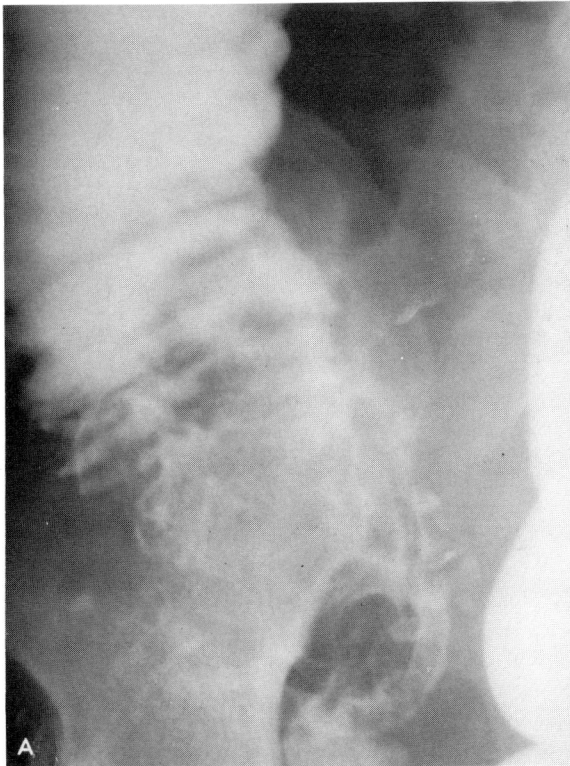

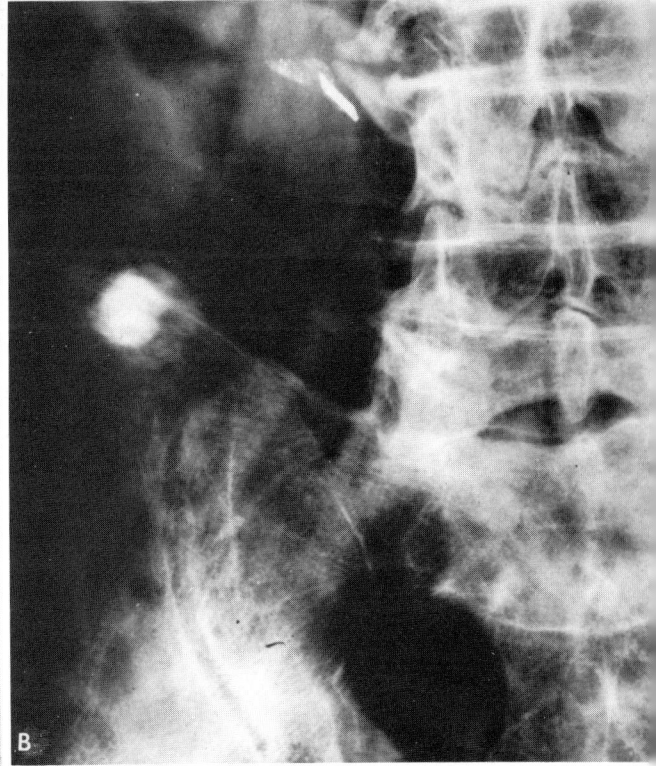

Figure 4-45. (A) Cecal diverticulitis, in which a large inflammatory mass is causing distortion and compression of the cecum, but the mucosa is intact. Several cecal diverticula are present. The appendix is filled with barium. (B) Plain abdominal radiograph in another case showing calcification in a cecal diverticulum. (From Felson, B., and Wiot, J. F.: Some interesting right lower quadrant abnormalities. Radiol. Clin. N. Amer., 7:83, 1969.)

Figure 4–46. Barium enema examination showing several small diverticula of the terminal ileum.

the diverticula develop at the site of vascular entry into the bowel wall. The majority are in the terminal portion of the ileum (Fig. 4–46). Parulekar reported 46 cases of acquired terminal ileal diverticula in 63 cases of small bowel diverticulosis.[31] Felson was only able to find three cases in the literature.[32] Parulekar claims that the difference is careful routine spot film examination of all patients having barium studies.[31] The ratio of jejunal to ileal diverticula in Parulekar's series was 1 to 8, whereas Felson states that the ratio is 5 to 1.

The size of the diverticula ranges from slightly less than 2 mm. to 15 mm. They are generally small, resemble sigmoid diver-

ticula and may be single or multiple.

Most ileal diverticula lie near the ileocecal valve. Meckel's diverticulum is usually more proximally situated and arises on the antimesenteric border. It is generally larger and longer than the acquired type.

Complications of terminal ileal diverticula are rare despite their similarity to sigmoid diverticula. Acute diverticulitis is probably the result of irritation or occlusion of the diverticulum by food particles or a foreign body (Figs. 4–47 and 4–48). Local swelling occurs and causes further stasis. The inflammation tends to extend intramurally, but perforation may supervene, with formation of a localized abscess or general-

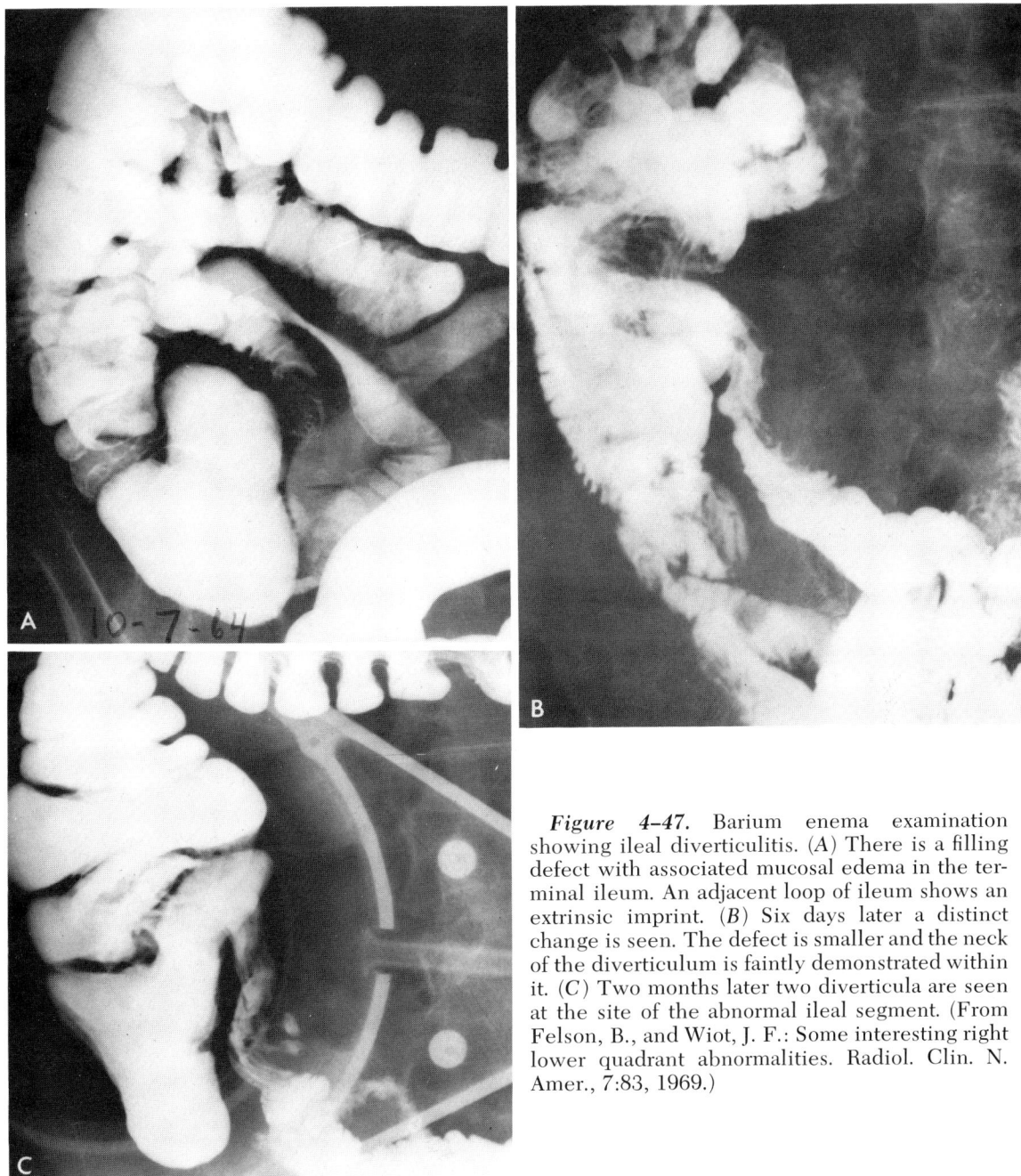

Figure 4-47. Barium enema examination showing ileal diverticulitis. (A) There is a filling defect with associated mucosal edema in the terminal ileum. An adjacent loop of ileum shows an extrinsic imprint. (B) Six days later a distinct change is seen. The defect is smaller and the neck of the diverticulum is faintly demonstrated within it. (C) Two months later two diverticula are seen at the site of the abnormal ileal segment. (From Felson, B., and Wiot, J. F.: Some interesting right lower quadrant abnormalities. Radiol. Clin. N. Amer., 7:83, 1969.)

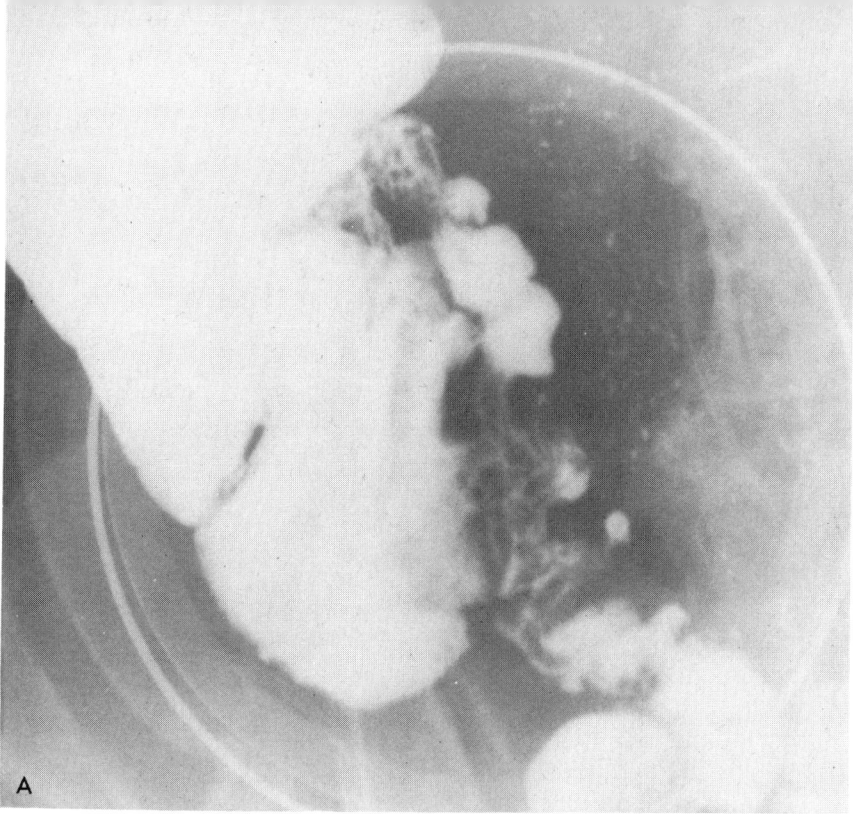

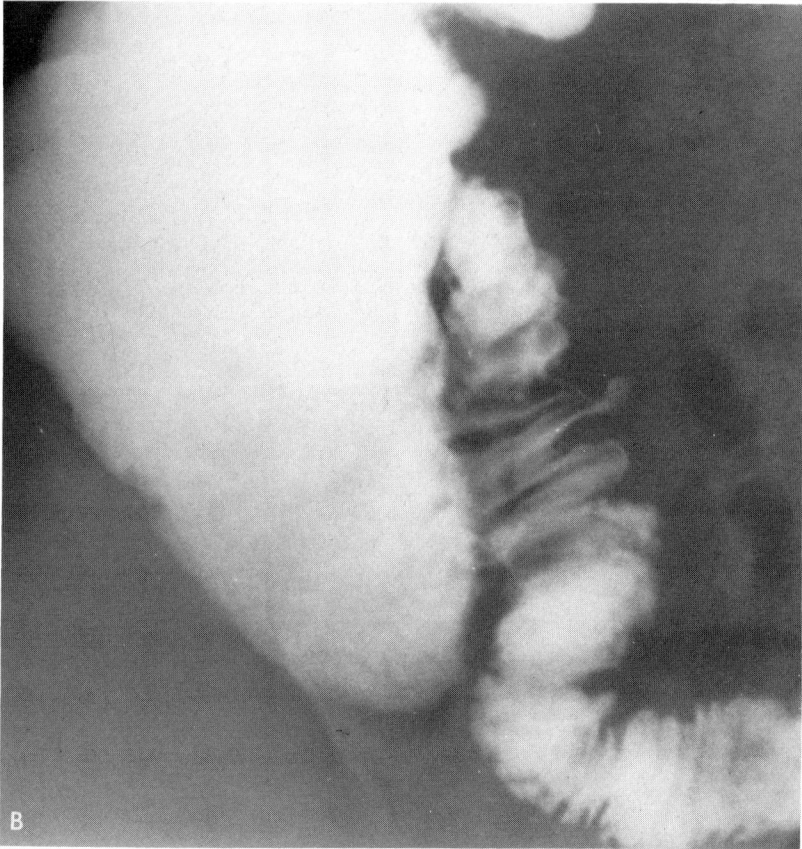

Figure 4-48. Barium enema examination (A and B) and postevacuation study (C) showing diverticulitis of the terminal ileum. Edema, spasm and numerous diverticula of the ileum are evident. The postevacuation radiograph (C) shows localized edema of the cecum and a mass due to the adjacent inflammatory reaction. At surgery, there was an inflammatory mass involving the ileum and cecum caused by a perforated ileal diverticulitis.

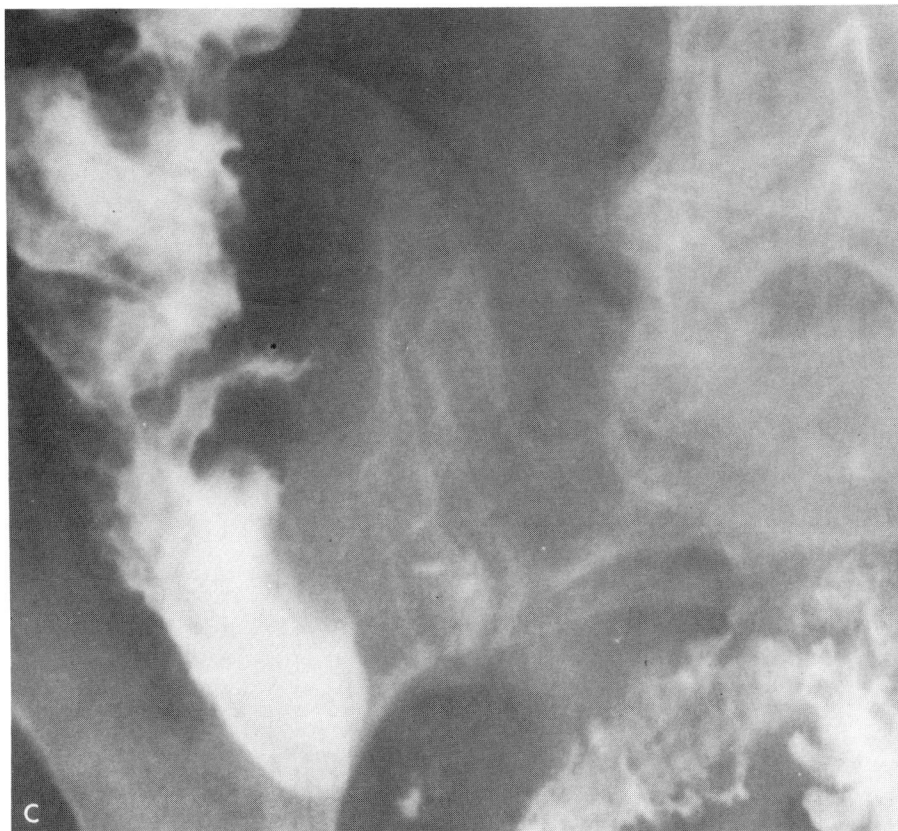

Figure 4–48. (*Continued*)

ized peritonitis. When the inflammatory process subsides, fibrosis may cause partial obstruction of the small intestine.

There are no characteristic clinical manifestations of ileal diverticulitis. Symptoms are usually those of acute appendicitis.

Like diverticulitis of the colon, edema may occlude the orifices of the diverticula. When this happens, a radiologic diagnosis is impossible. Suspicion should be raised if there are spasm and mucosal edema of a short segment of the terminal ileum. As the inflammatory process subsides, the diverticula may become evident, and the diagnosis is then established.

URTICARIA OF THE COLON

The colon is rarely affected by allergic reactions. Clemett reported a patient with radiologic manifestations who was assumed to have urticaria of the colon on the basis of a penicillin reaction.[33] The colon returned to normal as the reaction subsided and no histologic studies were possible.

Berk and Millman reported a case with identical radiographic findings on barium enema examination[33] (Fig. 4–49). Postmortem examination showed that the patient died of severe allergic myocarditis. No cause for the hypersensitivity response could be determined.

The histologic abnormalities in the colon in urticaria are distinctive. Flat, reticulated, discrete plaques can be identified on barium enema examination. Lymphoid hyperplasia, colitis cystica profunda, air bubbles in the barium mixture and eosinophilic gastroenteritis might be considered in the differential diagnosis, but none of these closely resembles the pattern of urticaria.

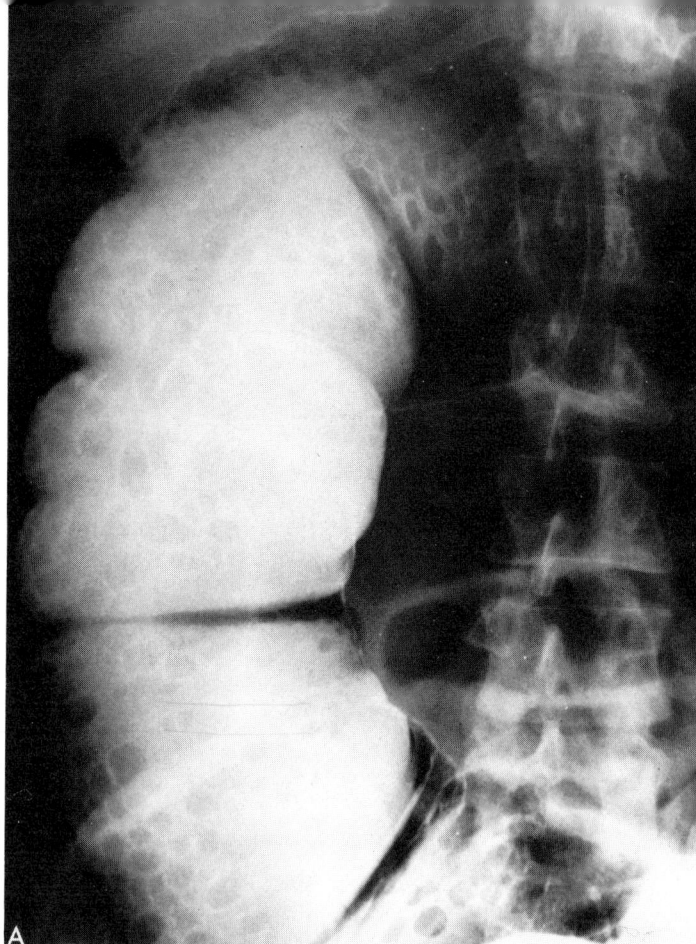

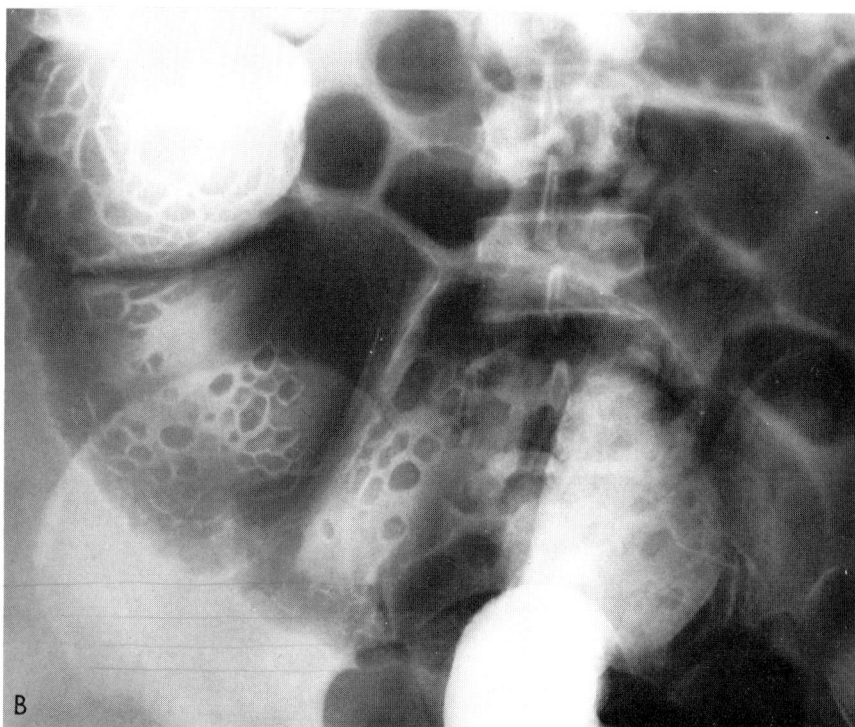

Figure 4-49. (A, B and C) Barium enema and air contrast examination showing flat, raised plaques on the surface of the cecum and ascending colon due to urticaria. (D) Gross specimen showing the mucosal surface of the ascending colon. (From Berk, R. N., and Millman, S. J.: Urticaria of the colon. Radiology, 99:539, 1971.)

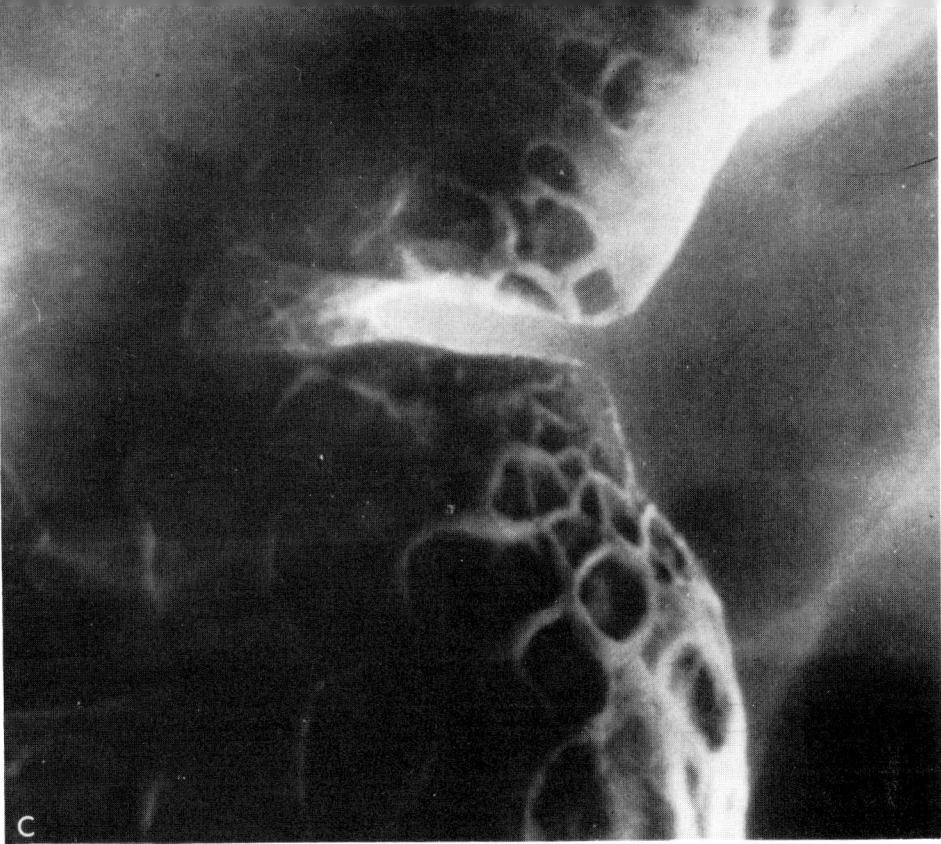

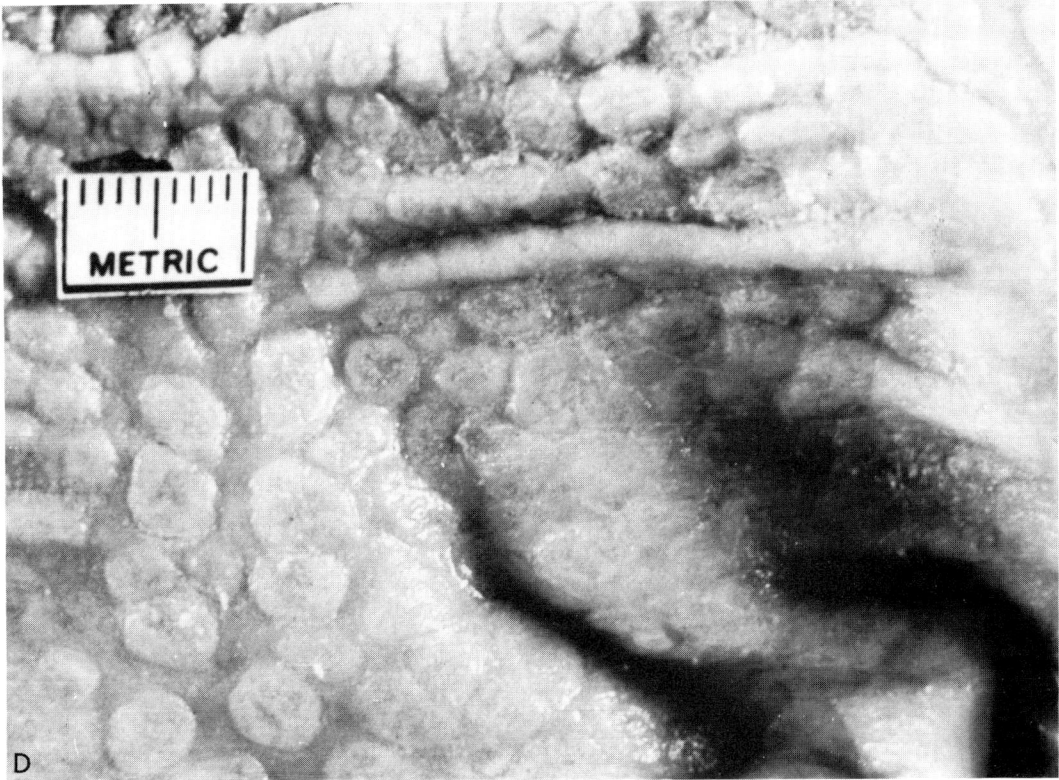

Figure 4–49. (Continued)

NONSPECIFIC ULCER OF THE CECUM

Solitary benign ulceration of the cecum is a rare disease of uncertain etiology. The clinical, radiologic and pathologic findings are nonspecific, so that preoperative diagnosis, except by colonoscopy, is difficult. The disease was first described by Cruveilhier in 1832 and since then less than 100 cases have been reported.[34]

The ulcer is usually single, punched out, close to the ileocecal valve, 0.5 cm. to 4 cm. in size, variable in shape and deep enough to involve the muscle coats of the cecum.[35] The histologic changes are nonspecific and consist of a benign ulcer covered with necrotic granulation tissue and surrounded by inflammation. The adjacent inflammatory reaction is often intense and simulates a tumor mass.

Various theories as to the etiology of the disease have been proposed.[34] The most popular is that the ulcer is the result of localized ischemia. Others have suggested that the lesion results from stercoraceous trauma, viral infection or diet containing excessive cereal. Several of the patients have been on steroids, which may play a role in the development of the ulcer.

The clinical manifestations of the disease are nonspecific and usually simulate acute appendicitis or tumor. Pain is variable and may be either acute or insidious in onset in patients ranging in age from 18 to 67. Complications include perforation with localized abscess formation or generalized peritonitis. Sutherland et al. reported three patients with bleeding, probably from cecal ulcers, which were diagnosed by angiography and treated by vasopressin infusion through the arterial catheter. All three pa-

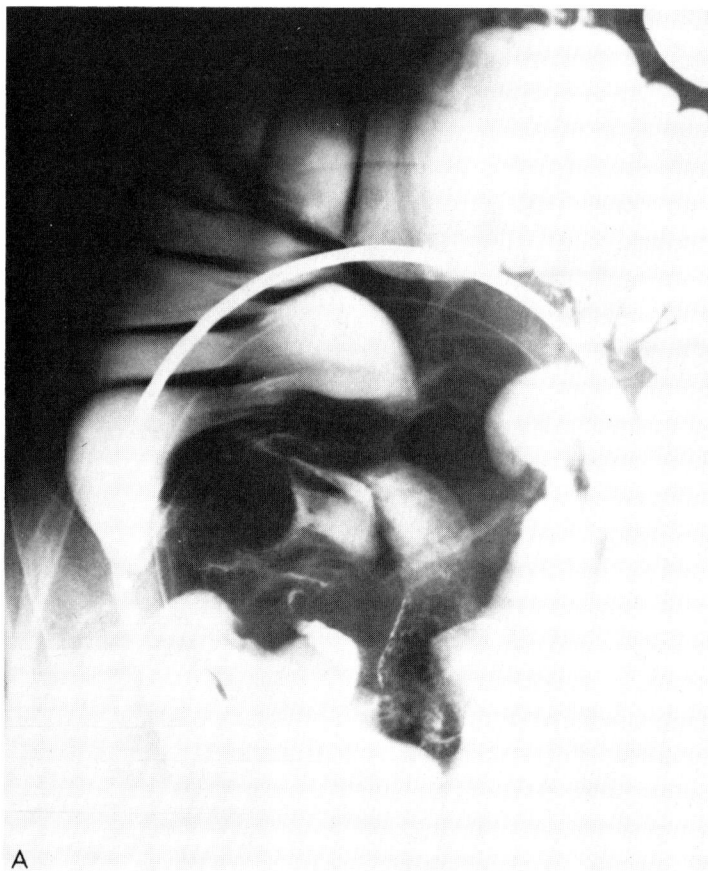

A

Figure 4–50. Barium enema examination (A and B) and air contrast study (C) showing a smooth mass in the base of the cecum due to granulation tissue caused by healing of a solitary ulcer of the cecum.

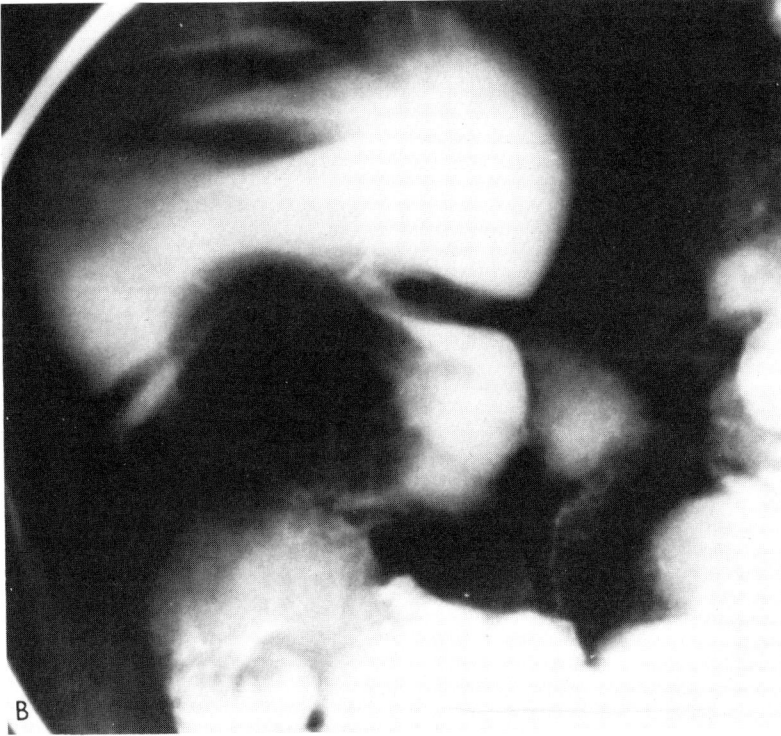

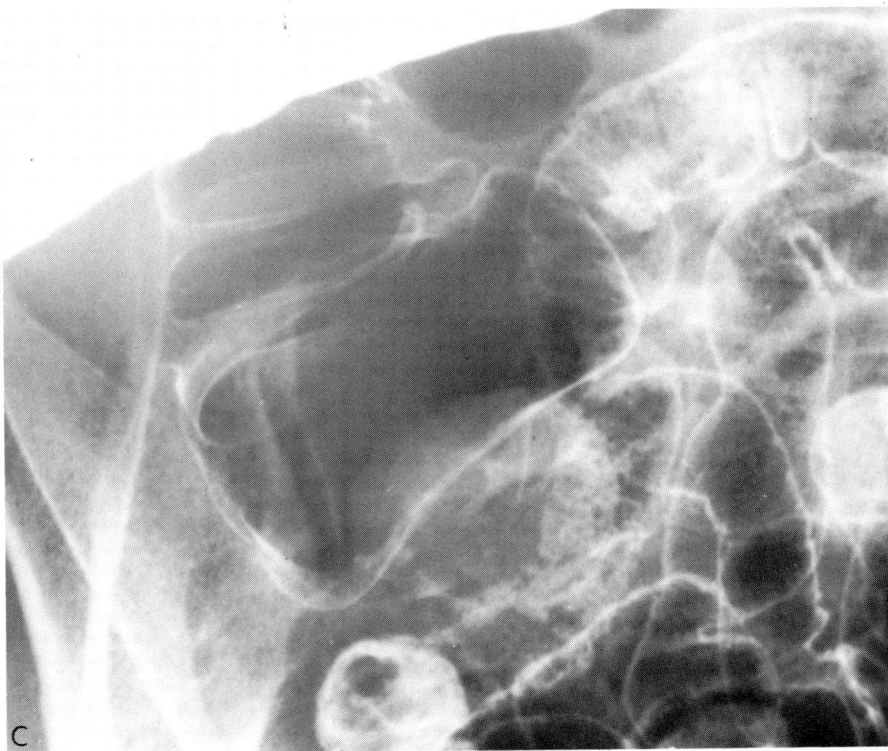

Figure 4–50. (Continued)

tients were taking steroids.[36] Nagasako et al. made a correct preoperative diagnosis by direct visualization of the ulcer through the colonoscope.[37]

The ulcer is rarely visible on barium enema examination, even with air contrast studies.[38] Usually a nodular swelling is present in the cecum owing to the inflammatory mass associated with ulcer (Fig. 4–50). Localized irritability, hypermotility, spasm and stricture may be present. The lesion cannot be differentiated radiographically from a cecal carcinoma, tuberculosis or amebiasis.

Because of the danger of perforation, treatment is surgical resection of the cecum.

LEUKEMIC TYPHLITIS

Leukemic typhlitis is a necrotizing colitis localized to the cecum and found in leukemic children who have been receiving treatment and are in the terminal stage of their disease.[39] Prolla and Kirsner have categorized the pathology of the lesion into four main types: hemorrhagic, leukemic infiltration, agranulocytic and fungal.[40] Clinically, differentiation from acute appendicitis is difficult, if not impossible.

Involvement of the cecum may not necessarily represent a specific form of inflammation. The lesion may accompany agranulocytosis from any cause and present as a

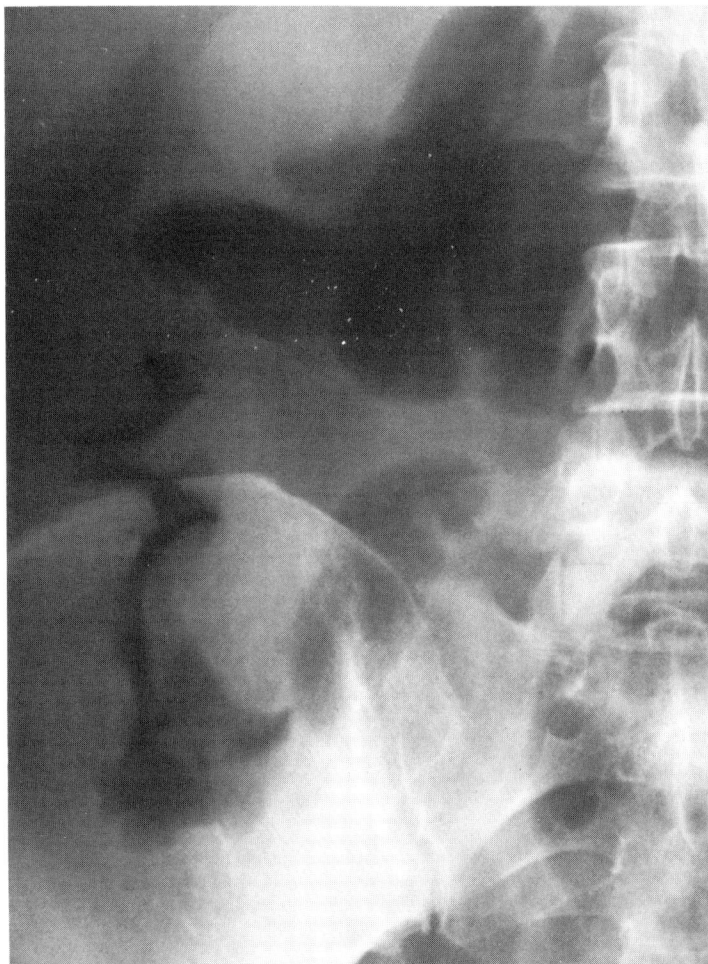

Figure 4-51. Plain abdominal radiograph showing thickening and nodularity of the cecum and ascending colon due to leukemic typhlitis. The patient was in an advanced stage of chronic lymphatic leukemia when she developed right lower quadrant pain. Postmortem examination several days later showed necrosis of the cecum and ascending colon with hemorrhage into the bowel wall.

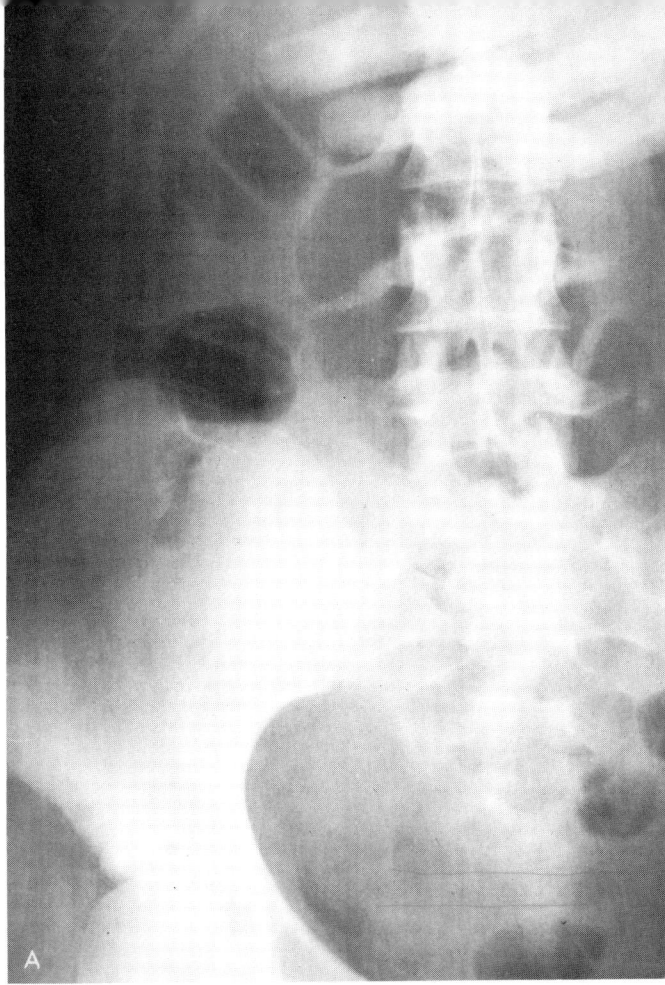

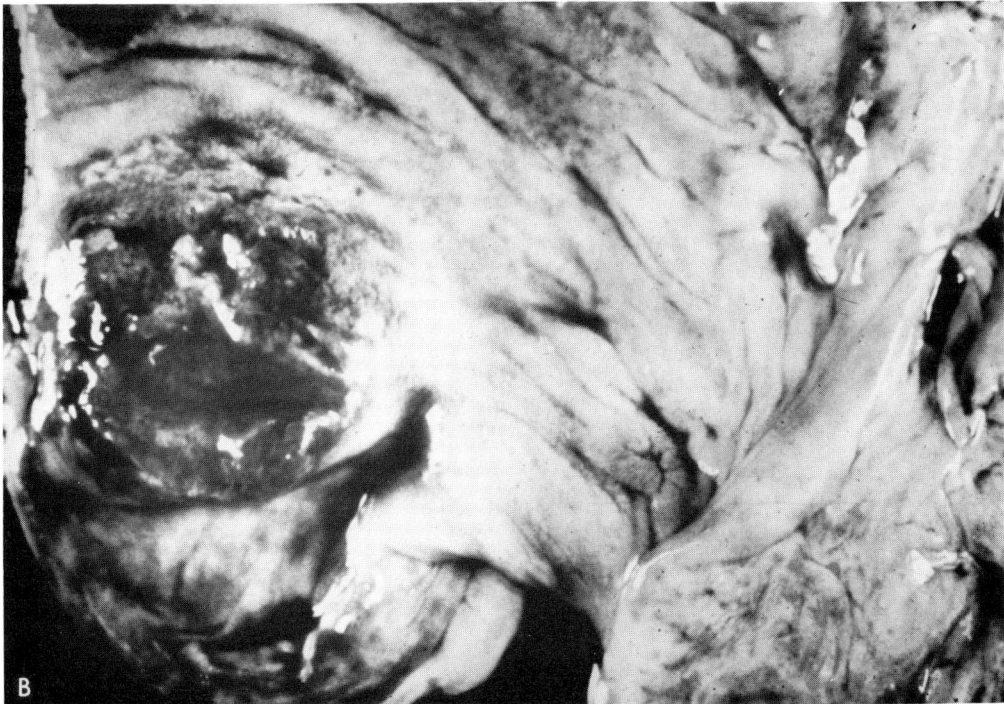

Figure 4–52. (A) Plain abdominal radiograph in a cachectic patient with acute myelocytic leukemia showing leukemic typhlitis. There is a paucity of bowel gas and a poorly defined density in the right lower quadrant. (B) Gross specimen of the cecum showing the mucosal surface of the cecum. An area of necrosis and hemorrhage is present in the wall.

sharply localized area of necrosis in the cecum, as an extensive lesion involving other portions of the colon and small intestine or as a cecal ulcer with similar ulcers elsewhere in the intestine. Histologically, edema, dilated blood vessels, necrosis and masses of organisms may be identified. Leukemic infiltration is rare. The usual microscopic pattern is one of hemorrhage and agranulocytic cellulitis. Bacteria or fungi can usually be identified.

Pathogenic possibilities include leukemic infiltration, localized hemorrhage or inflammation.[41] The most frequent of these is inflammation, but the underlying pathway of cecal involvement is uncertain. One possibility is that some alteration of the intestinal mucosa occurs with subsequent and progressive ulceration, hemorrhage and invasion by enteric organisms.

The clinical differentiation between typhlitis and appendicitis is difficult. Patients with typhlitis complain of abdominal pain and distention. Abdominal rigidity is absent in both typhlitis and acute appendicitis in leukemic patients.

The radiographic findings are nonspecific[39] (Figs. 4–51 and 4–52). Barium studies are rarely performed because of the moribund condition of most of the patients. An early finding on plain abdominal radiographs is a relative lack of bowel gas in the right lower quadrant. Air-fluid levels are usually lacking. Progressive distention of the small bowel occurs with absence of large bowel gas. An ill-defined density fills the right lower quadrant and is most likely due to the dilated atonic cecum. This soft tissue density and absence of air-fluid levels are helpful in differentiating typhlitis from appendicitis. When the diagnosis of typhlitis is made, the prognosis is grave.

REFERENCES

1. Marshak, R. H., and Lindner, A. E.: Radiology of the Small Intestine. Philadelphia, W. B. Saunders Co., 1970, pp. 158–274.
2. Lennard-Jones, J. E., and Morson, B. C.: Changing concepts in Crohn's disease. Disease-A-Month, August, 1969, pp. 1–37.
3. Norlen, B. J., Krause, U., and Bergman, L.: An epidemiological study of Crohn's disease. Scand. J. Gastroent., 5:385, 1970.
4. Silverman, F. N., and Lee, C. M., Jr: Chronic regional enteritis and growth retardation. Amer. J. Dis. Child., 103:569, 1962.
5. Berk, R. N., and Chamovitz, R.: Recurrent acute small bowel obstruction due to a narrow ileocolic anastomosis, a surgical complication with distinctive features. Radiology, 93:117, 1969.
6. Cohen, S., Kaplan, M., Gottlieb, L., and Patterson, J.: Liver disease and gallstones in regional enteritis. Gastroenterology, 60:237, 1971.
7. Wolf, B. S., and Marshak, R. H.: Granulomatous colitis: roentgen features. Amer. J. Roent., 88:662, 1962.
8. Lockhart-Mummery, H. E., and Morson, B. C.: Crohn's disease of the large intestine. Gut, 5:493, 1964.
9. Margulis, A. R., Goldberg, H. I., Lawson, T. L., Montgomery, C. K., Rambo, O. N., Noonan, C. D., and Amberg, J. R.: The overlapping spectrum of ulcerative and granulomatous colitis. Amer. J. Roent., 113:325, 1971.
10. Mitchell, R. S., and Bristol, L. J.: Intestinal tuberculosis: an analysis of 346 cases diagnosed by routine intestinal radiography on 5,529 admissions for pulmonary tuberculosis, 1924–49. Amer. J. Med. Sci., 227:241, 1954.
11. Moss, J. D., and Knauer, C. M.: Tuberculous enteritis. Gastroenterology, 65:959, 1973.
12. Tandon, H. D., and Prakash, A.: Pathology of intestinal tuberculosis and its distinction from Crohn's disease. Gut, 13:260, 1972.
13. Werbeloff, L., Novis, B. H., Bank, S., and Marks, I. N.: The radiology of tuberculosis of the gastrointestinal tract. Brit. J. Radiol., 46:329, 1973.
14. Doxiadis, T.: Amebiasis, a non-tropical disease. Med. Times, 97:86, 1969.
15. Weinfeld, A.: Roentgen appearance of intestinal amebiasis. Amer. J. Roent., 96:311, 1966.
16. Hill, M. C., and Goldberg, H. I.: Roentgen diagnosis of intestinal amebiasis. Amer. J. Roent., 99:77, 1967.
17. Pittman, F. E., Pittman, J. C., and El-Hashimi, W. K.: Studies of human amebiasis. III Ameboma: a radiologic manifestation of amebic colitis. Amer. J. Dig. Dis., 18:1025, 1973.
18. Reeder, M. M., and Hamilton, L. C.: Radiologic diagnosis of tropical diseases of the gastrointestinal tract. Rad. Clin. N. Amer., 7:57, 1969.
19. Barrett-Connor, E.: Human intestinal nematodiasis in the United States. Calif. Med., 117:8, 1972.
20. Richman, R. H., and Lewicki, A. M.: Right ileocolitis secondary to anisakiasis. Amer. J. Roent., 119:329, 1973.
21. Beeson, P. B., and McDermott, W.: Cecil-Loeb Textbook of Medicine. 13th ed. Philadelphia, W. B. Saunders Co., 1971, pp. 574–578.
22. DeBusscher, G.: Roentgen study of the small intestine in various disorders. Acta. Gastroent. Belg., 13:295, 1950.
23. Miller, W. T., DePoto, D. W., Scholl, H. W., and Raffensperger, E. C.: Evanescent colitis in the young adult: A new entity? Radiology, 100:71, 1971.
24. Ellis, K., and McConnell, D. J.: Hereditary angioneurotic edema involving small intestine. Radiology, 92:518, 1969.
25. Pearson, K. D., Buchignani, J. S., Shimkin, P. M., and Frank, M. M.: Hereditary angioneurotic edema. Amer. J. Roent., 116:256, 1972.
26. Roswit, B.: Complications of radiation therapy:

The alimentary tract. Semin. Roent., 9:51, 1974.

27. Colcock, B. P., and Braasch, J. W.: Surgery of the Small Intestine in the Adult. Philadelphia, W. B. Saunders Co., 1968, pp. 161–165.

28. Bosniak, M. A., Hardy, M. A., Quint, J., and Ghossein, N. A.: Demonstration of the effect of irradiation on canine bowel using in vivo photographic magnification angiography. Radiology, 93:1361, 1969.

29. Mason, G. R., Dietrich, P., Friedland, G. W., and Hanks, G. E.: The radiological findings in radiation-induced enteritis and colitis, a review of 30 cases. Clin. Radiol., 21:232, 1970.

30. Nissenbaum, J., Sparks, A. J., and Ellison, G. R.: Cecal diverticulum, report of a case diagnosed by roentgen ray and surgically proved: review of the original literature. Amer. J. Roent., 73:596, 1955.

31. Parulekar, S. G.: Diverticulosis of the terminal ileum and its complications. Radiology, 103:283, 1972.

32. Felson, B., and Wiot, J. F.: Some interesting right lower quadrant entities. Radiol. Clin. N. Amer., 7:83, 1969.

33. Berk, R. N., and Millman, S. J.: Urticaria of the colon. Radiology, 99:539, 1971.

34. Cromar, C. D. L.: Benign ulcer of the cecum. Amer. J. Dig. Dis., 13:230, 1946.

35. Benninger, G. W., Honig, L. J., and Fein, H. D.: Nonspecific ulceration of the cecum. Amer. J. Gastroent., 55:594, 1971.

36. Sutherland, D., Frech, R. S., Weil, R., Najarian, J. S., and Simmons, R. L.: The bleeding cecal ulcer; pathogenesis, angiographic diagnosis and nonoperative control. Surgery 71:290, 1972.

37. Nagasako, K., Ikezawa, H., Gyo, S., and Takemoto, T.: Preoperative diagnosis of nonspecific ulcer of the cecum by colonofiberscopy. Dis. Colon Rectum, 15:413, 1972.

38. Feldman, M.: Clinical roentgenology of the digestive tract. 3rd ed. Baltimore, Williams and Wilkins Co., 1948, p. 901.

39. Wagner, M. L., Rosenberg, H. S., Fernbach, D. J., and Singleton, E. B.: Typhlitis, a complication of leukemia in childhood. Amer. J. Roent., 109:341, 1970.

40. Prolla, J. C., and Kirsner, J. B.: Gastrointestinal lesions and complications of leukemias. Ann. Intern. Med., 61:1084, 1964.

41. Sherman, N. J., and Wooley, M. M.: The ileocecal syndrome in acute childhood leukemia. Arch. Surg., 107:39, 1973.

Chapter 5

TUMORS OF THE ILEOCECAL AREA

BENIGN LESIONS

Benign tumors of the cecum, ascending colon and terminal ileum often have characteristic radiologic features that are sufficiently specific to permit a definite diagnosis or at least to suggest that a lesion is benign rather than malignant.[1] This is of considerable practical importance since, in some patients, surgery may be avoided. In those cases in which a confident diagnosis cannot be made on the basis of the radiographic characteristics of the lesion, inspection and biopsy of the lesion by colonoscopy may establish a definite diagnosis.

Tumors of the cecum must be differentiated from a normal or enlarged ileocecal valve. It is important, therefore, to identify the valve with compression spot radiographs made at fluoroscopy. When this cannot be done, it is helpful to be certain that the terminal ileum is filled with barium during the barium enema examination in order to locate the site of the valve. A small bowel examination with compression spot radiographs of the cecum made when the cecum has filled with barium is useful in the rare instance in which the terminal ileum will not fill during the barium enema examination. Air introduced, per rectum, during the small bowel study will produce air contrast of the cecum and may be helpful in delineating a cecal mass. In many cases the use of glucagon will relax the ileocecal valve and permit filling of the terminal ileum with barium, when it is not otherwise possible during the barium enema examination.

Adenomatous Polyps

The most common benign tumors of the colon are adenomatous polyps. These epithelial lesions present on radiographic examination as sessile or pedunculated intraluminal masses of varying size with a moderately irregular surface (Fig. 5–1). Differentiation from carcinoma may be difficult or impossible unless the tumor is on a stalk greater than 2 cm. in length. Broad-based masses cannot be considered to be benign without histologic verification. It is essential, therefore, that meticulous radiographic technique be used to demonstrate the exact gross morphology of the tumor. Compression spot radiographs made under fluoroscopic control in multiple projections with both full-column barium and air contrast techniques are helpful in identifying the base of the lesion and in determining the presence and length of the stalk. A second examination at a later date is necessary to ensure that the polyp actually exists and that the defect noted on the barium enema examination is not due to fecal material or to an air bubble.

Villous Adenomas

Villous adenomas, less common than adenomatous polyps, are premalignant lesions which often have a characteristic radiologic appearance.[2] These epithelial polypoid tumors, also termed villous papillomas and papillary adenomas, have multiple frondlike outgrowths of epithelium over their surface that produce a shaggy irregular appearance on barium studies (Fig. 5–2). Barium in the interstices of the surface of the tumor produces a characteristic reticulated or lacelike radiographic appearance. This typical radiographic appearance is more frequent in villous adenomas occurring on the left side of the colon. In fact, most villous adenomas in the cecum and ascending colon are indistinguishable from adenomatous polyps and adenocarcinoma.[3] Stalks are unusual in villous adenomas and are rarely more than 1 cm. in length. Hence, a polypoid lesion on a long stalk is far more likely to be an adenomatous polyp.

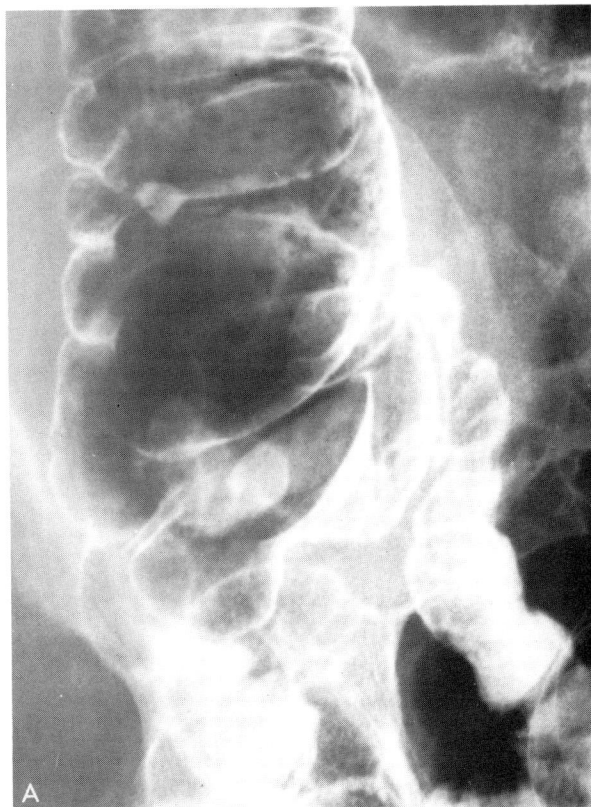

Figure 5–1. Air contrast barium enema examination in two patients (*A* and *B*) showing adenomatous polyps in the cecum. Broad-based masses with an irregular surface of this type cannot be considered to be benign without histologic verification.

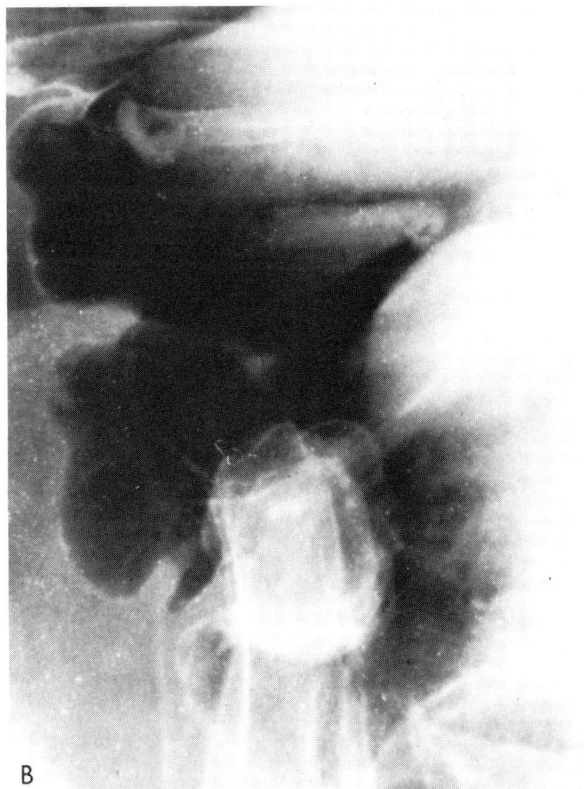

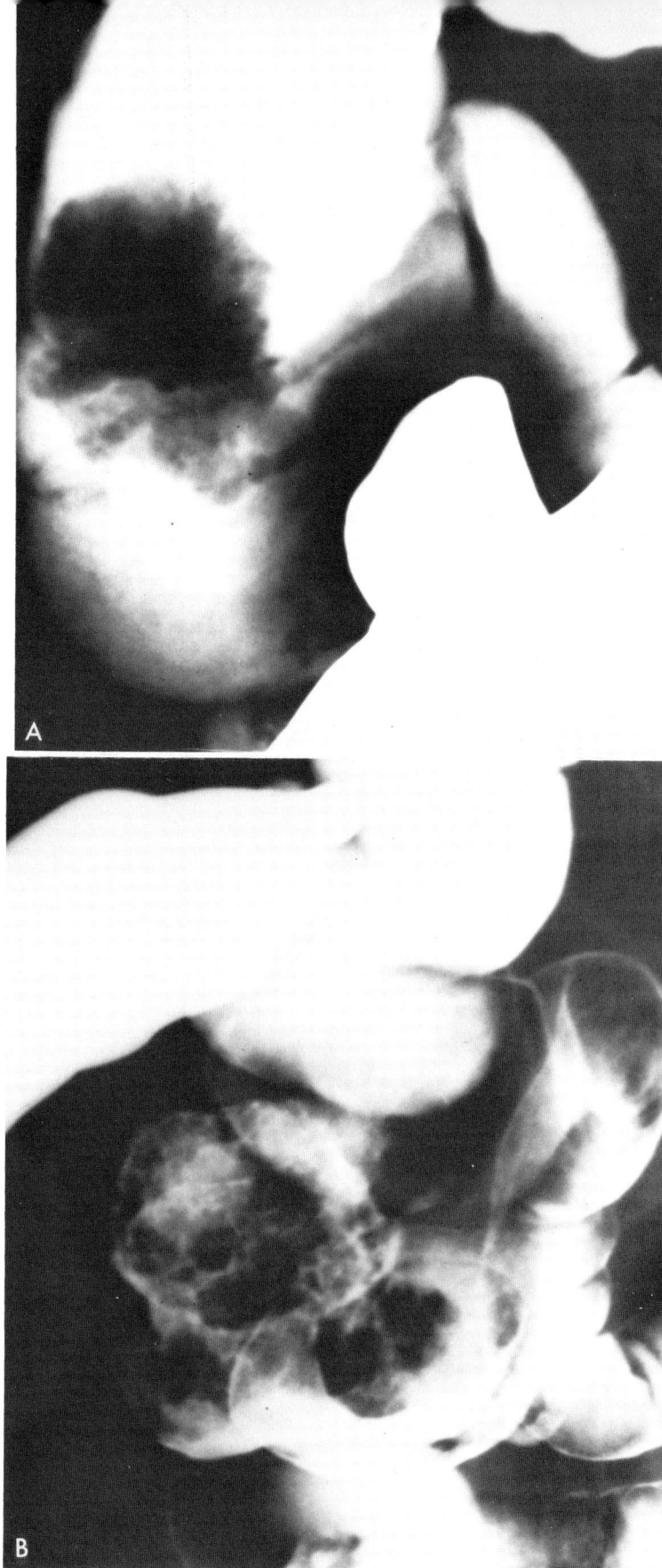

Figure 5–2. Barium enema examination (A) and air contrast study (B) showing a villous adenoma in the cecum. The shaggy frondlike surface is typical.

Lipomas

Although lipomas are second only to adenomatous polyps in frequency, they are still uncommon. Comfort and Stearns reported an incidence of 0.5 per cent in a series of 3924 autopsies.[4] The age of patients with lipomas ranges from 40 to 81 years, with a mean of 62 years. Males and females are equally involved. Although lipomas may occur in any portion of the gastrointestinal tract, almost two-thirds occur in the colon and, of these, most are in the cecum and ascending colon. Castro reported that 69 per cent of 45 lipomas involve the right half of the colon, especially in the cecal region.[5] Seventy-five per cent of the tumors were single, and the average diameter was 3.0 cm. with a range of 0.5 to 8.0 cm.

On gross section, 90 per cent of colonic lipomas are submucosal in origin and are covered with intact mucosa. The remainder begin in the subserosa, probably related to appendices epiploicae. After a period of years, a pseudopedicle may be produced by repeated contraction of the bowel and the mass then may present as an intraluminal polypoid filling defect. Histologic examination shows well-encapsulated adult fat cells with varying amounts of fibrous stroma. Ulceration, necrosis and cystic degeneration are frequent. Calcification is occasionally present. Sarcomatous degeneration does not occur.

Lipomas of the colon are usually asymptomatic, unless they ulcerate and bleed or become large enough to produce obstruction, intussusception or a palpable mass. Complete obstruction and severe bleeding, however, are rare. According to Schatzki, intussusception of the colon in an adult should suggest the possibility of a lipoma.[6]

The diagnosis of lipoma can be suspected when radiographic examination discloses a single, sharply defined smooth intramural lesion of the colon[7] (Fig. 5–3). When the tumor is on a pseudopedicle, the smooth mass is intraluminal. The shape and sharpness of the tumor mass and the completely smooth surface with normal pliability and distensibility of the bowel wall are often sufficient to suggest the diagnosis of a lipoma. The defect is usually spherical or oval, although occasionally it is irregular due to lobulations in the mass. The pedicle is short and broad, and the lesion is quite immobile. Lipomas may reach large size without developing a pseudopedicle and becoming an intraluminal mass. They then produce an intramural defect with sharp demarcation, a broad base and a smooth surface.

With proper spot radiographs, the soft consistency of the tumor can be detected by noting that the tumor changes in shape with compression. The relative radiolucency of the tumor is rarely appreciated on conventional barium enema examinations. This feature can best be evaluated by giving a water enema and making low kilovoltage radiographs of the abdomen, a technique devised by Margulis and Jovanovich.[8] When the lipoma is surrounded with water, the relative radiolucency of the tumor can be detected. Margulis and Jovanovich were able to make a definite diagnosis of lipoma in five successive cases by this technique. Stevens was successful in two more.[9]

Simple surgical excision of uncomplicated colonic lipomas is usually recommended. Removal is necessary because, although the tumors are benign, they are not harmless. None of the tumors becomes malignant, but excision establishes a definite diagnosis and avoids the hazards of colonic obstruction, intussusception and bleeding.

Leiomyomas

Leiomyomas are uncommon in the colon. When they occur they tend to maintain their intramural position more consistently than lipomas and therefore remain asymptomatic until they reach large size.[1] Mucosal folds overlying the lesion persist for a longer period than in a lipoma and, as a result, the characteristic sharp demarcation of mural lesions may be absent.[1] Specific differentiation from other extramucosal lesions is usually impossible. The broad-based, lobulated smooth mass indicates only that the lesion arises in the bowel wall beneath the mucosa and that it is intramural.

Multiple Familial Polyposis Syndromes

The cecum, ascending colon and terminal ileum may be involved in any of the familial

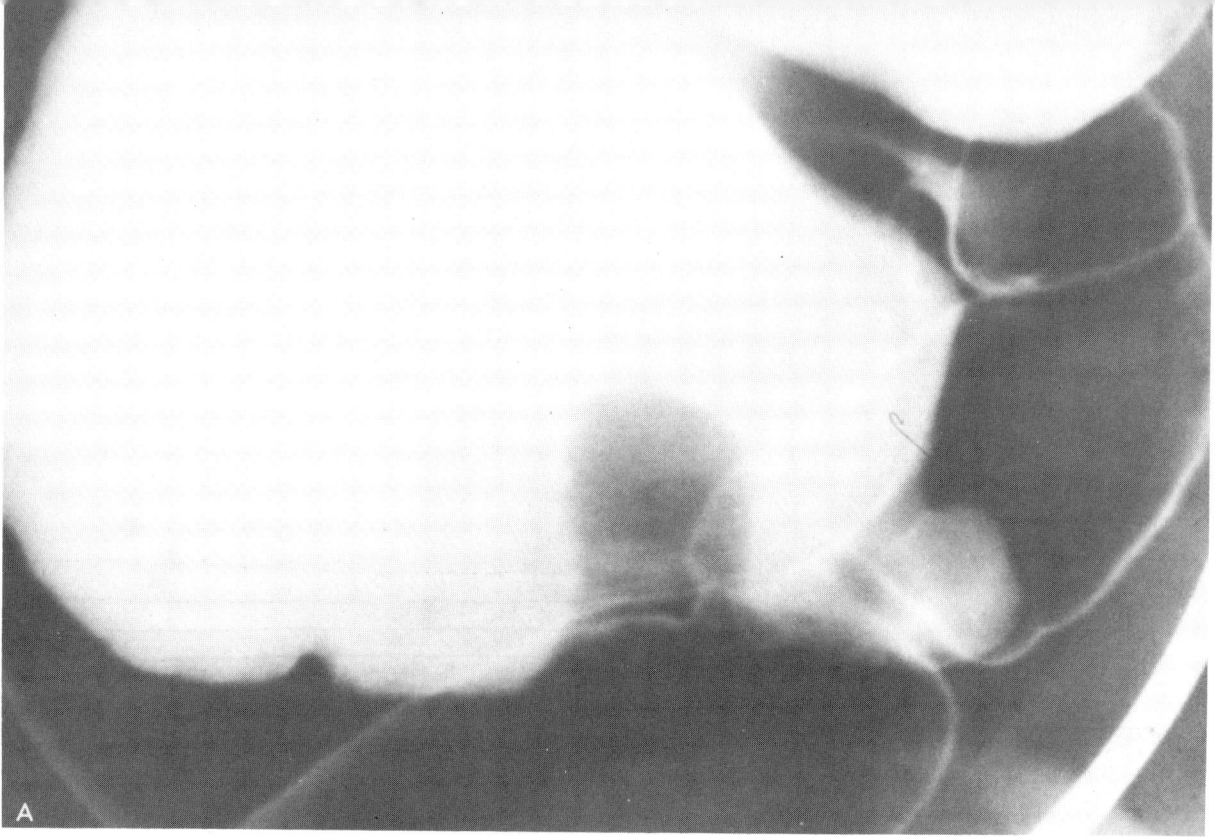

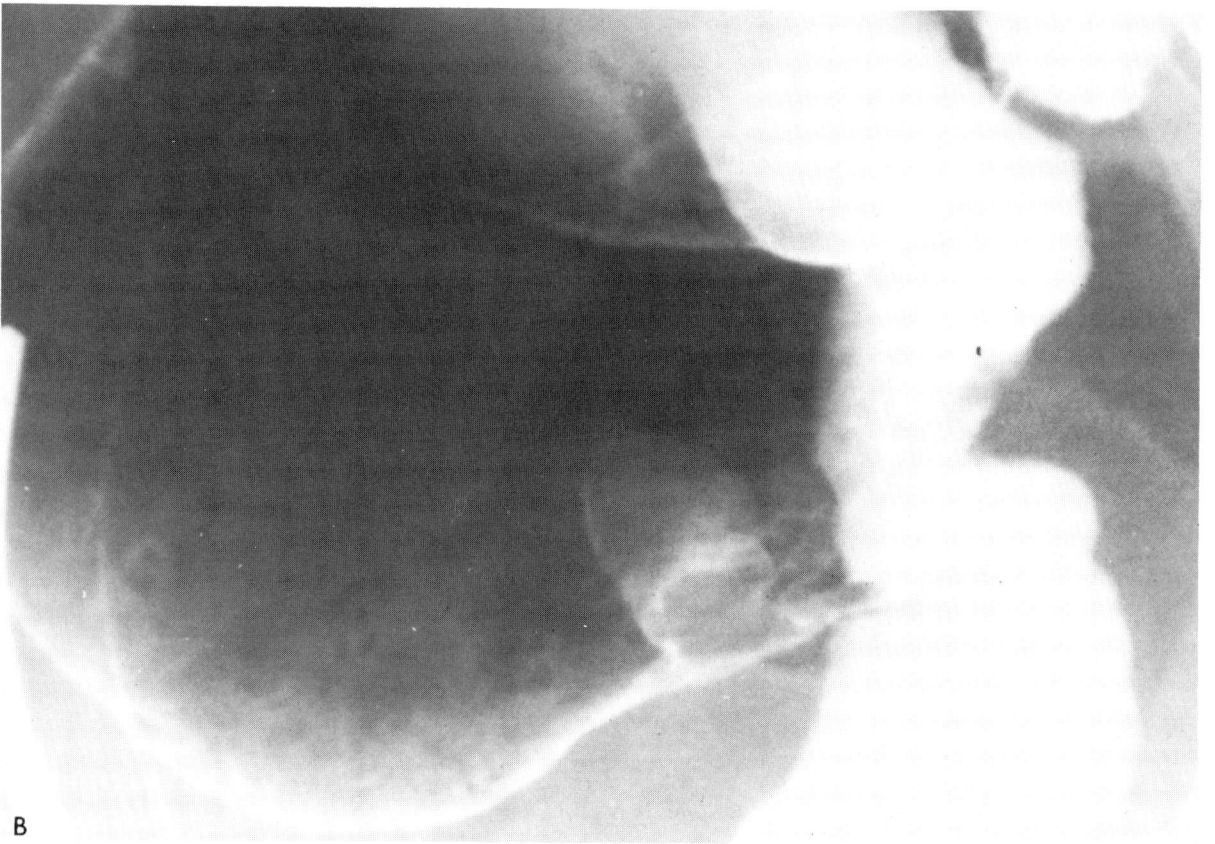

Figure 5–3. Barium enema examination (A) and air contrast study (B) showing a smooth sessile lipoma in the cecum. The lesion has grown into the lumen, simulating an adenomatous polyp or an inverted appendiceal stump. The relative radiolucency of the lipoma is not apparent.

polyposis syndromes that affect other portions of the small and large intestine. Multiple adenomatous polyps occur in the colon in familial polyposis.[10] These usually appear at the age of puberty and generally vary in size. The high malignant potential of this disease is widely recognized. Adenomatous polyps also occur in the colon in Gardner's syndrome, which is also a premalignant condition.[11, 12] These patients have multiple osteomas, usually about the paranasal sinuses and mandible; dentition abnormalities; and cutaneous fibromas. Care should be taken to consider the possibility of the syndrome in patients who present with multiple osteomas and delayed dentition before the onset of gastrointestinal symptoms. Peutz-Jeghers syndrome is a hereditary disease in which polyps of the small bowel or colon occur in association with melanin pigmentation of the lips and buccal mucosa.[13] The small intestine is usually involved and the polyps are hamartomas. Malignant complications are rare. Ectodermal abnormalities including alopecia, nail dystrophy and hyperpigmentation occur with the polyposis in the Cronkhite-Canada syndrome.[14] In these patients the polyps are inflammatory rather than neoplastic and are of the juvenile variety. Ruymann reported a patient with the manifestations of Cronkhite-Canada syndrome who also had devastating protein-losing enteropathy, cachexia and protein depletion.[15] We saw a case of inflammatory polyps of the small intestine in a 20-month-old child who had severe protein loss but no ectodermal changes[16] (Fig. 5–4).

Rare Tumors

Benign tumors such as neurofibromas, lymphangiomas, hemangiomas and teratomas are rare. The diagnosis of hemangioma may be made on the basis of the presence of phleboliths. Teratomas may be detected by the presence of characteristic calcification or ossification. Neurofibromas may be suspected in patients with von Recklinghausen's neurofibromatosis who have mural intestinal lesions.[17] However, in the majority of instances in these rare types of tumors a specific diagnosis cannot be made, and operative intervention is required.

MALIGNANT LESIONS

Adenocarcinoma of the Cecum and Ascending Colon

Despite the well-known clinical and roentgenologic features of carcinoma of the cecum and ascending colon, the disease is frequently unsuspected clinically and missed roentgenologically. Schutt and Walker found that 50 per cent of their patients had been examined by a physician for a significant complaint one month or more prior to the establishment of a correct diagnosis.[18] Cooley et al. recognized that the highest number of errors in the diagnosis of carcinoma of the colon on barium enema examination, exclusive of the rectum, occurred in the cecum.[19] In their series of 240 consecutive cases of carcinoma of the colon, false negative results on barium enema occurred in 27 per cent in the cecum compared to only 8 per cent in the sigmoid colon. Allcock found that 65 per cent of incorrect diagnoses were in the cecum while 20 per cent were in the sigmoid colon.[20]

In 1962 the clinical and radiologic findings in 91 cases of carcinoma of the cecum or ascending colon seen at the Pondville Hospital, Pondville, Massachusetts, between 1939 and 1961 were reviewed.[21] The average age of the patients was 67, with a range of 29 to 79 years. Abdominal pain was clearly the most common presenting complaint (80 per cent). The average duration of symptoms on admission was 8 months. Physical examination showed an abdominal mass in nearly three-fourths of the patients. Nine patients (10 per cent) had simultaneous carcinoma of the colon in sites other than the cecum or ascending colon. Twenty-three per cent had a second separate primary carcinoma of the colon or another organ either before the diagnosis of carcinoma of the cecum and ascending colon was made or after the disease was treated. Preoperative complications directly related to the carcinoma were found in 34 per cent of the patients. These included obstruction, hemorrhage, appendicitis, abscess, colonic perforation, intussusception and fistulas. The overall five-year survival rate was 35 per cent. Of the 91 patients, 38 per cent had carcinoma of the cecum, 15 per cent had carcinoma at the ileocecal valve, 39 per cent had a tumor in the ascending colon and 8

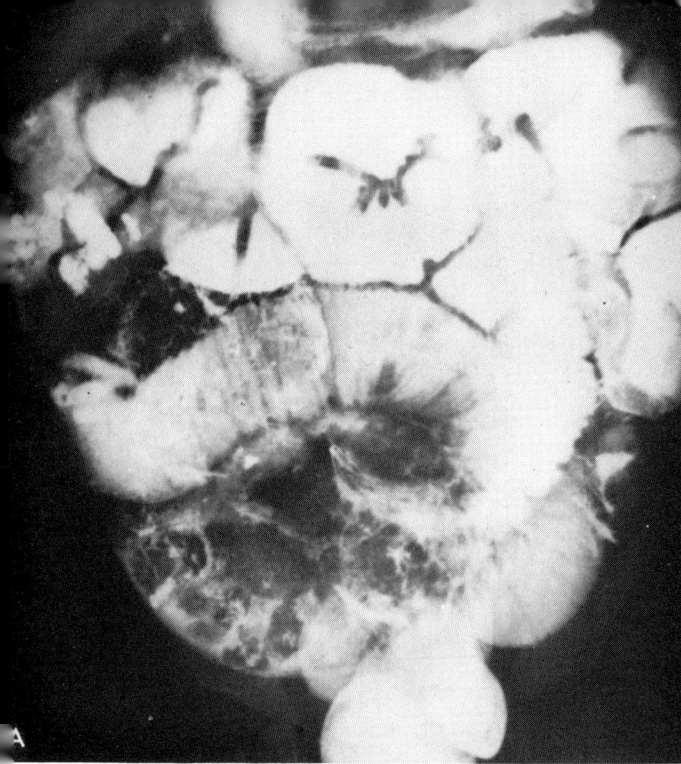

Figure 5–4. Small bowel examination (A) in a 20-month-old child, showing multiple inflammatory polyps. The child had severe protein-losing enteropathy. Gross specimen (B) showing the mucosal surface of a portion of the ileum. Multiple coalescing inflammatory polyps are present. Low power microscopic section (C) of a representative polyp showing numerous retention cysts with inflamed stroma. (From Berk, R. N., Rush, J. L., and Elson, E. C.: Multiple inflammatory polyps of the small intestine with cachexia and protein losing enteropathy. Radiology, 95:611, 1970.)

B

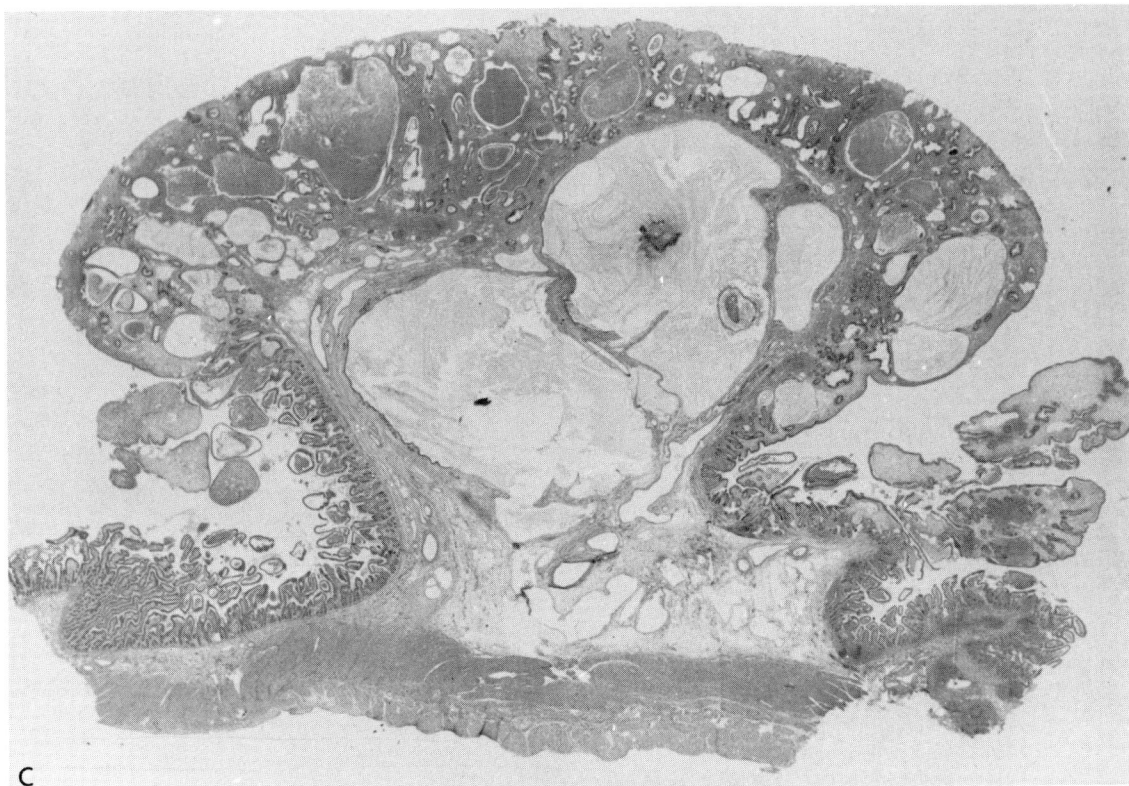

C

Figure 5–4. (Continued)

per cent had the lesion at the hepatic flexure.

In these 91 cases, 70 per cent had a polypoid tumor mass visible on the barium enema examination, while 50 per cent had an annular lesion (Figs. 5–5 to 5–11). The high incidence of annular features is somewhat at variance with the notion that lesions of the right colon are characteristically polypoid. The cecum was irregular and contracted in two-thirds of the patients with cecal lesions. It was clear that compression spot radiographs taken at fluoroscopy when the colon was filled with barium were the most helpful in recognizing an abnormality or clarifying one of uncertain cause. Without adequate spot film examination it was often impossible to differentiate a lesion from the ileocecal valve or from fecal material.

All patients with carcinoma in the ascending colon presented roentgenographic findings of an obvious abnormality on the barium enema examination, and in most instances the abnormality was identifiable as carcinoma. Of the patients in whom the

tumor was in the cecum or in the region of the ileocecal valve, approximately one-half showed changes that legitimately could be confused with fecal material; one-fourth presented findings thought to be compatible with various inflammatory processes about the appendix, ileum and cecum; and one-tenth had tumors that resembled a prominent ileocecal valve. In any series selected on the basis of positive diagnosis, it is impossible to be certain of the number of cases in which the diagnosis was missed. Of this group of 91 patients at least 16 per cent were diagnosed incorrectly by radiographic study before surgery. The correct diagnosis in many of these became known following surgery for associated disease. In two of 13 undiagnosed patients, the correct diagnosis was not made because of technical problems in performing a satisfactory examination. Five patients were interpreted as having an inflammatory process involving the appendix or terminal ileum, with carcinoma of the cecum suggested as a second possibility. All five had involvement of the terminal

(Text continued on page 265)

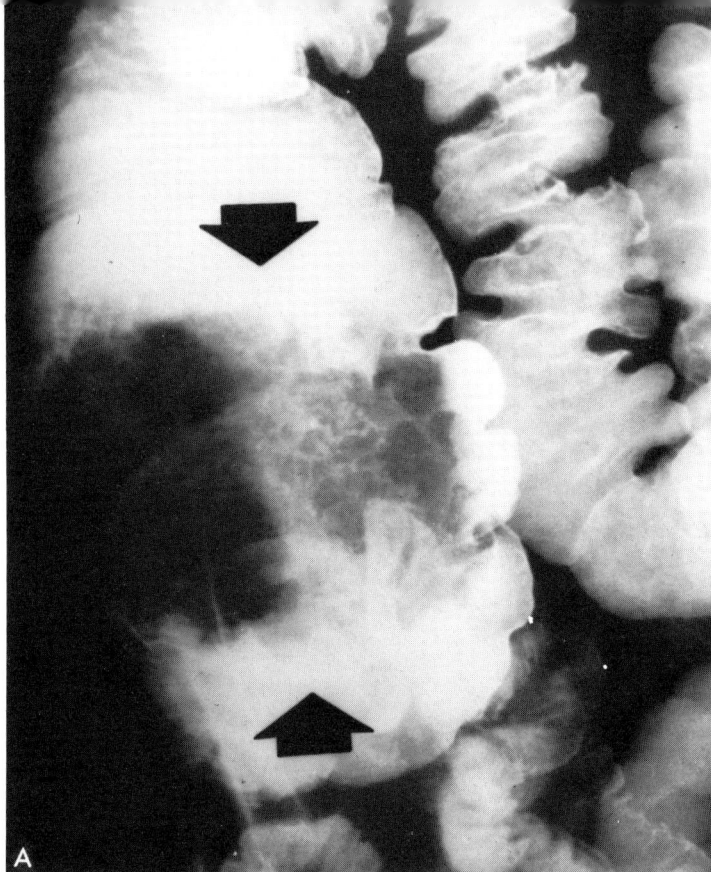

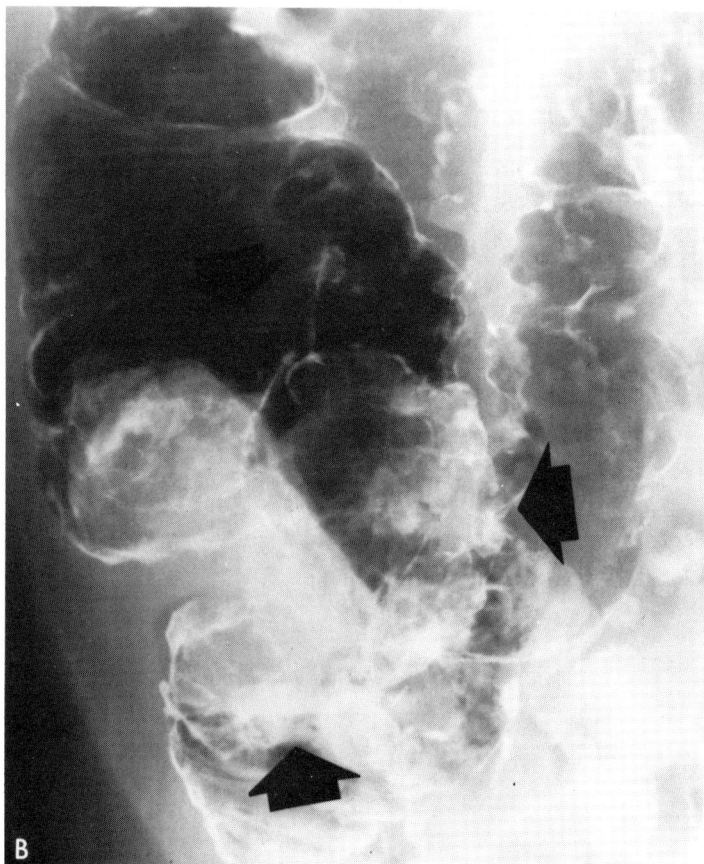

Figure 5–5. Barium enema examination (A) and air contrast study (B) showing a large polypoid adenocarcinoma of the cecum (arrows). A fungating mass with an irregular surface is typical of adenocarcinoma.

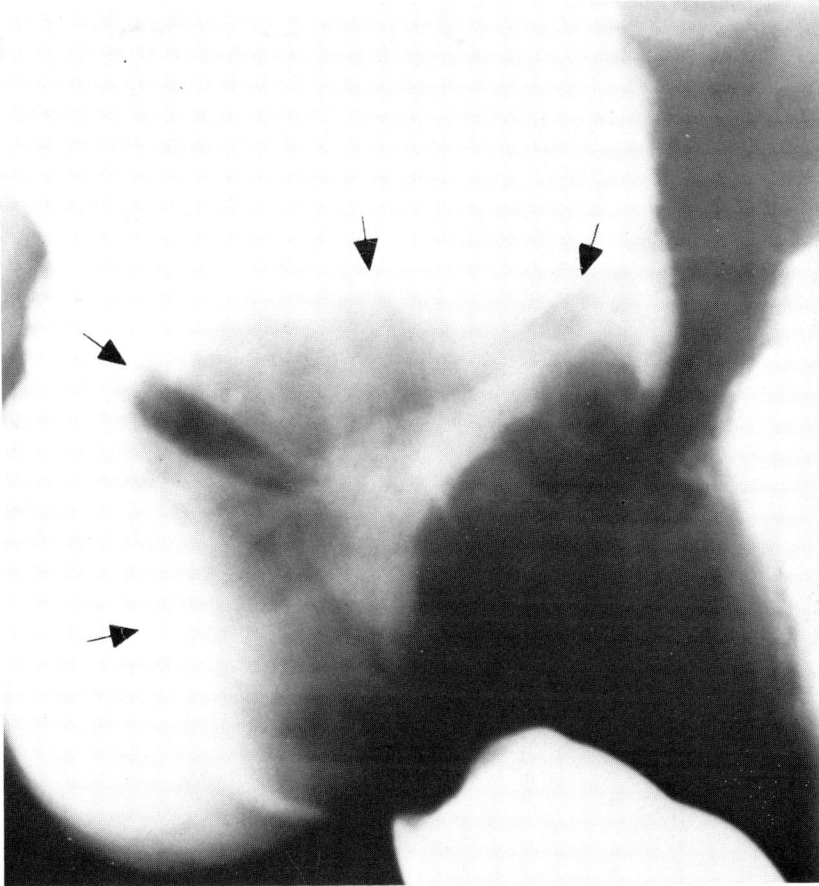

Figure 5–6. Barium enema examination showing a polypoid adenocarcinoma of the cecum (arrows). Compression spot radiographs optimally define the gross features of the lesion.

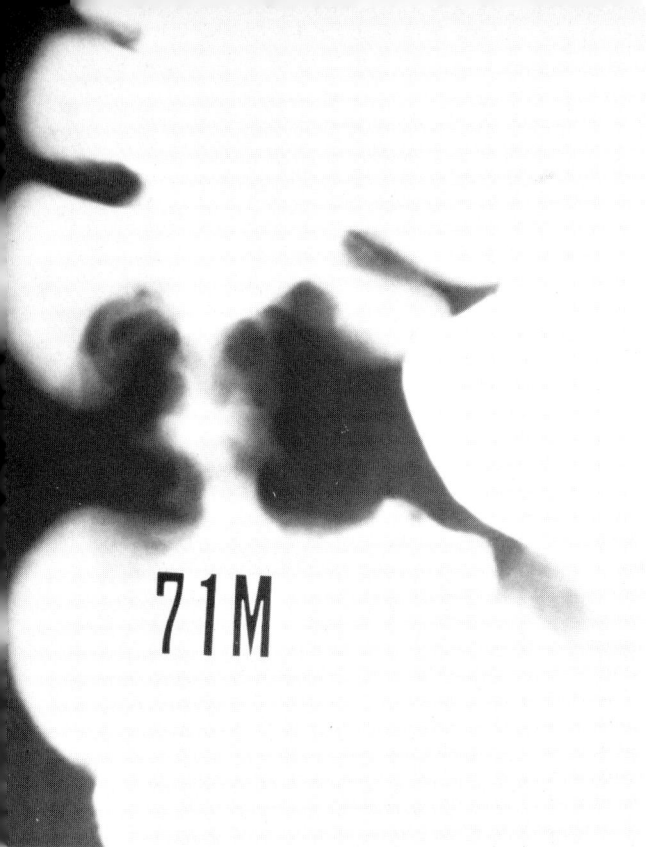

Figures 5–7 to 5–10. Barium enema examinations in four patients showing annular adenocarcinomas of the ascending colon. The sharp overhanging edge of the tumor and the mucosal destruction are typical. Annular carcinomas are not uncommon in the cecum and ascending colon.

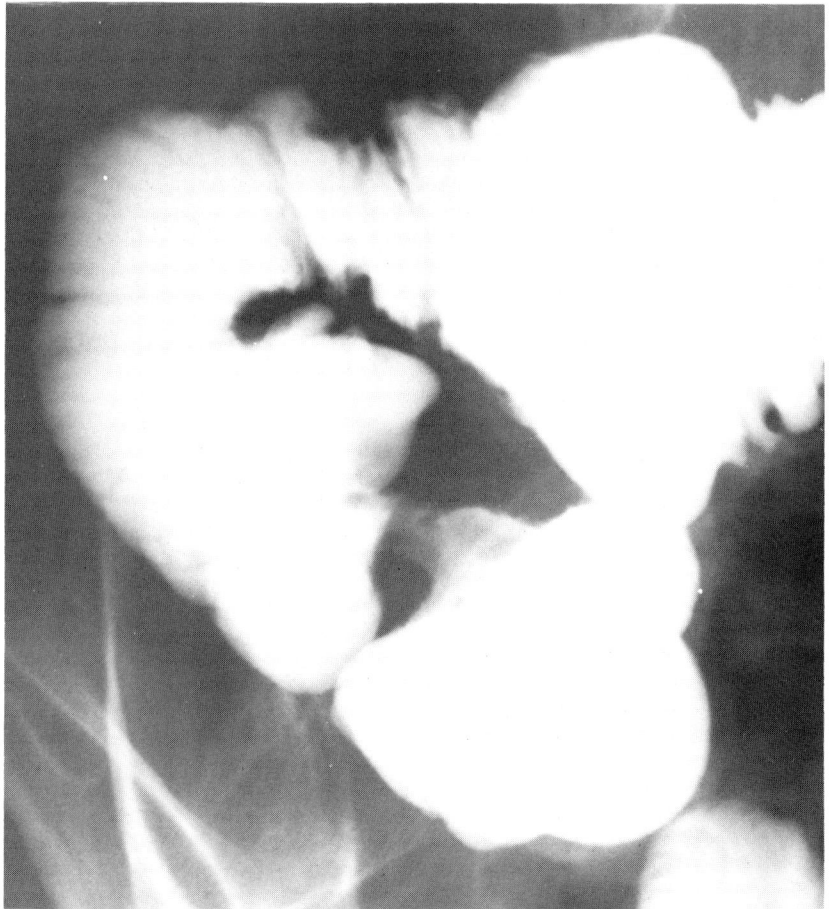

Figure 5–8. (See legend above.)

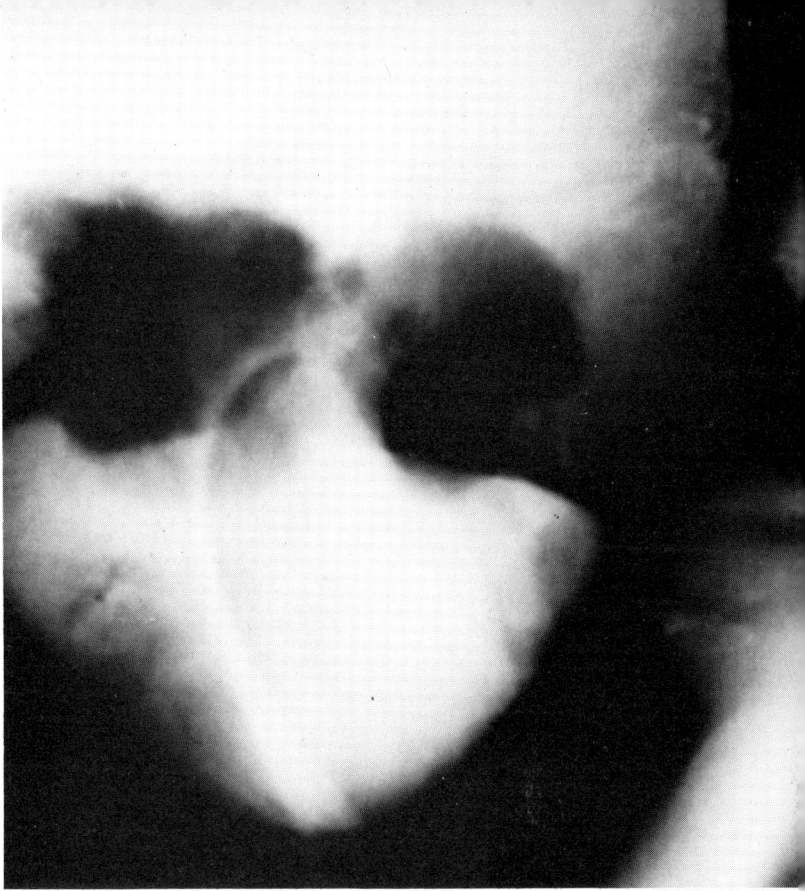

Figure 5–9. (See legend on p. 258.)

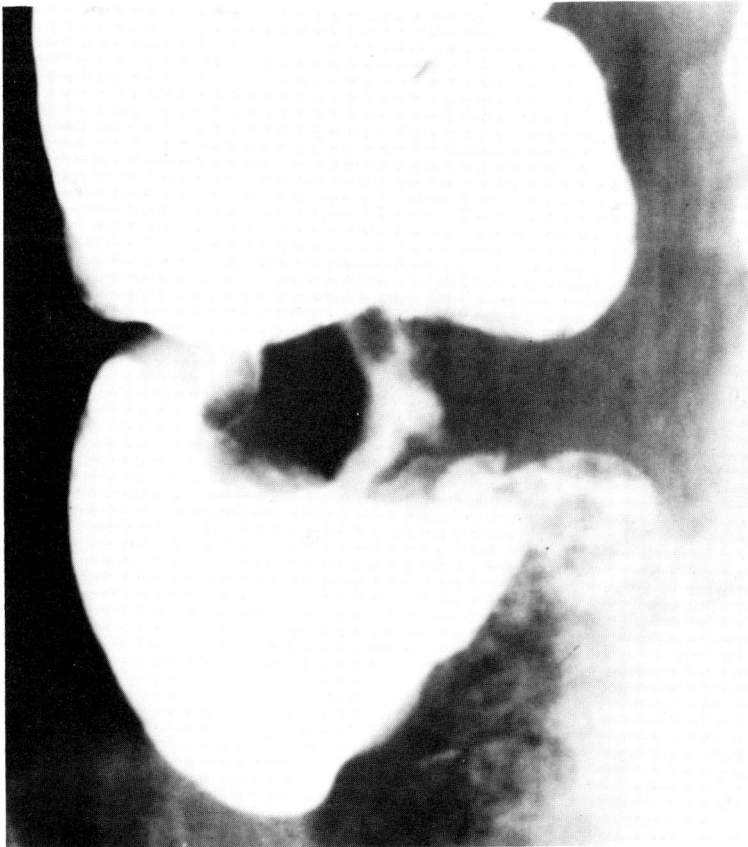

Figure 5–10. (See legend on p. 258.)

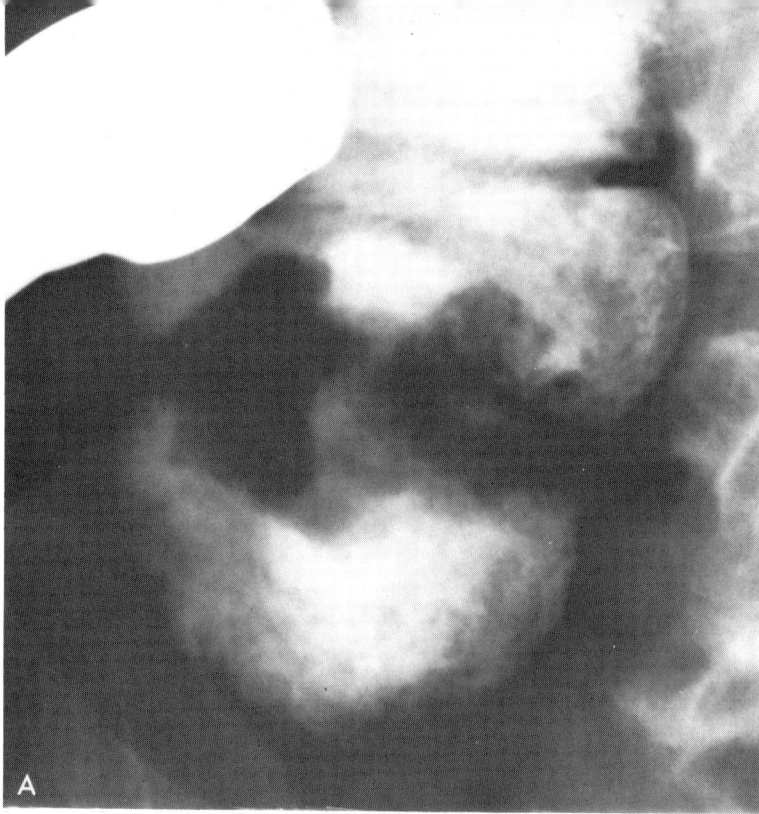

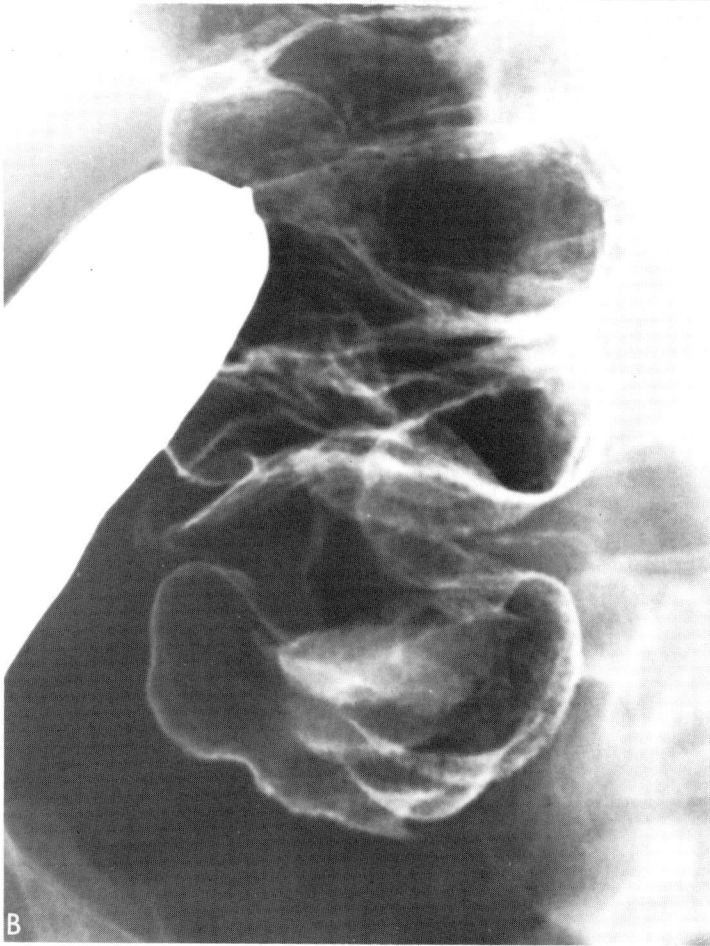

Figure 5–11. Barium enema examination (A) and air contrast study (B) showing an annular adeno-carcinoma of the ascending colon. Persistent narrowing with overhanging margins is characteristic.

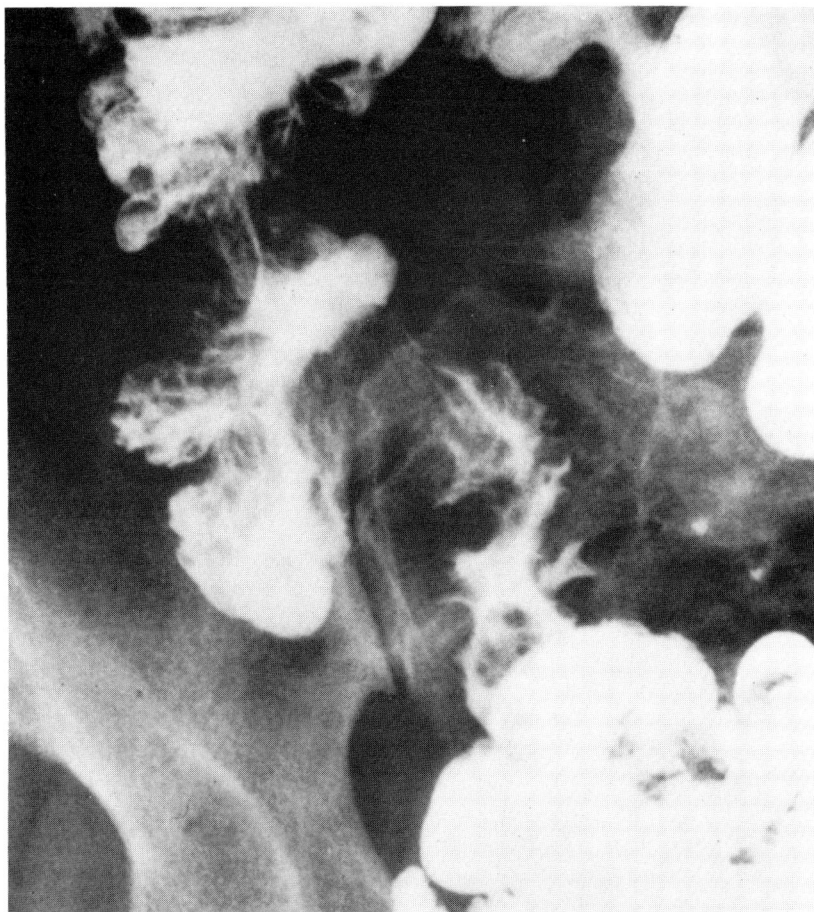

Figure 5–12. Small bowel examination showing a carcinoma of the cecum extending into the terminal ileum. Cecal carcinoma rarely crosses the ileocecal valve, and when it does it simulates regional enteritis.

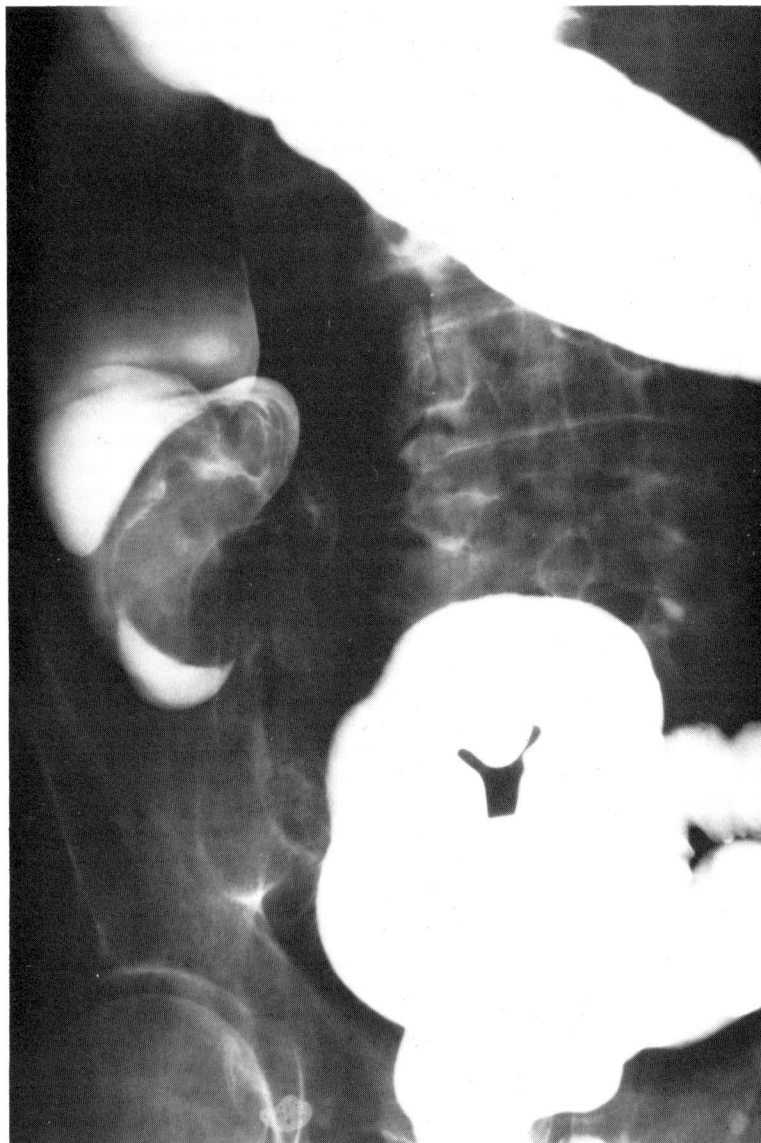

Figure 5–13. Barium enema examination showing a large polypoid mass in the cecum due to colocolic intussusception of a cecal carcinoma.

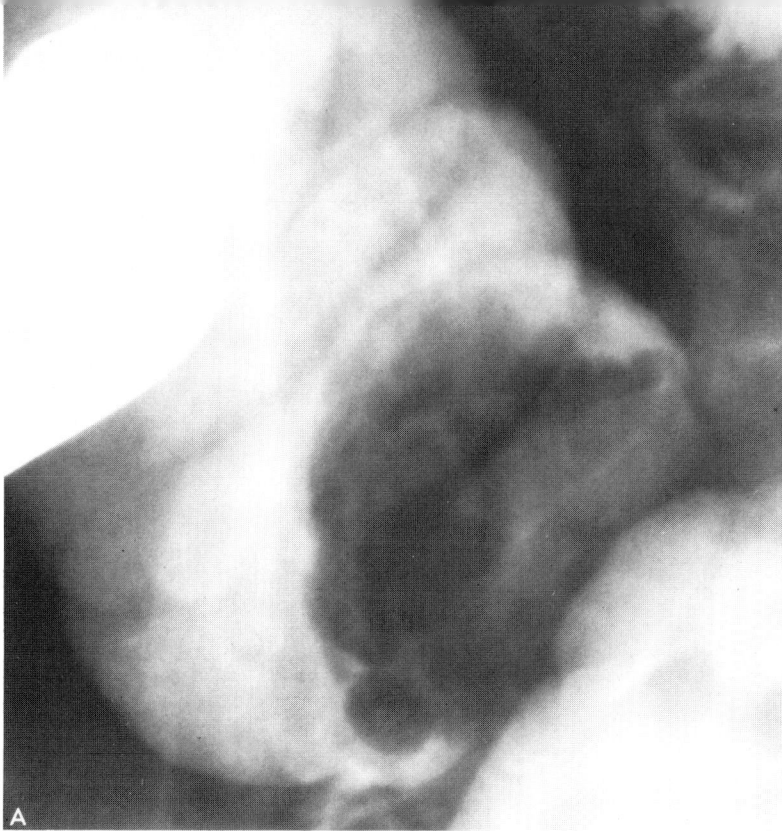

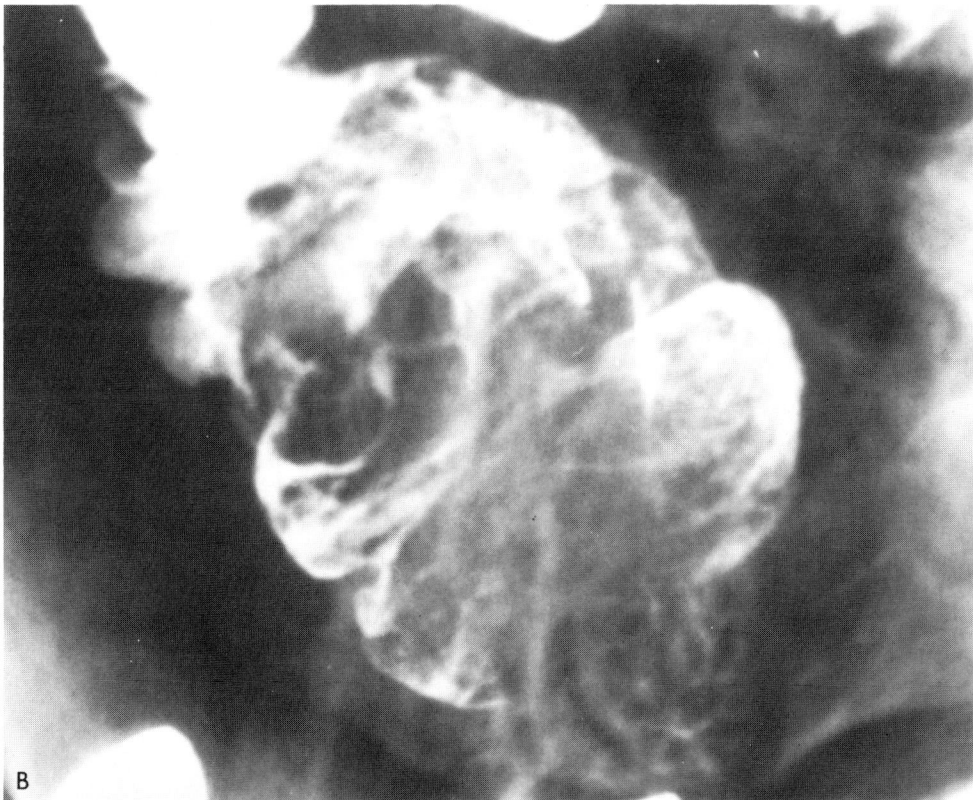

Figure 5-14. Barium enema examination (A) and postevacuation radiograph (B) showing a polypoid adenocarcinoma of the cecum which intussuscepts into the ascending colon on the postevacuation examination.

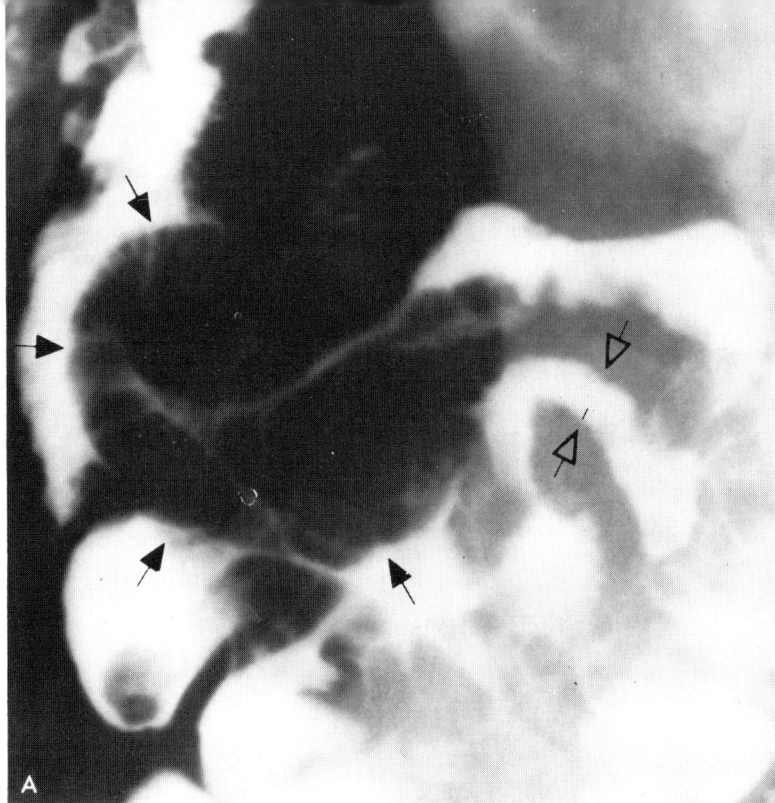

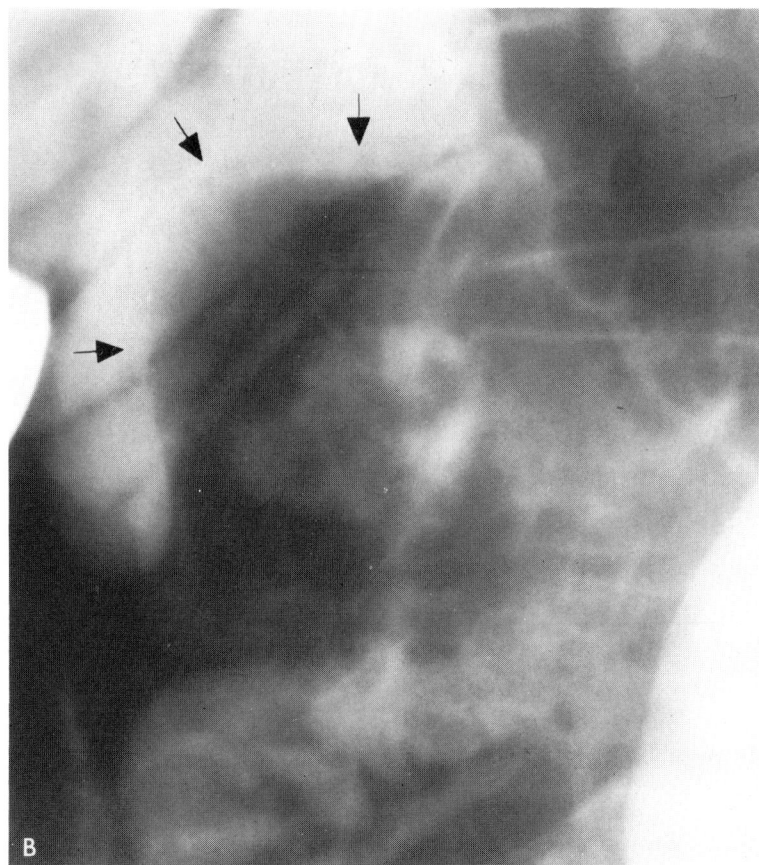

Figure 5–15. Small bowel examination (A) and barium enema study (B) showing a carcinoma of the cecum with a fistula into the terminal ileum (open arrows). The well-defined mass (closed arrows) is the distal margin of an edematous ileocecal valve.

ileum with tumor. Hence, while it may be generally true that carcinoma of the colon rarely crosses the ileocecal valve in comparison to lymphosarcoma, which does so regularly, involvement of the terminal ileum does not exclude the diagnosis of colonic carcinoma (Fig. 5–12).

Occasionally, radiographic studies demonstrate obstruction, perforation, intussusception or fistulas (Figs. 5–13 to 5–15).

Adenoacanthoma

Adenoacanthoma is a rare, malignant tumor of the gastrointestinal tract characterized by the presence of squamous carcinoma in combination with the ordinary glandular elements of adenocarcinoma.[22] Special histologic stains show the presence of keratin and mucin. The identification of keratin confirms the presence of squamous epithelium, while mucin indicates the presence of glandular elements. When mucin is absent in a tumor containing squamous epithelium, the lesion is classified as an epidermoid carcinoma.

Adenoacanthoma is not unusual in the rectum or the esophagus, where the source of the squamous elements is apparent. However, elsewhere in the gastrointestinal tract, the pathogenesis of the tumor is more obscure. The origin of the squamous cells may be from inclusion rests of ectoderm, the transformation of glandular cells or the malfunction of genetic replication leading to differentiation of adenocarcinoma into squamous carcinoma.[23]

Since the first case described by Herxheimer in 1908, a total of 18 adenoacan-

(Text continued on page 268)

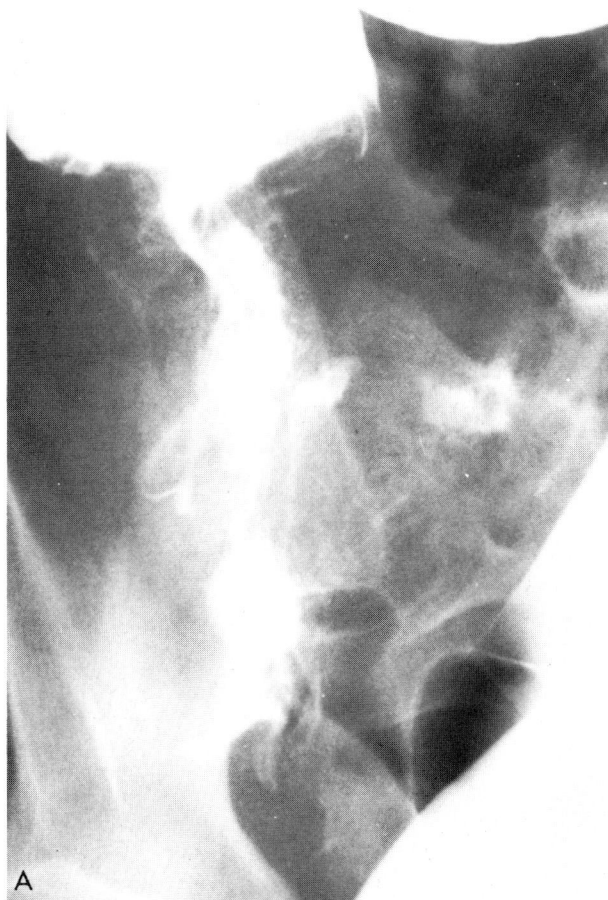

Figure 5–16.   Barium enema examination (A, B and C) showing an annular adenoacanthoma of the cecum. Note the smooth distal surface of the lesion apparent in C. Surgical specimen looking into the cecum from above (D) showing necrotic tumor. The distal margin of the lesion projecting into the ascending colon is smooth, accounting for the appearance in C. The ileum is toward the bottom of the illustration. Postevacuation barium enema radiograph (E) showing the tumor in the cecum 8 months earlier. The tumor on the earlier examination performed elsewhere was mistaken for fecal material.

(Figure 5–16 continued on following pages)

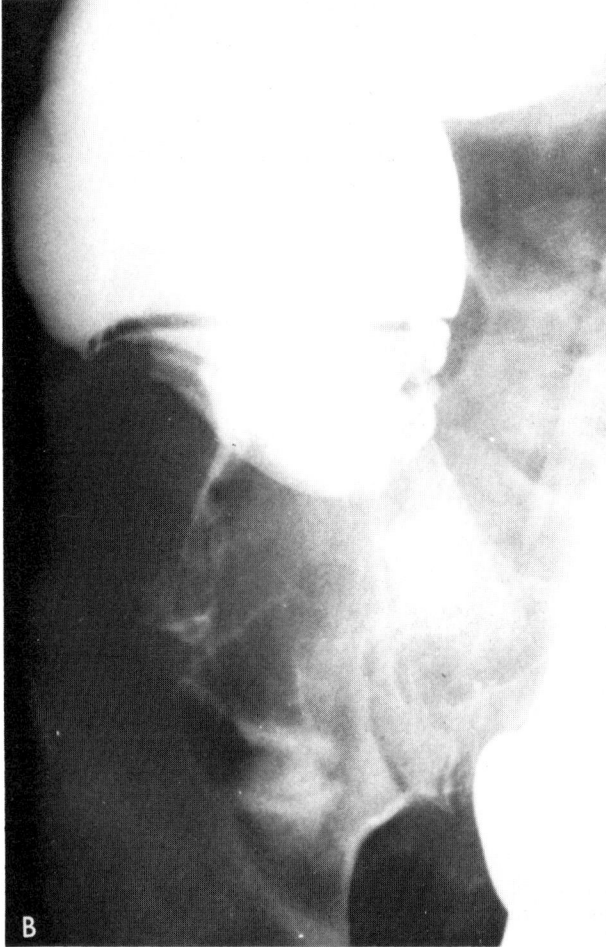

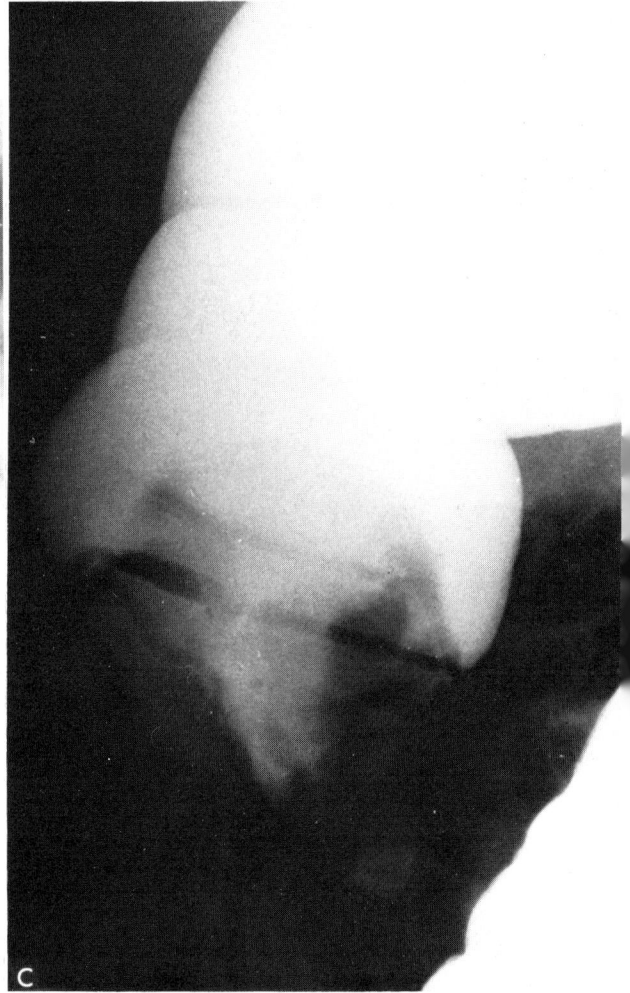

Figure 5–16. (Continued)

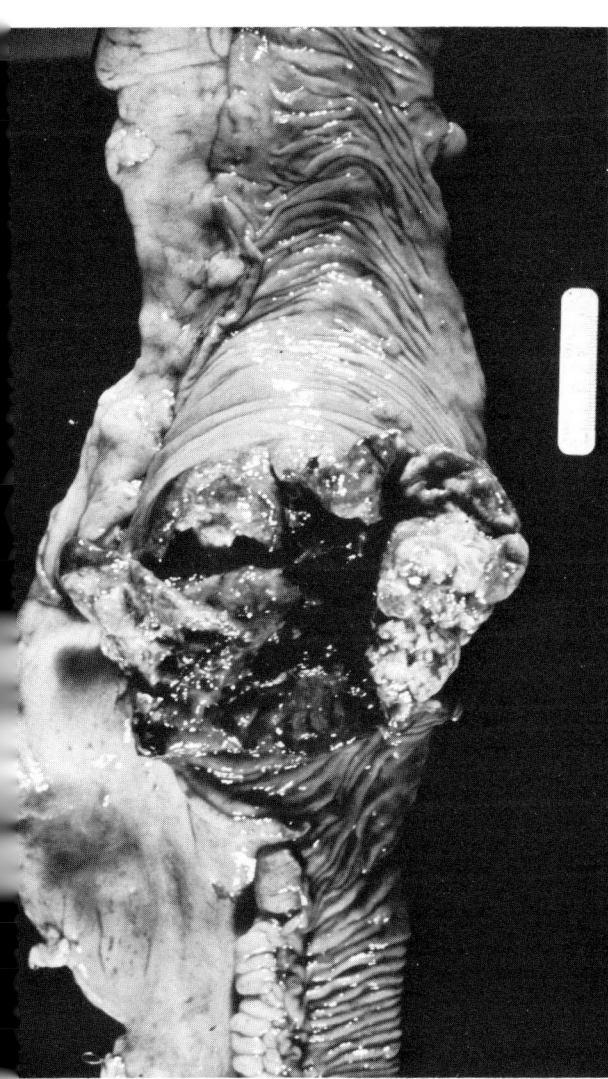

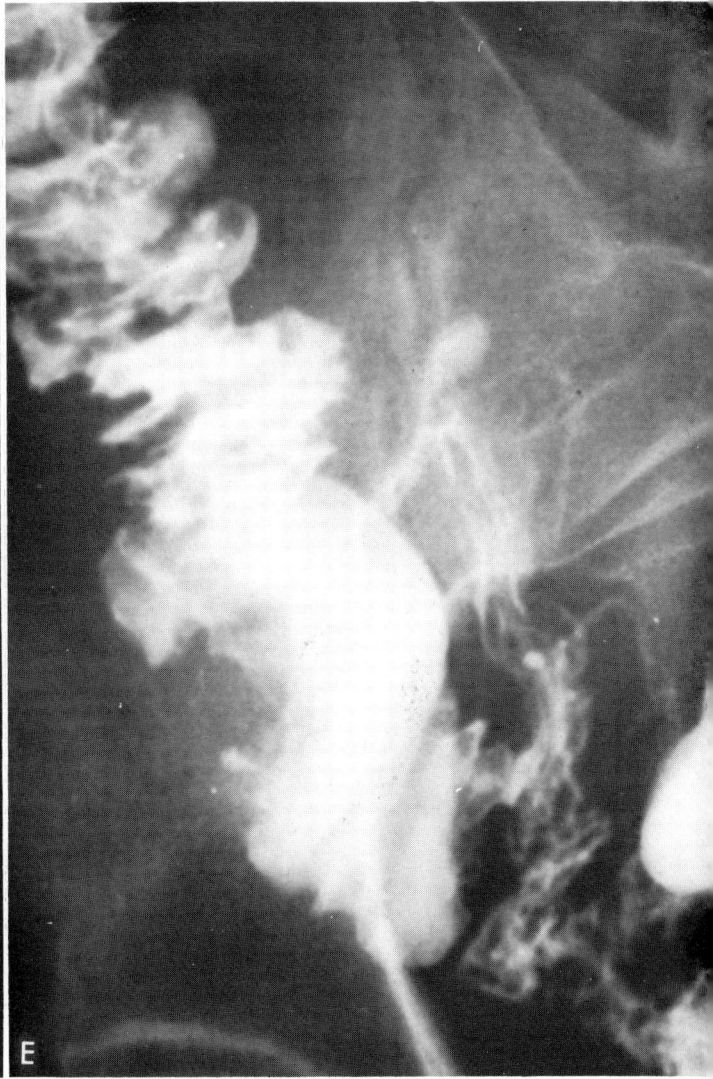

Figure 5-16. (Continued)

thomas of the cecum and ascending colon have been reported.[24] Up until 1967, 66 cases have been described in the stomach, where the lesion is more frequent.[25]

Clinically, there is little to distinguish adenoacanthoma of the cecum from the more common adenocarcinoma in the same location. The age and sex incidence are identical. Like other cecal tumors, adeno-acanthoma may remain silent for long periods, often reaching large size before being recognized. Anemia and fatigue frequently precede the gastrointestinal symptoms.

Like adenocarcinoma, adenoacanthoma of the cecum is usually polypoid, producing obstruction late in the course of the disease, although annular and infiltrating varieties are not unusual.

The prognosis of adenoacanthoma of the cecum is difficult to determine because of the limited number of cases reported. However, like squamous cell carcinoma of the esophagus and rectum, the five-year survival is lower than with adenocarcinoma.

The roentgen features of adenoacanthoma of the cecum are indistinguishable from the more conventional adenocarcinoma (Fig. 5–16). Both lesions produce polypoid masses with or without annular features and obstruction. Differentiation from adenomatous polyps, villous adenoma and ameboma is required.

Carcinoid

Carcinoids are slowly growing tumors that arise in the Kulchitsky or argentaffin cells in the crypts of Lieberkühn of the intestinal epithelium.[26] Although histologically benign, most pathologists agree that all carcinoids, except those arising in the appendix, are potentially malignant. The tumor usually produces single or multiple, well-circumscribed nodules in the small bowel. As the lesion grows, it infiltrates the bowel wall and extends into the mesentery and the mesenteric lymph nodes, often inciting an intense fibrotic response.

The tumor is most common in the appendix, where it is relatively benign and rarely metastasizes. Diagnosis is usually made incidentally at surgery or at postmortem examination. In the intestine, the ileum is most commonly involved. Eighty per cent of small bowel carcinoids occur in the last 60 cm. of the ileum. In one series of 56 patients with carcinoid of the gastrointestinal tract, the age range was from 14 to 86 years, with a mean of 51 years.[27] Over two-thirds of the patients were females.

When carcinoids metastasize to the liver, they may produce the carcinoid syndrome, which consists of cutaneous flushing, diarrhea, wheezing and intestinal colic due to the production of serotonin and other products.[28] The syndrome is not produced by in-

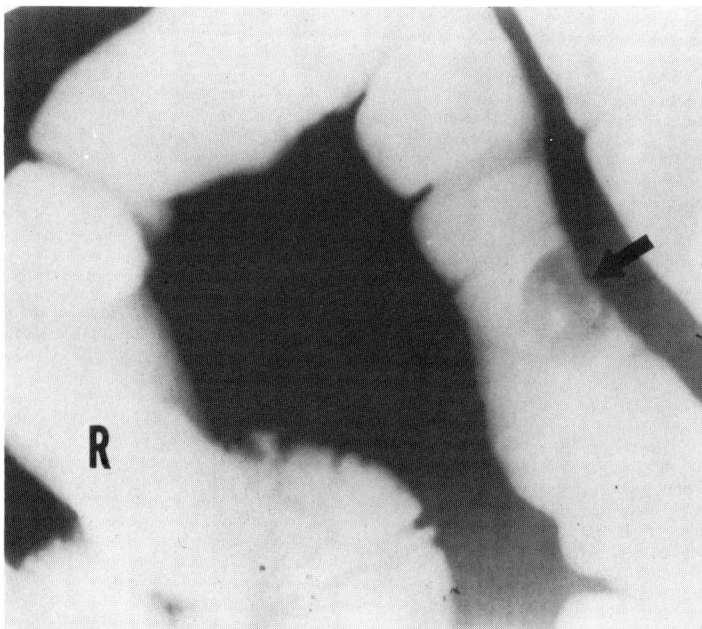

Figure 5–17. Small bowel examination showing a smooth nodule in the terminal ileum due to a carcinoid (arrow). Single or multiple masses of this type are characteristic.

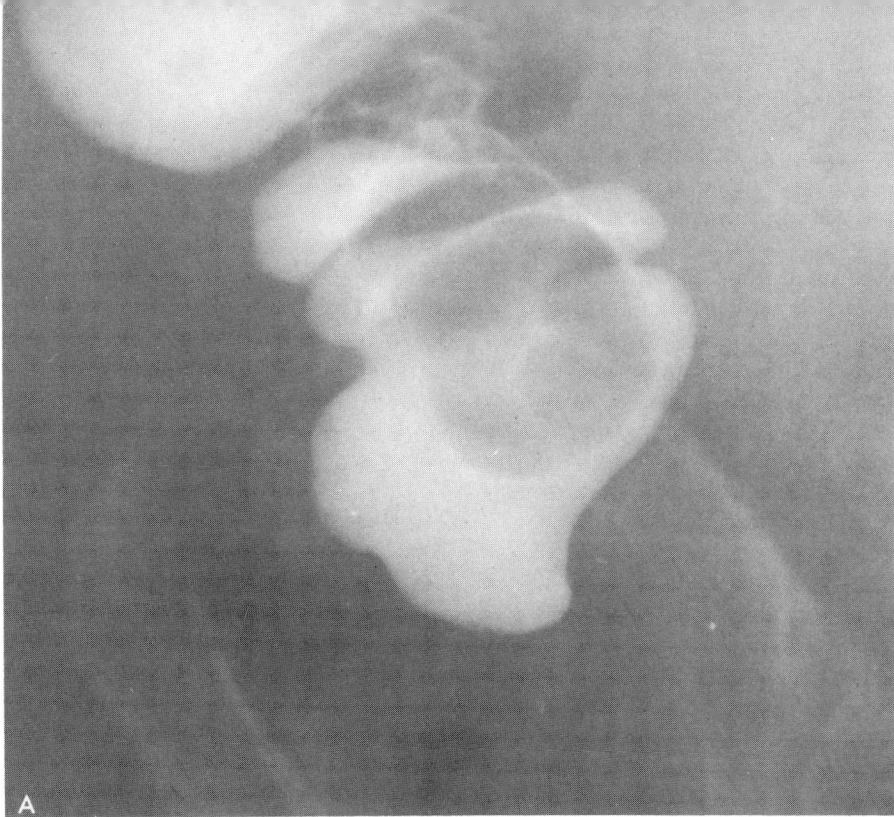

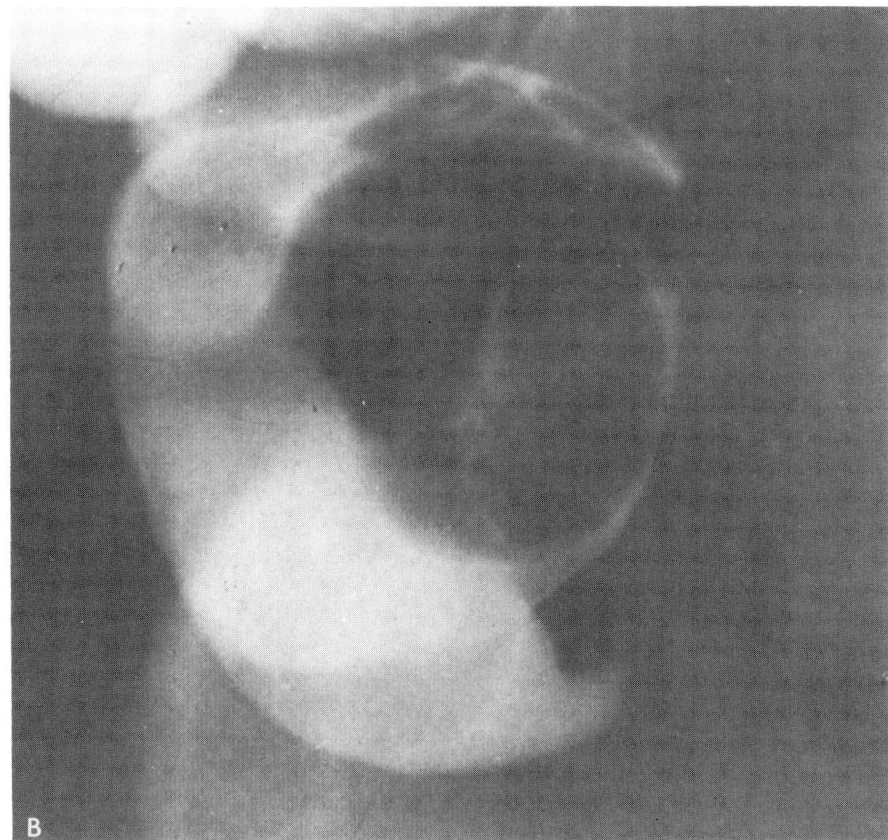

Figure 5-18. Barium enema examination (*A* and *B*) showing a large carcinoid of the ileum prolapsing into the cecum.

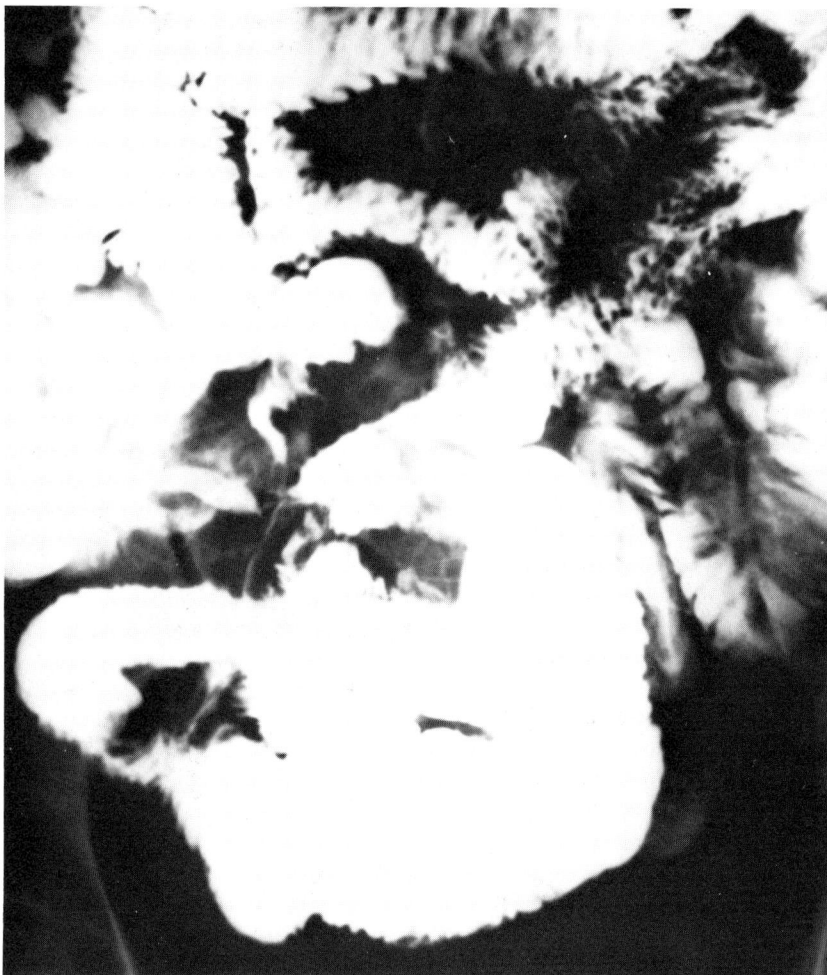

Figure 5–19.   Small bowel examination showing marked angulation and kinking of small bowel loops due to the desmoplastic reaction in the mesentery incited by a carcinoid of the terminal ileum. Unless nodules of tumor are identified, differentiation from adhesions is difficult.

testinal carcinoids unless liver metastases occur, because the serotonin in the portal circulation is degraded by the liver.

The tumors are not often diagnosed radiographically because they are usually small[29] (Fig. 5–17). Single or multiple submucosal nodules are characteristic. Intussusception may occur (Fig. 5–18). The smooth well-circumscribed lesions are difficult to distinguish from other submucosal lesions. However, as the tumor extends into the mesentery, the associated desmoplastic reaction causes kinking and angulation of the small bowel[30] (Fig. 5–19). Obstruction and ulceration are unusual but may occur (Fig. 5–20A). Large masses of lymph nodes in the mesentery may displace adjacent small bowel loops.

Nodules in the small bowel due to carcinoid must be differentiated from lymphosarcoma and metastatic disease. Calcification in carcinoid tumors is rare, but when it occurs the lesion must be differentiated from appendicoliths, mucoceles of the appendix and enteroliths associated with obstructing lesions of the ileum or Meckel's diverticula.[31] When narrowing of the intestinal lumen is associated with the tumor, the differential diagnosis must include regional enteritis, primary adenocarcinoma and ischemic ulcers of the intestine.

The angiographic findings of ileal carcinoid tumors consist of a characteristic stellate arterial pattern with irregular distal intramesenteric and arcade vessels and vasa recta, narrowing of the deep arterial mesenteric branches, variable tumor stain and nonvisualization of draining veins. The striking stellate arterial arrangement of irregular small arteries is due to the retraction of the mesentery associated with the desmoplastic reaction. Shimkin et al.[32] noted an intense tumor stain in their patient with carcinoid of the ileum compared to the poor to moderate tumor blush of the cases reported by Reuter and Boijsen[33] (Fig. 5–20B and C).

Lymphoma

Lymphomas can be conveniently divided into lymphosarcoma (which also includes reticulum cell sarcoma) and Hodgkin's disease because of the tendency of the former to incite little scirrhous reaction compared to the latter.[34] The intestine may be the sole site of involvement or may be involved as part of widespread disease. The terminal ileum is most often affected because of the great amounts of lymphoid tissue in the ileum compared to other segments of the gastrointestinal tract. The small bowel is involved more frequently in children than in adults.

Lymphoma arises in the lymphoid tissue in the submucosa of the intestine and extends along the long axis of the bowel. Growth may be into the lumen or outside the bowel. Exophytic extension produces large tumor masses outside of the bowel, extending into the mesentery. Other tumors are multiple. Because lymphosarcoma is a medullary tumor, the lesion may grow to large size before obstruction occurs which produces symptoms. Hemorrhage, perforation, intussusception, fistula and malabsorption may occur.

Marshak divides lymphosarcoma into five morphologic categories: nodular, infiltrating, polypoid, exo-endoenteric and mesenteric.[35] Many cases have features of several forms of the disease.

In the nodular variety of lymphosarcoma there are multiple intraluminal or intramural tumor masses of varying size. The lesions are sharply defined and frequently extend across the ileocecal valve into the cecum. The lumen of the bowel is normal or increased in diameter. Differentiation from metastatic and inflammatory disease may be difficult. In regional enteritis the cobblestone mucosal pattern may simulate lymphosarcoma, although ulcerations, edema, narrowing of the lumen and exudate in the lumen are present in regional enteritis. The nodules in lymphosarcoma are usually larger and better defined and the caliber of the bowel is normal or increased. There is no mucosal edema and no exudate in the intestinal lumen.

When lymphosarcoma infiltrates the bowel wall there may be varying degrees of segmental narrowing and dilatation. Diffuse infiltration without nodularity or a tumor mass may be impossible to differentiate from regional enteritis (Fig. 5–21). Marshak points out that the narrowing in lymphosarcoma is never marked, and when stricture occurs it is associated with ulceration and infection.[35] Segmental dilatation may be marked, producing bizarre aneurysmal dilatation of the bowel lumen with nodular, thickened walls.

In the polypoid form of the disease, the tumor grows into the lumen from its submucosal origin. The lesions are usually smooth and sharply demarcated and are typical of mural lesions of any etiology (Fig. 5–22). Peristalsis may produce a pseudopedicle or the mass may form the lead point for intussusception. The tumors may reach large size without producing narrowing of the intestinal lumen and are often multiple. Intussusception is not uncommon (Figs. 5–23 and 5–24).

When ulceration and necrosis occur, lymphosarcoma may assume the exo-endoenteric form described by Marshak.[34] Bizarre, irregular cavities fill with barium and displace adjacent loops of intestine (Fig. 5–25). Numerous complex fistulas are often present. Fistula and abscess formation occur in regional enteritis, but the absence of narrowing of the lumen and the presence of nodular defects in the wall of the cavity are typical of lymphosarcoma.

When lymphosarcoma begins in the mesentery or grows exophytically from the bowel wall, large extrinsic tumor masses displace adjacent loops of intestine. This variety of lymphosarcoma must be differentiated from other tumors of the mesentery such as liposarcoma and mesenteric cysts.

Hodgkin's disease may produce a poly-

(Text continued on page 278)

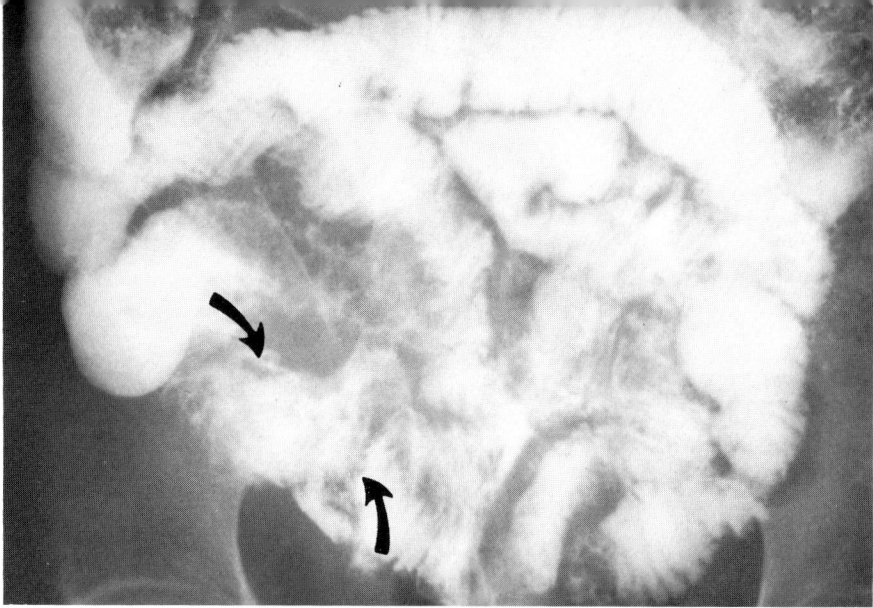

A

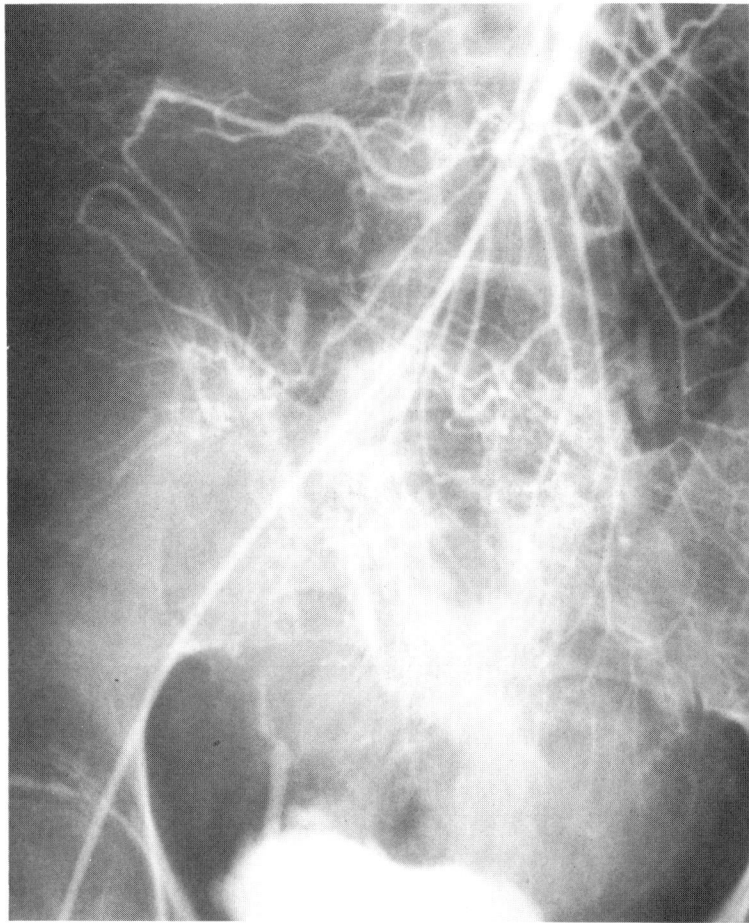

B

Figure 5–20. A small bowel examination (A) showing an ulcerating carcinoid of the ileum (arrows). Superior mesenteric angiograph in the arterial phase (B) shows irregular small arteries arranged in a stellate manner. The capillary phase (C) reveals an intense tumor stain in the right lower quadrant and pelvis. One year after the angiogram was performed, autopsy showed the ileum and cecum to be fixed by fibrous adhesions infiltrated with tumor cells. Multiple submucosal and subserosal tumor nodules were present in the ileum. A 7-mm. mucosal ulceration was present 10 cm. from the ileocecal valve. (From Shimkin, P. M., DeVita, V. T., and Doppman, J. L.: Arteriography of an ileal carcinoid tumor. J. Canad. Assoc. Radiol., 22:259, 1971.)

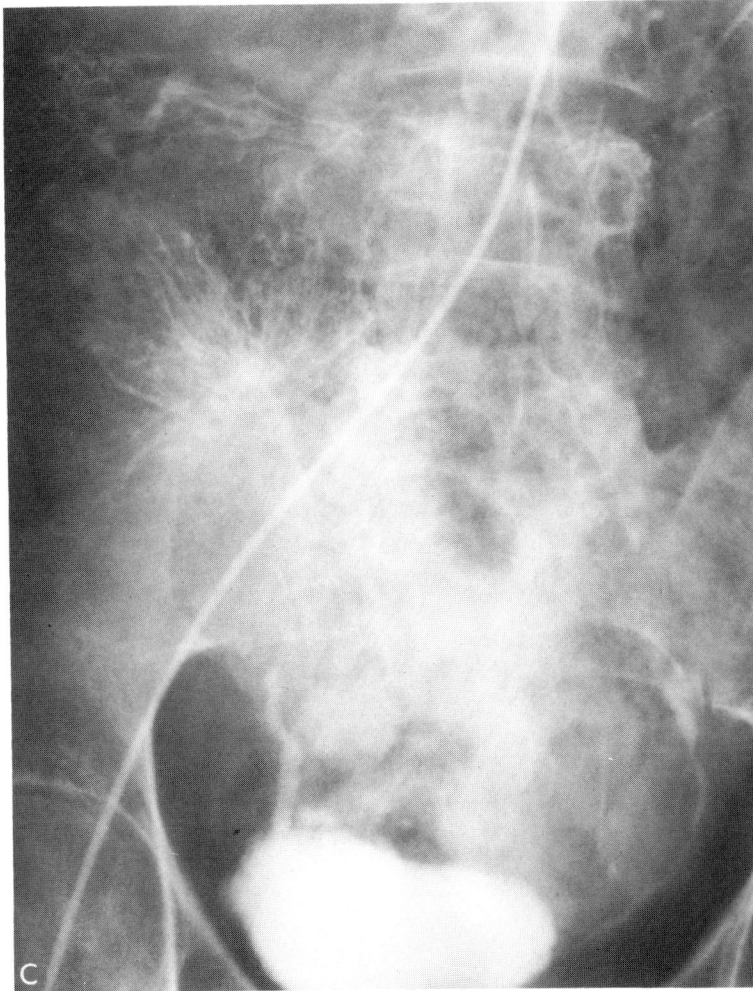

Figure 5–20. (*Continued*)

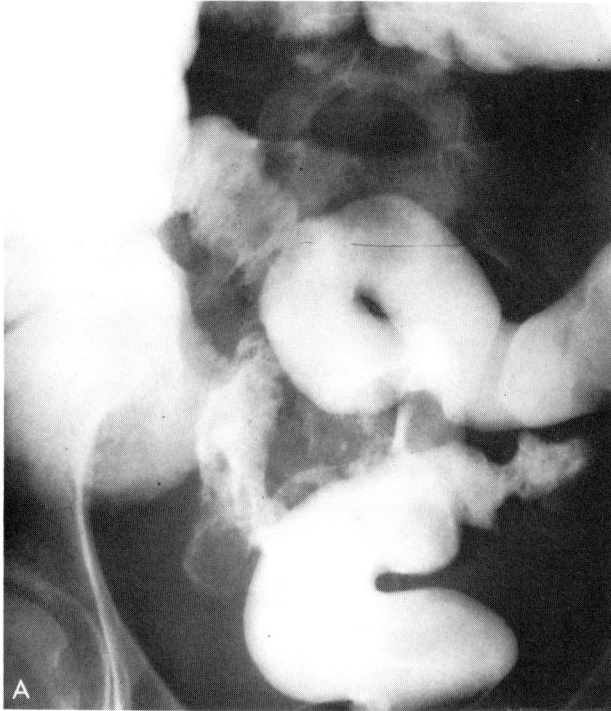

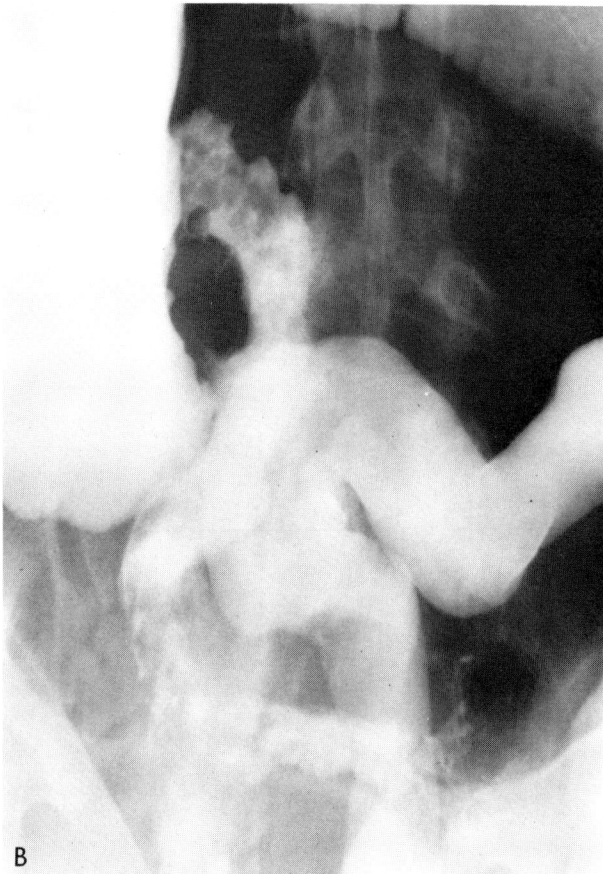

Figure 5–21. Anterior-posterior (*A*) and "angle" (*B*) views from a barium enema examination showing the infiltrating variety of lymphosarcoma involving the terminal ileum. This is a common location for intestinal involvement, which is particularly frequent in children. The ulcerations, exudate, eccentric involvement, spasm, narrowing and cobblestone pattern of regional enteritis are absent.

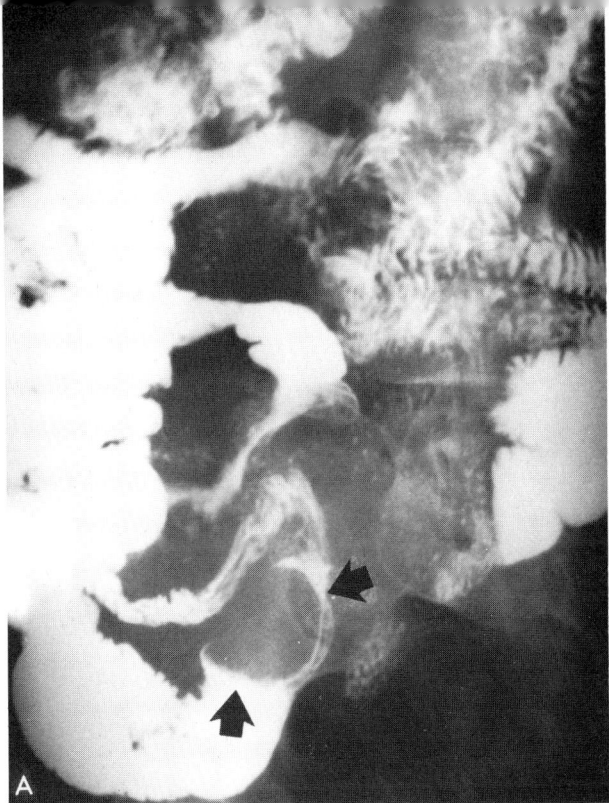

Figure 5-22. Small bowel examination showing the polypoid variety of lymphosarcoma of the small intestine. Multiple large smooth masses are present (arrows). Five separate tumors were found at surgery.

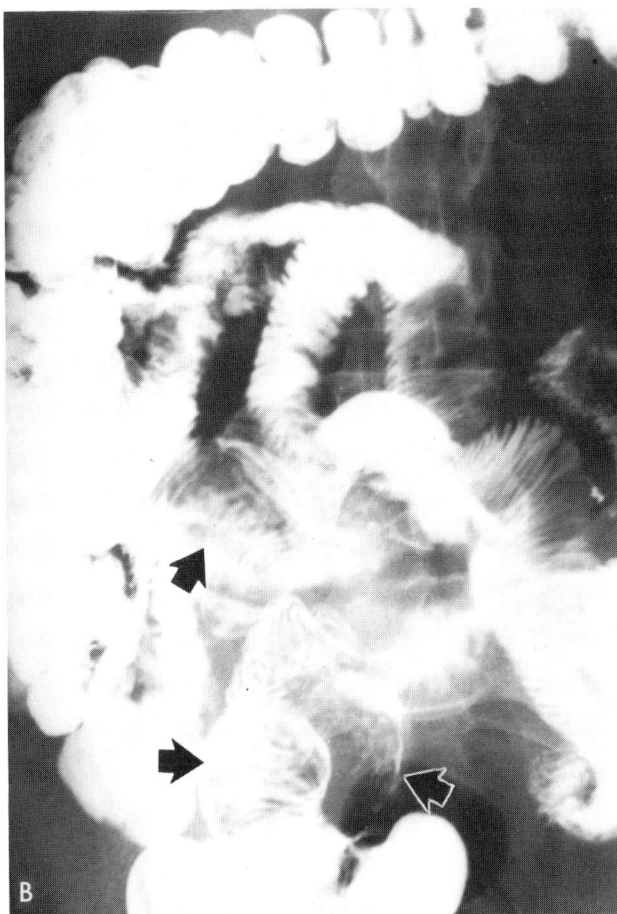

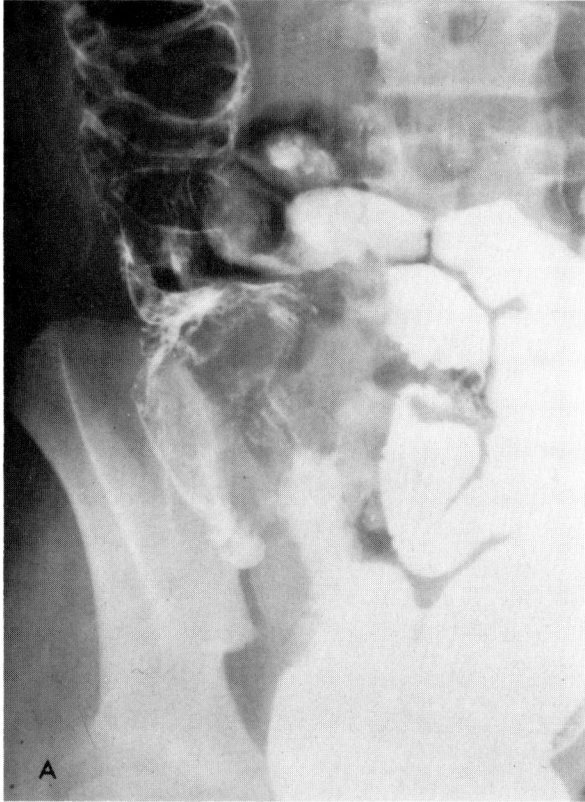

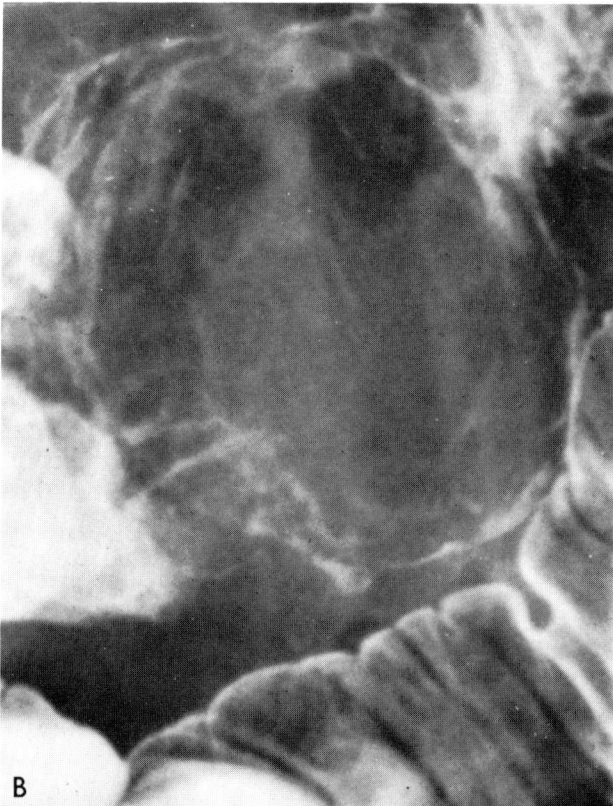

Figure 5–23. Small bowel examination (A) showing a polypoid lymphoma of the terminal ileum. The large, relatively smooth mass within the cecum is contiguous with the terminal ileum. The mass is better seen on a spot radiograph (B).

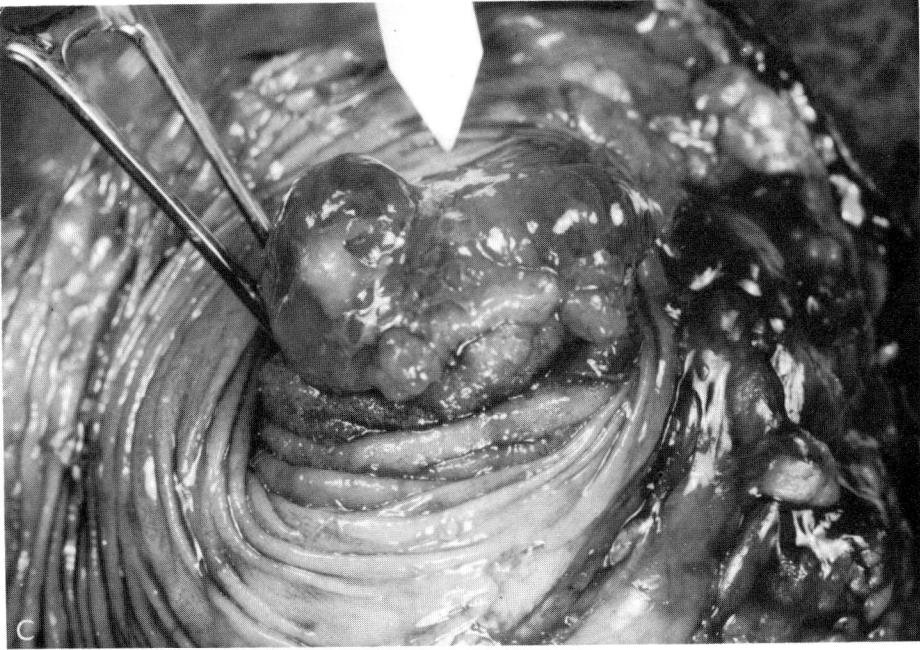

Figure 5-23. (Continued) The lymphomatous mass is identified in the open cecum at operation (*C*). The coil-spring appearance and edema of the cecal folds are evident. (From Felson, B., and Wiot, J. F.: Some interesting right lower quadrant entities. Radiol. Clin. N. Amer., 7:83, 1969.)

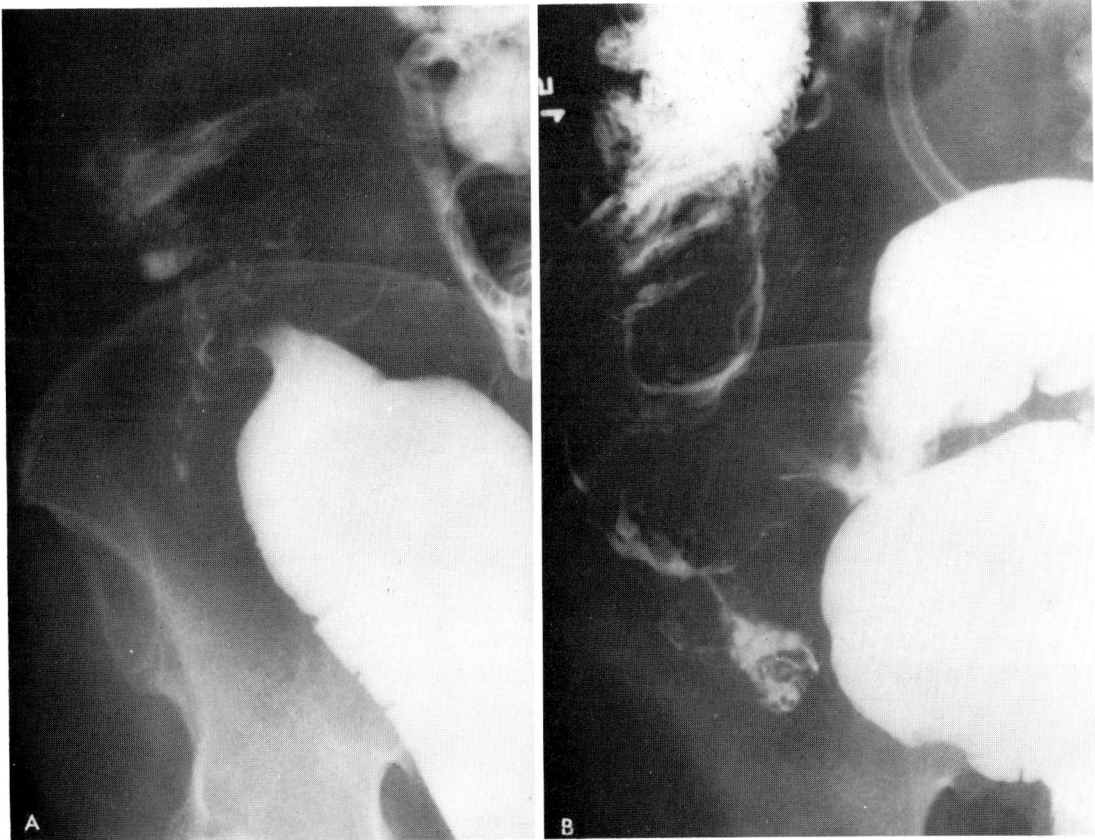

Figure 5-24. Small bowel examination showing polypoid lymphosarcoma of the terminal ileum intussuscepting into the colon. The markedly dilated terminal ileum tapers distally (*A*). There is a large mass in the ascending colon due to ileocolic intussusception (*B*). (From Felson, B., and Wiot, J. F.: Some interesting right lower quadrant entities. Radiol. Clin. N. Amer., 7:83, 1969.)

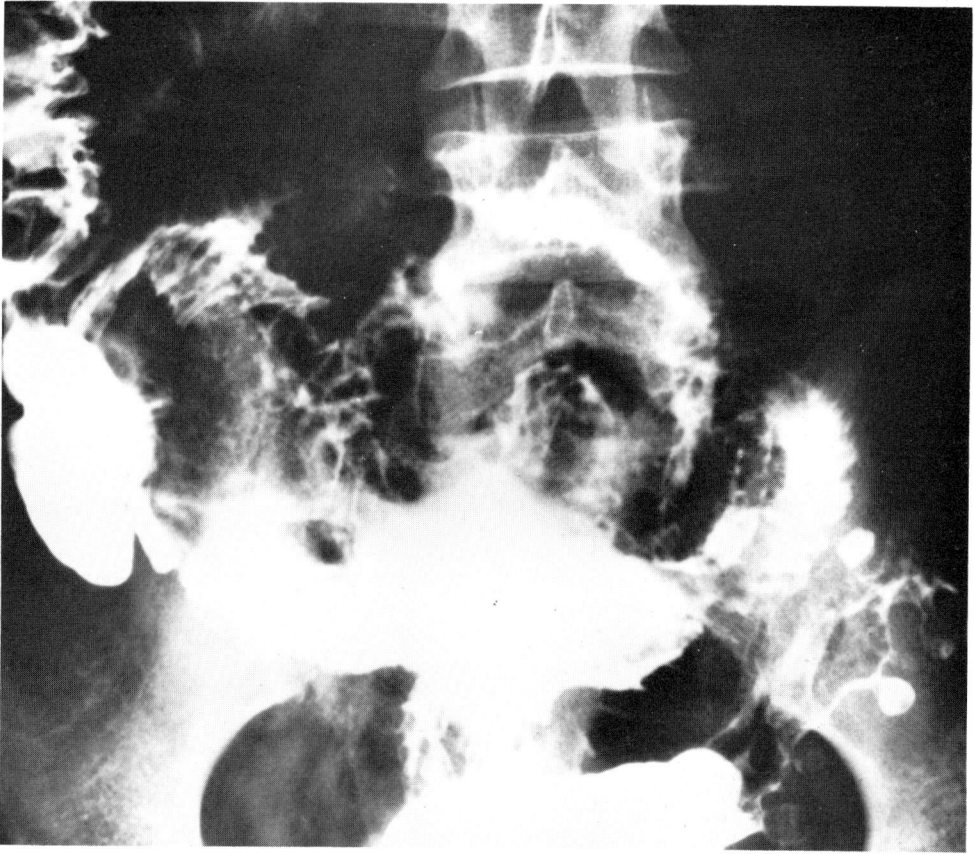

Figure 5-25. The small bowel examination showing the exoenteric form of lymphosarcoma. A bizarre irregular cavity is filled with barium and connects to other intestinal segments by numerous fistulas.

poid intraluminal mass, an annular constriction or a mesenteric lesion, depending on the direction of its growth. However, because it is often associated with an intense desmoplastic reaction, narrowing of the intestinal lumen is frequent. Because of this, Hodgkin's disease is difficult to differentiate from primary adenocarcinoma of the bowel and from cicatrizing inflammatory diseases. Like lymphosarcoma, Hodgkin's disease is common in the terminal ileum and in this location it commonly crosses the ileocecal valve to involve both the valve and the cecum. Carcinoma of the cecum, contrarily, rarely extends into the ileum.

Leiomyosarcoma

Leiomyosarcomas of the small intestine often present with bizarre radiologic features consisting of large irregular necrotic cavities which fill with air or barium. The tumor is usually a bulky, polypoid mass which may extend into the lumen or become largely exophytic.[36] Exophytic lesions may be attached to the bowel by a small pedicle, producing little or no deformity of the intestine. In distinction to the lymphomas, which involve the terminal ileum more frequently than other portions of the intestine, leiomyosarcomas are uniformly distributed throughout the small bowel. The tumor is usually single, although multiple tumors have been described.

Plain abdominal radiographs may show a large irregular cavity filled with air. Barium studies show that the cavity communicates with the bowel and demonstrate that the crater has thick, nodular walls (Fig. 5-26). Differentiation from the exo-endoenteric form of lymphosarcoma may be difficult, but the excavation in leiomyosarcoma tends to be larger and more bizarre.

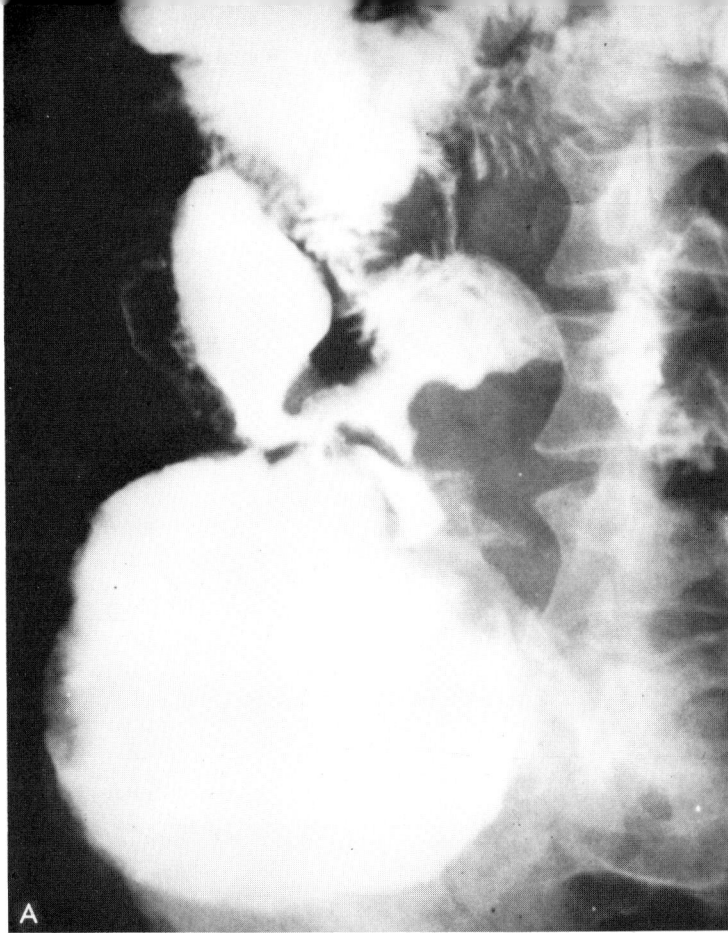

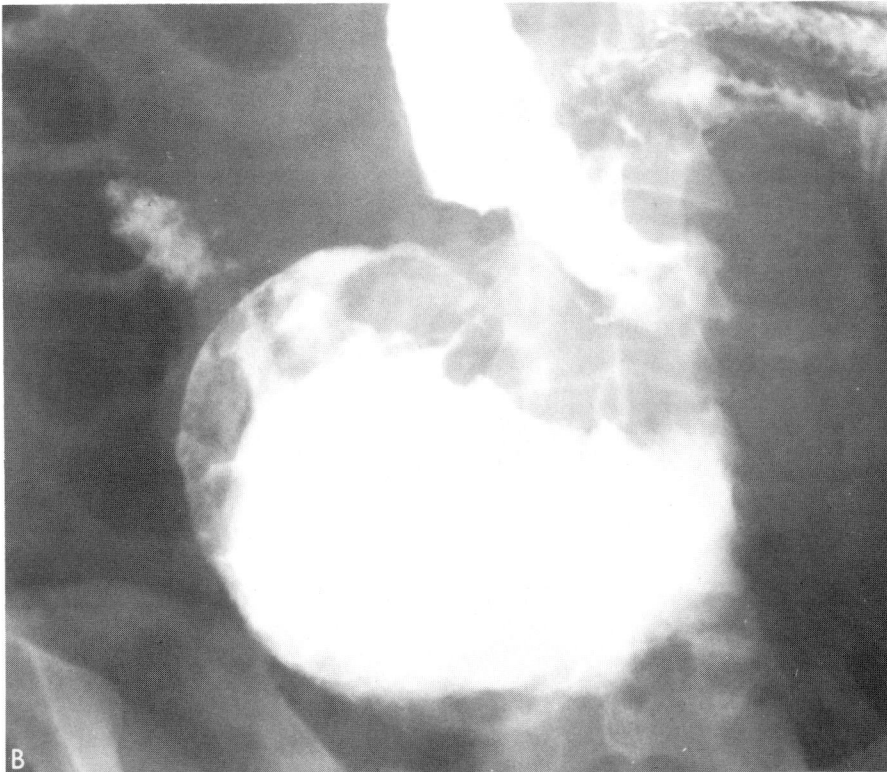

Figure 5–26. Small bowel examination showing a large cavitating leiomyosarcoma. The irregular cavity (A) with a thick, nodular wall (B) is characteristic of this type of tumor.

Primary and Metastatic
Adenocarcinoma of the Small Bowel

Primary adenocarcinoma of the small intestine is rare, particularly in the ileum.[36] The lesion usually is a short, sharply demarcated annular lesion causing destruction of the mucosa. Polypoid and exophytic primary tumors are unusual. Adenocarcinoma of the small intestine may be associated with regional enteritis and with sprue.

The tumors which most frequently metastasize to the small intestine are carcinomas of the ovary, pancreas, stomach, colon and breast. The metastatic deposit originates in the mesentery or beneath the serosa of the small bowel. As the tumor grows, peristalsis may create a pseudopedicle and produce an intraluminal tumor mass. More often the metastasis grows into the bowel wall, producing a mural defect with a smooth surface and sharp demarcation.

Marshak divides metastatic disease of the small intestine into three morphologic categories which include a single or multiple lesion involving primarily the wall of the small bowel, lesions involving primarily the mesentery (with or without involvement of adjacent bowel wall) and lesions that are multiple and involve both the bowel wall and the mesentery.[37] Ascites is common in the last group.

Discrete tumor masses in the first category produce well-defined mural lesions in the bowel wall that are difficult to distinguish from other extramucosal tumors (Fig. 5–27). As the lesion grows, the lumen of the bowel may become irregularly narrowed.

(Text continued on page 284)

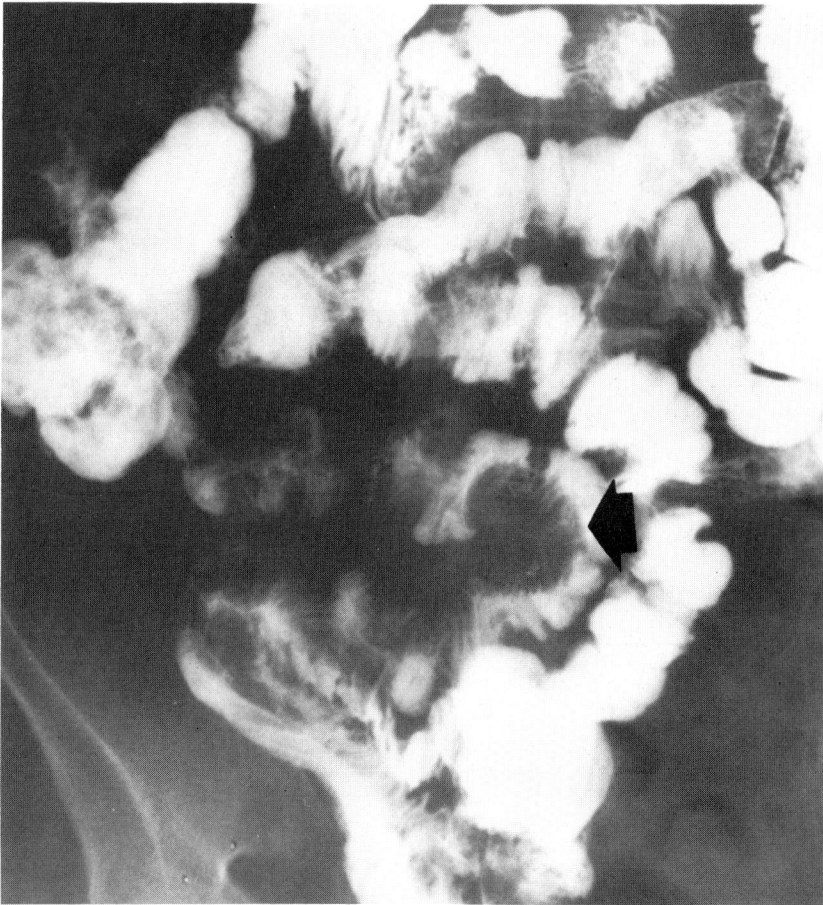

Figure 5–27. Small bowel examination showing a metastasis to the small intestine from carcinoma of the ovary (arrow). The discrete tumor mass produces a mural lesion in the bowel wall.

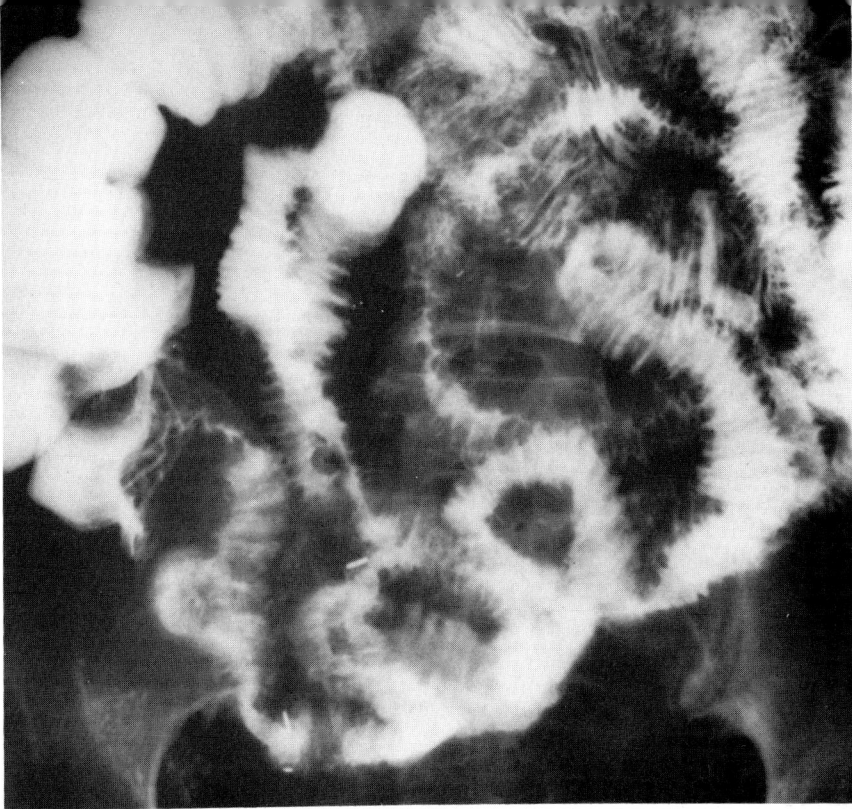

Figure 5-28. Small bowel examination showing multiple metastases to the small intestine and mesentery from carcinoma of the ovary. Metastatic disease in the mesentery produces pleating, angulation and separation of the bowel loops.

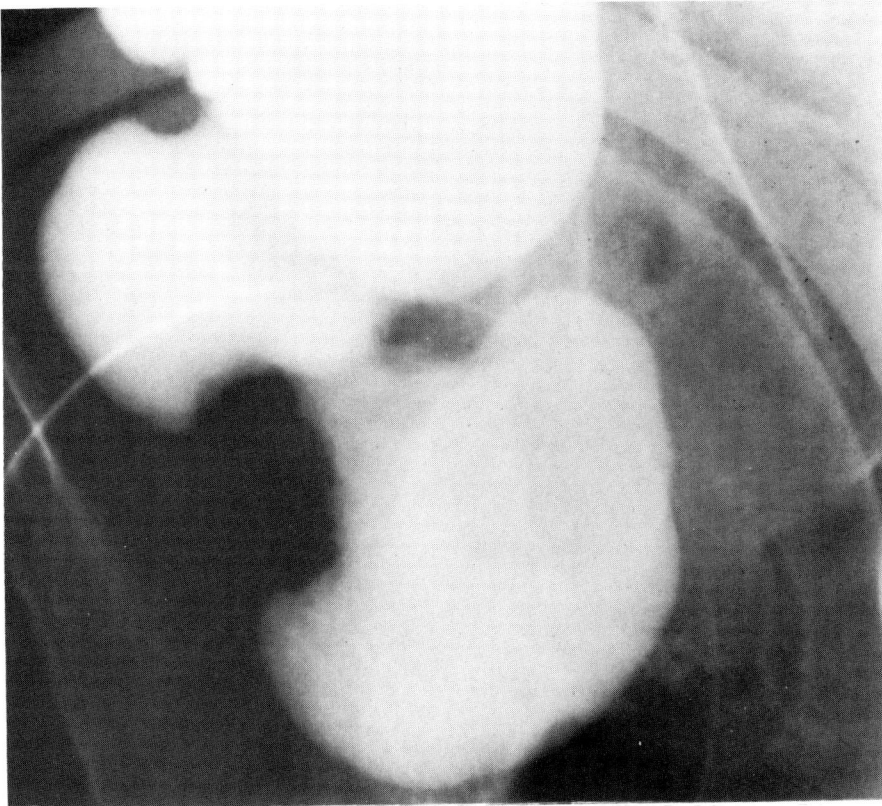

Figure 5-29. Barium enema examination showing an annular constriction of the cecum due to a metastasis from carcinoma of the pancreas. A mural implant with pleating of the bowel wall is characteristic of metastases to the bowel. When the metastasis involves the entire circumference of the intestine, differentiation from a primary carcinoma is difficult unless the intact mucosa is identified.

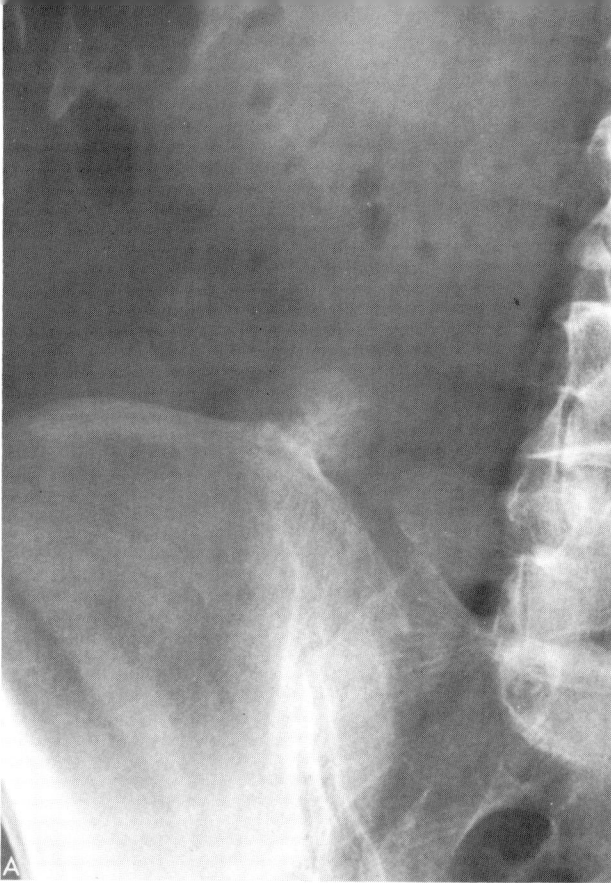

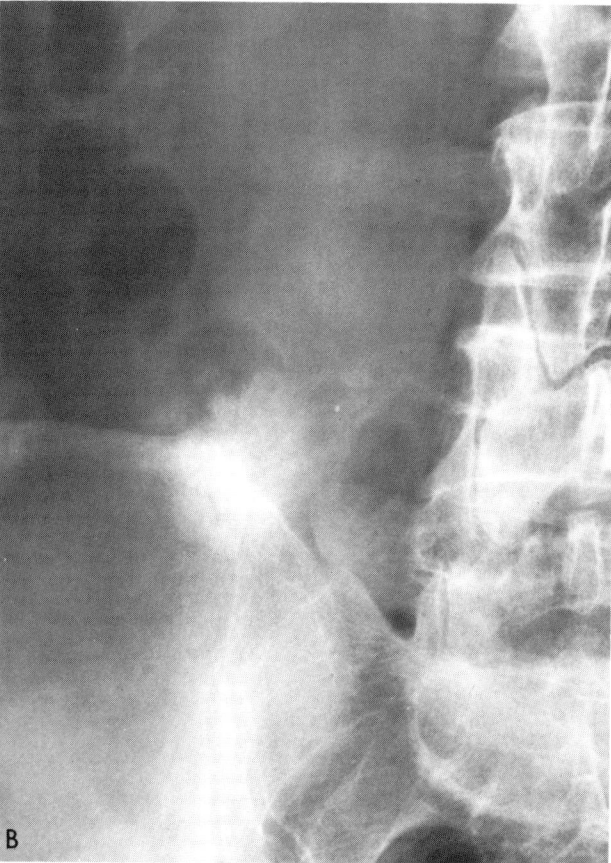

Figure 5-30. Plain abdominal radiographs showing a calcification in the retroperitoneum due to a calcifying metastasis from carcinoma of the cecum. (A) The calcification was first noted 3 years after resection of a carcinoma of the cecum. (B) Four months later the calcification has markedly increased in size. (C) Small bowel examination showing displacing of bowel loops around the mass. At operation, a retroperitoneal mass was present owing to metastatic adenocarcinoma with calcification. (D) Barium enema examination showing the original adenocarcinoma of the cecum at the ileocecal valve (arrows).

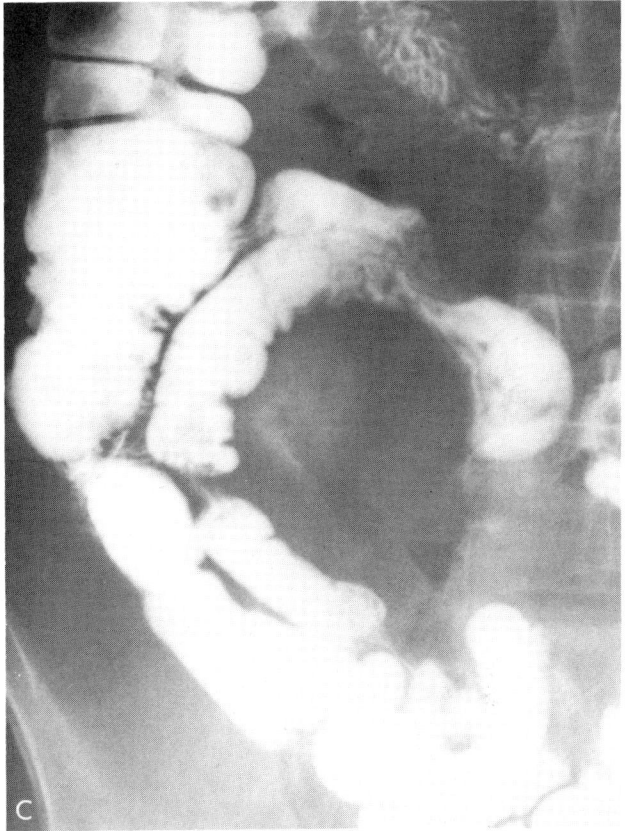

Figure 5–30. (Continued)

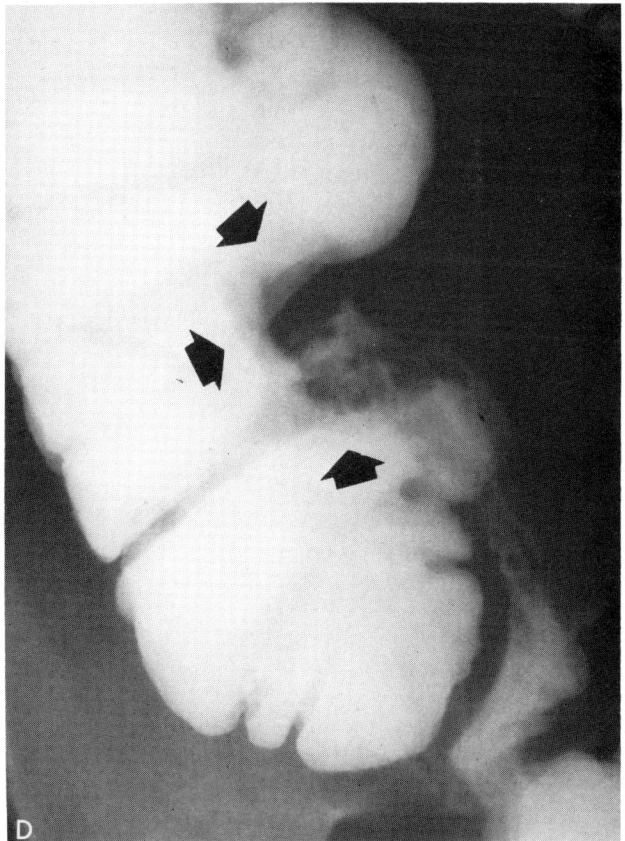

Metastatic foci arising in the mesentery produce smooth indentations of the bowel loops. The tumor may infiltrate the wall of the bowel and produce angulation, fixation and obstruction. In the third category—lesions involving both the bowel wall and the mesentery—multiple areas of eccentric or concentric narrowing with intervening areas of dilatation occur. Shortening of the mesenteric root causes crowding of the valvule and produces characteristic pleating in the contour of the mural tumor (Fig. 5–28).

When the metastasis is single, differentiation from a primary tumor of the bowel may be impossible. According to Marshak, primary tumors tend to produce obstruction earlier and ulceration later than metastatic disease.[37] When the metastatic tumor involves the mesentery, differentiation from other masses in the mesentery may be impossible. The presence of ascites is helpful in indicating the presence of metastatic disease. Regional enteritis rarely involves such short, localized segments of the bowel and is associated with edema of the bowel wall and exudate in the lumen. Hodgkin's disease and ischemic strictures may narrow the lumen and are often more asymmetrical than metastatic disease.

Meyers noted that the distribution of intraperitoneal malignant seeding correlates with the pathways of flow of ascitic fluid.[38] He noted four predominant sites, including the pouch of Douglas at the rectosigmoid level, the right lower quadrant at the lower end of the small bowel mesentery, the left lower quadrant along the superior border of the sigmoid mesocolon and the right paracolic gutter lateral to the cecum and ascending colon. Meyers concluded that growth of secondarily seeded neoplasms in the abdomen is dependent upon the natural flow of ascites within the peritoneal space.[38]

Metastases to the colon produce single or multiple eccentric mural defects with pleating of the bowel wall. Occasionally the tumor may encircle the bowel and simulate an annular primary adenocarcinoma (Fig. 5–29).

Calcification in metastatic disease in the liver, retroperitoneum or intestine may occur (Fig. 5–30). This is usually associated with metastases from mucinous adenocarcinoma of the stomach or colon and it may occur after radiation therapy, particularly after radiation for carcinoma of the ovary.[34]

REFERENCES

1. Wolf, B. S.: Roentgen features of benign tumors of the colon. Surg. Clin. N. Amer., 45:1141, 1965.
2. Wolf, B. S.: Roentgen diagnosis of villous tumors of the colon. Amer. J. Roent., 84:1093, 1960.
3. Kaye, J. J., and Bragg, D. G.: Unusual roentgenologic and clinicopathologic features of villous adenomas of the colon. Radiology, 91:799, 1968.
4. Comfort, M. W., and Stearns, M. W.: Submucous lipomata of the gastrointestinal tract. Surg. Gynec. Obstet., 52:101, 1931.
5. Castro, E. B., and Stearns, M. W.: Lipoma of the large intestine. Dis. Colon Rectum, 15:441, 1972.
6. Schatzki, R.: The roentgenologic appearance of intussuscepted tumors of the colon with and without barium examination. Amer. J. Roent., 41:549, 1939.
7. Wychulis, A. R., Jackman, R. J., and Mayo, C. W.: Submucous lipomas of the colon and rectum. Surg. Gynec. Obstet., 118:337, 1964.
8. Margulis, A. R., and Jovanovich, A.: Roentgen diagnosis of submucous lipomas of the colon. Amer. J. Roent., 84:1114, 1960.
9. Stevens, G. M.: Water enema to verify lipoma of the colon. Amer. J. Roent., 96:292, 1966.
10. Bussey, H. J. R.: Gastrointestinal polyposis. Gut, 11:970, 1970.
11. Gardner, E. J., and Richards, R. C.: Multiple cutaneous and subcutaneous lesions occurring simultaneously with hereditary polyposis and osteomatosis. Amer. J. Hum. Genet., 5:139, 1953.
12. Thomas, K. E., Watne, A. L., Johnson, J. G., Roth, E., and Zimmerman, B.: Natural history of Gardner's syndrome. Amer. J. Surg., 115:218, 1968.
13. Dormandy, T. L.: Gastrointestinal polyposis with mucocutaneous pigmentation (Peutz-Jeghers syndrome). New Eng. J. Med., 256:1093, 1957.
14. Cronkhite, L. W., and Canada, W. J.: Generalized gastrointestinal polyposis: An unusual syndrome of polyposis, pigmentation, alopecia and onychotrophia. New Eng. J. Med., 252:1011, 1955.
15. Ruymann, F. B.: Juvenile polyps with cachexia. Report of an infant and comparison with Cronkhite-Canada syndrome in adults. Gastroent., 57:431, 1969.
16. Berk, R. N., Rush, J. L., and Elson, E. C.: Multiple inflammatory polyps of the small intestine with cachexia and protein-losing enteropathy. Radiology, 95:611, 1970.
17. Davis, G. B., and Berk, R. N.: Intestinal neurofibromatosis in von Recklinghausen's disease. Amer. J. Gastroent., 60:410, 1973.
18. Schutt, R. P., and Walker, J. H.: The problem of early diagnosis in right colon carcinoma. West. J. Surg., 64:29, 1956.
19. Cooley, R. N., Agnew, C. H., and Rios, G.: Diagnostic accuracy of the barium enema study in carcinoma of the colon and rectum. Amer. J. Roent., 84:316, 1960.
20. Allcock, J. M.: An assessment of the accuracy of the clinical and radiological diagnosis of carcinoma of the colon. Brit. J. Radiol., 31:272, 1958.

21. Berk, R. N., Sadowsky, N. L., and Levenson, S. S.: Carcinoma of the cecum and ascending colon. Amer. J. Surg., *104*:27, 1962.

22. Berk, R. N., Scher, G. S., and Bode, D. F.: Unusual tumors of the gastrointestinal tract. Amer. J. Roent., *113*:159, 1971.

23. Rabson, S. M.: Adenosquamous cell carcinoma of intestine. Arch. Path., *21*:308, 1936.

24. Birnbaum, W.: Squamous cell carcinoma and adenoacanthoma of the colon. J.A.M.A., *212*:1511, 1970.

25. Parks, R. E.: Squamous neoplasms of the stomach. Amer. J. Roent., *101*:447, 1967.

26. Smith, F. H., and Murphy, R.: Carcinoid tumors. Med. Clin. N. Amer., *44*:465, 1960.

27. Swensen, S. R., Snow, E., and Gaisford, W. D.: Carcinoid tumors of the gastrointestinal tract. Amer. J. Surg., *126*:818, 1973.

28. Grahame-Smith, D. G.: Progress report, the carcinoid syndrome. Gut *11*:189, 1970.

29. Marshak, R. H., and Lindner, A. E.: Radiology of the Small Intestine. Philadelphia, W. B. Saunders Co., 1970, p. 316.

30. Miller, E. R., and Herrmann, W. W.: Argentaffin tumors of the small bowel; roentgen sign of malignant change. Radiology, *39*:214, 1942.

31. Case records of the Massachusetts General Hospital, Case 34-1973. New Eng. J. Med., *289*:419, 1973.

32. Shimkin, P. M., DeVita, V. T., and Doppman, J. L.: Arteriography of an ileal carcinoid tumor. J. Canad. Assoc. Radiol., *22*:259, 1971.

33. Reuter, S. R., and Boijsen, E.: Angiographic findings in two ileal carcinoid tumors. Radiology, *87*:836, 1966.

34. Marshak, R. H., and Lindner, A. E.: Radiology of the Small Intestine. Philadelphia, W. B. Saunders Co., 1970, p. 355.

35. Marshak, R. H., Wolf, B. S., and Eliasoph, J.: Roentgen findings in lymphosarcoma of the small intestine. Amer. J. Roent., *86*:682, 1961.

36. Good, C. A.: Tumors of the small intestine. Amer. J. Roent., *89*:685, 1963.

37. Marshak, R. H., Khilnani, M. T., Eliasoph, J., and Wolf, B. S.: Metastatic carcinoma of small bowel. Amer. J. Roent., *94*:385, 1965.

38. Meyers, M. A.: Distribution of intra-abdominal malignant seeding; dependency on dynamics of flow of ascitic fluid. Amer. J. Roent., *119*:198, 1973.

Chapter 6

MISCELLANEOUS LESIONS OF THE ILEOCECAL AREA

LYMPHOID HYPERPLASIA

Nodular lymphoid hyperplasia occurring in Peyer's patches in the terminal ileum is a normal finding in children and adolescents.[1] Multiple, discrete nodules 0.5 to 1.0 cm. in diameter are visible on compression radio-graphs of the terminal ileum (Figs. 6–1 and 6–2). The nodules are sharply defined, and there is no indication of associated inflammation, such as edema, ulceration, exudate or spasm. The wall of the bowel is distensible, pliable and nontender. Differentiation from lymphosarcoma and regional enteritis

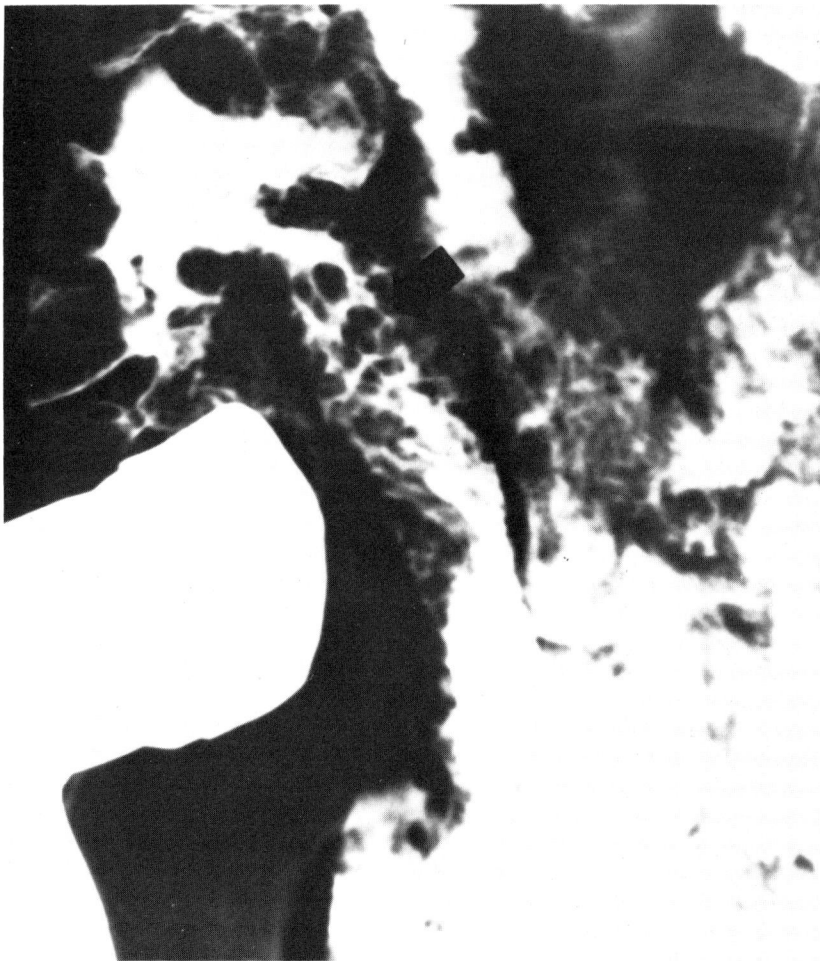

Figure 6–1. Small bowel examination showing lymphoid hyperplasia of the terminal ileum in a 12-year-old girl. Prominent, sharply defined nodules are apparent.

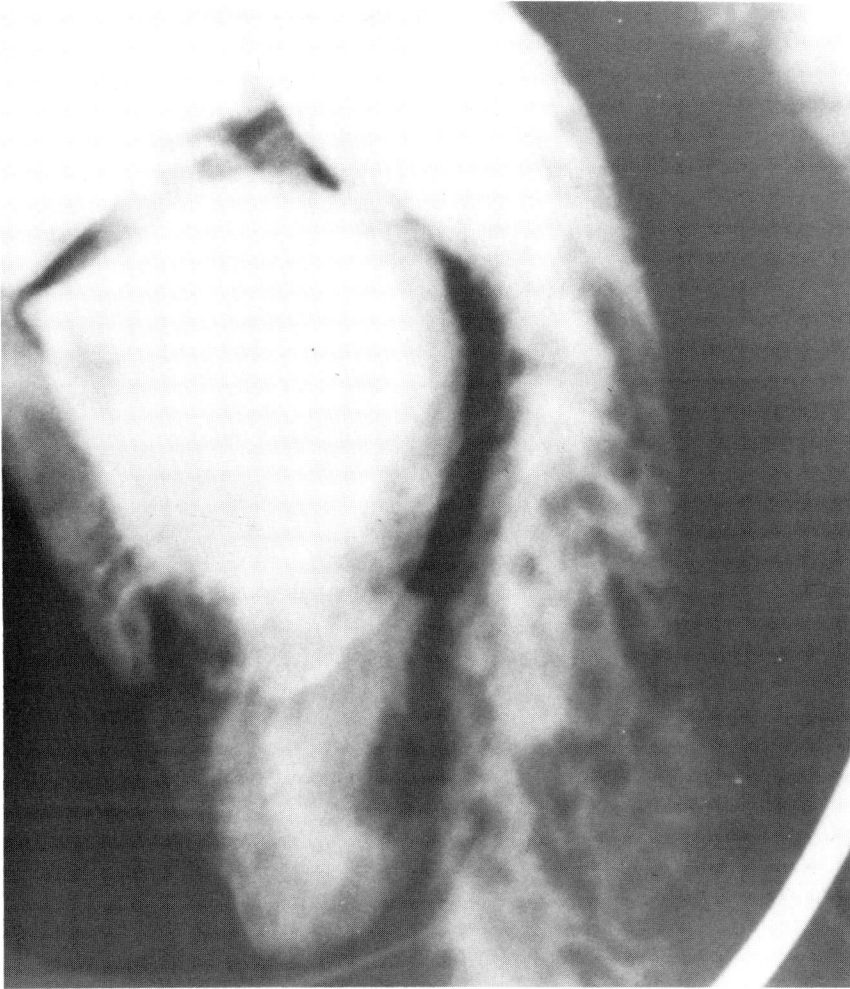

Figure 6-2. Barium enema examination showing lymphoid hyperplasia of the terminal ileum in a 17-year-old girl. Uniform small nodules are visible.

may occasionally be difficult. Thickening and blunting of the mucosal folds and ulceration are often present in lymphosarcoma, and the changes of inflammation are prominent in cases of regional enteritis.

Lymphoid hyperplasia may also occur in the colon in infants and children.[2] Older patients are rarely involved (Fig. 6–3). Characteristic small, uniform umbilicated polypoid lesions may be identified throughout the colon. A fleck of barium in the center of the small mass is often identified and represents a characteristic umbilication at the apex of the lymphoid nodule. The small masses are often best seen on air contrast studies of the colon, but they may also be noted as undulations in the wall of the colon when it is distended with barium. The le-

sions vary in size from 1 to 4 mm. and are relatively uniform in size, round and well demarcated. They are present throughout the colon from the cecum to the rectum, although some cases have a segmental distribution. Histologically, the lesions are solitary follicles of lymphoid tissue in the submucosa and lamina propria bulging into the intestinal lumen. Hyperplasia is a normal response to a variety of stimuli. Both infection and allergy have been implicated. Since the lesions are entirely benign, it is essential to differentiate lymphoid hyperplasia from familial polyposis, for which a total colectomy is indicated. Sigmoidoscopy and biopsy of the rectal lesions may be required.

In 1966 Hermans et al. described a syn-

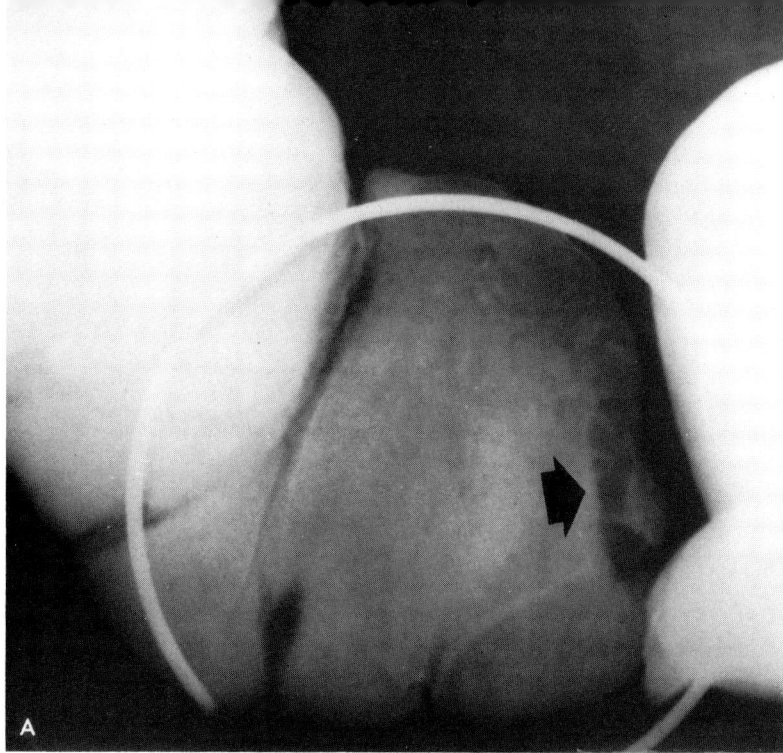

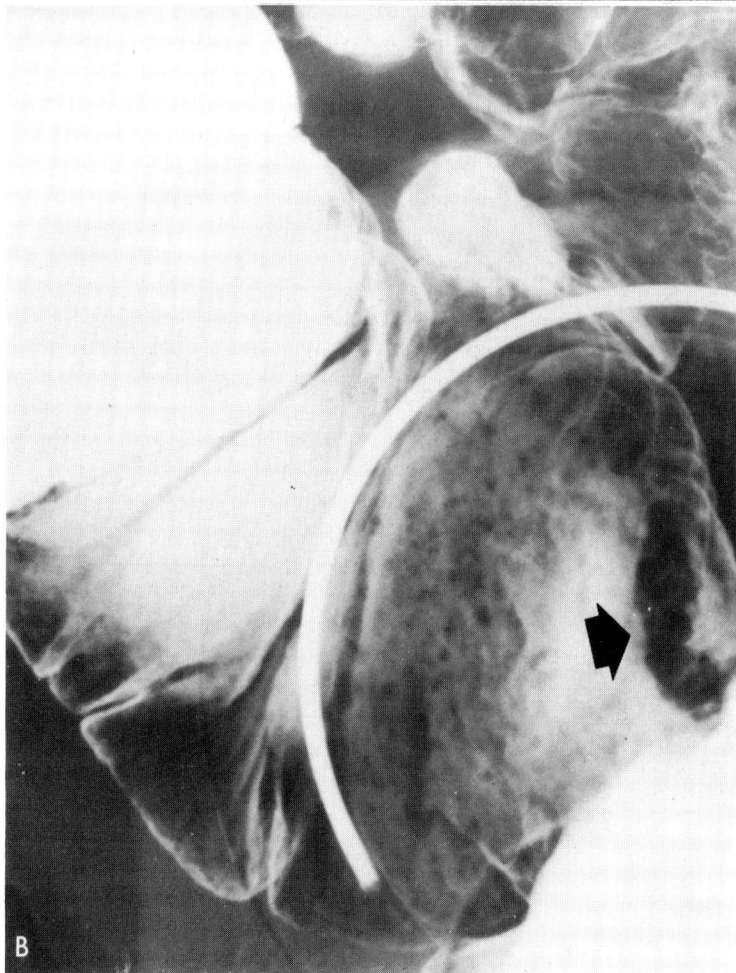

Figure 6–3. Barium enema (A) and air contrast examination (B) showing a villous adenoma (arrow) in the cecum of a 45-year-old man. The air contrast projections show multiple, small, uniform nodules in the cecum. At surgery, there was a villous adenoma at the tip of the cecum and lymphoid hyperplasia.

drome of diffuse nodular lymphoid hyperplasia, increased susceptibility to respiratory infections, malabsorption and hypogammaglobulinemia often associated with intestinal infestation with *Giardia lamblia*.[3] It has been postulated that the lymphoid hypertrophy may be the morphologic response of functionally inadequate tissue. The patients are almost always beyond the age at which lymphoid hyperplasia is present as a normal variant and range from 15 to 62 years of age in one series. The diffuse nodular hyperplasia often extends throughout the colon and small bowel, although jejunal involvement may be most striking. When giardiasis is present, thickening and irregularity of the mucosal folds in the duodenum and jejunum due to edema can be identified. Differentiation from the nodules that occur in mastocytosis is usually possible on a clinical basis. Nodules in early lymphosarcoma con-

tinue to enlarge, so that differentiation from lymphoid hyperplasia may require follow-up examination.

FOREIGN BODIES

Most foreign bodies pass through the gastrointestinal tract without complications. Occasionally, large objects such as fruit pits or conglomerates of persimmon seeds form bezoars and cause intestinal obstruction. Sharp objects such as chicken bones, needles and pins may penetrate the intestinal wall and produce an abscess or free perforation and generalized peritonitis. The ileocecal area is particularly prone to this, presumably because of the delay in transit of the foreign body proximal to the ileocecal valve. Foreign body abscesses in this location are difficult to differentiate from an appendiceal abscess, unless the foreign

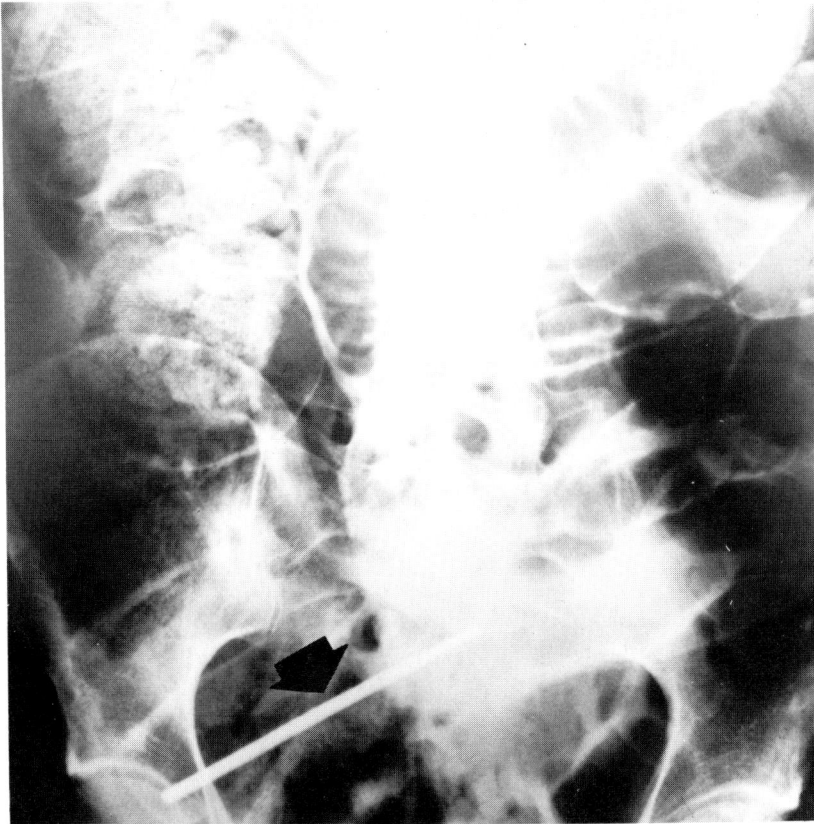

Figure 6–4. Plain abdominal radiograph in a 34-year-old demented man showing a foreign body (thermometer) in the right lower quadrant of the abdomen (arrow). Free intraperitoneal air is visible between the loops of intestine. At surgery, there was a perforation of cecum caused by the thermometer, with free intraperitoneal air and generalized peritonitis.

body can be identified on the plain abdominal radiograph. Although not as obvious as metallic foreign bodies, chicken bones are sometimes visible, permitting a specific diagnosis.[4] Bizarre objects are often swallowed by demented individuals. These may pass through the intestinal tract, remain in the stomach without causing symptoms or perforate the intestine, producing an abscess or generalized peritonitis (Fig. 6–4).

ENDOMETRIOSIS OF THE ILEOCECAL AREA

Intestinal endometriosis consists of implants of tissue composed of endometrial glands and stroma in the bowel wall. The implant and the intense desmoplastic reaction it incites invade the bowel wall extending into the muscularis. The result is a typical intramural lesion with a smooth surface and sharp margins. The mucosa remains intact over the lesion. The fibrous reaction and dense scarring may produce angulation and kinking of the bowel wall, leading to intestinal obstruction or intussusception. The pathogenesis of the implants is not certain. Several theories have been proposed including transtubal reflux, bloodstream and lymphatic embolization and coelomic metaplasia.[5]

The lesion is not rare. Over 800 cases have been reported since Rokitansky's first description in 1860.[6] Twelve per cent of patients with endometriosis in one series had intestinal implants.[7] Of these, intestinal involvement was in the sigmoid in 85 per cent, the small bowel in 6 per cent, the cecum in 5 per cent and the appendix in 4 per cent. Forty-three cases of cecal endometriosis had been reported by 1954.

It is not commonly appreciated that intestinal endometriosis may produce symptoms long after the menopause. This is well substantiated by the histories in four out of five cases reported by Tedeschi and Masand.[8] Two of the patients with intestinal endometriosis in Colcock and Lamphier's series were more than 64 years old.[9] In a case of McKittrick's, the patient was 64 years old and the clinical manifestations of intestinal obstruction occurred 14 years after the menopause.[10] Fallon reviewed 225 patients with endometrial implants and noted that 4 per cent were in girls under 20 years of age.[11] The youngest case reported is Sutton's patient, who was 14 years old.[12]

The clinical manifestations are chiefly those of crampy abdominal pain due to partial intestinal obstruction. Many of these women have undergone extensive diagnostic investigation with negative findings and have been dismissed as having psychosomatic disease. The association of pain and rectal bleeding with menstruation has been overemphasized. Since the endometrial implant rarely involves the intestinal mucosa, bleeding is uncommon. Abdominal pain may occur at any time.

In the sigmoid colon, serosal fibrosis causing compression and kinking of the colon which simulates diverticulitis is the most common radiologic manifestation. In the small bowel, a nodular implant may be present in association with obstruction[13] (Fig. 6–5). In the cecum, constriction and kinking seldom occur and the presence of an intramural, extramucosal mass is characteristic. The overlying mucosa is spared[14] (Fig. 6–6). The differential diagnosis includes cecal carcinoma, appendiceal abscess, mucocele of the appendix and a variety of intrinsic inflammatory lesions of the cecum.

CATHARTIC COLON

The cathartic colon is the term applied to changes in the colon that can be identified after prolonged and excessive use of laxative medications.[15] Most patients have used cathartics in large doses daily for over 15 years. Heilbrun reported the first case in 1943, noting that the localization of the radiologic findings in the colon and small intestine corresponds to the site of action of the cathartic used.[16] The emodins, such as cascara sagrada and aloes, are mild and act on the large intestine, irritating the mucosa and reflexly stimulating increased muscular activity and propulsion. The drastic cathartics podophyllin and jalap are more potent and act in the small intestine, where they cause increased motility. Phenolphthalein has its effect largely on the colon, while calomel increases small bowel motility. Hence, patients taking more irritant resins and chemicals show changes in the small bowel as well as in the colon. Those taking the milder drugs have abnormalities in the

(Text continued on page 295)

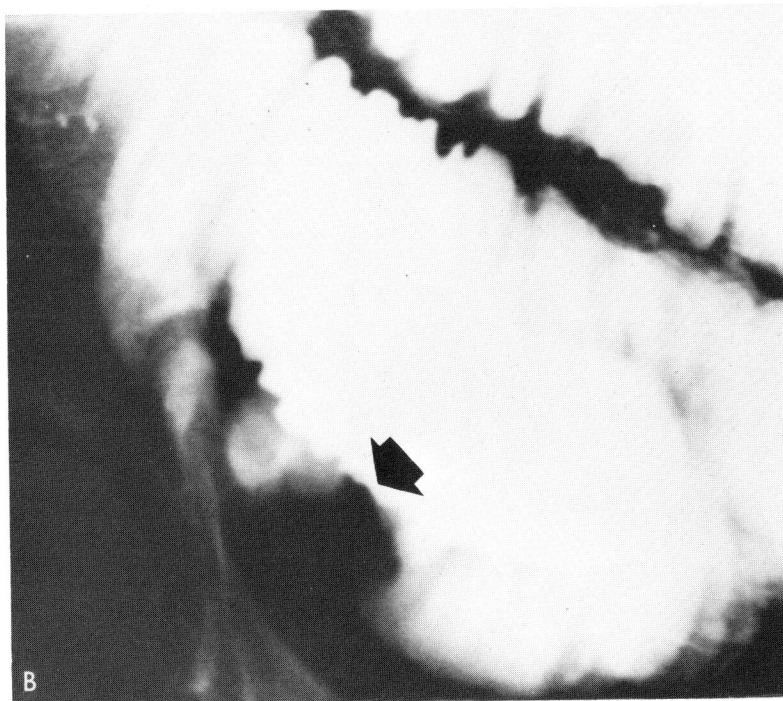

Figure 6–5. Small bowel examination (A) and spot radiograph (B) showing small bowel obstruction due to endometriosis. A mural lesion is visible in the wall of the intestine (arrow), with dilatation proximally. The obstruction is due to kinking of the bowel. An implant of endometriosis in the small bowel may be difficult to differentiate from primary and metastatic carcinoma, lymphosarcoma, and carcinoid of the intestine.

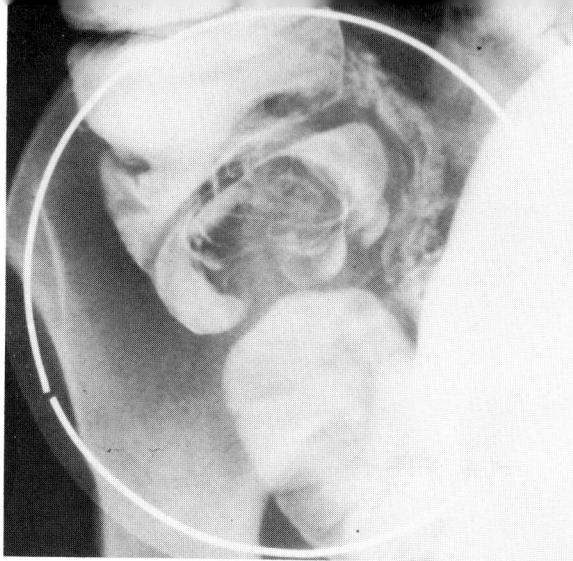

Figure 6–6. Barium enema examination showing endometriosis of the cecum. The combination of an intramural, extramucosal lesion of the cecum without spasm suggests endometriosis. (From Felson, B., and Wiot, J. F. Some interesting right lower quadrant entities. Radiol. Clin. N. Amer., 7:83, 1969.)

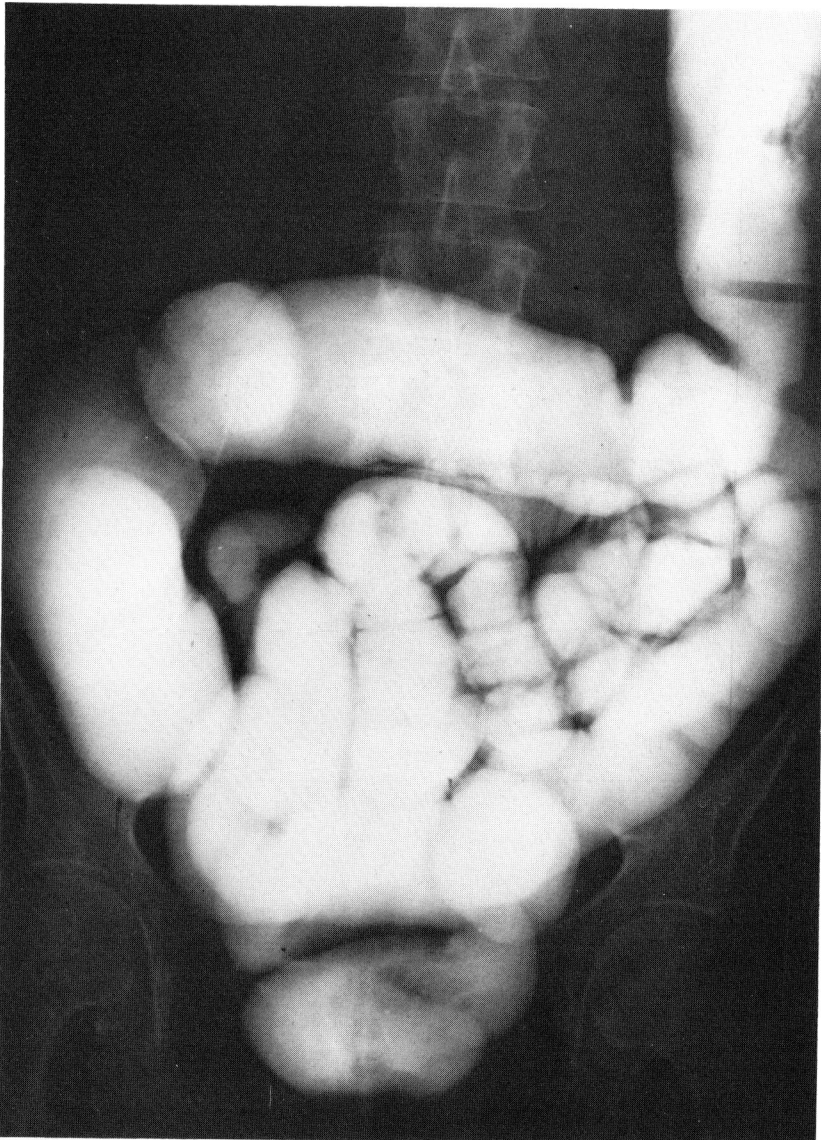

Figure 6–7. Barium enema examination showing cathartic colon. The right colon is smooth and the haustra are absent. The patient has been taking large doses of cathartics daily for 10 years.

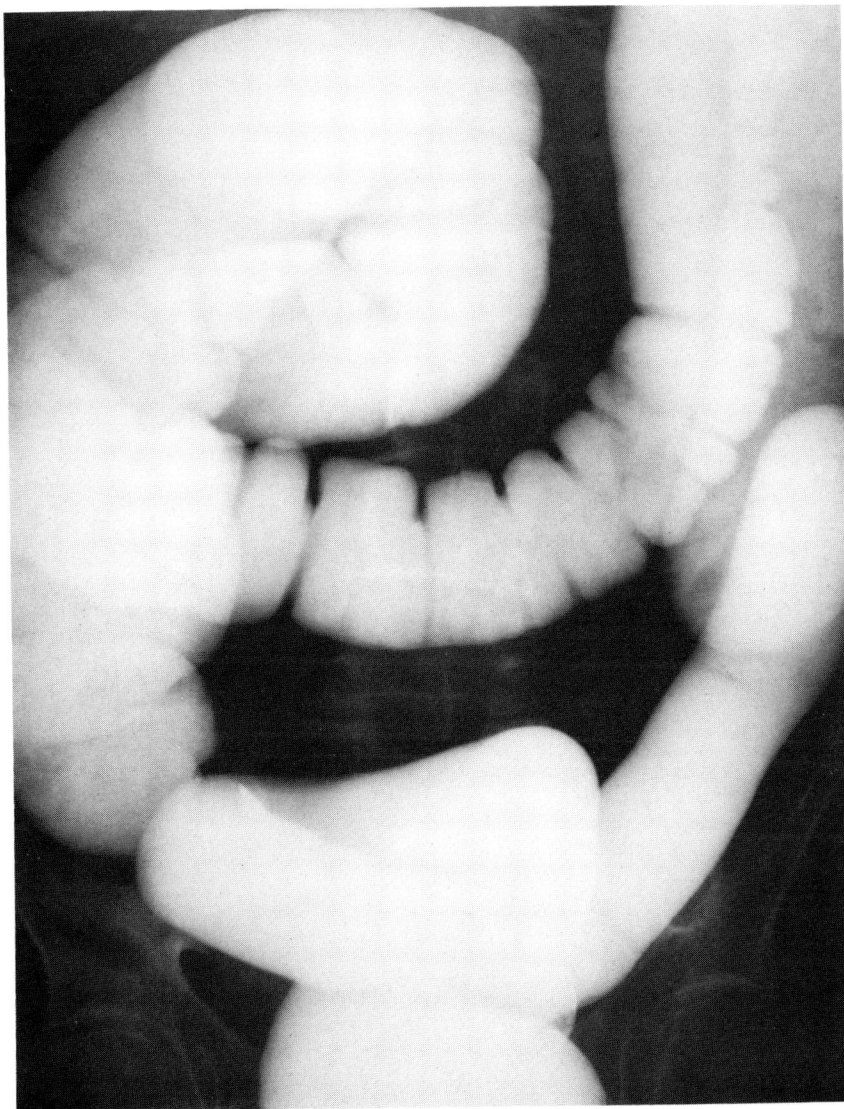

Figure 6–8. Barium enema examination showing cathartic colon. The colon is dilated and the haustra are deficient in the cecum, ascending colon and rectosigmoid. The patient has a long history of cathartic abuse.

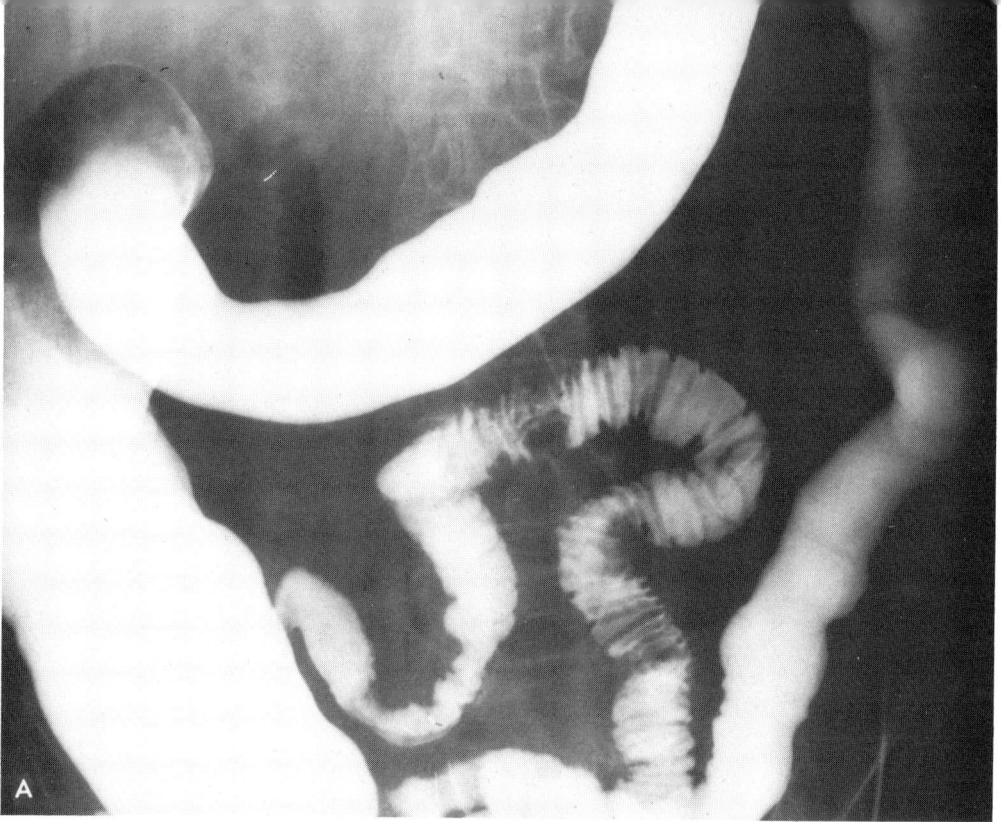

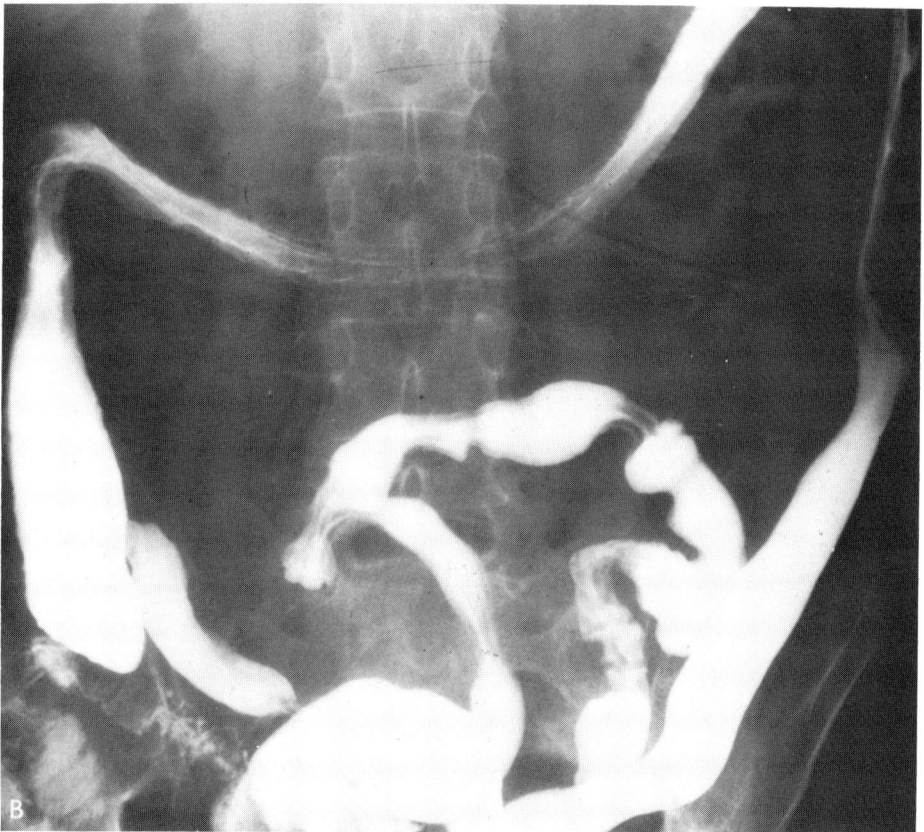

Figure 6–9. Barium enema examination (A) and postevacuation radiograph (B) showing cathartic colon. The haustra are lost and the mucosa is thickened. The patient has a long history of cathartic abuse.

cecum, proximal colon and ileocecal valve. The bulk or saline cathartics do not cause radiologic abnormalities.

The most severe changes occur in the proximal colon and consist of absent or diminished haustral markings, bizarre contractions, inconstant areas of narrowing and an appearance of atrophy[17] (Figs. 6–7 to 6–9). The bowel remains moderately distensible or even dilated in contrast to the rigid tubular colon in chronic ulcerative colitis. The colon empties poorly on the postevacuation examination. The mucosal pattern is linear or absent with no ulceration, edema or flocculation of barium. In more severe cases the left side of the colon is also involved, but the sigmoid and rectum remain distensible. The terminal ileum may show similar changes for varying lengths with narrowing and loss of the normal mucosal pattern. The ileocecal valve is shortened, flattened, gaping and incompetent. Shortening of the colon especially on the right side occurs, but the flexures are usually normally placed. Areas of narrowing are inconstant and may disappear during a single examination or vary in length.

The radiologic features of cathartic colon resemble a variety of inflammatory diseases of the colon and small bowel including ulcerative colitis, granulomatous colitis, amebiasis and tuberculosis. The changes most closely simulate chronic ulcerative colitis, but ulcerative colitis almost always involves the rectum and sigmoid. Isolated involvement of the right side is rare. Edema, ulceration and pseudopolyps are present in acute ulcerative colitis. In the chronic stage, the colon is often more rigid and narrow than in cathartic colon, but the differential diagnosis may require knowledge of the clinical history. When the changes are limited to the cecum, amebiasis must be considered. In amebiasis, spasm irritability and deformity are usually present and the terminal ileum is spared. Tuberculosis involves the terminal ileum and cecum. Irritability, irregular deformity and edema are usually prominent and the ascending colon is less likely to be involved. In tuberculosis the chest radiograph is usually positive.

The diagnosis often depends on the clinical features which are typical of either inflammatory bowel disease or cathartic colon. Patients with inflammatory bowel disease have abdominal cramps and diarrhea, and sigmoidoscopy shows friability, erythema and ulceration. The history of constipation and prolonged cathartic abuse in patients with cathartic colon permits an accurate distinction.

OBSTRUCTION OF THE TERMINAL ILEUM

Obstruction of the terminal ileum differs in a number of ways from more proximal small bowel obstruction. Distal ileal obstruction is much more common and has certain distinguishing clinical and radiologic features. Vomiting occurs later in the course of distal ileal obstruction and is less forceful. The plain abdominal radiograph shows more dilated intestinal loops and more numerous air-fluid levels. Certain causes of obstruction tend to be more common in the ileum. Adhesions from previous pelvic surgery, obstruction due to incarcerated inguinal and femoral hernias and obturation obstructions tend to involve the distal small bowel.

In this section, the radiologic features of simple distal ileal obstruction will be contrasted with those of strangulated obstruction. Various causes of ileal obstruction will be discussed and illustrated. These include extrinsic abnormalities occluding the bowel (hernia and adhesions), obturation obstruction of the intestinal lumen (gallstone ileus, meconium ileus equivalent and postgastrectomy bezoars) and primary disease of the bowel wall (ischemic and caustic ulcers).

The radiologic manifestations of small bowel obstruction on plain abdominal radiographs depend on the location, the degree and the duration of the obstruction. The characteristic findings are the presence of dilated loops of small bowel filled with air and fluid (Fig. 6–10). Air-fluid levels are visible on upright or decubitus radiographs. Early in the course of the obstruction, gas and fecal material may be identified in the colon, but later, after the colon has had a chance to empty, a striking disparity between the dilated small intestine and the empty colon becomes evident. In partial small bowel obstruction, air and fecal material may persist in the colon, but the disparity is usually still apparent. In cases of

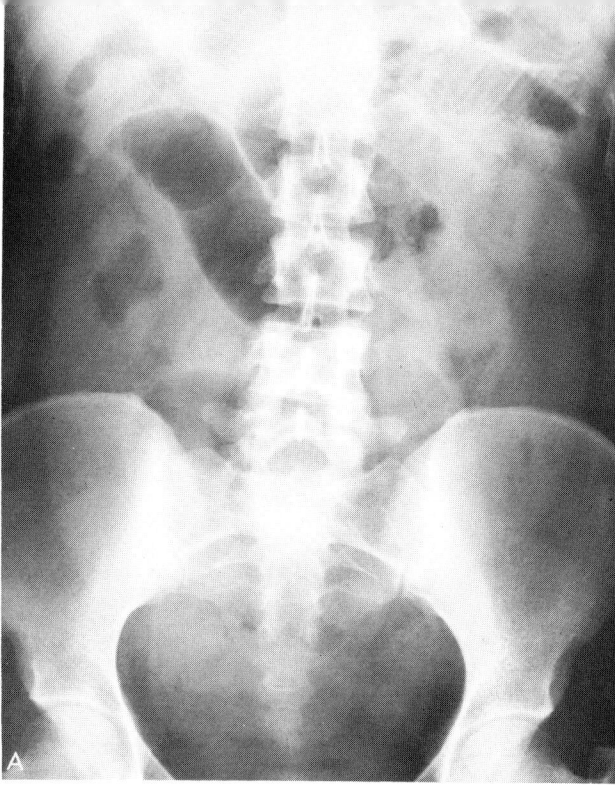

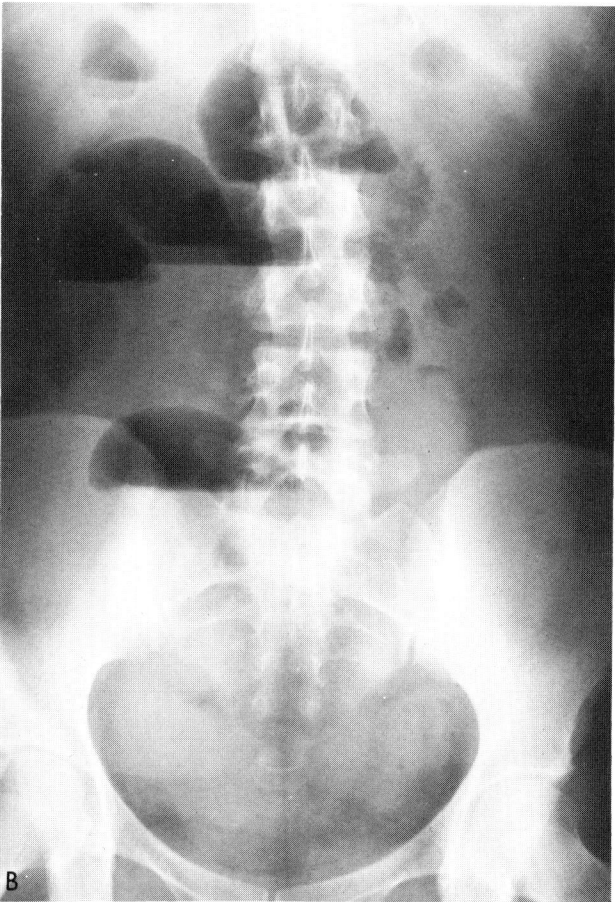

Figure 6–10. Plain abdominal radiograph showing dilated air and fluid-filled loops of small intestine due to obstruction of the ileum caused by adhesions. (A) Supine; (B) upright.

paralytic ileus, proportional distention of both the small and large bowel is the typical finding.

In small bowel obstruction, the dilated, air-filled small intestinal loops have a hoop-shaped, radial or "tonic" appearance with short fluid levels compared to the flaccid, atonic appearance of the small bowel with long fluid levels that are present in paralytic ileus. The dilated small bowel tends to be centrally located in the abdomen compared to the colon, which lies in a vertical position parallel to the flanks. Dilatation of the small bowel is never as great as it is in the colon. The small bowel diameter is rarely capable of exceeding 5 cm.

When the obstruction is in the distal ileum, the dilated loops of small bowel are more numerous and the number of air-fluid levels is greater than when the obstruction occurs more proximally. The most dilated portion of the bowel is the segment adjacent to the obstruction, a feature which may distinguish mechanical obstruction from paralytic ileus when the ileus is localized to the small bowel. Paralytic ileus involving only the small bowel is frequent in mesenteric infarction and acute pancreatitis. In these cases the degree of small bowel distention is uniform throughout the length of the dilated bowel or is even more marked in the proximal segments.

Small bowel obstruction occurs most often in the ileum, where the obstruction is frequently due to adhesions in the lower abdomen from previous surgery. Obstruction may be due to incarceration of the bowel in an inguinal or a femoral hernia. In these cases air or barium may be visible in the incarcerated loop below the inguinal ligament on the plain abdominal radiograph, indicating the presence of a hernia (Fig. 6–11). Obturation obstruction may be due to gallstone ileus, meconium ileus equivalent, food bezoars after partial gastrectomy, masses of ascarids, impacted medications and other situations in which the intestinal lumen becomes occluded because of its abnormal content.

The classic example of obturation obstruction is gallstone ileus in which a large gallstone erodes into the duodenum from the gallbladder or common duct.[18] The gallstone usually obstructs the bowel in the terminal ileum proximal to the ileocecal valve. Occasionally, the gallstone is passed per rectum without causing obstruction (Fig. 6–

12). The manifestations on the plain abdominal radiograph consist of the triad of air in the biliary tree, a gallstone in an ectopic location and small bowel dilatation. While gallstone ileus accounts for only 2 per cent of the cases of small bowel obstruction, the radiographic findings are striking and permit a precise diagnosis. Elderly women are particularly prone to this complication.

Although small bowel obstruction is a well-known feature of cystic fibrosis in newborn infants, intestinal obstruction in children and young adults with the disease is generally less well appreciated. This complication has been termed meconium ileus equivalent.[19] At least 36 cases have been reported in patients ranging in age from a few months to 25 years. As in the newborn group, the putty-like, inspissated, tenacious fecal material becomes impacted in the cecum and terminal ileum, producing mechanical obstruction. In most cases the obstruction is due to abnormal stool associated with pancreatic insufficiency, whereas in the newborn it results from thick intestinal mucus. If meconium ileus equivalent is properly diagnosed, surgery for intestinal obstruction can often be avoided by appropriate medical treatment with detergent enemas and pancreatic enzymes given orally and rectally. Detection of the syndrome often depends on the recognition of characteristic plain film radiographic findings (Fig. 6–13). Large amounts of fecal material in the cecum and terminal ileum associated with proximal small bowel distention are typical. A fecal bolus in the terminal ileum is almost pathognomonic.

Obturation obstruction also occurs in patients who have had gastric surgery[20] (Fig. 6–14). The complication usually occurs long after the surgery and is frequently associated with the ingestion of large amounts of fibrous food, particularly oranges. Occasionally, the proximal convex margin of the bezoar or the bezoar itself can be identified on plain abdominal radiographs in patients with small bowel obstruction. Primary and metastatic neoplastic diseases of the small bowel and inflammatory diseases such as regional enteritis, ischemia and strictures due to ingested caustic agents may narrow the lumen of the ileum and produce intestinal obstruction. Primary carcinoma is relatively rare in the ileum compared to its incidence in the duodenum. In fact, the incidence of carcinoma of the duodenum equals

(*Text continued on page 305*)

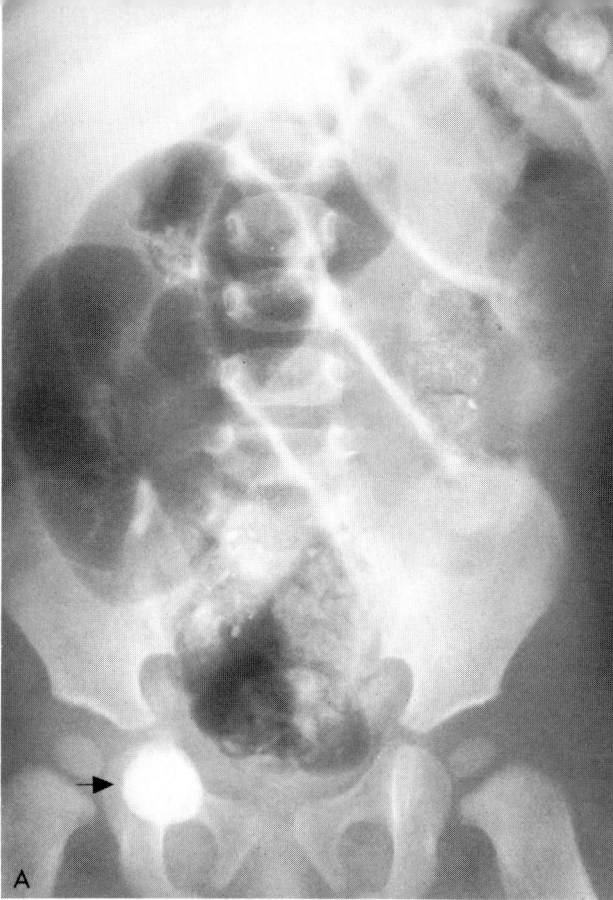

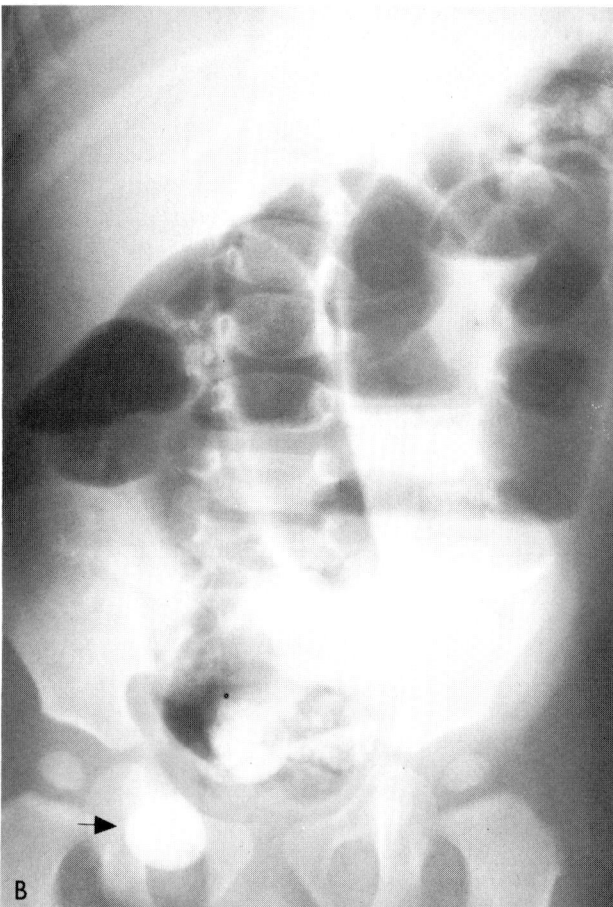

Figure 6-11. Plain abdominal radiograph showing dilated air and fluid-filled loops of small intestine in a 14-month-old girl with small bowel obstruction due to an incarcerated femoral hernia. Opaque medication in the incarcerated bowel shows the intestinal segment below the inguinal ligament (arrow).

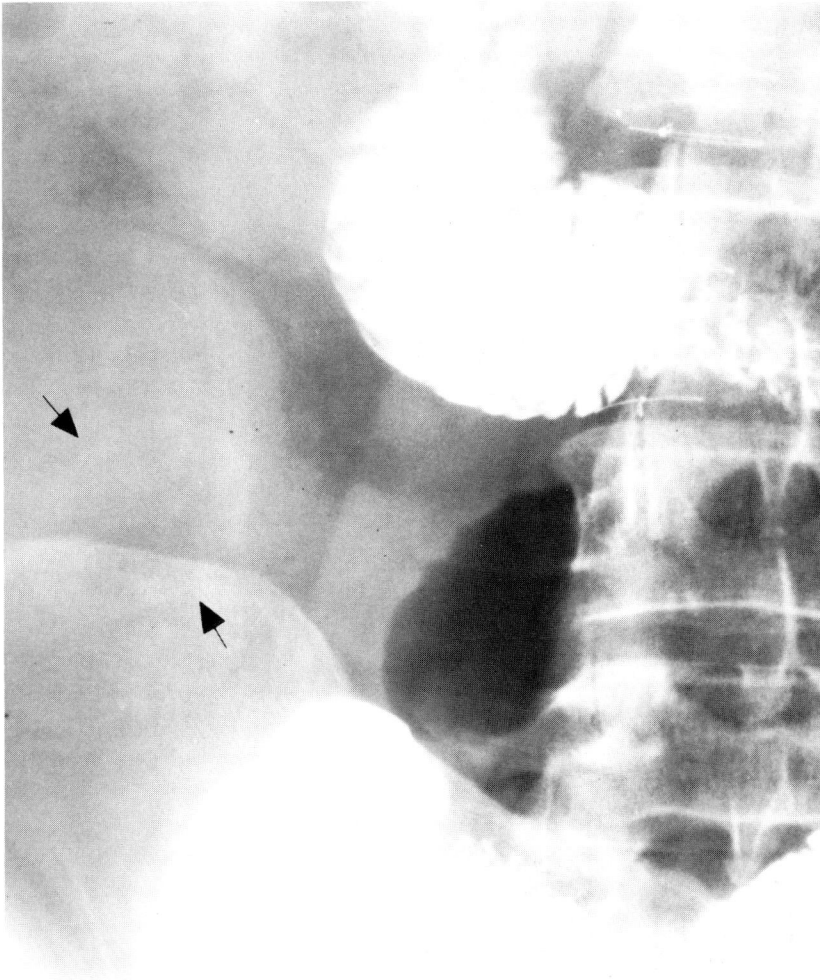

Figure 6–12. Small bowel examination showing a large calcified gallstone in the small intestine (arrows). The stone eroded into the duodenum, creating a cholecystoduodenal fistula, and passed spontaneously per rectum without producing intestinal obstruction or air in the biliary tree.

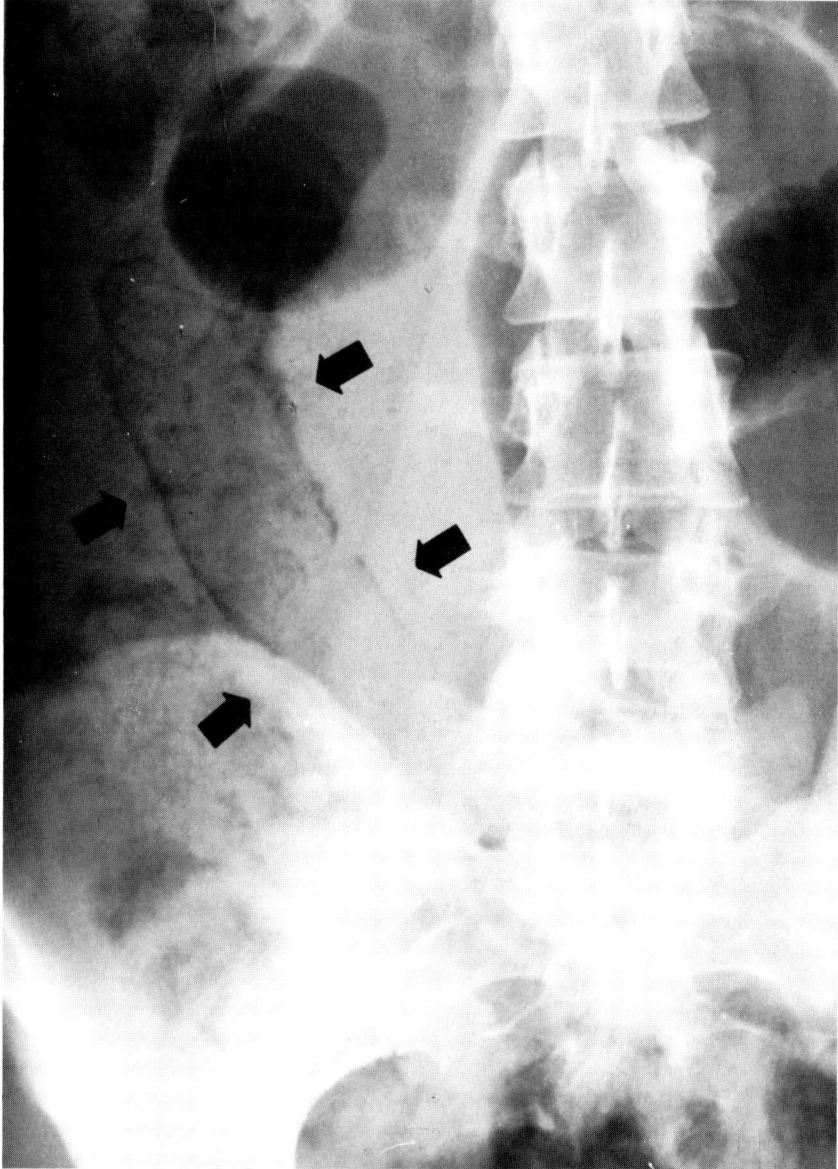

Figure 6–13. Plain abdominal radiograph showing abnormal intestinal content in the terminal ileum (arrows). The material was a bezoar consisting of fecal-like material which obstructed the bowel. Studies for pancreatic fibrosis were positive, establishing the diagnosis of meconium ileus equivalent in this 19-year-old girl who was not previously suspected of having the disease. (From Berk, R. N., and Lee, F. A.: Late gastrointestinal manifestations of cystic fibrosis. Radiology, *106*:377, 1973.)

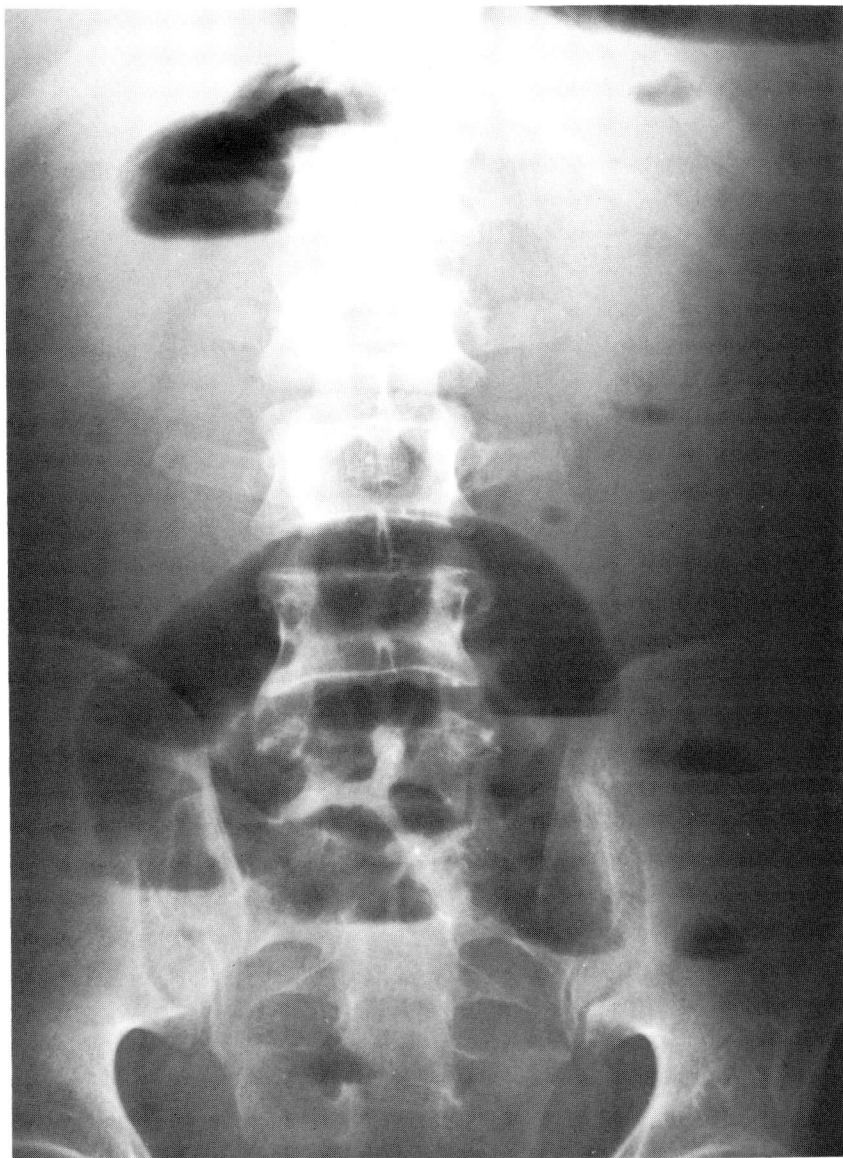

Figure 6–14. Upright plain abdominal radiograph in a 47-year-old man who had a hemigastrectomy 12 years earlier. Dilated air and fluid-filled loops of small intestine are evident. At surgery, there was an obturation obstruction of the terminal ileum due to a bezoar.

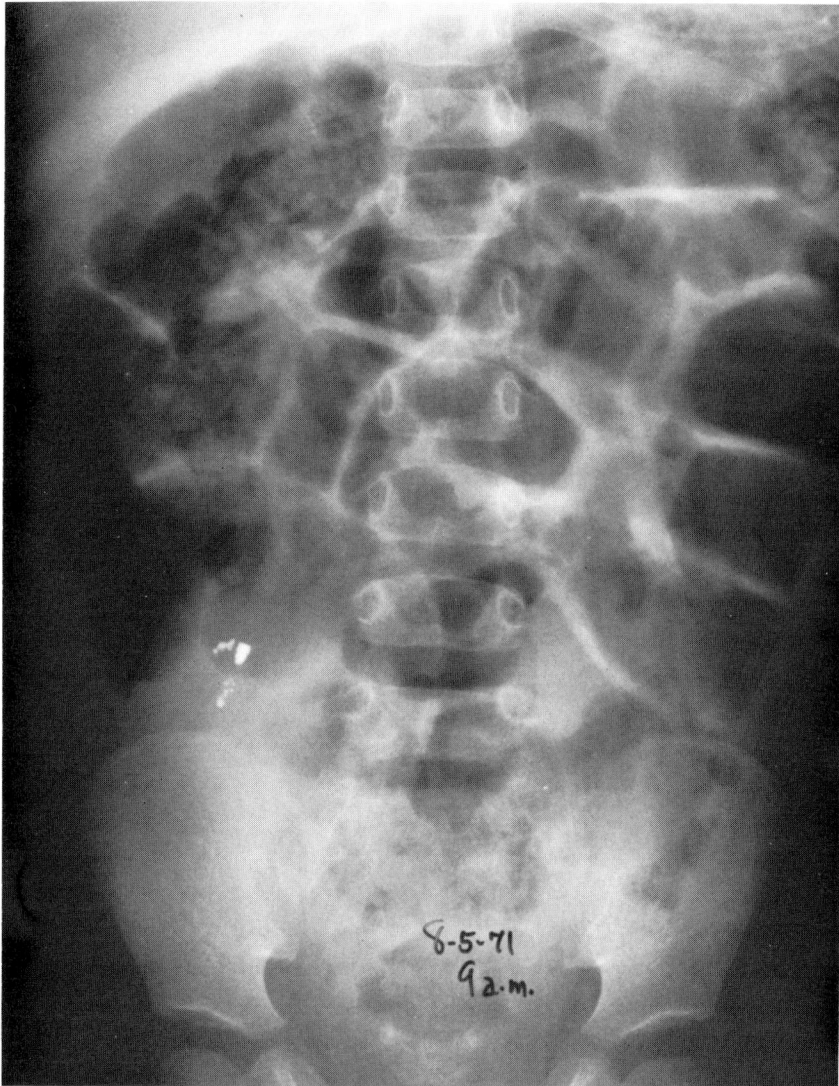

Figure 6–15. Plain abdominal radiograph in a 2-year-old boy who four months earlier had ingested a massive amount of iron. There is a mass in the right lower quadrant. Barium from a previous barium enema examination is present in the appendix. At surgery, there was a right lower quadrant abscess due to perforation of the terminal ileum secondary to a stricture of the ileum induced by the iron.

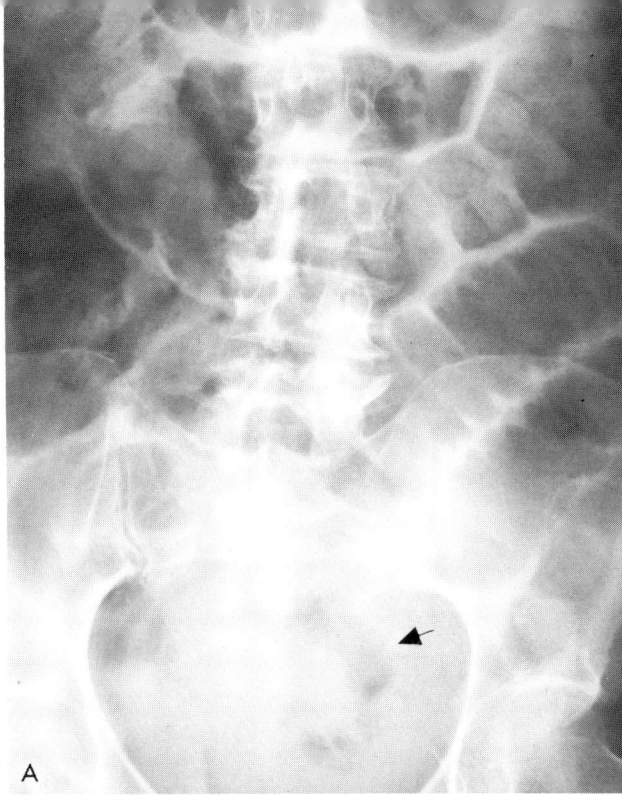

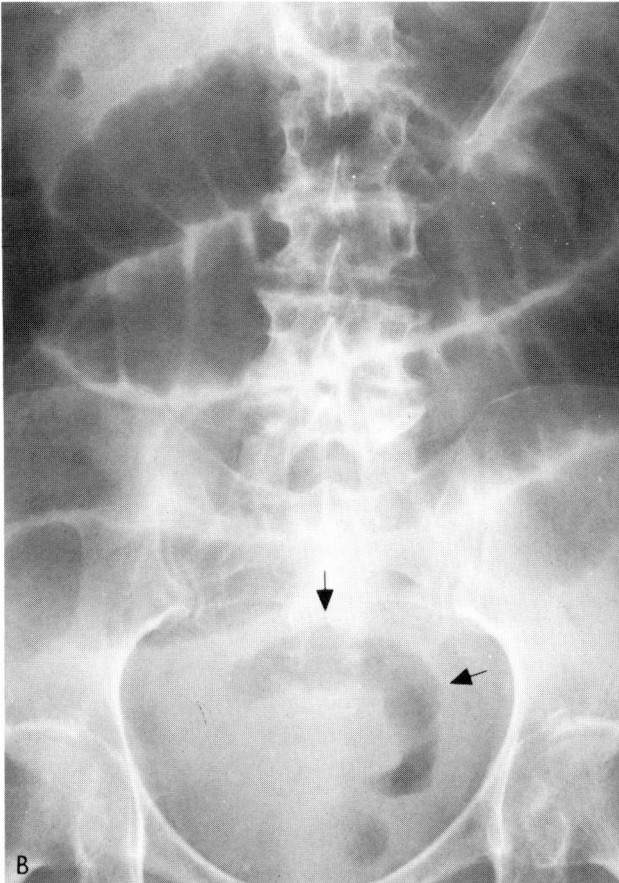

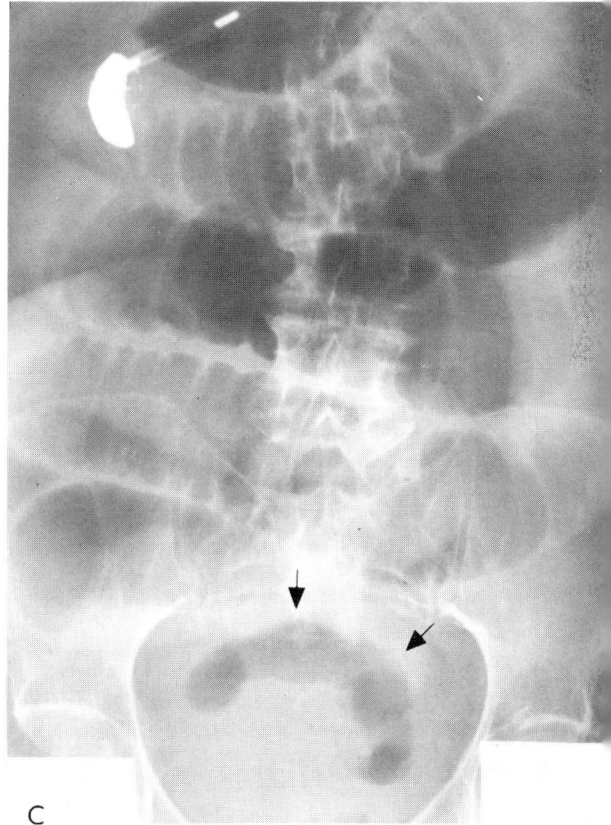

Figure 6–16. Supine plain abdominal radiograph showing a strangulated small bowel obstruction due to adhesions which produced a volvulus of the terminal ileum. Abdominal radiographs were made on admission (*A*), one day later (*B*) and 3 days later (*C*). A fixed, rigid thickened loop of intestine is visible in the pelvis on all three examinations (arrows). At surgery, there was infarction of a segment of the terminal ileum due to a volvulus of the ileum, with strangulation caused by an adhesion with small bowel obstruction.

303

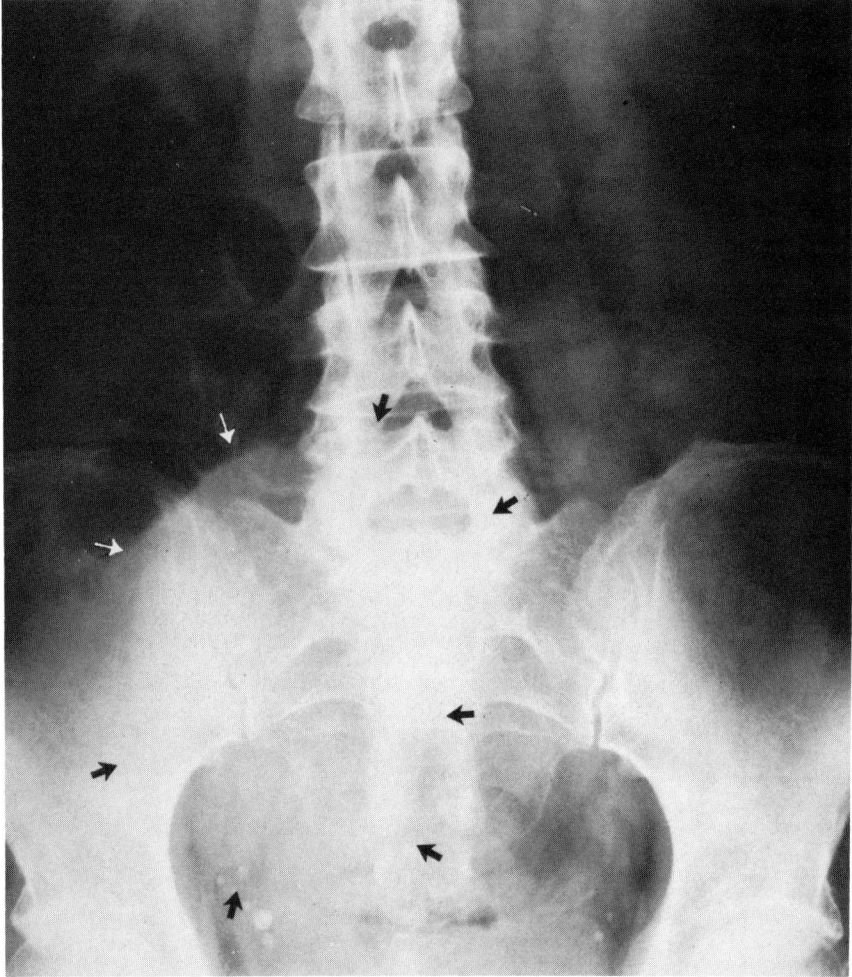

Figure 6–17. Supine abdominal radiograph in an elderly woman showing a large abdominal mass (arrows). At surgery, there was a strangulated small bowel obstruction due to an adhesion with a volvulus. The mass was due to a segment of infarcted bowel which was dilated and filled with fluid (pseudotumor).

that of the entire ileum and jejunum combined. Lymphosarcoma frequently occurs in the terminal ileum but because of the absence of an associated desmoplastic reaction the lumen is rarely sufficiently occluded to produce obstruction until late in the course of the disease. Metastatic disease involves the serosal surface of the intestine or the mesentery. When obstruction occurs, it is usually because of kinking and angulation of the small bowel. Progressive fibrosis of the bowel wall occurs in regional ileitis, so that chronic intestinal obstruction is a frequent complication of the disease. Cicatrizing small bowel ulcers occur in cases of localized ischemia of the small bowel or in patients taking enteric-coated potassium chloride. Fibrotic strictures may occur in young children who ingest ferrous sulfate tablets[21] (Fig. 6–15). Small intestine ulceration occurs owing to the caustic effect of the iron. The loss of mucosal integrity allows the ferrous sulfate to gain entrance into the veins and lymphatics, causing necrosis. Healing is by fibrosis, which eventually leads to stricture and obstruction.

Strangulation obstruction of the small bowel occurs when a loop of bowel becomes obstructed at both ends. The vessels in the mesentery are compromised and a hemorrhagic infarction of the bowel occurs. When the intestinal obstruction is incomplete at the proximal end, air may enter the ischemic loop and produce a fixed, air-filled loop of bowel on plain abdominal radiographs (Fig. 6–16). When the loop is distended, the thick line between the halves of the loop resembles the crease in a coffee bean, so that the presence of a fixed gas-filled segment of the small bowel has been termed the "coffee bean" sign.[22] When the ischemic loop is completely obstructed, edema of the wall and transudation of fluid into the lumen produces a fixed, fluid-filled mass, termed a pseudotumor[23] (Fig. 6–17). The ischemic segment may simulate an ovarian cyst or other mass lesions. Linear collections of gas in the bowel wall or gas in the portal venous system are direct radiologic signs of strangulation.

ADHERENT FECALITH IN THE ILEOCECAL AREA

The sticky fecal material of patients with cystic fibrosis may form a persistent mass in the colon, particularly in the cecum.[19] The adherent fecalith may be palpable and simulate a colonic neoplasm or suggest acute appendicitis when there is associated tenderness. Occasionally, the fecal bolus forms the lead point of an intussusception. The tenacious scybalum produces a tumor-like defect on barium enema examination which often persists on repeated examinations, even after several weeks (Fig. 6–18). Such masses should be managed by conservative medical therapy, because in most cases the fecal material is eventually passed spontaneously. When a persistent fecal bolus is identified on barium enema examination in a young patient who is not known to have cystic fibrosis, a review of the history and determination of the sweat electrolytes are indicated to consider this diagnosis. Occasionally, an adherent barolith occurs in the cecum as a complication of a barium enema examination in an apparently normal patient. In these cases inspissated barium becomes adherent to the mucosa and persists in the cecum for weeks or months before it is passed.

PNEUMATOSIS OF THE ILEOCECAL AREA

Gas in the wall of the terminal ileum and cecum may occur in association with infarction or gas-producing infection of the bowel, in association with intestinal obstruction or in a benign condition most commonly termed pneumatosis cystoides intestinalis.[24]

When intestinal ischemia leads to necrosis of the bowel, the integrity of the mucosa is lost, allowing gas in the lumen to dissect between the layers of the bowel wall. Plain abdominal radiographs disclose linear or ringlike collections of gas in the involved segments. The gas may have a bubbly or foamy appearance which may be confused with feces in the lumen of the bowel. However, in these cases gas can actually be identified in the wall of the bowel when the bowel wall is seen in profile. In severe cases the intramural air may enter the portal venous system where it can be identified as gas within the liver.[25]

Intestinal infarction is usually seen in elderly patients in association with vascular occlusion due to atherosclerosis or in relation to low intestinal perfusion states. It

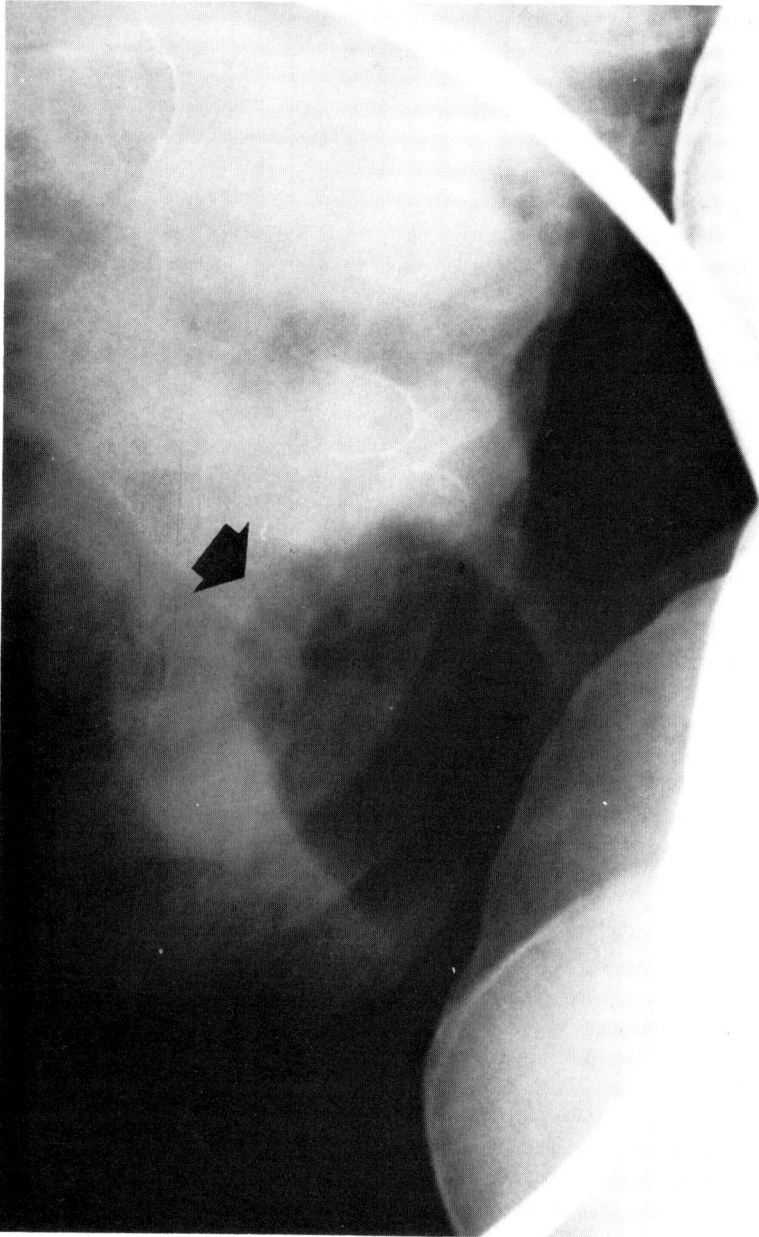

Figure 6–18. Spot film radiograph of the cecum taken during a barium enema examination in a 28-year-old man with proved cystic fibrosis. An adherent fecalith is present (arrow). The mass persisted for 10 days before it passed spontaneously. (From Berk, R. N., and Lee, F. A.: Late gastrointestinal manifestations of cystic fibrosis. Radiology, *106*:377, 1973.)

Figure 6–19. Supine (A) and oblique (B) abdominal radiographs showing acute bacterial emphysematous colitis. The patient is a 26-year-old female with morbid obesity who one month earlier had an ileal bypass operation. Extensive air in the wall of the right side of the colon is apparent. At surgery, there was an abscess caused by bacterioides adjacent to the hepatic flexure and the liver and in the rectus sheath. Air was present in the colon causing a foamy appearance over the surface. The abscesses were drained. The pneumatosis subsided on subsequent radiographs over the next week with intensive antibiotic therapy. Air in the bowel wall is not an absolute indication of intestinal infarction, even in the absence of idiopathic pneumatosis cystoides intestinalis.

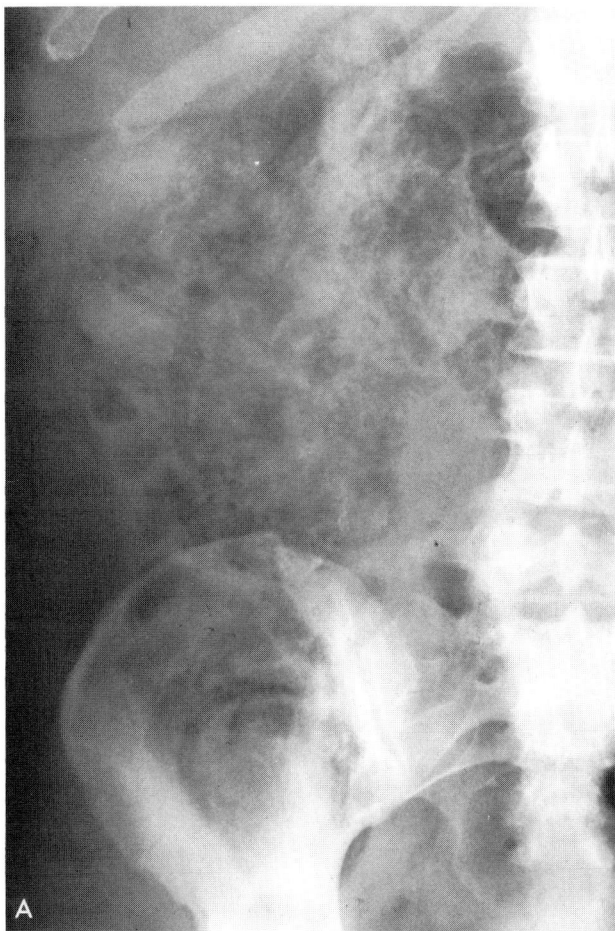

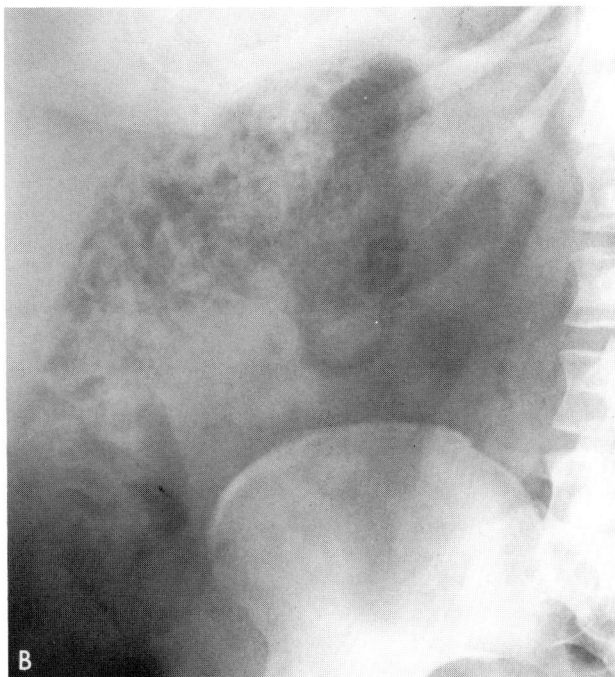

may also occur in younger patients with strangulating obstruction of the bowel, emboli or vasculitis. Necrotizing enterocolitis occurs in premature and debilitated infants with poor resistance to infection.[26] Gas-forming bacteria from the lumen invade the bowel wall through the intestinal mucosa and produce a fulminating necrotizing cellulitis. As in adults, the intramural gas has the appearance of fecal material radiographically. However, feces are not observed on abdominal radiographs in premature infants.

When a gas-forming organism enters the bowel wall by continuity with adjacent infections, acute bacterial inflammation may result. In these cases gas can be identified in the bowel wall in the absence of intestinal infarction (Fig. 6–19).

Intramural gas collections may occur without infection or infarction in cases of intestinal obstruction. The intramural gas collections occur distal to the obstructing lesion.[24] It seems likely that ulcerations in the mucosa proximal to an obstruction allow gas in the lumen to enter the wall of the bowel and to dissect into the distal portions of the intestine.

Pneumatosis cystoides intestinalis is a benign, relatively rare condition affecting both the large and small intestine and is characterized by multiple gas-filled cysts in the subserosa and submucosa.[27] The colon is involved more often than the small intestine. Many patients with pneumatosis have chronic obstructive pulmonary disease. In these cases gas may enter the pulmonary interstitial spaces from ruptured alveoli, dissect along peribronchial and perivascular planes into the mediastinum and pass through the diaphragm into the retroperitoneum and mesentery into the bowel wall. Radiologically, the gas-filled cysts may simulate multiple polyps or the thumbprinting that occurs in intestinal ischemia. Differentiation depends on recognition that the cysts are of air density. Pneumoperitoneum occurs when the gas cysts rupture into the peritoneal space. The free air may persist if the rate of air entering the peritoneal cavity equals or is greater than the rate of absorption.[24] Persistent pneumoperitoneum with large quantities of air may produce a distressing sensation of abdominal fullness and distention in patients with pneumatosis who are otherwise asymptomatic.

ILEAL PROLAPSE

Pouting or eversion of the ileal mucosa through the lips of the ileocecal valves is a rare occurrence, although Fleischner noted that a small rosette of ileal mucosa at the mouth of the ileocecal valve is not an uncommon finding in normal anatomic specimens.[28] The abnormality is probably asymptomatic and may be related to lipomatous infiltration of the ileocecal valve, which is a frequent associated finding. When the prolapsing tissue reaches large size, it may produce a smooth defect in the cecum. The

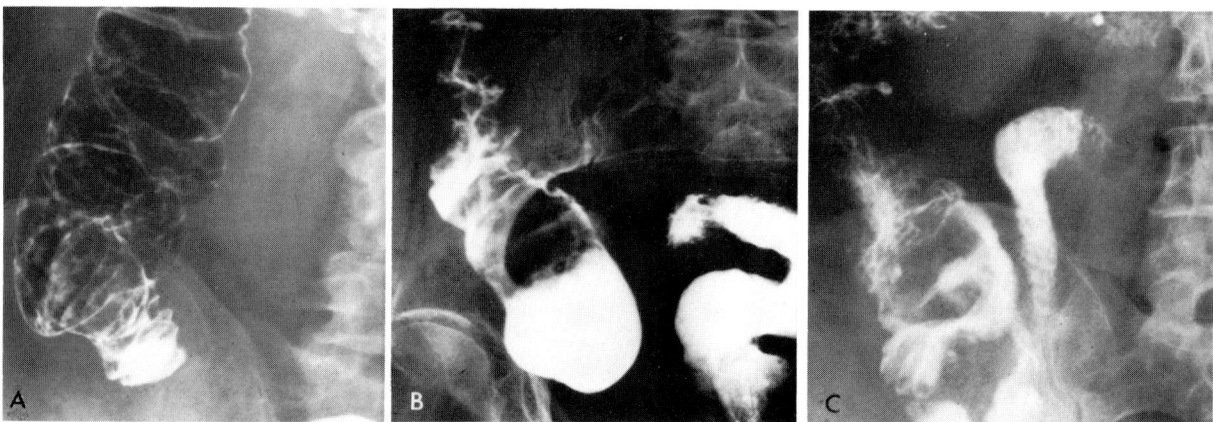

Figure 6–20. Barium enema and air contrast colon examination showing ileal prolapse. (A and B) A smooth intracecal mass is present. (C) The terminal ileum is filled with barium and can be traced to the cecal mass. Barium in the ileal lumen is seen in the center of the mass. (From Felson, B., and Wiot, J. F.: Some interesting right lower quadrant entities. Radiol. Clin. N. Amer., 7:83, 1969.)

surface is smooth, and barium in the center may identify the lumen of the bowel. If careful compression spot radiographs are made, the origin of the mass can be traced to the ileocecal valve.

The most difficult differential diagnosis involves carcinoma of the cecum. Felson et al. reported a patient who was operated on twice, each time at a different institution, with the mistaken diagnosis of carcinoma of the cecum, when the cecal mass noted on barium enema examination was actually due to ileal prolapse[14] (Fig. 6–20). The proper diagnosis is possible if it is recognized that the mass arises between the lips of the ileocecal valve. Carcinoma of the ileocecal valve or cecum shows mucosal destruction. Differentiation of ileal prolapse from ileocolic intussusception may be impossible when the prolapse is large.

INTRAMURAL HEMORRHAGE IN THE ILEOCECAL AREA

Submucosal bleeding into the ileum and cecum may occur after trauma or in association with any bleeding diatheses such as those associated with Dicumarol therapy, hemophilia, leukemia, and Henoch-Schönlein's purpura.[29] Hematomas after trauma are usually more localized than bleeding in association with coagulation defects, because after trauma the blood is still able to clot.

Plain abdominal radiographs may show paralytic ileus and an isolated, fixed, thickened loop of bowel with an irregular wall and narrowing of the lumen (Fig. 6–21). Barium studies disclose segmental involvement of a small bowel loop with striking thickening of the folds and a distorted, rigid, spikelike contour which resembles a stack of thick coins or platters[30] (Fig. 6–22). Spasm, irritability, edema, exudate and obstruction are usually absent, and thumbprinting in the involved segment is not uncommon. Complete resolution is the rule.

Differential diagnosis includes intestinal infarction and edema of the bowel wall from a variety of causes including nephrosis, cirrhosis, hypoproteinemia, cardiac failure and Whipple's disease. In these cases, the sharply defined, parallel transverse mucosal folds are not as prominent, nor is the bowel as rigid. The changes are generally more diffuse, and evidence of mucosal edema and spasm is present.

ILEOCOLIC INTUSSUSCEPTION

Intussusception results from the invagination of one portion of the intestine into another. It ranks second only to appendicitis as the most common acute abdominal emergency in children and is the most frequent cause of intestinal obstruction in the pediatric age group.[31] In children less than 2 years of age the cause of the intussusception is unknown in 95 per cent of the cases. Other cases are due to hyperplasia of Peyer's patches in the terminal ileum, inversion of Meckel's diverticulum and enlarged lymph nodes. In the absence of peritonitis, hydrostatic reduction by means of a barium enema examination is indicated for intussusception occurring in infancy.

Less than 10 per cent of the recognized cases of intussusception occur in adults.[32] In ileocolic intussusception, the cecum remains normally orientated as compared to ileocecal intussusception, in which the cecum also invaginates into the colon along with the terminal ileum and ileocecal valve. Ninety per cent of adults with intussusception have associated pathologic conditions, including benign and malignant tumors, inflammatory lesions and Meckel's diverticulum. Colocolic intussusceptions are frequently associated with colonic lipomas, especially those that grow intraluminally and develop a stalk.

Plain abdominal radiographs may show intestinal obstruction with dilated loops of small bowel.[33] Occasionally, the intussuscepting mass or its distal convex edge can be identified. Barium enema examination may show a "coiled spring" pattern of barium around the intussusceptum or the intussusception may appear as a mass in the cecum (Fig. 6–23). When barium is given orally, only the narrow central channel of the intussusception may be visible as a thin irregular thread of barium.[34]

VOLVULUS OF THE CECUM

Volvulus of the cecum occurs when the cecum undergoes an axial twist or torsion which causes intestinal obstruction with or

(Text continued on page 313)

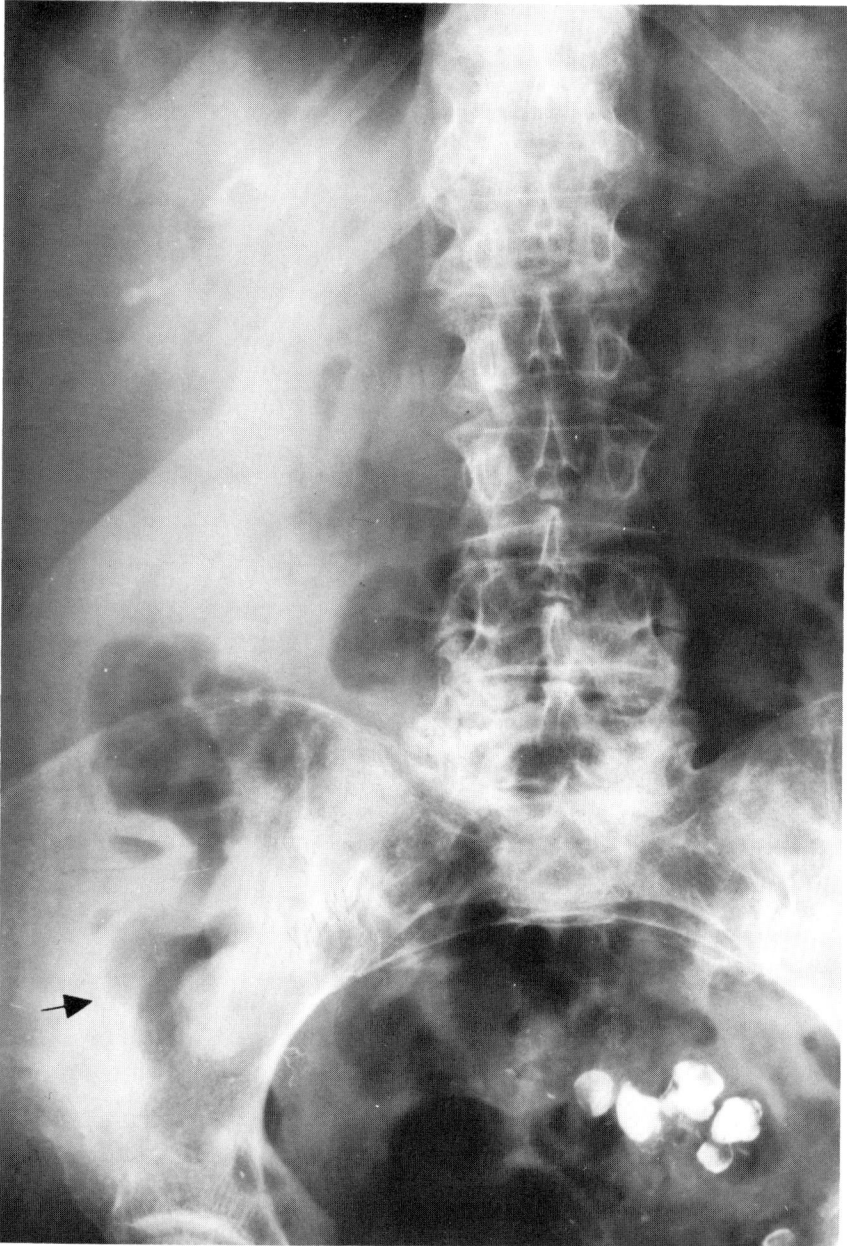

Figure 6-21. Supine plain abdominal radiograph showing a thickened nodular abnormality in the ascending colon with a distorted lumen (arrow). The patient had hemorrhage into the cecum and ascending colon due to a hemorrhagic diathesis associated with leukemia.

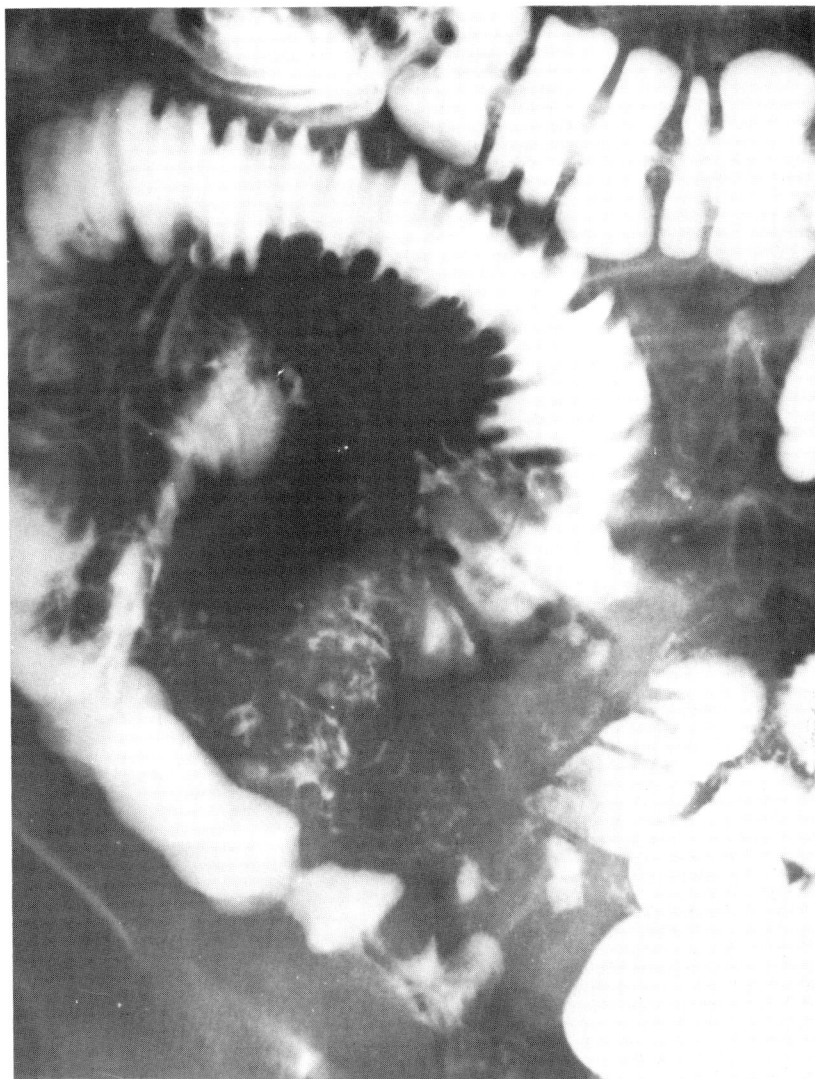

Figure 6–22. Small intestine examination showing a segment of small bowel with a stacked-coin or picket-fence appearance due to submucosal hemorrhage caused by excessive Dicumarol therapy.

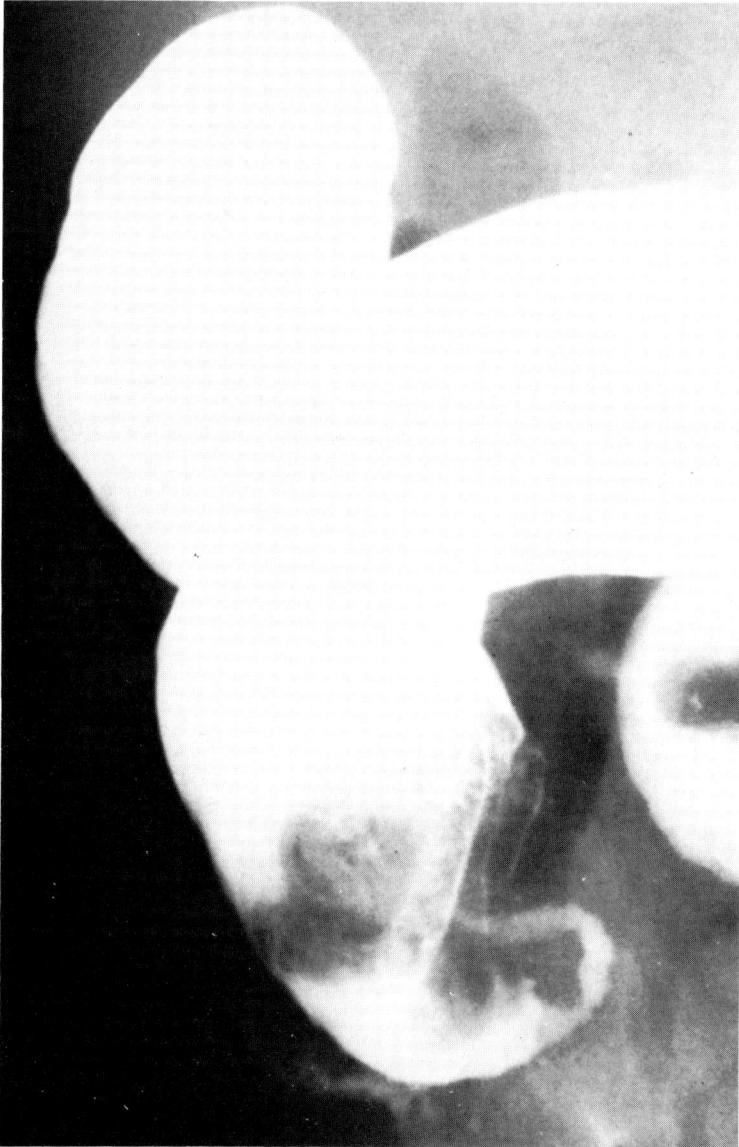

Figure 6–23. Barium enema examination in a 2-year-old boy showing an ileocecal intussusception caused by an inverted Meckel's diverticulum.

without occlusion of the mesenteric vessels. The term is inaccurate, since the terminal ileum and a portion of the ascending colon are usually involved. The volvulus is possible only when the cecum is free to move on a long mesentery. A cecal bascule occurs when the cecum is folded on the ascending colon without a twist or torsion.[35] This is often associated with adhesions. Intestinal obstruction occurs at the point where the colon is folded.

There are several important differences between a cecal and a sigmoid volvulus.[36] Cecal volvulus is less common, occurs in younger patients and is more often seen in females. Cecal volvulus causes small bowel obstruction, whereas a sigmoid volvulus leads to colonic obstruction. The rotation is most often clockwise in cecal volvulus and counterclockwise in sigmoid volvulus. Distal obstructing lesions in the colon are sometimes found in association with a cecal volvulus, whereas patients with sigmoid volvuli are often bedridden, senile or paraplegic but have no other colonic lesions.

Plain abdominal radiographs show the distended cecum and ascending colon displaced in the abdomen[37] (Figs. 6–24 and 6–25). Gas in the normal location of the cecum may be absent, and dilated loops of small bowel may indicate the presence of associated small bowel obstruction. The dilated

(Text continued on page 317)

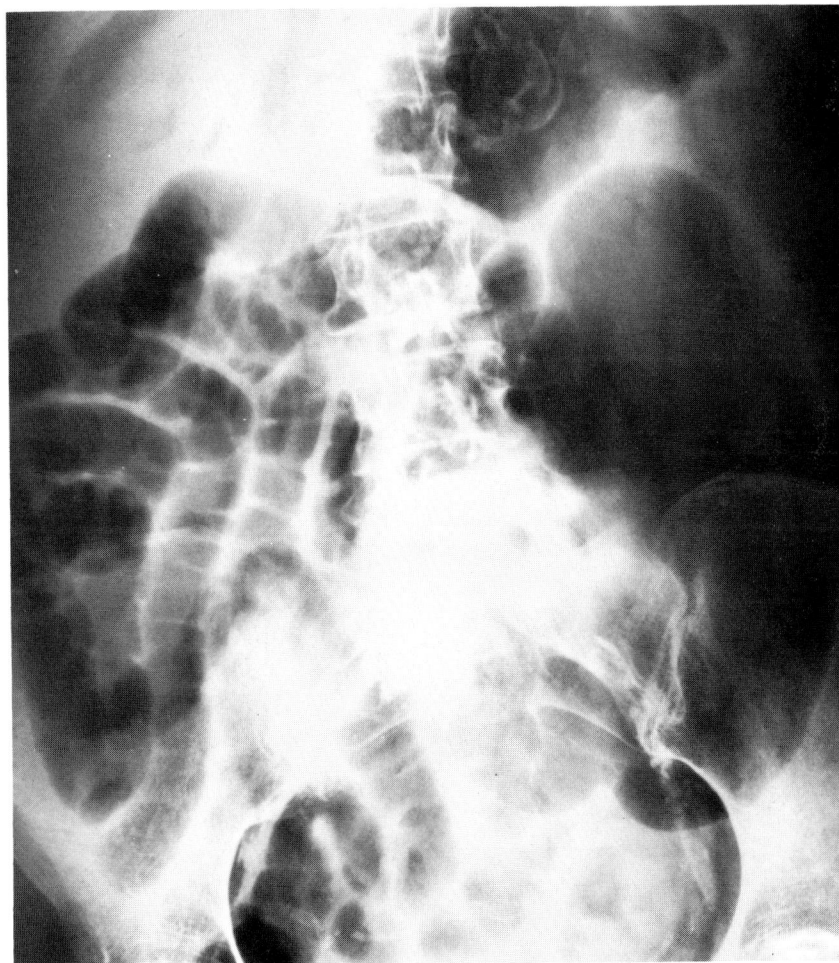

Figure 6–24. Plain abdominal radiograph in an elderly female showing a large collection of gas on the left side of the abdomen due to a cecal volvulus. Numerous dilated loops of small bowel present on the right are due to the small bowel obstruction caused by the volvulus. Free intraperitoneal air was present on the upright chest radiograph indicating that the volvulus had perforated. Operation confirmed the findings.

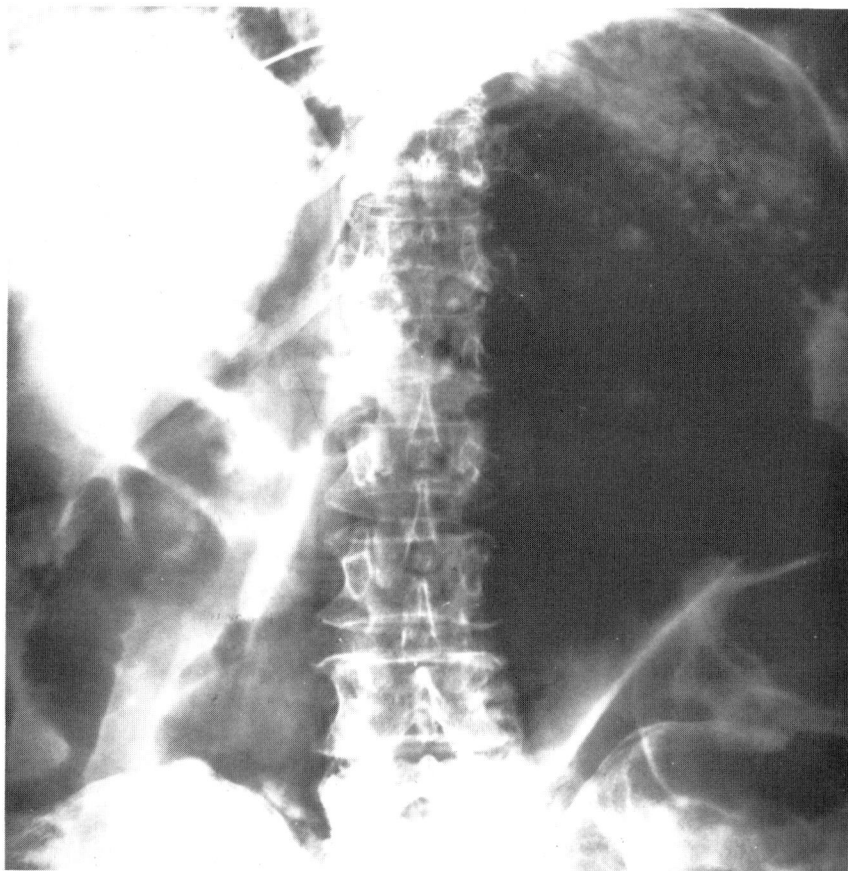

Figure 6–25. Plain abdominal radiograph showing a huge gas collection in the left upper quadrant due to a cecal volvulus with marked distention of the cecum.

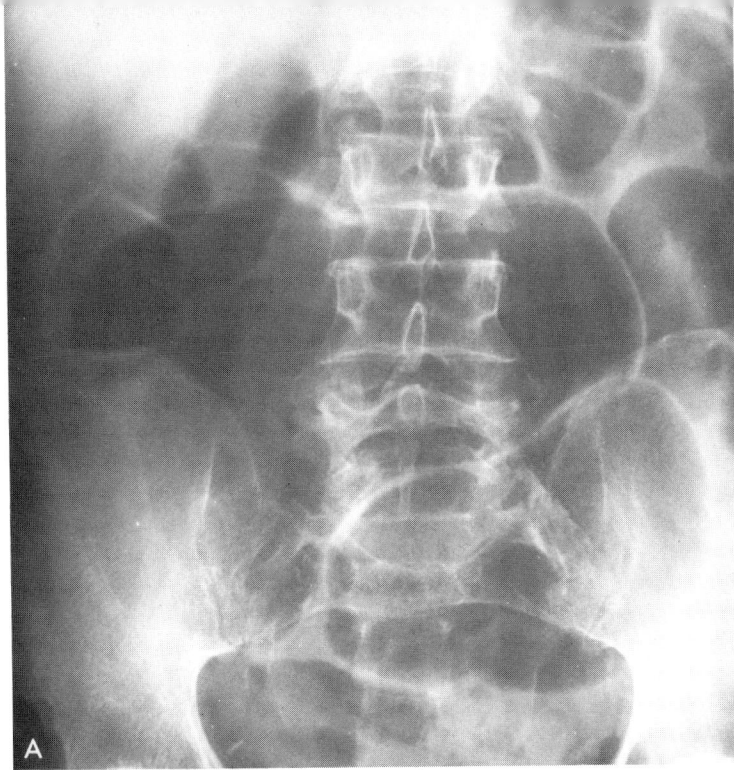

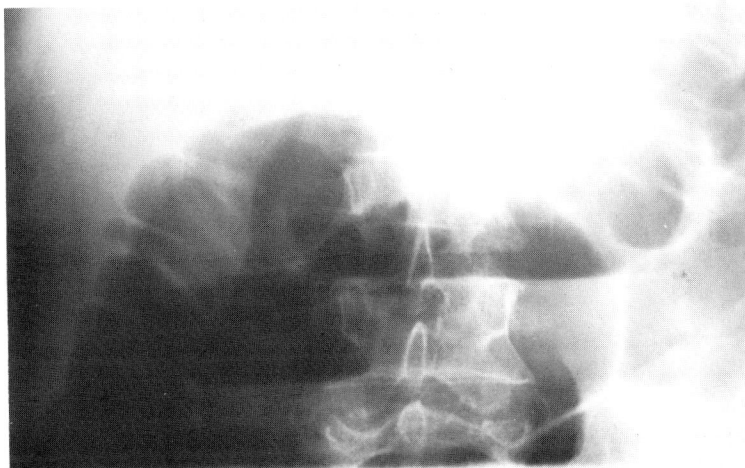

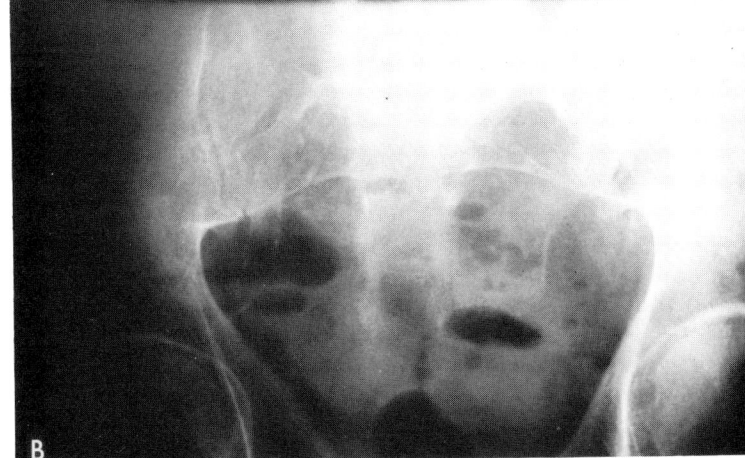

Figure 6-26. Supine (*A*) and upright (*B*) abdominal radiograph showing marked dilatation of the cecum due to localized ileus. At surgery, the patient had acute cholecystitis. The cecum was dilated but was not otherwise abnormal.

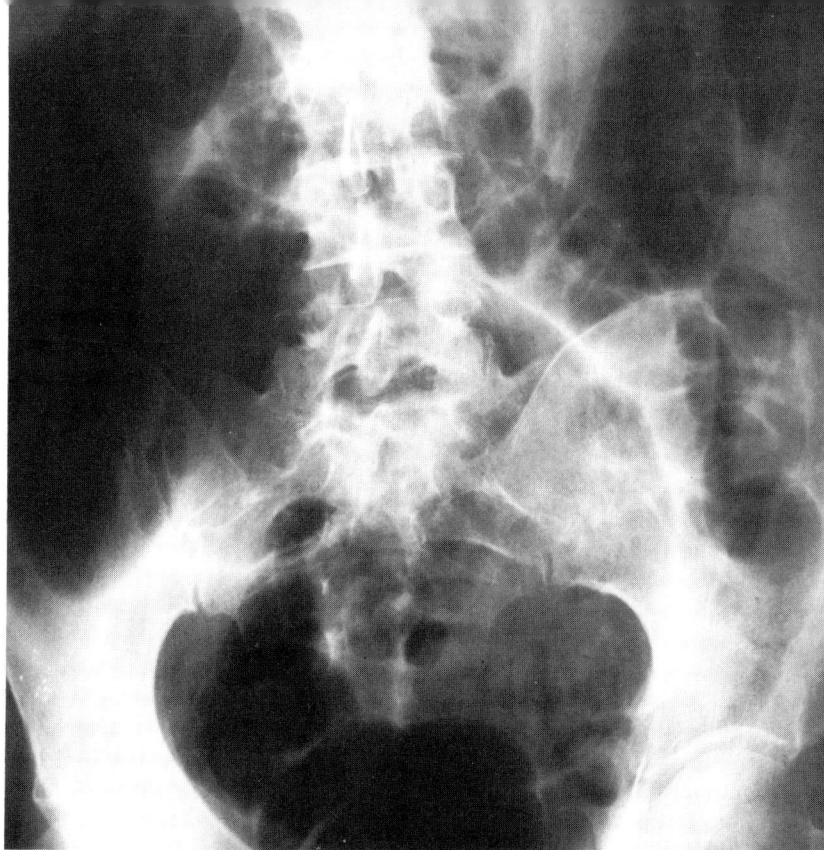

Figure 6–27. Supine abdominal radiograph showing marked dilatation of the cecum. The patient was an elderly women who was found lying on the floor under a bed with the bed compressing her abdomen. She had apparently been helpless in this condition for 3 or 4 days. At surgery, there was infarction of the cecum due to extensive venous thrombosis.

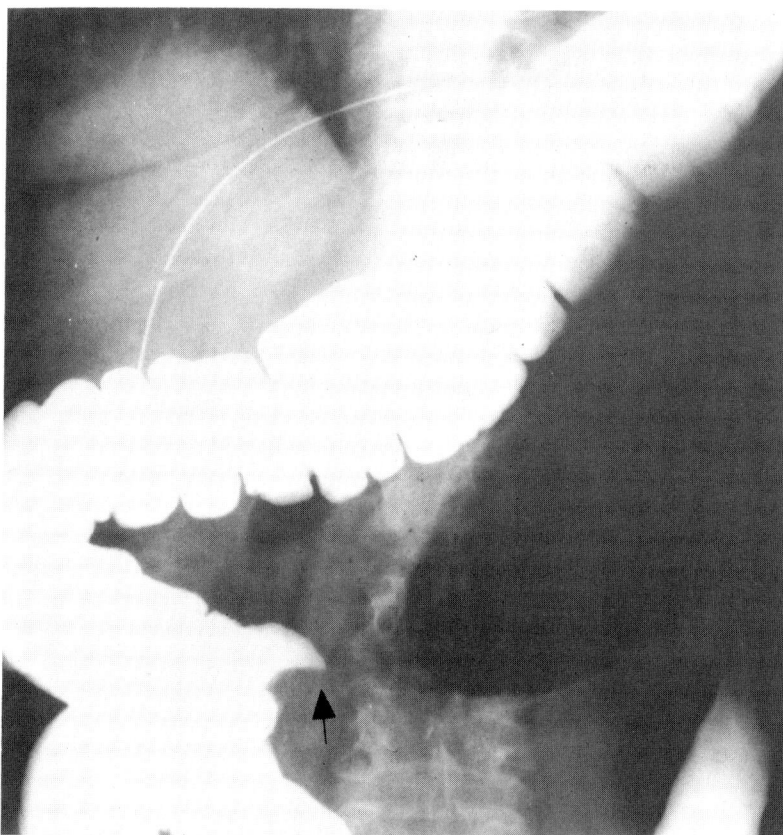

Figure 6–28. Barium enema examination showing a cecal volvulus. The "bird beak" configuration of the termination of the barium column is typical of a volvulus (arrow).

cecum most often lies in the upper abdomen, frequently in the left upper quadrant where it must be differentiated from the stomach. Fluid levels in the volvulus are sharp and straight compared to the fluid levels in the stomach which are often irregular or fuzzy owing to the presence of gastric mucus.[38] Differentiation of cecal volvulus from isolated cecal ileus and cecal infarction may require barium enema studies (Figs. 6–26 and 6–27).

Barium enema examination is usually characteristic in cases of cecal volvulus. The point of obstruction in the ascending colon is identified with a typical "bird's beak" configuration (Fig. 6–28). Reduction during the barium enema examination is rare.

EXTRINSIC INVOLVEMENT OF THE CECUM AND ASCENDING COLON

The retroperitoneal position of the cecum and ascending colon makes these structures

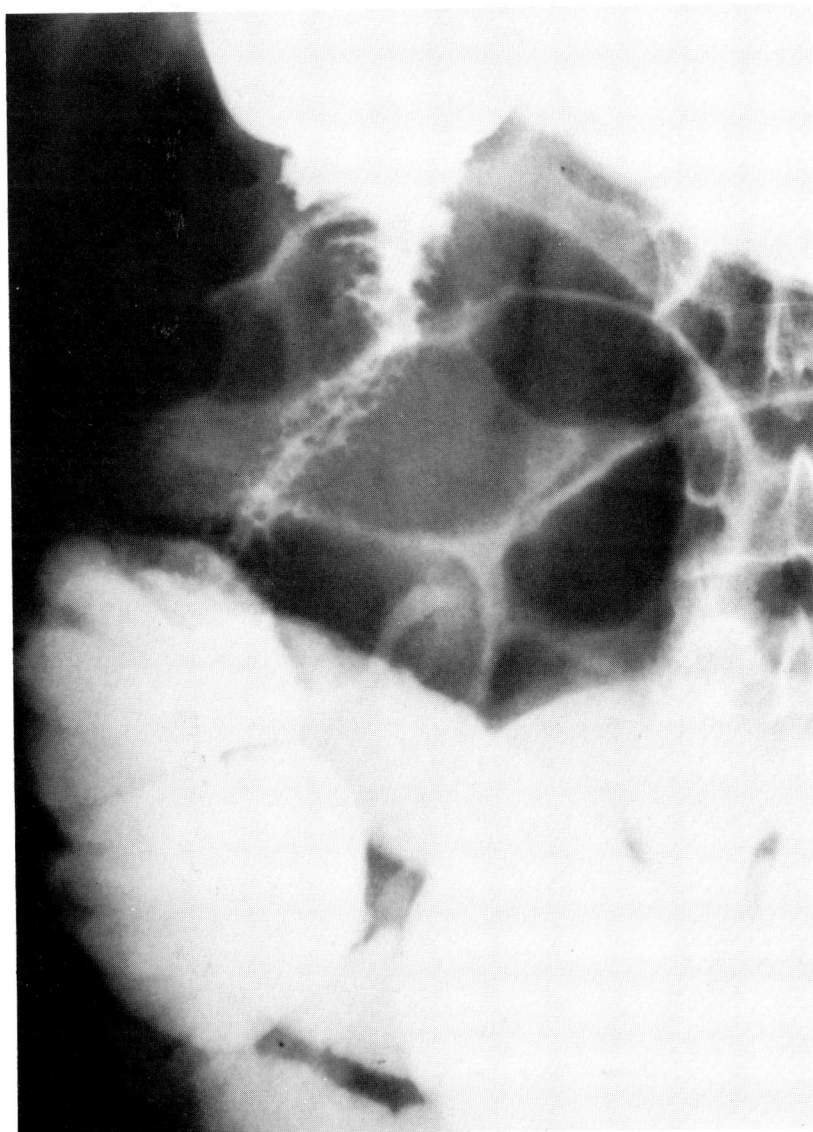

Figure 6–29. Barium enema examination showing extramucosal compression of the ascending colon due to retroperitoneal hemorrhage in a patient with leukemia. The blood dissected into the short mesentery of the colon and surrounded the bowel.

Figure 6–30. Postevacuation radiograph from the barium enema examination of a patient with acute pancreatitis. Spasm, irritability and edema of the ascending colon are due to the inflammatory reaction in the retroperitoneum dissecting into the right lower quadrant (arrow).

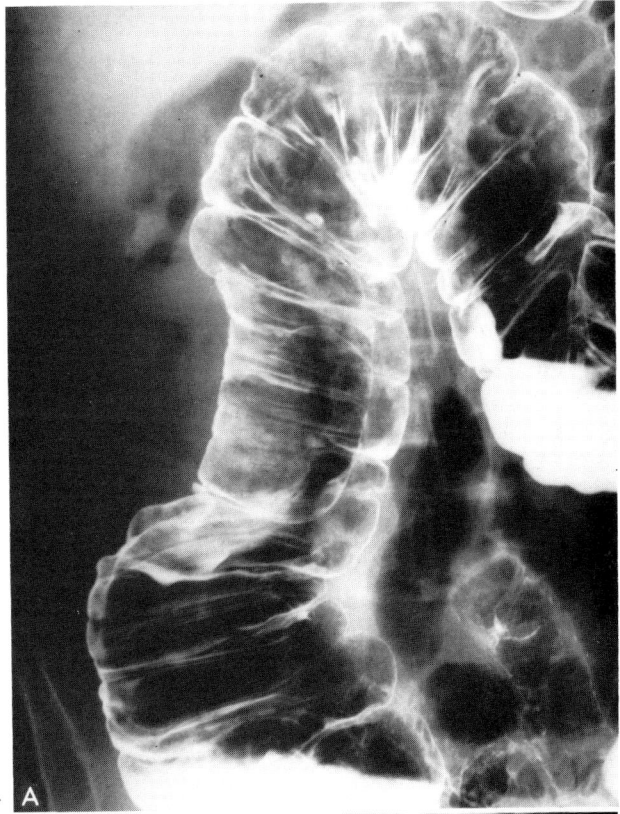

Figure 6–31. Air contrast enema examination showing extramucosal compression of the posterior-lateral aspect of the ascending colon due to direct extension from carcinoma of the right kidney. (*A*) Frontal projection. (*B*) Oblique projection.

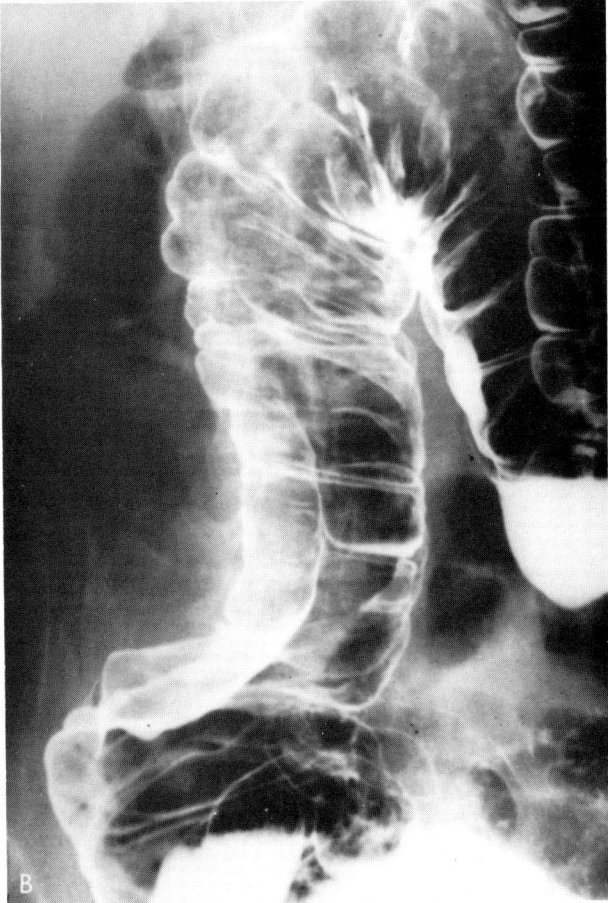

particularly prone to contiguous involvement from diseases in the retroperitoneal space. Thus, retroperitoneal hemorrhage from various causes, including trauma or a bleeding diathesis of any type, may surround the ascending colon and produce an extramucosal defect on barium enema examination (Fig. 6–29). The inflammatory reaction associated with acute pancreatitis may extend down the retroperitoneum into the right lower quadrant and into the root of the small bowel mesentery.[39] Barium studies in this situation show irritability, deformity and edema in the cecum and ascending colon (Fig. 6–30). The radiographic findings may suggest primary inflammatory disease of the bowel. Proximity of the ascending colon to the right kidney permits involvement of the colon with primary or recurrent hypernephroma. The renal tumor may involve only the wall of the ascending colon, producing a smooth, extramucosal intramural mass defect or it may grow into and through the bowel wall to include the mucosa (Fig. 6–31). Rarely, extension of retroperitoneal tumors into the ascending colon may create a fistula into the colon, causing extravasation of barium from the colon into the retroperitoneal space on barium studies.

REFERENCES

1. Marshak, R. H., and Lindner, A. E.: Radiology of the Small Intestine. Philadelphia, W. B. Saunders Co., 1970, p. 461.
2. Capitanio, M. A., and Kirkpatrick, J. A.: Lymphoid hyperplasia of the colon in children. Radiology, 94:323, 1970.
3. Hermans, P. E., Huizenga, K. A., Hoffman, H. N., Brown, A. L., and Markowitz, H.: Dysgammaglobulinemia associated with nodular lymphoid hyperplasia of the small intestine. Amer. J. Med., 40:78, 1966.
4. Berk, R. N., and Reit, R. J.: Intra-abdominal chicken bone abscess. Radiology, 101:311, 1971.
5. Rio, F. W., Edwards, D. L., and Regan, J. F.: Endometriosis of the small intestine. Arch. Surg., 101:403, 1970.
6. Macafee, C. H., and Greer, H. L.: Intestinal endometriosis: a report of 29 cases and a survey of the literature. J. Obstet. Gynec. Brit. Comm., 67:539, 1960.
7. Elliott, G. B., Christensen, R. M., and Elliott, K. A.: Invasive endometriosis of the intestine, report of 21 cases. Canad. J. Surg., 13:387, 1970.
8. Tedeschi, L. G., and Masand, G. P.: Endometriosis of the intestines, a report of 7 cases. Dis. Colon Rectum, 14:360, 1971.
9. Colcock, B. P., and Lamphier, T. A.: Endometriosis of the large and small intestine. Surgery, 28:997, 1950.
10. McKittrick, L. S.: Discussion. New Eng. J. Med., 217:17, 1937.
11. Fallon, J., Brosnan, J. T., and Moran, W. G.: Endometriosis, 200 cases considered from the viewpoint of the practitioner. New Eng. J. Med., 235:669, 1946.
12. Sutton, L. A.: The clinical features of endometriosis. N.Y. State J. Med., 41:1343, 1941.
13. LiVolsi, V. A., and Perzin, K. H.: Endometriosis of the small intestine, producing intestinal obstruction or simulating neoplasm. Amer. J. Dig. Dis., 19:100, 1974.
14. Felson, B., and Wiot, J. F.: Some interesting right lower quadrant entities. Radiol. Clin. N. Amer., 7:83, 1969.
15. Heilbrun, N., and Bernstein, C.: Roentgen abnormalities of the large and small intestine associated with prolonged cathartic ingestion. Radiology, 65:549, 1955.
16. Heilbrun, N.: Roentgen evidence suggesting enterocolitis associated with prolonged cathartic abuse. Radiology, 41:486, 1943.
17. Marshak, R. H., and Gerson, A.: Cathartic colon. Amer. J. Dig. Dis., 5:724, 1960.
18. Eisenman, J. I., Finck, E. J., and O'Loughlin, B. J.: Gallstone ileus, a review of the roentgenographic findings and report of a new roentgen sign. Amer. J. Roent., 101:361, 1967.
19. Berk, R. N., and Lee, F. A.: The late gastrointestinal manifestations of cystic fibrosis of the pancreas. Radiology, 106:377, 1973.
20. Rogers, L. F., Davis, E. K., and Harle, T. S.: Phytobezoar formation and food boli following gastric surgery. Amer. J. Roent., 119:280, 1973.
21. Vuthibhagdee, A., and Harris, N. F.: Antral stricture as a delayed complication of iron intoxication. Radiology, 103:163, 1972.
22. Mellins, H. Z., and Rigler, L. G.: Roentgen findings in strangulating obstructions of small intestine. Amer. J. Roent., 71:404, 1954.
23. Rigler, L. G., and Pogue, W. L.: Roentgen signs of intestinal necrosis. Amer. J. Roent., 94:402, 1965.
24. Nelson, S. W.: Extraluminal gas collections due to diseases of the gastrointestinal tract. Amer. J. Roent., 115:225, 1972.
25. Wiot, J. F., and Felson, B.: Gas in the portal venous system. Amer. J. Roent., 86:920, 1961.
26. Stevenson, J. K., Graham, C. B., Oliver, T. K., and Goldenberg, V. E.: Neonatal necrotizing enterocolitis, report of 21 cases with 14 survivors. Amer. J. Surg., 118:260, 1969.
27. Keyting, W. S., McCarver, R. R., Kovarik, J. L., and Daywitt, A. L.: Pneumatosis intestinalis: new concept. Radiology, 76:733, 1961.
28. Fleischner, F. G., and Bernstein, C.: Roentgen anatomic studies of the normal ileocecal valve. Radiology, 54:43, 1950.
29. Wiot, J. F.: Intramural small intestinal hemorrhage—a differential diagnosis. Semin. Roentgenol., 1:219, 1966.
30. Khilnani, M. T., Marshak, R. H., Eliasoph, J., and Wolf, B. S.: Intramural intestinal hemorrhage. Amer. J. Roent., 92:1061, 1964.
31. Benson, C. D., Lloyd, J. R., and Fischer, H.: Intussusception in infants and children. Arch. Surg., 86:745, 1963.
32. Weilbaecher, D., Bolin, J. A., Hearn, D., and Ogden, W.: Intussusception in adults, review of 160 cases. Amer. J. Surg., 121:531, 1971.

33. LeVine, M., Schwartz, S., Katz, I., Burko, H., and Rabinowitz, J.: Plain film findings in intussusception. Brit. J. Radiol., 37:678, 1964.

34. Carlson, H. C.: Small intestinal intussusception, an easily misunderstood sign. Amer. J. Roent., 110:338, 1970.

35. Bobroff, L. M., Messinger, N. H., Subbarao, K., and Beneventano, T. C.: The cecal bascule. Amer. J. Roent., 115:249, 1972.

36. Meyers, J. R., Heifetz, C. J., and Baue, A. E.: Cecal volvulus, a lesion requiring resection. Arch. Surg., 104:594, 1972.

37. Frimann-Dahl, J.: Volvulus of the right colon. Acta Radiol. (Stockholm), 41:141, 1954.

38. Caruso, R. D., and Berk, R. N.: The fuzzy fluid level sign. Radiology, 98:369, 1971.

39. Meyers, M. A., and Evans, J. A.: Effects of pancreatitis on the small bowel and colon: spread along mesenteric planes. Amer. J. Roent., 119:151, 1973.

INDEX

Note: Page numbers in *italics* refer to illustrations.